METHODS IN MOLECULAR BIOLOGY™

Series Editor
John M. Walker
School of Life Sciences
University of Hertfordshire
Hatfield, Hertfordshire, AL10 9AB, UK

For further volumes:
http://www.springer.com/series/7651

Homology Modeling

Methods and Protocols

Edited by

Andrew J.W. Orry

Molsoft L.L.C., San Diego, CA, USA

Ruben Abagyan

Skaggs School of Pharmacy and Pharmaceutical Sciences, University of California, San Diego, La Jolla, CA, USA;
San Diego Supercomputer Center, University of California, San Diego, La Jolla, CA, USA

Humana Press

Editors
Andrew J.W. Orry, Ph.D.
Molsoft L.L.C.
San Diego, CA, USA
andy@molsoft.com

Ruben Abagyan, Ph.D.
Skaggs School of Pharmacy
and Pharmaceutical Sciences
University of California, San Diego
La Jolla, CA, USA
and
San Diego Supercomputer Center
University of California, San Diego
La Jolla, CA, USA

ISSN 1064-3745 e-ISSN 1940-6029
ISBN 978-1-4939-5880-1 ISBN 978-1-61779-588-6 (eBook)
DOI 10.1007/978-1-61779-588-6
Springer New York Dordrecht Heidelberg London

Printed on acid-free paper

Humana Press is part of Springer Science+Business Media (www.springer.com)

Preface

Knowledge about protein tertiary structure can guide mutagenesis experiments, help in the understanding of structure–function relationships, and aid the development of new therapeutics for diseases. Homology modeling is an in silico method that predicts the tertiary structure of a query amino acid sequence based on a homologous experimentally determined template structure. The method relies on the observation that the tertiary structure of a protein is better conserved than sequence and therefore two proteins that are not fully conserved at the sequence level may still share the same fold. Structures solved by X-ray crystallography and NMR are deposited in the Protein Data Bank (PDB) and form the templates for homology modeling. The human proteome has approximately 20,000 annotated human proteins and only 4,900 human protein fragments and domains can be found in the PDB.

The main steps in a homology modeling experiment are template selection, alignment, backbone and side-chain prediction, and structure optimization, including ligand-guided optimization and evaluation. Errors at the template selection step will result in an incorrect model and so care is needed to identify a template structure that has significant homology with the query sequence. The template sequence is aligned to the query sequence and the alignment is adjusted to ensure optimal correspondence between the homologous regions. The backbone atoms of the model are mapped onto the three-dimensional template structure and nonconserved side-chain orientations are predicted. Optimization of the model in a force field removes steric clashes and improves the hydrogen-bonding network between atoms. Evaluation of the final model highlights regions where there are errors in the model, for example, nonconserved loops, which may need to be modeled independently of the conserved regions. While the ability of models to predict ligand binding is still limited as evaluated recently in a GPCR DOCK 2010 competition, there is noticeable progress.

Energy sampling methods used in the homology modeling optimization step also have application for predicting how ligands bind to the model. Modeling methods are required even when an X-ray or NMR structure is available because the number of possible ligand–receptor combinations is extremely high and experimentally solving all of them is not practical.

In this book, experts in the field describe each homology modeling step from first principles, highlighting the pitfalls to avoid and providing first-hand solutions to common modeling problems. In addition, the book contains chapters from colleagues who model particularly challenging proteins such as membrane proteins where template structures are scarce or large macromolecular assemblies. The book also describes methods that can be applied once the initial model is complete, such as those which can be used to optimize the ligand-binding pocket of the model and predict protein–protein interactions.

We would like to express our sincere thanks to all the authors who so generously contributed their time and knowledge to this book.

San Diego, CA, USA | *Andrew J.W. Orry, Ph.D.*
La Jolla, CA, USA | *Ruben Abagyan, Ph.D.*

Contents

Contributors

RUBEN ABAGYAN • *Skaggs School of Pharmacy and Pharmaceutical Sciences, University of California, San Diego, La Jolla, CA, USA; San Diego Supercomputer Center, University of California, San Diego, La Jolla, CA, USA*
ANTONINA ANDREEVA • *MRC Laboratory of Molecular Biology, Cambridge, UK*
ANDREW J. BORDNER • *Mayo Clinic, Scottsdale, AZ, USA*
LORENZA BORDOLI • *SIB Swiss Institute of Bioinformatics, Biozentrum University of Basel, Basel, Switzerland*
JONAS CARLSSON • *IFM Bioinformatics and SeRC (Swedish e-Science Research Centre), Linköping University, Linköping, Sweden*
STEFANO COSTANZI • *Laboratory of Biological Modeling, National Institute of Diabetes and Digestive and Kidney Diseases, National Institutes of Health, DHHS, Bethesda, MD, USA*
ANTOINE DAINA • *School of Pharmaceutical Sciences, University of Geneva, University of Lausanne, Geneva, Switzerland*
MICHAEL A. DOLAN • *Bioinformatics and Computational Biosciences Branch, National Institute of Allergies and Infectious Diseases, National Institutes of Health, Bethesda, MD, USA*
THOMAS E. FERRIN • *Resource for Biocomputing, Visualization, and Informatics, Department of Pharmaceutical Chemistry, University of California, San Francisco, San Francisco, CA, USA*
MARKUS GRUBER • *Center of Applied Molecular Engineering, Division of Bioinformatics, University of Salzburg, Salzburg, Austria*
DARRELL HURT • *Bioinformatics and Computational Biosciences Branch, National Institute of Allergies and Infectious Diseases, National Institutes of Health, Bethesda, MD, USA*
KEEHYOUNG JOO • *Center for In Silico Protein Science, Center for Advanced Computation, Korea Institute for Advanced Study, Seoul, Korea*
VSEVOLOD KATRITCH • *Department of Molecular Biology, The Scripps Research Institute, La Jolla, CA, USA*
IRINA KUFAREVA • *Skaggs School of Pharmacy and Pharmaceutical Sciences, University of California, San Diego, La Jolla, CA, USA*
KEREN LASKER • *Department of Bioengineering and Therapeutic Sciences, University of California, San Francisco, San Francisco, CA, USA; Department of Pharmaceutical Chemistry, University of California, San Francisco, San Francisco, CA, USA; California Institute for Quantitative Biosciences (QB3), University of California, San Francisco, San Francisco, CA, USA; The Blavatnik School of Computer Science, Tel-Aviv University, Ramat Aviv, Israel*

JINWOO LEE • *Department of Mathematics, Kwangwoon University, Seoul, Korea*
JOOYOUNG LEE • *Center for In Silico Protein Science, Center for Advanced Computation, School of Computational Sciences, Korea Institute for Advanced Study, Seoul, Korea*
NIR LONDON • *Department of Microbiology and Molecular Genetics, Institute for Medical Research Israel-Canada, Hadassah Medical School, The Hebrew University, Jerusalem, Israel*
JAMES W. NOAH • *Southern Research Institute, Birmingham, AL, USA*
ALESSANDRA NURISSO • *School of Pharmaceutical Sciences, University of Geneva, University of Lausanne, Geneva, Switzerland*
ANDREW J.W. ORRY • *Molsoft L.L.C., San Diego, CA, USA*
BENGT PERSSON • *IFM Bioinformatics and SeRC (Swedish e-Science Research Centre), Linköping University, Linköping, Sweden; Science for Life Laboratory, Department of Cell and Molecular Biology, Karolinska Institutet, Stockholm, Sweden*
BARAK RAVEH • *Department of Microbiology and Molecular Genetics, Institute for Medical Research Israel-Canada, Hadassah Medical School, The Hebrew University, Jerusalem, Israel; The Blavatnik School of Computer Science, Tel-Aviv University, Ramat Aviv, Israel*
AINA WESTRHEIM RAVNA • *Medical Pharmacology and Toxicology, Department of Medical Biology, Faculty of Health Sciences, University of Tromsø, Tromsø, Norway*
MANUEL RUEDA • *Skaggs School of Pharmacy and Pharmaceutical Sciences, University of California, San Diego, La Jolla, CA, USA; San Diego Supercomputer Center, University of California, San Diego, La Jolla, CA, USA*
ANDREJ SALI • *Department of Bioengineering and Therapeutic Sciences, University of California, San Francisco, San Francisco, CA, USA; Department of Pharmaceutical Chemistry, University of California, San Francisco, San Francisco, CA, USA; California Institute for Quantitative Biosciences (QB3), University of California, San Francisco, San Francisco, CA, USA*
ORA SCHUELER-FURMAN • *Department of Microbiology and Molecular Genetics, Institute for Medical Research Israel-Canada, Hadassah Medical School, The Hebrew University, Jerusalem, Israel*
TORSTEN SCHWEDE • *SIB Swiss Institute of Bioinformatics, Biozentrum University of Basel, Basel, Switzerland*
MANFRED J. SIPPL • *Center of Applied Molecular Engineering, Division of Bioinformatics, University of Salzburg, Salzburg, Austria*
AROOP SIRCAR • *EMD Serono Research Center, Inc., Billerica, MA, USA*
STEFAN J. SUHRER • *Center of Applied Molecular Engineering, Division of Bioinformatics, University of Salzburg, Salzburg, Austria*
INGEBRIGT SYLTE • *Medical Pharmacology and Toxicology, Department of Medical Biology, Faculty of Health Sciences, University of Tromsø, Tromsø, Norway*

MAXIM TOTROV • *Molsoft L.L.C., San Diego, CA, USA*
JAVIER A. VELÁZQUEZ-MURIEL • *Department of Bioengineering and Therapeutic Sciences, University of California, San Francisco, San Francisco, CA, USA; Department of Pharmaceutical Chemistry, University of California, San Francisco, San Francisco, CA, USA; California Institute for Quantitative Biosciences (QB3), University of California, San Francisco, San Francisco, CA, USA*
ČESLOVAS VENCLOVAS • *Institute of Biotechnology, Vilnius University, Vilnius, Lithuania*
ROSS C. WALKER • *Department of Chemistry and Biochemistry, University of California, San Diego, La Jolla, CA, USA; San Diego Supercomputer Center, University of California, San Diego, La Jolla, CA, USA*
BENJAMIN M. WEBB • *Department of Bioengineering and Therapeutic Sciences, University of California, San Francisco, San Francisco, CA, USA; Department of Pharmaceutical Chemistry, University of California, San Francisco, San Francisco, CA, USA; California Institute for Quantitative Biosciences (QB3), University of California, San Francisco, San Francisco, CA, USA*
MARKUS WIEDERSTEIN • *Center of Applied Molecular Engineering, Division of Bioinformatics, University of Salzburg, Salzburg, Austria*
ZHENG YANG • *Resource for Biocomputing, Visualization, and Informatics, Department of Pharmaceutical Chemistry, University of California, San Francisco, San Francisco, CA, USA*

Chapter 1

Classification of Proteins: Available Structural Space for Molecular Modeling

Antonina Andreeva

Abstract

The wealth of available protein structural data provides unprecedented opportunity to study and better understand the underlying principles of protein folding and protein structure evolution. A key to achieving this lies in the ability to analyse these data and to organize them in a coherent classification scheme. Over the past years several protein classifications have been developed that aim to group proteins based on their structural relationships. Some of these classification schemes explore the concept of structural neighbourhood (structural continuum), whereas other utilize the notion of protein evolution and thus provide a discrete rather than continuum view of protein structure space. This chapter presents a strategy for classification of proteins with known three-dimensional structure. Steps in the classification process along with basic definitions are introduced. Examples illustrating some fundamental concepts of protein folding and evolution with a special focus on the exceptions to them are presented.

Key words: Protein domain, Protein motif, Protein repeat, Oligomeric complex, Protein classification, Conformational changes, Chameleon sequences, Fold decay, Fold transitions, Circular permutation

1. Introduction

Over five decades have passed from the time when the first three-dimensional structure of globular protein, myoglobin, was solved (1). Since this pioneering work, the determination of protein structures has seen tremendous increase. The largest repository of structural data, the Protein Data Bank (2), currently holds more than 70,000 protein structures. This wealth of structural data provides unprecedented opportunity to study and better understand the molecular mechanisms of protein function and evolution. A key to achieving this lies in the ability to analyse these data and organize them in a coherent classification scheme.

Andrew J.W. Orry and Ruben Abagyan (eds.), *Homology Modeling: Methods and Protocols*, Methods in Molecular Biology, vol. 857, DOI 10.1007/978-1-61779-588-6_1, © Springer Science+Business Media, LLC 2012

The notion of protein structure classification has emerged from early studies aiming to elucidate the basic principles of protein folding and protein structure evolution. In the late 1970s, Chothia and coworkers pioneered the division of protein structures into four major classes, based on their secondary structure composition and demonstrated that simple geometrical principles govern their mutual arrangement into distinct architectures (3–5). In the early 1980s, in the "Anatomy and Taxonomy of Protein Structure," Jane Richardson has provided the first general classification scheme for protein structures founded on their architecture and topological details (6, 7).

Several protein structure classifications were developed in the 1990s. Liisa Holm and Chris Sander established the Families of Structurally Similar Proteins (FSSP), a fully automatic classification based on structural alignments generated using Dali algorithm (8). FSSP explored the concept of structural neighbourhood and thus creating continuum rather than discrete view of protein structure space. Similarly, the Molecular Modeling DataBase (MMDB) developed at National Center for Biotechnology Information (NCBI) provided a look at the structural neighbourhood but based on the VAST structure comparison algorithm (9). Nearly at the time of the FSSP and MMDB development, the Structural Classification of Proteins (SCOP) database was created at LMB Cambridge by Alexey Murzin, Steven Brenner, Tim Hubbard, and Cyrus Chothia (10). The notion of protein evolution, embodied in SCOP, allowed to create discrete groupings of proteins based not only on their structural similarity but also on their common evolutionary origin. Like in the Linnaean taxonomy, discrete units (domains) were grouped hierarchically on the basis of their common structural and evolutionary relationships. Soon after the release of SCOP, another protein structural classification, Class, Architecture, Topology, Homology (CATH), was developed at UCL London by Orengo et al. (11, 12). Similar to SCOP, the CATH database organized protein domains into hierarchical levels but in contrast to SCOP, used a semi-automatic, rather than manual approach for classification. Each of these classifications remains widely used today and became invaluable resource in many areas of protein structure research.

This chapter discuses a methodology for classification of proteins with known structure. Steps in the classification process along with basic definitions are introduced. Examples illustrating some fundamental concepts of protein folding and evolution, with a special focus on the exceptions to them, are presented. At the end, an overview of the widely used classifications is given.

2. Materials

Automated methods for sequence and structure comparison are indispensible part of protein structure classification process. The most commonly used comparison tools along with the sequence and structural data resources are listed in Table 1. The reader is directed to the references therein for more details about algorithms and descriptions of databases.

3. Units of Protein Classification

Structural similarities between proteins can arise at different levels of protein structure organization. These similarities can be local, comprising only a few secondary structural elements, or global, extending to the entire tertiary or quaternary structure. Each of these structural similarities can indicate biologically relevant relationships between proteins and thus provide important insights into protein function and structure evolution.

This section aims to describe basic units of protein structure classification. Beside protein domain that is most commonly used, additional units of classification, namely motif, repeat, and protein complex are introduced.

3.1. Protein Domain

Domain, as a general feature of protein three-dimensional structure, was primary described by Wetlaufer in terms of regions of polypeptide chain that can enclose in a compact volume and fold autonomously (13). Wetlaufer also introduced the concept of continuous and discontinuous structural regions and proposed an approach for defining domains. Later on, Rossmann based on his observations on dehydrogenases proposed that domains represent genetic units which in the course of evolution have been transferred and combined with other structurally distinct domains to produce functionally different but related proteins (14). These, in essence, conceptually different approaches to delineate domains have evolved in a broad definition of domain as a unit of folding, structure, function, and evolution.

Generally, one or more of the following criteria can be used to define protein domain:

1. A compact, globular region of structure that is semi-independent of the rest of the polypeptide chain (structural domain); this region can consist of one or more segments of the polypeptide chain, the entire polypeptide chain or several polypeptide chains.

Table 1
Databases and tools for protein analysis

Sequence databases	
Uniprot (141)	http://www.uniprot.org
NCBI (142)	http://www.ncbi.nlm.nih.gov/
Structure databases	
PDB (2)	http://www.pdb.org
Protein structure classifications	
SCOP (10)	http://scop.mrc-lmb.cam.ac.uk/scop/
CATH (12)	http://www.cathdb.info/
SISYPHUS (28)	http://sisyphus.mrc-cpe.cam.ac.uk/
3D complex (27)	http://www.3Dcomplex.org
Structural neighbourhoods	
MMDB (142)	http://www.ncbi.nlm.nih.gov/sites/entrez?db=structure
FSN (137)	http://fatcat.burnham.org/fatcat-cgi/cgi/FSN/fsn.pl
Dali DB (135, 143)	http://ekhidna.biocenter.helsinki.fi/dali/start
COPS (136)	http://cops.services.came.sbg.ac.at/
Tools for analysis	
Tools for sequence comparison and similarity searches	
BLAST & PSIBLAST (85)	http://www.ncbi.nlm.nih.gov/blast
FASTA3 (144)	http://www.ebi.ac.uk/Tools/fasta33
HMMER (86)	http://selab.janelia.org/
Tools for structure comparison and similarity searches	
Dali (143)	http://ekhidna.biocenter.helsinki.fi/dali_server/
VAST (145)	http://www.ncbi.nlm.nih.gov/Structure/VAST/vastsearch.html
SSAP (146)	http://www.cathdb.info
FATCAT (147)	http://fatcat.burnham.org/
CE (148)	http://cl.sdsc.edu/
Mammoth (149)	http://ub.cbm.uam.es/mammoth/mult/
Topmatch (150)	http://topmatch.services.came.sbg.ac.at/TopMatchFlex.php
TM-align (151)	http://zhanglab.ccmb.med.umich.edu/TM-align/
Other resources	
DisProt (84)	http://www.disprot.org/
PROSITE (26)	http://www.expasy.org/prosite
Consurf (140)	http://consurf.tau.ac.il/
Database of membrane proteins (152)	http://blanco.biomol.uci.edu/Membrane_Proteins_xtal.html
Pratt (38)	http://www.ebi.ac.uk/Tools/pratt/index.html
Jalvew (139)	http://www.jalview.org/

2. A region of protein that occurs in nature either in isolation or in more than one context of multidomain proteins (evolutionary domain).
3. A region of protein structure that is associated with a particular function (functional domain).

Often when dividing a protein structure into domains not all of these criteria can simultaneously be satisfied. Structural domains, for instance, may not be associated with a particular function or evolutionary domains can consists of two or more structural domains. Similarly, some protein functional domains can contain more than one structural domain. One example of functional domain composed of two structural domains is the structure of D-aminopeptidase DppA that consists of an N-terminal 5-stranded $\alpha/\beta/\alpha$ domain and a C-terminal 5-stranded β/α domain (Fig. 1) (15). The active site of this enzyme is located in a cleft between the two domains that comprises the most conserved part of the protein. The functionally active protein requires the presence of two domains. None of these domains exists on its own or in combination with other domains and therefore the evolutionary domain spans over the two structural domains.

The selection of criteria used for defining domains should depend on the type of analysis for which domains will be used. For protein structure analysis and structure comparison searches, the domain defined as a structural unit is more appropriate. Some structural domains, however, might not be suitable for sequence

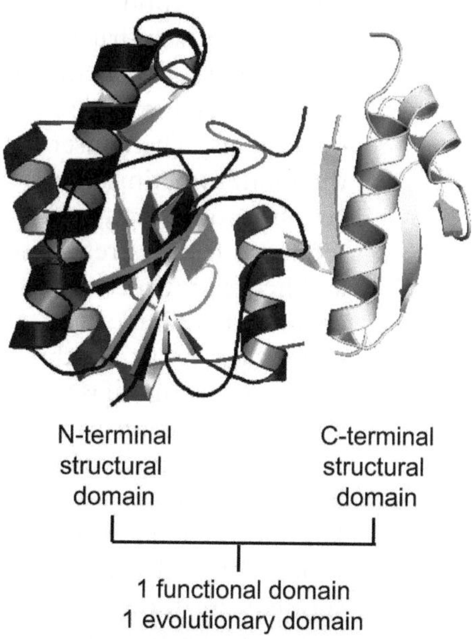

Fig. 1. Domains in the structure of D-aminopeptidase DppA (pdb 1hi9).

analysis particularly when the domain consists of two or more discontinuous segments or the domain boundaries disrupt a highly conserved sequence motif that can be crucial for detection of proteins' homologs.

Assignment of novel domains can be done by visual inspection or by using automated methods. Over the past years, several methods for automatic detection of domains have been devised (16–25). Many of them, however, disagree in their domain definitions. The problem with these methods arises from the fact that there is no simple quantitative definition of protein domain. One approach to tackle with this problem is by combining the results of several independent automatic domain definition programmes with visual inspection. This strategy has been implemented by the authors of CATH, in which domains are assigned by using the results of three different methods PUU (18), Domak (20), and DETECTIVE (22) in combination with manual validation. Domains can also be assigned by similarity to already known domains by using either sequence or structure comparison tools.

3.2. Other Units of Classification

Most classifications use the protein domain as classification unit. Within the classification scheme, domains are usually organized hierarchically depending on their structural and evolutionary relationships. The units described here, add extra complexity to the hierarchical presentation of relationships between proteins. They can be classified either separately (as in refs. 26, 27) or as interrelationships within the hierarchical scheme (as in ref. 28).

3.2.1. Protein Motifs

Protein motif is a local, relatively small, contiguous region within a protein polypeptide chain that can be distinguish by a well-defined set of properties (structural and/or functional). There are two types of motifs: sequence and structural. Sequence motif represents a conserved amino acid sequence pattern that is common to a group of proteins. The conservation of the amino acid residues within the motif sometimes can be strict and also may be defined within a certain group, e.g., hydrophobic, polar, or charged. The unique sequence features reflect structural and/or functional constraints and hence sequence motifs usually reside in regions of polypeptide chain that are important for the protein either to perform its tasks or to adopt particular three-dimensional conformation.

Structural motif is regarded as a combination of a few secondary structural elements with a specific geometric arrangement. In contrast to protein domain, it lacks compactness and a well-defined hydrophobic core. Typical examples for structural motifs are Greek-key motif found in β-sandwiches (29), helix-turn-helix (HTH) motif (30), helix-hairpin-helix (HhH) motif (31), etc. Structural motifs were thought that cannot fold independently if they are expressed separately from the rest of the protein. However, recently the HTH motif of engrailed homeodomain was found to fold independently in solution and having essentially the same structure as in the full-length protein (32). This finding allows arguing that some structural motifs may act as a folding template and increase the likelihood for a successful non-homologous recombination (reviewed in ref. 33).

Quite often, but not always a local sequence motif resides in a local structural motif. Some sequence motifs, however, can span over dissimilar structural motifs. For instance, a number of cytochrome c proteins contain a sequence motif defined by C-X2-C-H pattern that binds heme via two invariant Cys residues and coordinates heme iron via conserved His residue. This heme-binding sequence motif spans over regions that have different conformations as shown in Fig. 2. Similarly, (pro)aerolysin and α-hemolysin share a common sequence motif described with [KT]-X2-N-W-X2-T-[DN]-T pattern. Both proteins have globally distinct structures and the sequence motif resides in structurally dissimilar regions.

Similar sequence and structural motifs can be found in structurally distinct proteins. This can result in significant sequence hits between proteins which structures are globally dissimilar. Some of these motifs, however, are of particular interest since they are frequently related to function. Some examples of such motifs are KH motif (34), HTH motif (30), nucleotide-binding motif (35), Ca-binding (DxDxDG) motif (36), P-loop motif (37), etc. The P-loop motif, for instance, is a Gly-rich sequence motif that comprises a flexible loop between a β-strand and an α-helix. This motif is involved in binding of mononucleotides, e.g., ATP, GTP, and directly interacts with one of the phosphate groups. Detection of this motif by sequence analysis tools is relatively straightforward. Several topologically different structures are found to contain the P-loop motif. Another example is the "nucleophile elbow and

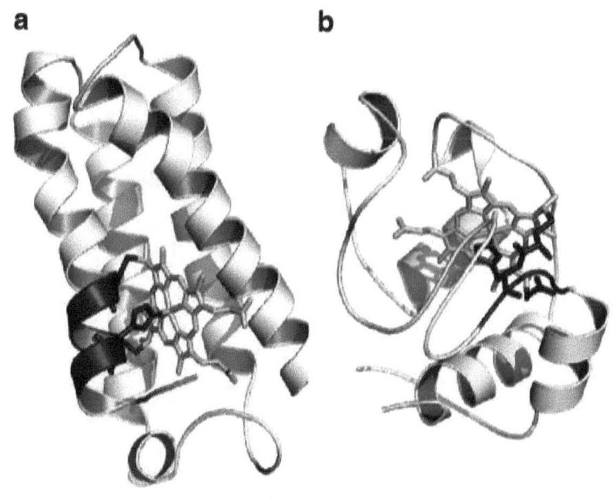

Fig. 2. The structures of (**a**) cytochrome c′ (pdb 1a7v) and (**b**) cytochrome c (pdb 1fhb). The sequence motif common to both proteins is shown in *black*.

oxyanion hole" structural motif that encompasses a discontinuous $\beta/\beta\alpha$ motif and harbours the nucleophilic and the oxyanion-hole amino acid residues that constitute the catalytic site in different enzymes. The nucleophile (Ser, Asp, or Cys) is located in a sharp turn between a β-strand and an α-helix, the so-called nucleophile elbow. The oxyanion-hole is usually formed by mainchain NH groups of two Gly, one of which frequently follows the nucleophile. The conserved $\beta/\beta\alpha$ structural motif is found in a number of α/β catalytic domains with different β-sheet topologies (Fig. 3).

The presence of common sequence motifs in proteins with dissimilar structures can create challenges for protein structure prediction (see Note 6). Knowledge of the occurrence of these motifs and the structural context in which they are observed is essential for protein modeling.

Sequence motifs can be easily identified within a multiple sequence alignment or by sequence comparisons. One widely used

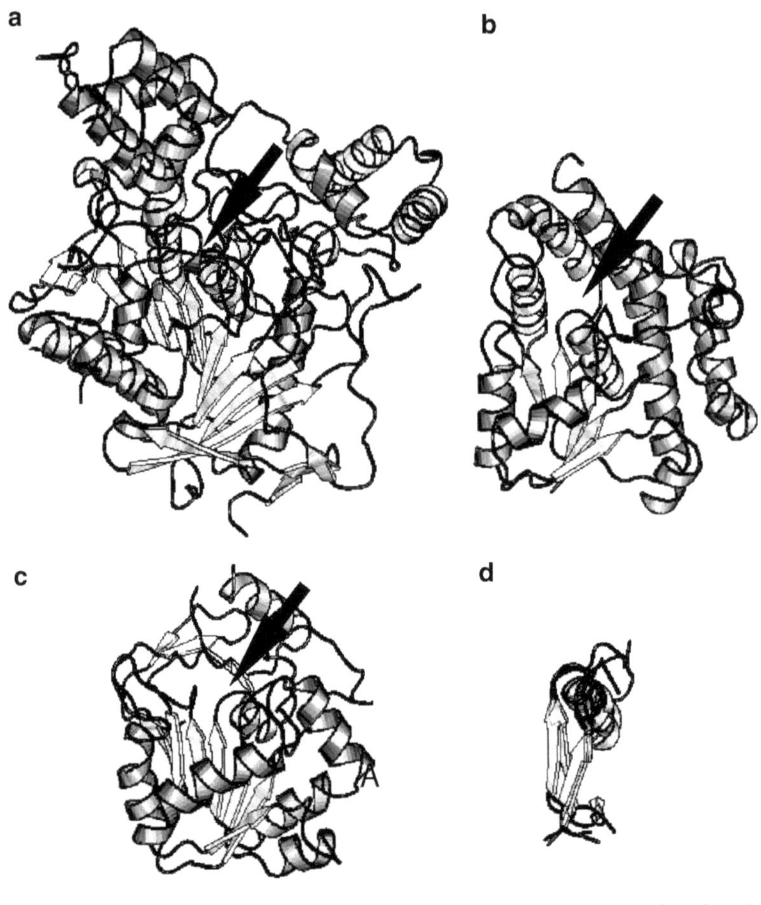

Fig. 3. The structures of (**a**) acetylcholinesterase (pdb 2ack), (**b**) malonyl-CoA:acyl carrier protein transacylase (pdb 1mla), (**c**) aspartyl dipeptidase (pdb 1fye), and (**d**) the "Nucleophile elbow and oxyanion hole" structural motif. *Arrows* indicate the location of the motif in the structures.

resource is PROSITE that contains a collection of protein sequence motifs along with tools for protein sequence analysis and motif detection (26). Programmes are available for automatic generation of sequence patterns (38–41). Detection of structural motifs, particularly in the absence of sequence similarity, is not straightforward. SPASM/RIGOR are programmes that can be used for the detection of small structural motifs (42). Spatial arrangements of side chain and main chain (SPASM) uses a user-defined motif and compares it against a database of protein structures. RIGOR allows searches with entire protein structure using a database of predefined structural motifs.

3.2.2. Protein Repeats

Symmetry and structural duplication are widespread features of natural proteins. A vast number of protein structures with internal symmetry and/or regularly repeating structural units are known to date. These units, also called protein repeats, are usually arranged tandemly in a sequence and/or structure. They exist in multiplicity and thus differ from domains that can exist on their own. Two types of repeats can be distinguish: sequence and structural repeats. Sequence repeat can be defined as any sequence of the same amino acid residue or group of similar amino acid residues repeated in a protein. Frequently, the sequence identity and the number of sequence repeats vary across protein homologs. Structural repeat is regarded as any arrangement of secondary structural elements repeated in a protein structure. The boundaries of sequence repeats frequently correlate with those of structural repeats but in some proteins, e.g., potII family of proteinase inhibitors (43) and WD40-containing proteins (44), the sequence and structural repeats do not coincide.

Protein repeats can fold into compact domains that have a different degree of complexity and shape; and are often symmetrical. Some homologous repetitive structures can bent and coil in different ways so that their global structural similarity can become negligible. These considerable structural variations are usually a result of distinct packing interactions between neighbouring repeats. Protein repeats can form fibrous domains, globular domains, solenoids, and toroids. Repeats in fibrous domains are usually small, comprising only a few residues [collagen, coiled coil (Fig. 4a)]. Some globular proteins contain interlocking repeats that are formed by supersecondary structural elements (Fig. 4b). Solenoids are formed by more simple secondary structural elements such as $\alpha\alpha$-hairpins [heat, armadillo, and tetratricopeptide repeats (Fig. 4c)], $\beta\beta$-hairpins and β-arches [β-superhelix (Fig. 4d)], $\alpha\beta$-hairpins [leucine-rich repeat (Fig. 4e)] and fold into open sometimes elongated repetitive structures. Similarly, toroids are built by simple secondary structural elements but in contrast to solenoids form closed structures [$\alpha\alpha$-toroids (Fig. 4f), β-propellers (Fig. 4g), $(\beta\alpha)8$-barrels (Fig. 4h)].

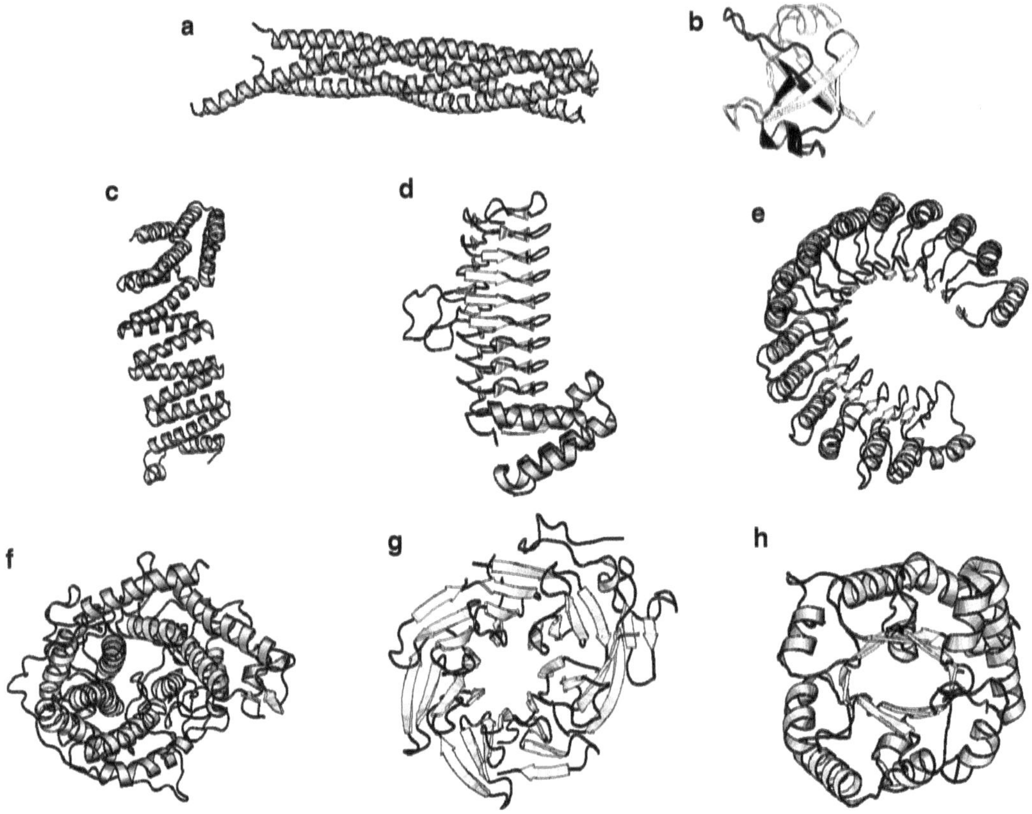

Fig. 4. Representative repetitive structures. (**a**) Coiled coil (pdb 1n7s), (**b**) structural repeats in globular domain (pdb 1cz4), (**c**) α-solenoid (pdb 1qqe), (**d**) β-solenoid (pdb 2jf2), (**e**) βα-solenoid (pdb 2bnh), (**f**) α-toroid (pdb 1gai), (**g**) β-toroid (pdb 1erj), and (**h**) βα-toroid (pdb 2jk2).

Methods for detecting repeats are available (45–48). Most of the methods for identification of sequence repeats utilize standard sequence comparison algorithms that are adapted for repeats. They usually perform well when the sequence similarity between repeats is substantial but fail to detect repeats with low sequence similarity or containing large insertions or deletions.

3.2.3. Protein Complexes

Majority of globular and membrane proteins assemble into oligomeric complexes consisting of two or more polypeptide chains. Within these oligomeric complexes two types can be distinguished, homomeric and heteromeric, that are composed of identical and non-identical chains, respectively. A large portion of protein complexes are homomeric with about 50–70% of proteins known to assemble into such structures (49). There are two different types of interfaces in oligomeric complexes: isologous (homologous) and heterologous. Isologous interface is formed by identical surfaces of the two subunits, whereas in heterologous interface, these surfaces are non-identical. Several studies in the past have addressed the structural properties of the oligomeric interfaces such as shape, size, packing, complementarity, etc. (50, 51) but these are beyond the scope of this chapter. Most of oligomeric structures posses symmetry. Dimers and trimers usually adopt cyclic symmetry, whereas dihedral symmetry is more common to tetramers (27, 52). Cubic symmetry is used in protein complexes such as ferritin and viral capsids to enclose vast cavities. Most oligomers adopt either cyclic or dihedral symmetry and only a small fraction of protein complexes have a cubic symmetry (53). Each of the features described above can be used as a criteria to organize and classify protein oligomeric complexes.

4. Classification Based on Protein Types

Proteins fall into four main groups each of which to large extent correlates with characteristic sequence and structural features. Given the striking differences between these groups, their organization and classification will be discussed separately.

4.1. Globular Proteins

Globular proteins are soluble in aqueous solutions. They tend to fold into compact units and their three-dimensional structure reflects their interaction with the solvent. Globular proteins are comparatively easy to analyse and crystallize and therefore, not surprisingly, this group of proteins is the best structurally characterized and comprises the largest fraction of protein structural space available for modeling. Their classification will be described in the next section of this chapter.

4.2. Fibrous Proteins

This group includes a number of structural proteins such as collagen, keratin, elastin, etc., most of which are insoluble. Depending on the secondary structure, fibrous proteins can be subdivided into three groups: triple helix, β-sheet fibres, and α-fibrous proteins. The former group is exemplified by collagen in which each individual polypeptide chain is folded into an extended polyproline type II helix. Three collagen chains coil around a central axis to form a right-handed triple helix. The second group of fibrous proteins tend to form β-sheet structures in which array of extended chains are stacked along the fibril axis. Besides β-keratin and silk proteins, this group includes amyloid fibres. The third group, also known as coiled-coil proteins, is becoming increasingly better understood in terms of sequence and structure. Typically, coiled coils are bundles of two, three, or more helices in which each helix is oriented parallel or antiparallel with respect to the adjacent one. These helices wrap around each other to form a supercoil which is usually left-handed. Although the formation of right-handed coiled-coils is less favourable, these are also observed in nature, e.g. in the structures of tetrabrachion (54), tetramerization domain of VASP

(55), IF regulatory subunitt of F-ATPase (56), and tetramerization domain of MNT repressor (57). Coiled-coil proteins can be homooligomeric or heterooligomeric.

A characteristic feature of the fibrous protein sequences is the presence of repetitive sequence motifs. Collagen, for instance, contains a short Gly-X-Y sequence motif where X is usually Pro and Y is Hyp. Characteristic for the canonical (left-handed, parallel) coiled-coil proteins are heptad repeats denoted as **a-b-c-d-e-f-g**, where **a** and **d** are hydrophobic residues located at the interface of the coiled-coil helices and **e** and **g** are polar residues exposed to the solvent. Nonheptad repeats result in non-canonical coiled-coils that lack left-handness or regular geometry. Right-handed coiled coils, for instance, contain an 11 residue repeat (undecatad repeat). The hydrophobic packing in these proteins substantially differs from the packing of the canonical coiled coils (54). Programmes for analysis of coils are Socket (58) and Twister (59). Socket identifies knobs-into-holes packing in coiled coils, whereas Twister determines the local structural parameters and detects local fluctuations in coiled-coil structures.

The first two subgroups of fibrous proteins are very poorly characterized and only few low resolution structures are available, e.g. the structure of collagen type I that has been recently determined by X-ray fibre diffraction (60). Coiled-coil proteins are difficult to crystallize due to aggregation problems and structures of fragments or relatively short coils are available. Classification of these proteins is usually based on the number of helices, their direction (parallel or antiparallel) and the handedness of the supercoil (left or right).

4.3. Membrane Proteins

Since the first low resolution structure of bacteriorhodopsin was determined by Henderson and Unwin in 1975 (61), much progress has been made in membrane crystallography. Currently, there are more than 200 high-resolution structures of unique membrane proteins. The majority of integral membrane proteins consist of transmembrane α-helices usually organized in bundles. Their topology can be defined on the basis of the number of transmembrane helices and their relative orientation with respect to the plane of the membrane bilayer. The geometry of the side-chains packing at the helix interfaces is reminiscent to knobs-into-holes packing observed in coiled coils (62). The transmembrane helices of proteins involved in proton and electron transport are highly hydrophobic, whereas transporter proteins such as lactose permease (63) have large hydrophilic cavities spanning along the membrane and their helices contain a number of polar and charged residues that are buried in the interior of the transmembrane domain. The transmembrane helices can have different length, different tilt with respect to the bilayer, and different type of distortions, e.g. kinks. Large dynamic changes in the helix orientation and packing interactions or local helix to coil transitions can occur in transmembrane proteins. This intrinsic dynamics of α-helical membrane proteins is a well-documented phenomenon and should be taken into account during structural analysis and classification (64–68).

Another architectural type observed mainly in outer membrane proteins is the β-sheet barrel. All known transmembrane β-barrels form closed structures in which their first strand is hydrogen bonded to the last. The number of strands in the barrel is even and all β-strands are antiparallel. Many barrels contain water filled channels and thus the interior residues are predominantly polar, whereas hydrophobic residues are exposed on the barrel surface. In some proteins, the barrel interior is occupied by additional secondary structural elements or domains. The barrel of autotransporter Nalp, for instance, is filled with an N-terminal helix (69), whereas the barrel of FhuA receptor is plugged by α/β domain (70).

Classification of membrane proteins is primary based on their typical architectural and topological features. Since some membrane proteins have evolved via duplication and fusion, it is important to examine the structure for the presence of internal repeats before it is compared to structures of other proteins. Structure comparison search with a repeat of this kind could reveal a similarity that can be missed if the entire structure is used.

4.4. Intrinsically Unstructured Proteins

Regions of proteins or even entire proteins at native conditions may lack ordered structure but in their functional state they can undergo disorder-to-order transition. These are known as natively unfolded, intrinsically disordered or intrinsically unstructured proteins (IUPs) (71–75). IUPs gained much interest over the last years particularly because they reside in functionally important regions in proteins and comprise a substantial fraction of eukaryotic proteome. Most importantly, these proteins or regions of proteins violate the classical sequence–structure–function paradigm of structural biology, that is, the protein sequence determines a unique 3D structure that in turn determines the proteins' function.

Intrinsic disorder offers several advantages such as binding of diverse ligands (functional promiscuity), provides a large interaction interface, rapid turnover in the cell, and allows high-specificity coupled with low-affinity interaction. IUPs exist in dynamic ensembles in which the backbone conformation varies over the time and which undergo non-cooperative conformational changes. Typically, the binding to their target (nucleic acid or protein) is accompanied with a shift in the conformational ensemble and a selection of "bound" conformation which is complementary to the binding partner. For example, a number of proteins such as VP16 and p53 contain acidic activation domains that are unstructured in a free state. Upon binding to different target proteins, they undergo disorder-to-order conformational change (76–79). Both electrostatic and hydrophobic interactions are attributed to this phenomenon.

While electrostatics is essential for the mutual attraction to the partner domain, the hydrophobic interactions are essential for the folding of the activation domain (78). Remarkably, although these activation domains bind to structurally distinct protein domains, in all instances they adopt α-helical conformation. Other IUPs, e.g. α-synuclein (80), the C-terminal regulatory domain of p53 (76), exhibit chameleon behaviour and can adopt different conformations (α-helical or β-structures) depending on the environment and the nature of their target domain.

When compared with globular proteins, sequences of IUPs are less conserved. In the absence of strong structural constraints, their sequences have change rapidly during the evolution. In general, IUPs lack the typical patterns of hydrophobic residues observed in globular proteins. Most of them have unusual sequences exhibiting low sequence complexity or high content of charged and low content of hydrophobic residues. This strong bias in their amino acid composition allows successful prediction of protein disorder from the sequence. Several programmes have been developed over the past years (81–83). Structures of quite a few intrinsically disordered regions of proteins bound to their partner proteins have been determined by X-ray crystallography and NMR. None of these, however, have been included in the scope of any of the current protein classifications. A recently developed database, DisProt, provides structural and functional information about disordered proteins (84).

5. Classification of Globular Proteins

The strategy for classifying protein structures, described here, concerns classification of globular proteins but it can be employed for other protein types such as membrane proteins. Steps in the classification procedure of protein domains will be outlined.

Classification of a new protein structure usually begins with analysis of the structure itself. This includes a search for any internal sequence and structural similarity; analysis of the proteins' oligomeric state (biological unit) and domain assignment. Detection of internal similarity can indicate duplication of domains in multidomain proteins or repeats in single domains. The constituent subunits of homooligomeric complexes can exchange equivalent core secondary structural elements (segment-swapping) and domains in these swapped structures should be defined by including corresponding parts of both polypeptide chains. Protein domains are usually consecutive in sequence, but in some proteins one domain can be inserted into another or in a more complex scenario, equivalent structural elements can be swapped between both domains. Because of the ambiguity in identifying domains on the basis of a single structure, it is usually best to start with preliminary domain assignment and tentatively to refine it during the classification process.

Classification of new protein structure depends on its relationship to other proteins with known 3D structure. This relationship can be structural arising from physics and chemistry of proteins favouring particular packing arrangements and topologies or evolutionary due to a descent from a common ancestral protein. Steps of classification aiming identification of these relationships are described below.

5.1. Assignment of Probable Evolutionary Relationships

Protein domains that have evolved from a common ancestor usually share common sequence, structural, and/or functional features. Significant global sequence similarity is considered to be a sufficient evidence for a common ancestry and usually defines close evolutionary relationships. Close evolutionary relationships are detectable with simple BLAST searches (85). More distant (remote) evolutionary relationships can be detected using PSI-BLAST or HMM-profile (86) searches or more sensitive profile–profile approaches such as PRC (87) and COMPASS (88). In the absence of sequence similarity, structural similarity along with commonality in function can also indicate a distant homology. In addition, conserved features such as rare or unusual topological details, conserved packing interactions, common binding/active sites can be used to support a confident conclusion for a common ancestry.

5.2. Assignment of Protein Fold

Assignment of fold is not trivial since there is no single universal definition of protein fold. The term "fold" was originally introduced to outline three major aspects of protein structure: the secondary structural elements of which it is composed, their spatial arrangement and their connectivity. The term "common fold" is used to describe the consensus subset of structural elements shared by a group of proteins. Proteins with the same common fold usually differ in their peripheral structural elements that may have distinct conformation or size. In extreme cases, particularly when homologous proteins are more divergent or have underwent events, such as deletions, insertions, etc (described in the next section), these differences may comprise more than a half of the domain.

Some folds are easy to recognize by eye, e.g. ($\beta\alpha$)8-barrel, β-propeller, and many others. For identification of a common fold, it is usually best to perform a structure comparison search against a database of proteins with known structures. Various structure comparison tools can be used to detect structural similarities and some of these are shown in Table 1. Frequently, different methods give different results. For interpretation of the structural similarities is recommended to use the results of several structure comparison algorithms (see Note 4).

5.3. Assignment of Protein Class

Depending on the secondary structure composition, globular protein domains can be divided into four major classes: all-α (predominantly α-helices), all-β (predominantly β-strands), α/β (alternating α-helices and β-strands, and α+β (segregated α-helices and β-strands) (see Note 5). A fifth class includes small proteins with little or no secondary structures. These are usually small proteins that are stabilized either by disulphide bonds or by metal coordination. The division into five classes is adopted by the SCOP classification scheme. Usually, the assignment of all-α and all-β protein classes is straightforward. The borderline between α/β and α+β classes is not always clear. For this reason, the authors of the CATH database, for instance, have merged these two classes into one, namely mixed αβ structures.

6. Dogmas, Principles and Rules, and Their Exceptions

The plethora of structural data accumulated over the past decade revealed numerous examples of atypical structural features and large structural variations that have challenged many longstanding tenets in protein science (33, 89–92). The central dogma of protein folding "one sequence–one structure" is increasingly being challenged as many structural variations are observed in protein families and their individual members. Many exceptions to the topological rules established by earlier protein structure analyses also become apparent. Knowledge of these is essential for both protein structure classification and modeling. Some examples are discussed in this section.

6.1. Sequence–Structure Relationships

In the early 1960s, Anfinsen proposed what he called a "thermodynamic hypothesis" of protein folding to explain the biologically active conformation of protein structure (93, 94). He theorized that the native structure of protein is thermodynamically the most stable under in vivo conditions. Anfinsen postulated that in a given environment, the protein structure is determined by the sum of interatomic interactions and hence by the amino acid sequence. While to a large extent this theory holds true for most proteins, there is a new growing phenomenon of proteins existing in multiple conformational states or adopting conformation that is not at the thermodynamic minimum. In addition, regions of some proteins exhibit chameleon behaviour and can fold into alternative secondary structures.

6.1.1. One Sequence: Many Folds

The most remarkable examples of proteins existing in equilibrium between two entirely different conformational states are Mad2 (95) and lymphotactin (96) (Fig. 5). The transition between the two conformations in both proteins involves a large rear-

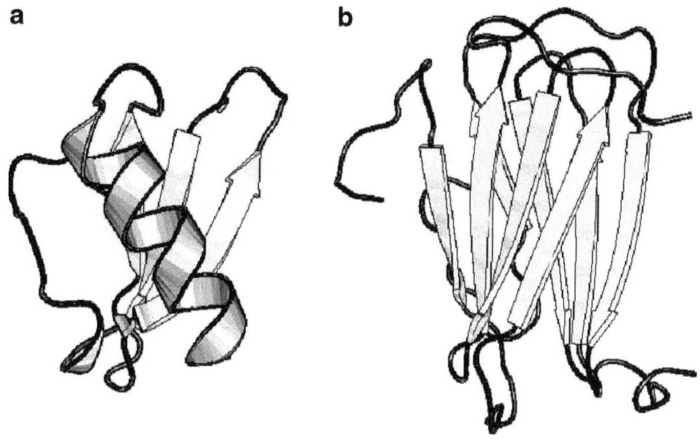

Fig. 5. The structures of two alternative folds of lymphotactin (Ltn10). (**a**) Monomeric Ltn10 (pdb 1j8i) and (**b**) dimeric Ltn10 (pdb 2jp1).

rangement of the hydrogen bonding network and many of the packing interactions.

Several proteins that assume multiple conformational states can adopt biologically active conformation that is not the thermodynamically most stable. This has been shown to play an important role for function. α-Lytic protease and α_1-antitrypsin, for instance, fold into metastable native state, while avoiding the stable but inactive conformation (reviewed in ref. 97). The formation of a metastable native state structure has been described for a number of proteins such as hemaglutinin (98), gp120 and gp41 from HIV (99), protein E from TBEV (100), and some heat shock transcription factors (101).

Depending on the environment some proteins can undergo dramatic conformational changes. The death domain of protein kinase Pelle (Pelle-DD), for example, adopts a six helical bundle characteristic for the death domain family. In the presence of MPD (2-methyl-2,4-pentanediol), the structure of Pelle-DD refolds into a single helix (102) (Fig. 6). Other factors such as pH, salt concentration, temperature are also known to induce conformational transitions. Lymphotactin, for instance, undergoes large structural rearrangement depending on temperature and salt concentration (103). In certain proteins, conformational transitions can be induced by changes in pH, as observed in influenza virus hemagglutinin (98) or pheromone-binding protein (104). Conformational switches can also be a result of experimental design. The design of truncated proteins, in which parts of the polypeptide chain is omitted, may result in dramatic changes of their fold or oligomeric state as observed in p73 (105), MinC (106), Kv7.1 (107), and more recently in human splicing protein PRP8 D4 domain (108).

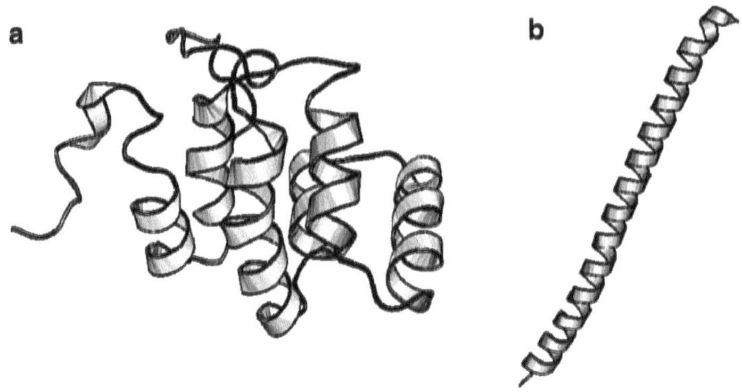

Fig. 6. The death domain of protein kinase Pelle (Pelle-DD) (**a**) solution structure, (**b**) crystal structure in MPD.

6.1.2. Chameleon Sequences

Strings of identical amino acid residues, the so-called chameleon sequences, can adopt alternative secondary structures (α-helix, β-strand, coil). Some chameleon sequences are found in structurally distinct proteins (109, 110). Others are present in individual proteins such as MAD2 (95), matα2 (111), elongation factor Tu (112, 113), p53 (76), Axh (114, 115), Radixin (116, 117), SecA (118), Lekti (119), etc. Most of these chameleon sequences undergo transitions from α-helix to β-strand. The conformational transitions in MAD2 and matα2 are particularly interesting since they are observed under identical conditions. In some proteins, these transitions occur upon oligomer formation. In isolated α-apical domain of thermosome, for instance, the crystal contacts involve a short helical segment resulting in the formation of a four helical bundle between symmetry-related molecules (Fig. 7a) (120, 121). In the closed thermosome, the same region participates in the formation of a β-barrel ring (Fig. 7b). Its conformation is stabilized by interactions provided by the equivalent regions of the adjacent subunits.

6.2. Topological Principles That Determine the Protein Structure

Several topological rules have been established during early analyses aiming to underline the basic principles that govern the protein structure (122–125). One of these postulates that secondary structures, α-helices, and β-sheets, closely pack to enclose hydrophobic core. Others describe preferences such as secondary structures adjacent in sequence are adjacent in structure, right-handedness of connections in β-X-β units, etc. Some topological features as knots and crossing connections were considered improbable and even prohibited. Nowadays, many exceptions of these rules have been found in protein structures. Some of these are shown in Fig. 8.

6.3. Evolution of Protein Structures

A common tenet of protein evolution is that the structure is more conserved than the protein sequence. While for many proteins that's true, steadily growing is the number of evolutionarily related proteins that revealed dramatic changes in their fold. These changes

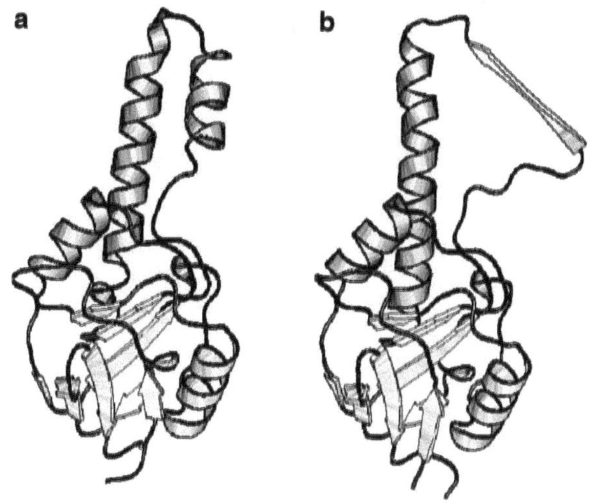

Fig. 7. α-Apical domain of thermosome. (**a**) Structure of isolated domain, (**b**) structure of a subunit in the closed thermosome.

affect not only the peripheral elements but the structural core as well (reviewed in refs. 33, 90, 92). Some examples are given below.

6.3.1. Fold Decay

Fold decay is a deletion event that affects the protein common fold. Fold decay is observed, for instance, in the family B of DNA polymerases. The exonuclease domain of prokaryotic DNA polymerases contains an additional five-stranded β-barrel subdomain with a canonical OB-fold. In the structures of archaeal polymerases, this domain has deletions of different size resulting in the formation of either a three-stranded curved β-sheet or an open β-barrel (Fig. 9).

6.3.2. Fold Transitions

Perhaps the most remarkable example of fold transition is observed in the structures of NusG and RfaH (126). The C-terminal domain of NusG is a SH3-like barrel that contains the so-called KOW motif. Despite the significant sequence similarity between this domain and the C-terminal domain of its homolog RfaH, the latter folds into α-helical domain instead of β-barrel (Fig. 10). Homology modeling of RfaH using the structure of NusG showed that the RfaH sequence can be easily tread on the NusG β-barrel while maintaining the hydrophobic core and avoiding steric clashes (126).

6.3.3. Architecture Transitions

Insertion of additional secondary structures to a common fold core can result in a novel architecture. YaeQ, for example, resembles the restriction endonucleases fold but it contains additional N- and C-terminal β-structures forming a five-stranded β-sheet (127) (Fig. 11). These extra secondary structural elements contribute to the formation of a distinct barrel-like architecture. Despite these

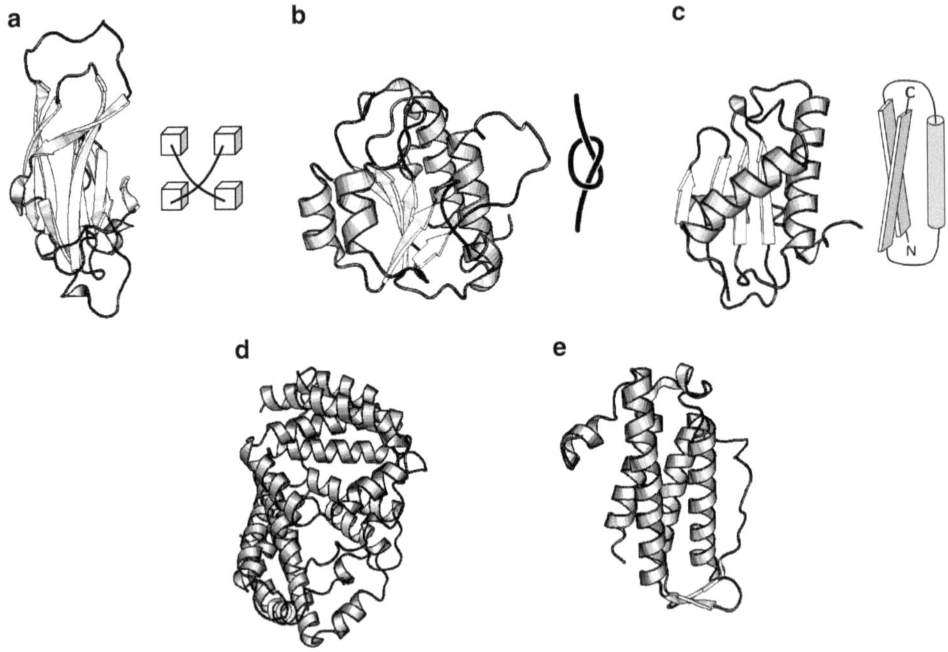

Fig. 8. Examples of exceptions to topological rules. **Rule**: connections between secondary structures neither cross each other nor make knots in the chain. **Exceptions**: (**a**) crossing connections in ecotin (pdb 1ifg) and (**b**) deep trefoil knot in the structure of YibK methyltransferase (pdb 1mxi); **Rule**: connections of β-X-β are right handed. Exception: (**c**) left-handed connection in the structure of Ribonuclease P (pdb 1a6f); **Rule**: the association of secondary structures, α-helices and β-sheets, close pack to form a hydrophobic core. **Exception**: (**d**) the structure of peridinin–chlorophyll–protein (pdb 1ppr) that does not have a core but instead enclosing ligand binding cavity; **Rule**: pieces of secondary structures that are adjacent in sequence are often in contact in three dimensions. **Exception**: (**e**) high contact order structure of representative of DinB-like family (pdb 2f22).

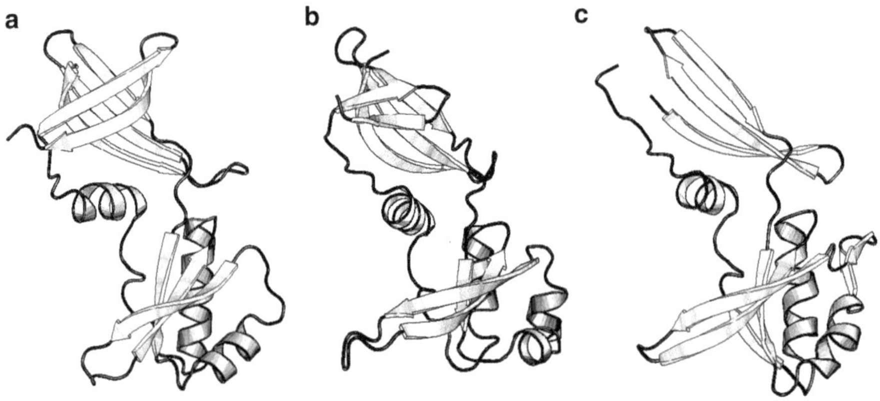

Fig. 9. Fold decay. Structures of exonuclease domains of (**a**) *Escherichia coli* DNA polymerase (pdb 1q8i), (**b**) *Sulfolobus solfataricus* DNA polymerase (pdb 1s5j), (**c**) *Thermococcus gorgonarius* DNA polymerase (pdb 1tgo).

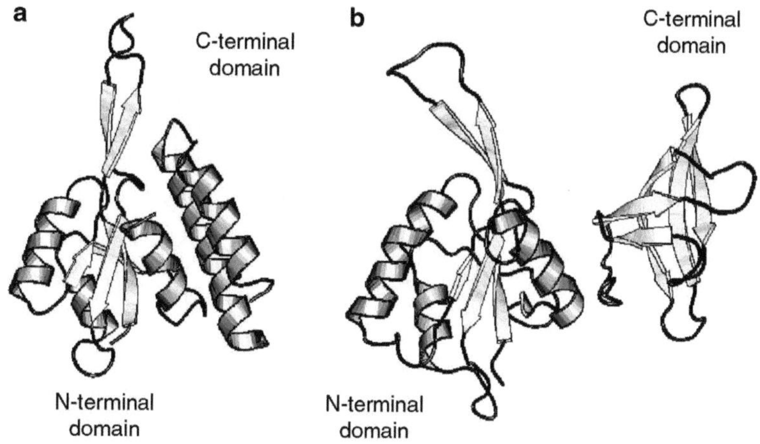

Fig. 10. Fold transition. Structures of (**a**) RfaH and (**b**) NusG.

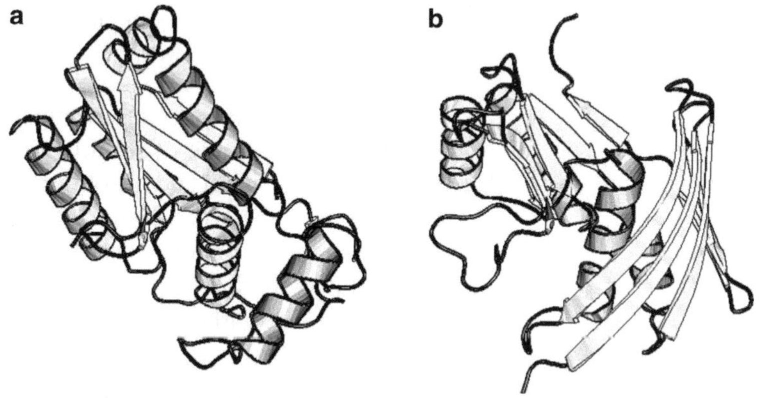

Fig. 11. Architecture transition. Structures of (**a**) restriction endonuclease *Bam*HI (pdb 1bam) and (**b**) *Yae*Q (pdb 2g3w).

differences, residues essential for catalysis in restriction endonucleases, are conserved in the YaeQ structure.

6.3.4. Circular Permutations

Circular permutation can be regarded as a change of the sequential order of the N- and C-terminal parts in protein structures. As such, it does not affect the relative spatial arrangement or packing interactions of the secondary structural elements. Numerous examples of circular permutations are known to date. One example is the structure of phospholipase CΔ C2-domain that has a circularly permuted topology of synaptotagmin I C2-domain (128, 129). The difference between the two topologies is in the first strand of synaptotagmin C2-domain that occupies the same spatial position as the last strand of the phospholipase CΔ C2-domain (Fig. 12).

6.3.5. Strand Flip and Swap

Strand flip is regarded as change of the orientation of the strand with respect to the core elements, whereas strand swap is an internal

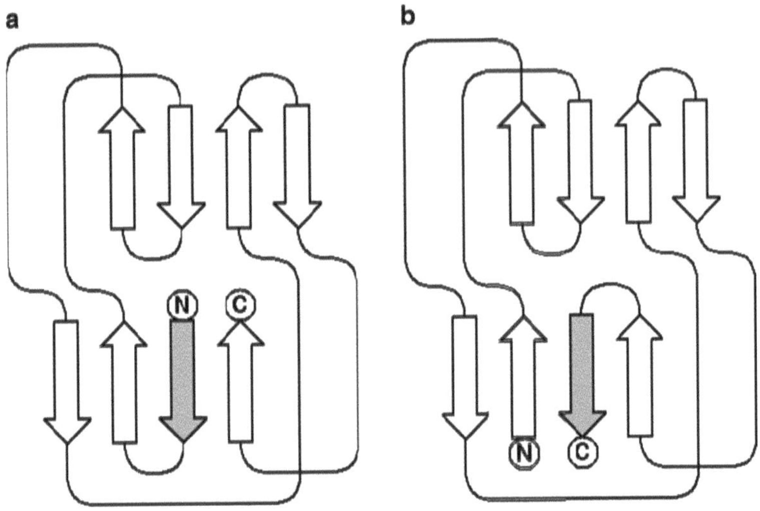

Fig. 12. Circular permutation. Topology diagram of (**a**) synaptotagmin C2-domain, (**b**) phospholipase CΔ C2-domain. Circularly permuted strand is shown in *grey*.

exchange of β-strands that occupy positions with similar environment. One well-known example of strand swap is triabin. The sequence similarity between triabin and nitrophorin is detectable with BLAST. The nitrophorin structure comprises an eight-stranded β-barrel in which all strands are antiparallel. The N-terminal region of triabin differs by swap of a β-hairpin, which results in a parallel arrangement of two pairs of β-strands (Fig. 13).

7. Protein Structure Classification Schemes

Two major manually curated classifications of protein structures are currently available, SCOP (10, 130, 131) and CATH (11, 19, 132). Both classifications have a hierarchical tree-like structure in which protein domains are arranged according to their structural and evolutionary relationships. While these classifications share some common philosophical underpinnings, they differ in several aspects such as domain definitions and classification assignments (133, 134). An overview of these classifications is given below.

A number of other resources that automatically cluster protein structures to build structural neighbourhoods are also available (8, 135–137) (see Table 1). The clustering in these databases depends on the structure comparison method that is employed and algorithm settings that are used. Since comparison methods differ in their results, particularly when the structural similarity between proteins is not significant, the resulting clusters are frequently very different.

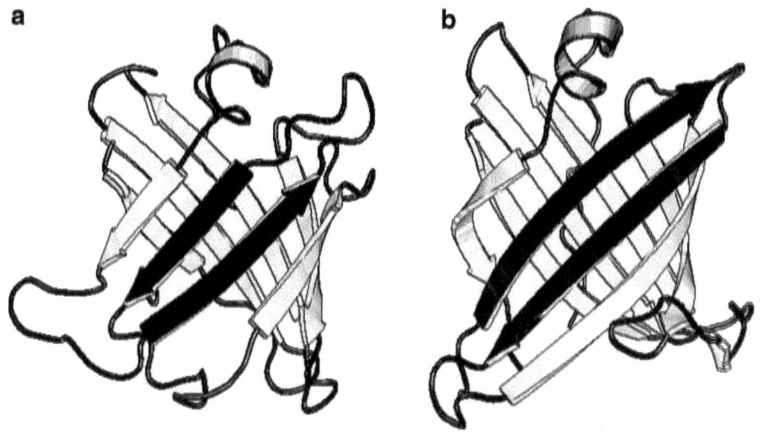

Fig. 13. Strand swap. Structures of (**a**) triabin (pdb 1avg) and (**b**) nitrophorin (pdb 1pee). Swapped β-hairpin is shown in *black*.

7.1. SCOP

SCOP is a database, in which the main focus is to place the proteins in a coherent evolutionary framework, based on their conserved sequence and structural features. It has been created as a hierarchy in which protein domains are arranged in different levels according to their structure and evolution. The SCOP hierarchy comprises the following seven levels: protein **Species**, representing a distinct protein sequence and its naturally occurring or artificially created variants; **Protein**, grouping together similar sequences of essentially the same functions that either originate from different biological species or present different isoforms within the same organism; **Family**, organizing proteins of related sequences but distinct functions; **Superfamily**, bringing together protein families with a common functional and structural features. Near the root of the SCOP hierarchy, structurally similar superfamilies are grouped into **Folds**, which are further arranged into **Classes** based on their secondary structural content.

The classification of proteins in SCOP is a bona fide research. During the classification process, the sequence and structural similarities between proteins are very carefully analysed and interpreted to achieve an optimal prediction of the proteins' evolutionary history. Thus, SCOP is an excellent resource to study the sequence and structural divergence of homologous proteins and the type of structural changes they underwent in the course of evolution.

Structural variations amongst homologous and individual proteins, and the existence of motifs common to structurally distinct proteins add extra complexity and create difficulties in their presentation on the SCOP hierarchy. A comprehensive annotation of these proteins is provided in SISYPHUS, a compendium of

SCOP database (28). The SISYPHUS design conceptually differs from the established classification schemes. In contrast to the latter that are domain-based, the database contains protein structural regions of different size that range from short fragments (motifs or repeats), domains to oligomeric biological units. These protein structural regions are organized in categories that are connected by complex non-hierarchical interrelationships. The relationships between these structural regions are evidenced by multiple alignments and annotated using controlled vocabulary (keywords) and Gene Ontology terms.

7.2. CATH

CATH is a hierarchical protein structure classification in which the protein domains are organized in nine levels. Lower levels of CATH comprise subfamilies of domains that are clustered based on their sequence similarity. Protein domains are merged in **Homologous superfamily** *(H-level)* if they share significant sequence, structure, and/or functional similarity. **Topology** *(T-level)* groups together proteins with a similar arrangement of their secondary structures and topology. Next level, **Architecture** *(A-level)* refers to the overall arrangement of the secondary structures regardless their connectivity. At the root of the hierarchy, **Class** *(C-level)* is defined according to the secondary structure composition. With the exception of A-level that is unique to CATH, the other levels have their equivalent in the SCOP database. The CATH classification protocol uses a highly automated system combined with manual curation (19). Supplementary resource to CATH is CATH-DHS (Dictionary of Homologous Structures) which contains multiple structural alignments, consensus information and functional annotations for proteins grouped at H-level in the classification (138).

7.3. 3D Complex

3D complex is a classification of protein complexes of known three-dimensional structure, representing their fundamental structural features as a graph (27, 52). Proteins are organized in 12 hierarchical levels by using one or more of the following criteria for comparison of the protein complexes: (1) topology of the complex, represented by the number of chains and their pattern of contacts; (2) domain architecture of each constituent chain in the complex according to SCOP classification; (3) number of non-identical chains per domain architecture within each complex; (4) sequence similarity between the constituent chains in the complex; (5) symmetry of the complex. The database allows browsing and analysis of both homomeric and heteromeric complexes and their evolutionary relationships.

8. Notes

1. Because of many structural variations observed amongst homologous proteins and exceptions to rules and definitions, any classification of protein structures will be approximate. The choice of classification scheme should depend on the applications for which it will be used.
2. Every group of related proteins has its own evolutionary history and may underwent events that may not be observed in other proteins. Case by case analysis of protein sequence and structural similarities is, therefore, recommended as it is more powerful way for the detection of protein evolutionary relationships.
3. Given a protein structure, perform sequence analysis of its close homologs with unknown structure. This is best done by search against a sequence database (see Table 1). The sequences of close homologs can be used to generate a multiple sequence alignment and project the sequence conservation on the structure. Best tools to use are Jalview (139) and Consurf (140). Analysis of this type can reveal strictly conserved structural features within the protein family some of which may be related to function.
4. Seek for peculiarities in protein structures such as unusual packing or topological details (knots, left-handed connections, crossing connections). These are characteristic features of folds and can assist in the decision making process during fold assignment.
5. During assignment of protein class, only the core elements of protein domain should be considered. The peripheral elements are usually less conserved and may contain additional structural elements.
6. A significant local sequence similarity between proteins does not necessarily indicate that their structures are globally similar. If a common sequence motif is identified in proteins with known structure, always analyse and compare their structures in order to classify them. If a local sequence match to a protein template structure is found, this not always means that the structure is a suitable template for homology modeling.

References

1. Kendrew, J. C., Bodo, G., Dintzis, H. M., Parrish, R. G., Wyckoff, H., and Phillips, D. C. (1958) A three-dimensional model of the myoglobin molecule obtained by x-ray analysis, *Nature 181*, 662–666.
2. Berman, H. M., Westbrook, J., Feng, Z., Gilliland, G., Bhat, T. N., Weissig, H., Shindyalov, I. N., and Bourne, P. E. (2000) The Protein Data Bank, *Nucleic Acids Res 28*, 235–242.
3. Chothia, C. (1984) Principles that determine the structure of proteins, *Annu. Rev. Biochem. 53*, 537–572.
4. Chothia, C., Levitt, M., and Richardson, D. (1977) Structure of proteins: packing of alpha-helices and pleated sheets, *Proc. Natl. Acad. Sci. USA 74*, 4130–4134.
5. Levitt, M., and Chothia, C. (1976) Structural patterns in globular proteins, *Nature 261*, 552–558.
6. Richardson, J. S. (1977) beta-Sheet topology and the relatedness of proteins, *Nature 268*, 495–500.
7. Richardson, J. S. (1981) The anatomy and taxonomy of protein structure, *Adv. Protein Chem. 34*, 167–339.
8. Holm, L., and Sander, C. (1994) The FSSP database of structurally aligned protein fold families, *Nucleic Acids Res 22*, 3600–3609.
9. Ohkawa, H., Ostell, J., and Bryant, S. (1995) MMDB: an ASN.1 specification for macromolecular structure, *Proc Int Conf Intell Syst Mol Biol 3*, 259–267.
10. Murzin, A. G., Brenner, S. E., Hubbard, T., and Chothia, C. (1995) SCOP: a structural classification of proteins database for the investigation of sequences and structures, *J Mol Biol 247*, 536–540.
11. Orengo, C. A., Pearl, F. M., Bray, J. E., Todd, A. E., Martin, A. C., Lo Conte, L., and Thornton, J. M. (1999) The CATH Database provides insights into protein structure/function relationships, *Nucleic Acids Res 27*, 275–279.
12. Orengo, C. A., Michie, A. D., Jones, S., Jones, D. T., Swindells, M. B., and Thornton, J. M. (1997) CATH – a hierarchic classification of protein domain structures, *Structure 5*, 1093–1108.
13. Wetlaufer, D. B. (1973) Nucleation, rapid folding, and globular intrachain regions in proteins, *Proc Natl Acad Sci USA 70*, 697–701.
14. Rossmann, M. G., Moras, D., and Olsen, K. W. (1974) Chemical and biological evolution of nucleotide-binding protein, *Nature 250*, 194–199.
15. Remaut, H., Bompard-Gilles, C., Goffin, C., Frere, J. M., and Van Beeumen, J. (2001) Structure of the Bacillus subtilis D-aminopeptidase DppA reveals a novel self-compartmentalizing protease, *Nat Struct Biol 8*, 674–678.
16. Alden, K., Veretnik, S., and Bourne, P. E. (2010) dConsensus: a tool for displaying domain assignments by multiple structure-based algorithms and for construction of a consensus assignment, *BMC Bioinformatics 11*, 310.
17. Alexandrov, N., and Shindyalov, I. (2003) PDP: protein domain parser, *Bioinformatics 19*, 429–430.
18. Holm, L., and Sander, C. (1994) Parser for protein folding units, *Proteins 19*, 256-268.
19. Redfern, O. C., Harrison, A., Dallman, T., Pearl, F. M., and Orengo, C. A. (2007) CATHEDRAL: a fast and effective algorithm to predict folds and domain boundaries from multidomain protein structures, *PLoS Comput Biol 3*, e232.
20. Siddiqui, A. S., and Barton, G. J. (1995) Continuous and discontinuous domains: an algorithm for the automatic generation of reliable protein domain definitions, *Protein Sci 4*, 872–884.
21. Sowdhamini, R., and Blundell, T. L. (1995) An automatic method involving cluster analysis of secondary structures for the identification of domains in proteins, *Protein Sci 4*, 506–520.
22. Swindells, M. B. (1995) A procedure for detecting structural domains in proteins, *Protein Sci 4*, 103–112.
23. Taylor, W. R. (1999) Protein structural domain identification, *Protein Eng 12*, 203–216.
24. Veretnik, S., Bourne, P. E., Alexandrov, N. N., and Shindyalov, I. N. (2004) Toward consistent assignment of structural domains in proteins, *J Mol Biol 339*, 647–678.
25. Zhou, H., Xue, B., and Zhou, Y. (2007) DDOMAIN: Dividing structures into domains using a normalized domain-domain interaction profile, *Protein Sci 16*, 947–955.
26. Sigrist, C. J., Cerutti, L., de Castro, E., Langendijk-Genevaux, P. S., Bulliard, V., Bairoch, A., and Hulo, N. (2010) PROSITE, a protein domain database for functional characterization and annotation, *Nucleic Acids Res 38*, D161–166.
27. Levy, E. D., Pereira-Leal, J. B., Chothia, C., and Teichmann, S. A. (2006) 3D complex: a structural classification of protein complexes, *PLoS Comput Biol 2*, e155.

28. Andreeva, A., Prlic, A., Hubbard, T. J., and Murzin, A. G. (2007) SISYPHUS – structural alignments for proteins with non-trivial relationships, *Nucleic Acids Res 35*, D253–259.
29. Hemmingsen, J. M., Gernert, K. M., Richardson, J. S., and Richardson, D. C. (1994) The tyrosine corner: a feature of most Greek key beta-barrel proteins, *Protein Sci 3*, 1927–1937.
30. Brennan, R. G., and Matthews, B. W. (1989) The helix-turn-helix DNA binding motif, *J Biol Chem 264*, 1903–1906.
31. Doherty, A. J., Serpell, L. C., and Ponting, C. P. (1996) The helix-hairpin-helix DNA-binding motif: a structural basis for non-sequence-specific recognition of DNA, *Nucleic Acids Res 24*, 2488–2497.
32. Religa, T. L., Johnson, C. M., Vu, D. M., Brewer, S. H., Dyer, R. B., and Fersht, A. R. (2007) The helix-turn-helix motif as an ultrafast independently folding domain: the pathway of folding of Engrailed homeodomain, *Proc Natl Acad Sci USA 104*, 9272–9277.
33. Andreeva, A., and Murzin, A. G. (2006) Evolution of protein fold in the presence of functional constraints, *Current Opinion in Structural Biology 16*, 399–408.
34. Grishin, N. V. (2001) KH domain: one motif, two folds, *Nucleic Acids Res 29*, 638–643.
35. Bellamacina, C. R. (1996) The nicotinamide dinucleotide binding motif: a comparison of nucleotide binding proteins, *FASEB J 10*, 1257–1269.
36. Rigden, D. J., and Galperin, M. Y. (2004) The DxDxDG motif for calcium binding: multiple structural contexts and implications for evolution, *J Mol Biol 343*, 971–984.
37. Saraste, M., Sibbald, P. R., and Wittinghofer, A. (1990) The P-loop – a common motif in ATP- and GTP-binding proteins, *Trends Biochem Sci 15*, 430–434.
38. Jonassen, I. (1997) Efficient discovery of conserved patterns using a pattern graph, *Comput Appl Biosci 13*, 509–522.
39. Jonassen, I., Collins, J. F., and Higgins, D. G. (1995) Finding flexible patterns in unaligned protein sequences, *Protein Sci 4*, 1587–1595.
40. Rigoutsos, I., and Floratos, A. (1998) Combinatorial pattern discovery in biological sequences: The TEIRESIAS algorithm, *Bioinformatics 14*, 55–67.
41. Ye, K., Kosters, W. A., and Ijzerman, A. P. (2007) An efficient, versatile and scalable pattern growth approach to mine frequent patterns in unaligned protein sequences, *Bioinformatics 23*, 687–693.
42. Kleywegt, G. J. (1999) Recognition of spatial motifs in protein structures, *J Mol Biol 285*, 1887–1897.
43. Lee, M. C., Scanlon, M. J., Craik, D. J., and Anderson, M. A. (1999) A novel two-chain proteinase inhibitor generated by circularization of a multidomain precursor protein, *Nat Struct Biol 6*, 526–530.
44. Neer, E. J., Schmidt, C. J., Nambudripad, R., and Smith, T. F. (1994) The ancient regulatory-protein family of WD-repeat proteins, *Nature 371*, 297–300.
45. Murray, K. B., Gorse, D., and Thornton, J. M. (2002) Wavelet transforms for the characterization and detection of repeating motifs, *J Mol Biol 316*, 341–363.
46. Heger, A., and Holm, L. (2000) Rapid automatic detection and alignment of repeats in protein sequences, *Proteins 41*, 224–237.
47. Andrade, M. A., Ponting, C. P., Gibson, T. J., and Bork, P. (2000) Homology-based method for identification of protein repeats using statistical significance estimates, *J Mol Biol 298*, 521–537.
48. Murray, K. B., Taylor, W. R., and Thornton, J. M. (2004) Toward the detection and validation of repeats in protein structure, *Proteins 57*, 365–380.
49. Levy, E. D., Boeri Erba, E., Robinson, C. V., and Teichmann, S. A. (2008) Assembly reflects evolution of protein complexes, *Nature 453*, 1262–1265.
50. Chothia, C., and Janin, J. (1975) Principles of protein-protein recognition, *Nature 256*, 705–708.
51. Jones, S., and Thornton, J. M. (1997) Analysis of protein-protein interaction sites using surface patches, *J Mol Biol 272*, 121–132.
52. Levy, E. D. (2007) PiQSi: protein quaternary structure investigation, *Structure 15*, 1364–1367.
53. Janin, J., Bahadur, R. P., and Chakrabarti, P. (2008) Protein-protein interaction and quaternary structure, *Q Rev Biophys 41*, 133–180.
54. Stetefeld, J., Jenny, M., Schulthess, T., Landwehr, R., Engel, J., and Kammerer, R. A. (2000) Crystal structure of a naturally occurring parallel right-handed coiled coil tetramer, *Nat Struct Biol 7*, 772–776.
55. Kuhnel, K., Jarchau, T., Wolf, E., Schlichting, I., Walter, U., Wittinghofer, A., and Strelkov, S. V. (2004) The VASP tetramerization domain is a right-handed coiled coil based on a 15-residue repeat, *Proc Natl Acad Sci USA 101*, 17027–17032.
56. Cabezon, E., Runswick, M. J., Leslie, A. G., and Walker, J. E. (2001) The structure of bovine IF(1), the regulatory subunit of mitochondrial F-ATPase, *EMBO J 20*, 6990–6996.
57. Nooren, I. M., Kaptein, R., Sauer, R. T., and Boelens, R. (1999) The tetramerization

domain of the Mnt repressor consists of two right-handed coiled coils, *Nat Struct Biol 6*, 755–759.

58. Walshaw, J., and Woolfson, D. N. (2001) Socket: a program for identifying and analysing coiled-coil motifs within protein structures, *J Mol Biol 307*, 1427–1450.
59. Strelkov, S. V., and Burkhard, P. (2002) Analysis of alpha-helical coiled coils with the program TWISTER reveals a structural mechanism for stutter compensation, *J Struct Biol 137*, 54–64.
60. Orgel, J. P., Irving, T. C., Miller, A., and Wess, T. J. (2006) Microfibrillar structure of type I collagen in situ, *Proc Natl Acad Sci USA 103*, 9001–9005.
61. Henderson, R., and Unwin, P. N. (1975) Three-dimensional model of purple membrane obtained by electron microscopy, *Nature 257*, 28–32.
62. Walters, R. F., and DeGrado, W. F. (2006) Helix-packing motifs in membrane proteins, *Proc Natl Acad Sci USA 103*, 13658–13663.
63. Guan, L., Mirza, O., Verner, G., Iwata, S., and Kaback, H. R. (2007) Structural determination of wild-type lactose permease, *Proc Natl Acad Sci USA 104*, 15294–15298.
64. Abramson, J., Smirnova, I., Kasho, V., Verner, G., Kaback, H. R., and Iwata, S. (2003) Structure and mechanism of the lactose permease of Escherichia coli, *Science 301*, 610–615.
65. Gupta, S., Bavro, V. N., D'Mello, R., Tucker, S. J., Venien-Bryan, C., and Chance, M. R. (2010) Conformational changes during the gating of a potassium channel revealed by structural mass spectrometry, *Structure 18*, 839–846.
66. Toyoshima, C., and Nomura, H. (2002) Structural changes in the calcium pump accompanying the dissociation of calcium, *Nature 418*, 605-611.
67. Olesen, C., Sorensen, T. L., Nielsen, R. C., Moller, J. V., and Nissen, P. (2004) Dephosphorylation of the calcium pump coupled to counterion occlusion, *Science 306*, 2251–2255.
68. Huang, Y., Lemieux, M. J., Song, J., Auer, M., and Wang, D. N. (2003) Structure and mechanism of the glycerol-3-phosphate transporter from Escherichia coli, *Science 301*, 616–620.
69. Oomen, C. J., van Ulsen, P., van Gelder, P., Feijen, M., Tommassen, J., and Gros, P. (2004) Structure of the translocator domain of a bacterial autotransporter, *EMBO J 23*, 1257–1266.
70. Locher, K. P., Rees, B., Koebnik, R., Mitschler, A., Moulinier, L., Rosenbusch, J. P., and Moras, D. (1998) Transmembrane signaling across the ligand-gated FhuA receptor: crystal structures of free and ferrichrome-bound states reveal allosteric changes, *Cell 95*, 771–778.
71. Dyson, H. J., and Wright, P. E. (2005) Intrinsically unstructured proteins and their functions, *Nat Rev Mol Cell Biol 6*, 197–208.
72. Dunker, A. K., Silman, I., Uversky, V. N., and Sussman, J. L. (2008) Function and structure of inherently disordered proteins, *Curr Opin Struct Biol 18*, 756–764.
73. Uversky, V. N., and Dunker, A. K. (2010) Understanding protein non-folding, *Biochim Biophys Acta 1804*, 1231–1264.
74. Uversky, V. N. (2002) Natively unfolded proteins: a point where biology waits for physics, *Protein Sci 11*, 739–756.
75. Tompa, P. (2002) Intrinsically unstructured proteins, *Trends Biochem Sci 27*, 527–533.
76. Joerger, A. C., and Fersht, A. R. (2010) The tumor suppressor p53: from structures to drug discovery, *Cold Spring Harb Perspect Biol 2*, a000919.
77. Rajagopalan, S., Andreeva, A., Rutherford, T. J., and Fersht, A. R. (2010) Mapping the physical and functional interactions between the tumor suppressors p53 and BRCA2, *Proc Natl Acad Sci USA 107*, 8587–8592.
78. Rajagopalan, S., Andreeva, A., Teufel, D. P., Freund, S. M., and Fersht, A. R. (2009) Interaction between the transactivation domain of p53 and PC4 exemplifies acidic activation domains as single-stranded DNA mimics, *J Biol Chem 284*, 21728–21737.
79. Jonker, H. R., Wechselberger, R. W., Boelens, R., Folkers, G. E., and Kaptein, R. (2005) Structural properties of the promiscuous VP16 activation domain, *Biochemistry 44*, 827–839.
80. Uversky, V. N. (2003) A protein-chameleon: conformational plasticity of alpha-synuclein, a disordered protein involved in neurodegenerative disorders, *J Biomol Struct Dyn 21*, 211–234.
81. Linding, R., Jensen, L. J., Diella, F., Bork, P., Gibson, T. J., and Russell, R. B. (2003) Protein disorder prediction: implications for structural proteomics, *Structure 11*, 1453–1459.
82. Romero, P., Obradovic, Z., Li, X., Garner, E. C., Brown, C. J., and Dunker, A. K. (2001) Sequence complexity of disordered protein, *Proteins 42*, 38–48.
83. Ward, J. J., Sodhi, J. S., McGuffin, L. J., Buxton, B. F., and Jones, D. T. (2004)

Prediction and functional analysis of native disorder in proteins from the three kingdoms of life, *J Mol Biol 337*, 635–645.

84. Sickmeier, M., Hamilton, J. A., LeGall, T., Vacic, V., Cortese, M. S., Tantos, A., Szabo, B., Tompa, P., Chen, J., Uversky, V. N., Obradovic, Z., and Dunker, A. K. (2007) DisProt: the Database of Disordered Proteins, *Nucleic Acids Res 35*, D786–793.
85. Altschul, S. F., Madden, T. L., Schaffer, A. A., Zhang, J., Zhang, Z., Miller, W., and Lipman, D. J. (1997) Gapped BLAST and PSI-BLAST: a new generation of protein database search programs, *Nucleic Acids Res 25*, 3389–3402.
86. Johnson, L. S., Eddy, S. R., and Portugaly, E. (2010) Hidden Markov model speed heuristic and iterative HMM search procedure, *BMC Bioinformatics 11*, 431.
87. Madera, M. (2008) Profile Comparer: a program for scoring and aligning profile hidden Markov models, *Bioinformatics 24*, 2630–2631.
88. Sadreyev, R. I., Tang, M., Kim, B. H., and Grishin, N. V. (2009) COMPASS server for homology detection: improved statistical accuracy, speed and functionality, *Nucleic Acids Res 37*, W90–94.
89. Andreeva, A., Prlic, A., Hubbard, T. J., and Murzin, A. G. (2007) SISYPHUS – structural alignments for proteins with non-trivial relationships, *Nucleic Acids Res. 35*, D253–259.
90. Grishin, N. V. (2001) Fold change in evolution of protein structures, *J Struct Biol 134*, 167–185.
91. Kinch, L. N., and Grishin, N. V. (2002) Evolution of protein structures and functions, *Curr Opin Struct Biol 12*, 400–408.
92. Alva, V., Koretke, K. K., Coles, M., and Lupas, A. N. (2008) Cradle-loop barrels and the concept of metafolds in protein classification by natural descent, *Curr Opin Struct Biol 18*, 358–365.
93. Anfinsen, C. B. (1973) Principles that govern the folding of protein chains, *Science 181*, 223–230.
94. Anfinsen, C. B., Haber, E., Sela, M., and White, F. H., Jr. (1961) The kinetics of formation of native ribonuclease during oxidation of the reduced polypeptide chain, *Proc Natl Acad Sci USA 47*, 1309–1314.
95. Luo, X., Tang, Z., Xia, G., Wassmann, K., Matsumoto, T., Rizo, J., and Yu, H. (2004) The Mad2 spindle checkpoint protein has two distinct natively folded states, *Nat Struct Mol Biol 11*, 338–345.
96. Tuinstra, R. L., Peterson, F. C., Kutlesa, S., Elgin, E. S., Kron, M. A., and Volkman, B. F. (2008) Interconversion between two unrelated protein folds in the lymphotactin native state, *Proc Natl Acad Sci USA 105*, 5057–5062.
97. Cabrita, L. D., and Bottomley, S. P. (2004) How do proteins avoid becoming too stable? Biophysical studies into metastable proteins, *Eur Biophys J 33*, 83–88.
98. Bullough, P. A., Hughson, F. M., Skehel, J. J., and Wiley, D. C. (1994) Structure of influenza haemagglutinin at the pH of membrane fusion, *Nature 371*, 37–43.
99. Chan, D. C., Fass, D., Berger, J. M., and Kim, P. S. (1997) Core structure of gp41 from the HIV envelope glycoprotein, *Cell 89*, 263–273.
100. Stiasny, K., Allison, S. L., Mandl, C. W., and Heinz, F. X. (2001) Role of metastability and acidic pH in membrane fusion by tick-borne encephalitis virus, *J Virol 75*, 7392–7398.
101. Orosz, A., Wisniewski, J., and Wu, C. (1996) Regulation of Drosophila heat shock factor trimerization: global sequence requirements and independence of nuclear localization, *Mol Cell Biol 16*, 7018–7030.
102. Xiao, T., Gardner, K. H., and Sprang, S. R. (2002) Cosolvent-induced transformation of a death domain tertiary structure, *Proc Natl Acad Sci USA 99*, 11151–11156.
103. Kuloglu, E. S., McCaslin, D. R., Markley, J. L., and Volkman, B. F. (2002) Structural rearrangement of human lymphotactin, a C chemokine, under physiological solution conditions, *J Biol Chem 277*, 17863–17870.
104. Zubkov, S., Gronenborn, A. M., Byeon, I. J., and Mohanty, S. (2005) Structural consequences of the pH-induced conformational switch in A. polyphemus pheromone-binding protein: mechanisms of ligand release, *J Mol Biol 354*, 1081–1090.
105. Joerger, A. C., Rajagopalan, S., Natan, E., Veprintsev, D. B., Robinson, C. V., and Fersht, A. R. (2009) Structural evolution of p53, p63, and p73: implication for heterotetramer formation, *Proc Natl Acad Sci USA 106*, 17705–17710.
106. Cordell, S. C., Anderson, R. E., and Lowe, J. (2001) Crystal structure of the bacterial cell division inhibitor MinC, *EMBO J 20*, 2454–2461.
107. Xu, Q., and Minor, D. L., Jr. (2009) Crystal structure of a trimeric form of the K(V)7.1 (KCNQ1) A-domain tail coiled-coil reveals structural plasticity and context dependent changes in a putative coiled-coil trimerization motif, *Protein Sci 18*, 2100–2114.
108. Schellenberg, M. J., Ritchie, D. B., Wu, T., Markin, C. J., Spyracopoulos, L., and Macmillan,

A. M. (2010) Context-Dependent Remodeling of Structure in Two Large Protein Fragments, *J Mol Biol 402*, 720–730.

109. Guo, J. T., Jaromczyk, J. W., and Xu, Y. (2007) Analysis of chameleon sequences and their implications in biological processes, *Proteins 67*, 548–558.
110. Mezei, M. (1998) Chameleon sequences in the PDB, *Protein Eng 11*, 411–414.
111. Tan, S., and Richmond, T. J. (1998) Crystal structure of the yeast MATalpha2/MCM1/DNA ternary complex, *Nature 391*, 660–666.
112. Abel, K., Yoder, M. D., Hilgenfeld, R., and Jurnak, F. (1996) An alpha to beta conformational switch in EF-Tu, *Structure 4*, 1153–1159.
113. Polekhina, G., Thirup, S., Kjeldgaard, M., Nissen, P., Lippmann, C., and Nyborg, J. (1996) Helix unwinding in the effector region of elongation factor EF-Tu-GDP, *Structure 4*, 1141–1151.
114. Chen, Y. W., Allen, M. D., Veprintsev, D. B., Lowe, J., and Bycroft, M. (2004) The structure of the AXH domain of spinocerebellar ataxin-1, *J Biol Chem 279*, 3758–3765.
115. de Chiara, C., Menon, R. P., Adinolfi, S., de Boer, J., Ktistaki, E., Kelly, G., Calder, L., Kioussis, D., and Pastore, A. (2005) The AXH domain adopts alternative folds the solution structure of HBP1 AXH, *Structure 13*, 743–753.
116. Hamada, K., Shimizu, T., Yonemura, S., Tsukita, S., and Hakoshima, T. (2003) Structural basis of adhesion-molecule recognition by ERM proteins revealed by the crystal structure of the radixin-ICAM-2 complex, *EMBO J 22*, 502–514.
117. Kitano, K., Yusa, F., and Hakoshima, T. (2006) Structure of dimerized radixin FERM domain suggests a novel masking motif in C-terminal residues 295-304, *Acta Crystallogr Sect F Struct Biol Cryst Commun 62*, 340–345.
118. Zimmer, J., Li, W., and Rapoport, T. A. (2006) A novel dimer interface and conformational changes revealed by an X-ray structure of B. subtilis SecA, *J Mol Biol 364*, 259–265.
119. Tidow, H., Lauber, T., Vitzithum, K., Sommerhoff, C. P., Rosch, P., and Marx, U. C. (2004) The solution structure of a chimeric LEKTI domain reveals a chameleon sequence, *Biochemistry 43*, 11238–11247.
120. Ditzel, L., Lowe, J., Stock, D., Stetter, K. O., Huber, H., Huber, R., and Steinbacher, S. (1998) Crystal structure of the thermosome, the archaeal chaperonin and homolog of CCT, *Cell 93*, 125–138.
121. Klumpp, M., Baumeister, W., and Essen, L. O. (1997) Structure of the substrate binding domain of the thermosome, an archaeal group II chaperonin, *Cell 91*, 263–270.
122. Chothia, C. (1984) Principles that determine the structure of proteins, *Annu Rev Biochem 53*, 537–572.
123. Chothia, C., and Finkelstein, A. V. (1990) The classification and origins of protein folding patterns, *Annu Rev Biochem 59*, 1007–1039.
124. Sternberg, M. J., and Thornton, J. M. (1976) On the conformation of proteins: the handedness of the beta-strand-alpha-helix-beta-strand unit, *J Mol Biol 105*, 367–382.
125. Sternberg, M. J., and Thornton, J. M. (1977) On the conformation of proteins: the handedness of the connection between parallel beta-strands, *J Mol Biol 110*, 269–283.
126. Belogurov, G. A., Vassylyeva, M. N., Svetlov, V., Klyuyev, S., Grishin, N. V., Vassylyev, D. G., and Artsimovitch, I. (2007) Structural basis for converting a general transcription factor into an operon-specific virulence regulator, *Mol Cell 26*, 117–129.
127. Guzzo, C. R., Nagem, R. A., Barbosa, J. A., and Farah, C. S. (2007) Structure of Xanthomonas axonopodis pv. citri YaeQ reveals a new compact protein fold built around a variation of the PD-(D/E)XK nuclease motif, *Proteins 69*, 644–651.
128. Essen, L. O., Perisic, O., Cheung, R., Katan, M., and Williams, R. L. (1996) Crystal structure of a mammalian phosphoinositide-specific phospholipase C delta, *Nature 380*, 595–602.
129. Sutton, R. B., Davletov, B. A., Berghuis, A. M., Sudhof, T. C., and Sprang, S. R. (1995) Structure of the first C2 domain of synaptotagmin I: a novel Ca2+/phospholipid-binding fold, *Cell 80*, 929–938.
130. Andreeva, A., Howorth, D., Brenner, S. E., Hubbard, T. J., Chothia, C., and Murzin, A. G. (2004) SCOP database in 2004: refinements integrate structure and sequence family data, *Nucleic Acids Res 32*, D226–229.
131. Andreeva, A., Howorth, D., Chandonia, J. M., Brenner, S. E., Hubbard, T. J., Chothia, C., and Murzin, A. G. (2008) Data growth and its impact on the SCOP database: new developments, *Nucleic Acids Res 36*, D419–425.
132. Cuff, A., Redfern, O. C., Greene, L., Sillitoe, I., Lewis, T., Dibley, M., Reid, A., Pearl, F., Dallman, T., Todd, A., Garratt, R., Thornton, J., and Orengo, C. (2009) The CATH hierarchy revisited-structural divergence in domain superfamilies and the continuity of fold space, *Structure 17*, 1051–1062.

133. Hadley, C., and Jones, D. T. (1999) A systematic comparison of protein structure classifications: SCOP, CATH and FSSP, *Structure 7*, 1099–1112.

134. Day, R., Beck, D. A., Armen, R. S., and Daggett, V. (2003) A consensus view of fold space: combining SCOP, CATH, and the Dali Domain Dictionary, *Protein Sci 12*, 2150–2160.

135. Holm, L., and Park, J. (2000) DaliLite workbench for protein structure comparison, *Bioinformatics 16*, 566–567.

136. Suhrer, S. J., Wiederstein, M., Gruber, M., and Sippl, M. J. (2009) COPS – a novel workbench for explorations in fold space, *Nucleic Acids Res 37*, W539–544.

137. Li, Z., Ye, Y., and Godzik, A. (2006) Flexible Structural Neighborhood – a database of protein structural similarities and alignments, *Nucleic Acids Res 34*, D277–280.

138. Bray, J. E., Todd, A. E., Pearl, F. M., Thornton, J. M., and Orengo, C. A. (2000) The CATH Dictionary of Homologous Superfamilies (DHS): a consensus approach for identifying distant structural homologues, *Protein Eng 13*, 153–165.

139. Waterhouse, A. M., Procter, J. B., Martin, D. M., Clamp, M., and Barton, G. J. (2009) Jalview Version 2 – a multiple sequence alignment editor and analysis workbench, *Bioinformatics 25*, 1189–1191.

140. Ashkenazy, H., Erez, E., Martz, E., Pupko, T., and Ben-Tal, N. (2010) ConSurf 2010: calculating evolutionary conservation in sequence and structure of proteins and nucleic acids, *Nucleic Acids Res 38 Suppl*, W529–533.

141. (2010) The Universal Protein Resource (UniProt) in 2010, *Nucleic Acids Res 38*, D142–148.

142. Sayers, E. W., Barrett, T., Benson, D. A., Bryant, S. H., Canese, K., Chetvernin, V., Church, D. M., DiCuccio, M., Edgar, R., Federhen, S., Feolo, M., Geer, L. Y., Helmberg, W., Kapustin, Y., Landsman, D., Lipman, D. J., Madden, T. L., Maglott, D. R., Miller, V., Mizrachi, I., Ostell, J., Pruitt, K. D., Schuler, G. D., Sequeira, E., Sherry, S. T., Shumway, M., Sirotkin, K., Souvorov, A., Starchenko, G., Tatusova, T. A., Wagner, L., Yaschenko, E., and Ye, J. (2009) Database resources of the National Center for Biotechnology Information, *Nucleic Acids Res 37*, D5–15.

143. Holm, L., and Rosenstrom, P. (2010) Dali server: conservation mapping in 3D, *Nucleic Acids Res 38 Suppl*, W545–549.

144. Pearson, W. R., and Lipman, D. J. (1988) Improved tools for biological sequence comparison, *Proc Natl Acad Sci USA 85*, 2444–2448.

145. Gibrat, J. F., Madej, T., and Bryant, S. H. (1996) Surprising similarities in structure comparison, *Curr Opin Struct Biol 6*, 377–385.

146. Orengo, C. A., and Taylor, W. R. (1996) SSAP: sequential structure alignment program for protein structure comparison, *Methods Enzymol 266*, 617–635.

147. Ye, Y., and Godzik, A. (2003) Flexible structure alignment by chaining aligned fragment pairs allowing twists, *Bioinformatics 19 Suppl 2*, ii246–255.

148. Shindyalov, I. N., and Bourne, P. E. (1998) Protein structure alignment by incremental combinatorial extension (CE) of the optimal path, *Protein Eng 11*, 739–747.

149. Ortiz, A. R., Strauss, C. E., and Olmea, O. (2002) MAMMOTH (matching molecular models obtained from theory): an automated method for model comparison, *Protein Sci 11*, 2606–2621.

150. Sippl, M. J., and Wiederstein, M. (2008) A note on difficult structure alignment problems, *Bioinformatics 24*, 426–427.

151. Zhang, Y., and Skolnick, J. (2005) TM-align: a protein structure alignment algorithm based on the TM-score, *Nucleic Acids Res 33*, 2302–2309.

152. Jayasinghe, S., Hristova, K., and White, S. H. (2001) MPtopo: A database of membrane protein topology, *Protein Sci 10*, 455–458.

Chapter 2

Effective Techniques for Protein Structure Mining

Stefan J. Suhrer, Markus Gruber, Markus Wiederstein, and Manfred J. Sippl

Abstract

Retrieval and characterization of protein structure relationships are instrumental in a wide range of tasks in structural biology. The classification of protein structures (COPS) is a web service that provides efficient access to structure and sequence similarities for all currently available protein structures. Here, we focus on the application of COPS to the problem of template selection in homology modeling.

Key words: Protein structure space, Protein structure comparison, Template selection, Structure alignment, Structure similarity search, Classification, Homology modeling, Ligand binding

1. Introduction

The repository of known protein structures contains a wealth of information about the relationships between protein sequences and protein structures. Many useful tools and databases have been developed to extract knowledge from this repository, but the appropriate organization of protein structure data remains a challenge.

The classification of protein structures (COPS) (1–3) provides access to the overwhelming number of structure and sequence relationships (4, 5) between all experimentally determined protein structures deposited in the Protein Data Bank (PDB) (6). COPS features a quantitative organization of protein structures according to a set of metric properties and principles. It includes methods for the automated decomposition of proteins into structural domains, pairwise structure comparison, and the instant visualization of structure similarities. Since COPS is updated weekly with every PDB release, it covers the complete set of publicly available protein structures.

Andrew J.W. Orry and Ruben Abagyan (eds.), *Homology Modeling: Methods and Protocols*, Methods in Molecular Biology, vol. 857, DOI 10.1007/978-1-61779-588-6_2, © Springer Science+Business Media, LLC 2012

In this chapter, we present and illustrate the usage of COPS with an emphasis on its use in homology modeling. Homology modeling builds on the observation that proteins of similar sequence frequently adopt similar structures (7). Proteins of unknown structure are modeled using the structures of other proteins as templates, given their sequences share significant similarity. In this procedure, the steps of template selection, template comparison, and evaluation for their use in model building are significantly affected by the way protein structure data is organized and accessible. Moreover, it is important to keep pace with the rapid growth of PDB which implies an ever increasing pool of template candidates. We discuss the key components of COPS and apply them to the step of template characterization in homology modeling.

2. Structure Mining with COPS

The COPS classification process includes the weekly download of structures from PDB, their decomposition into domains with TopDomain, the calculation of structural similarities with TopMatch (8), and the update of the COPS hierarchy with respect to the found similarities. The domains are organized in a tree similar to a file browser, where the domains correspond to tree nodes and pairwise structural similarities between domains correspond to tree edges. Currently, COPS provides five classification layers called *Distant* (30% relative structural similarity), *Remote* (40%), *Related* (60%), *Similar* (80%), and *Equivalent* (99%) (1, 9).

The graphical interface requires JavaScript to be enabled as well as a recent (version 10 or greater) Adobe® FlashPlayer® installation. For the proper three-dimensional (3D) visualization of protein structures and superimpositions, we recommend a modern workstation with a minimum display resolution of 1,024×768 pixels and a fast network connection. COPS is available online at http://cops.services.came.sbg.ac.at/.

At start up the first COPS page shows a widget where the main tools such as qCOPS, iCOPS, and ΔCOPS are listed. This tutorial is focused on the first application, quantitative COPS (qCOPS). A typical COPS query involves several steps (refer to Fig. 1 for a condensed view):

1. Main Query
   Enter a PDB four letter code (e.g., 2hhb) into the query input box (Fig. 2a) and press the button *Search* or the return/enter key on your keyboard. This queries the qCOPS server with the given PDB code. In this tutorial, we use 1z6t (10) as our query.
2. Selection Widget (Fig. 2b)
   The result of a query is listed in the *Selection Widget* which displays all COPS domains available for a given PDB code.

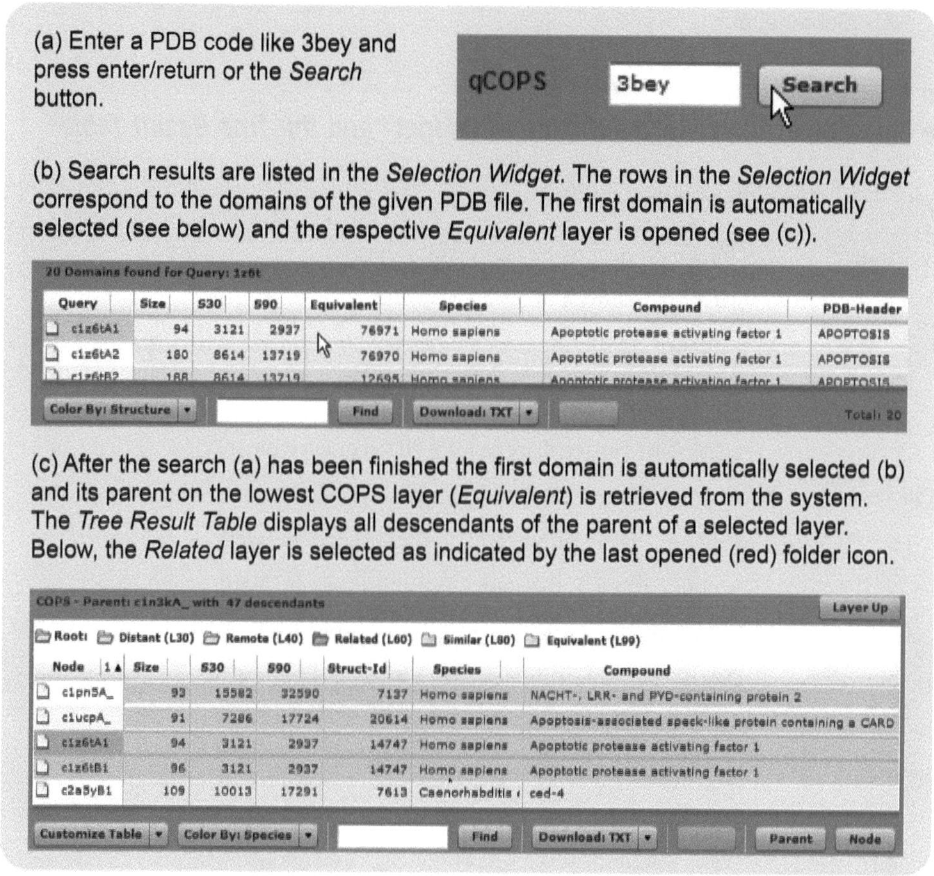

Fig. 1. The essential steps to use COPS.

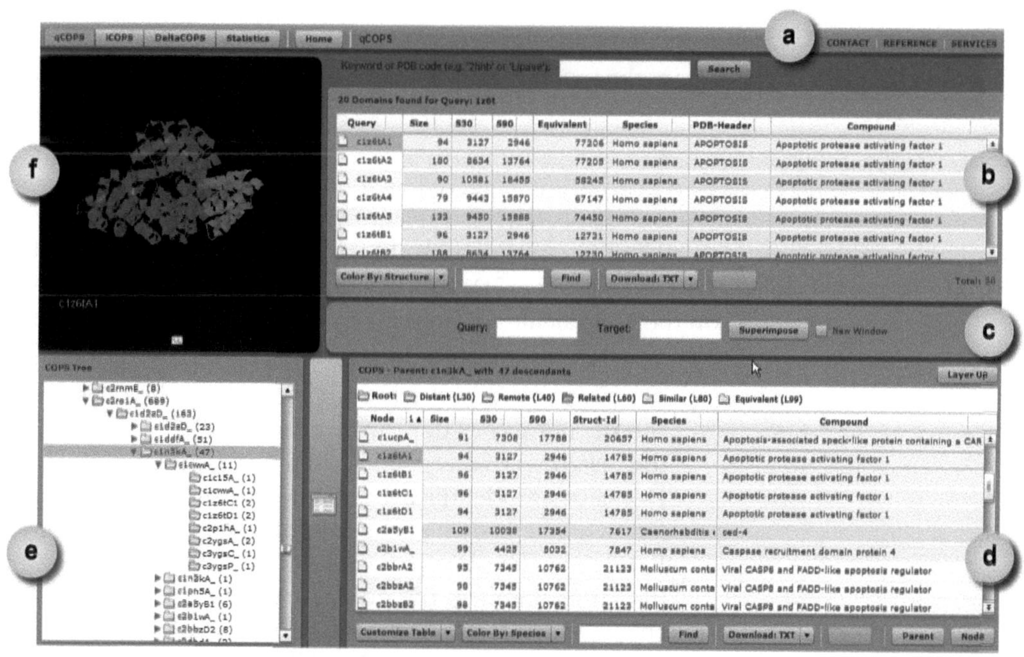

Fig. 2. COPS screen capture displaying the main sections of the interface: (**a**) Query input box, (**b**) Selection Widget, (**c**) Superimposition Box, (**d**) Tree Result Table, (**e**) Tree Widget, and (**f**) Jmol Widget.

Table 1
Table columns available in the *Selection Widget*[a] and the *Tree Result Table*[b]

Column	Description
Query/Node[a,b]	Unique domain name (see text for details)
Size[a,b]	Size of the domain in residues
S30[a,b]	Sequence classification code on layer S30. Domains with the same *S30* id are in the same sequence cluster and share at least 30% sequence identity
S90[a,b]	Sequence classification code on layer S90. Same as *S30*, but sequences within the same cluster share at least 90% sequence identity
Equivalent[a]	Structure classification code on the Equivalent layer (L90)
Struct-Id[b]	Structure classification code on the subsequent layer
Species[a,b]	Scientific name of the source organism used by UniProt and NCBI
PDB-Header[a,b]	HEADER classification record of the respective PDB file
Compound[a,b]	Describes the macromolecular contents of an entry
Method[b]	Experimental method
Resolution[b]	Resolution in Å
SG[b]	1 for Structural Genomics target, 0 otherwise
S-Kingdom[b]	Super Kingdom as defined in the NCBI taxonomy
Ligand Short[b]	Ligand short name
Ligand Long[b]	Ligand name
EC Number[b]	Enzyme classification number
Release Date[b]	Release date of the respective PDB file

Two actions are triggered as soon as the data of the *Selection Widget* has been loaded: First, the first domain is selected and visualized in context with the respective protein chain in the *Jmol Widget* (Fig. 2f), and second, the first domain is selected on the equivalent layer in the *Tree Result Table* (Fig. 2d) of the *Fold Space Navigator* (see below).

(a) The *Selection Widget* has a title bar where the query code and the number of domains are indicated. Every domain in the *Selection Widget* is annotated as described in Table 1. Domains are identified by a unique name constructed as follows: The first character is *c* followed by the four letter PDB code. The next letter specifies the PDB chain and the last letter numbers the domains within the chain. Single chain domains have an underscore as last character. For example, the code c1z6tB2 specifies domain two of chain B of PDB code 1z6t. Domains can be selected by clicking on the corresponding row in the table.

(b) The table rows are sorted by the domain names (*Query* column) by default. To sort the rows by any of the other columns just click on the respective column header. This is indicated by a small black triangle besides the column name which is visible when the column is sorted and the mouse pointer is placed over a column header. If the triangle points up the table is sorted in ascending order, if the triangle points down the sort order is descending. Additionally, a number is placed besides the triangle. This number indicates the sort order of the columns. For example, if the table rows are sorted by the *S30* column, a black triangle is visible in the *S30* column header together with the number one besides the column name. The number one indicates that column *S30* is the first sort criterion. We can now sort the table by a second criterion, e.g., the *Equivalent* column. This can be achieved by placing the mouse over the *Equivalent* column header and clicking on the number two appearing on the right side of the column name. Now the table rows are sorted or grouped firstly by the *S30* id and secondly by the *Equivalent* id. In other words, domains with more than 30% sequence identity are grouped together and these groups are then divided into subgroups of domains with more than 99% structural similarity. Other columns can be added to the sort criteria in the same fashion. To reset the sort criteria to the default sort order, just click on the column header of the *Query* column. More examples of useful sort combinations are given in the *Tree Result Table* paragraph of item 3.

You can also change the order of the columns in the table by dragging the column at the column header and dropping it at the desired position. To change a column width, place the mouse pointer over the grid lines separating two column headers and move the line with the appearing new mouse cursor to the desired width.

(c) Below the *Selection Widget* a toolbar is located that allows some customizations of the table. It is separated into three sections by pale vertical lines. With the drop-down list in the first section the table can be colored by different criteria. By default, the table is colored by *Structure*, which means all domains that share the same classification id on the *Equivalent* layer have the same color. In other words, domains in the same *Equivalent* layer are colored similarly. All columns (except *Query*) can be used for coloring the table. The coloring gives a quick overview of the domain composition of a protein and helps answering questions on the structural diversity of the domains. If we sort the domains of our example protein 1z6t by the *Equivalent* column and color by *Structure*, we instantly see that domains three, four, and five of chains A–D are structurally equivalent.

The next section of the toolbar is for searching the table with a domain name. For example, to get the third domain of chain C of 1z6t one can enter c1z6tC3 and click the *Search* button. The last section of the toolbar provides the data of the result table in different file formats such as CSV or XML.

3. Fold Space Navigator

   The *Fold Space Navigator* is a graphical representation of qCOPS and its design is largely equivalent to the structure of a file browser. Folder icons represent parent nodes (representative domain) on a given layer and the contents of a folder (i.e., the files) correspond to all child nodes (i.e., the complete subtree) of the respective family. The *Tree Widget* displays the path of the selected domain from the root (no structural similarities) of the hierarchical classification tree down to the equivalent layer (highest structural similarities). The structural relationship of all child nodes to the parent depends on the selected layer. On the equivalent layer, for example, all domains of a specific family have a structural similarity of ≥99% to the parent. The *Fold Space Navigator* contains three widgets: The *Tree widget*, the *Tree Result Table*, and the *Breadcrumb* for easy layer navigation. In the following, all three widgets are explained in detail.

   (a) Tree widget (Fig. 2e)

   The *Tree Widget* is hidden by default to maximize the *Tree Result Table* view. To uncover the *Tree Widget* just press the button on the left side of the *Tree Result Table*. The *Tree Widget* provides direct access to the nodes of the qCOPS hierarchy. Every icon folder corresponds to the parent domain on a specific layer. Besides an icon folder, the domain name of the representative domain (parent) is shown followed by the total number of child domains below the respective parent in parenthesis. Clicking on a folder icon loads the child domains into the *Tree Result Table*. The black arrows in front of the folder icons can be used to open or close a folder without loading the child nodes. Folder icons can be dragged and dropped into the *Superimposition Box* to get a structure alignment as we will see later (see item 4).

   (b) Tree Result Table (Fig. 2d)

   The *Tree Result Table* lists all child domains of a selected parent. The name of the parent and the number of descendants are displayed in the title bar of the table. The functionality of the table is similar to the result table of the *Selection Widget* (*see* item 2), but covers more columns and additional features. By default, the displayed columns are identical, except for the *Node* and the *Struct-Id* column. The *Node* column comprises domain names, too, but here it specifies the node names in the context of the classification tree. The *Struct-Id* column contains the layer id of a node on the subsequent layer (from root to leaf) or, if the current layer is the *Equivalent* layer, the id of the (leaf) node itself. As a consequence, nodes on the *Equivalent* layer have all unique *Struct-Id* values. The representative domain (parent) of the currently selected layer has a folder icon besides the *Node* name that distinguishes it from the other domains in the table. Clicking on a row in the *Tree Result Table* displays the TopMatch superimposition of the respective node and the selected domain in the *Selection Widget* and the *Jmol Widget*.

Using the sort combinations explained in item 2, it is easy to answer difficult questions with just a few clicks. For example, suppose we are interested in domains that have relative structural similarities of at least 60% but sequence identities below 30%. We use domain one (c1z6tA1) of chain A of our example structure 1z6t. We skip the *Equivalent* and *Similar* layers and directly select the *Related* layer in the *Breadcrumb* navigation (see item 3c). Sort the table by the *Struct-Id* column by clicking on the respective column header and add the *S30* column as the second sort criterion as explained in item 2. Now we only have to scroll through the table and search for domains with identical *Struct-Id* but different *S30* entries. This process can be simplified even more by additionally coloring the table by *Structure*; then we only have to search for table rows with identical color but different *S30* values. In our example, numerous pairs of domains fulfill these criteria. To check the results, e.g., c3lqrA1 and c2vgqA4, we simply superimpose the domains with TopMatch (see item 4). In fact, the domains have almost 80% relative structural similarity but less than 15% sequence identity.

The *Tree Result Table* has a toolbar, similar to the toolbar of the *Selection Widget* (item 2). The functionality is identical except for the *Customize Table* button. This button opens a menu that enables the user to add or remove columns from the *Tree Result Table* by checking or unchecking the corresponding check boxes, respectively (see Table 1 for a column description). The buttons *Parent* and *Node* at the right end side of the toolbar select the parent and the node row (the currently selected domain in the *Selection Widget*) in the *Tree Result Table*.

(c) Breadcrumb Navigation (Fig. 2d)
The *Breadcrumb Navigation* widget above the *Tree Result Table* displays the path of the selected domain from the root (no structural similarities) of the hierarchical classification tree down to the equivalent layer (highest structural similarities). Each node of a layer on the path is depicted as a folder icon (cf. *Tree Widget*) followed by the layer name and the layer shortcut in parenthesis. The currently selected layer is highlighted red. A click on one of the folder icons

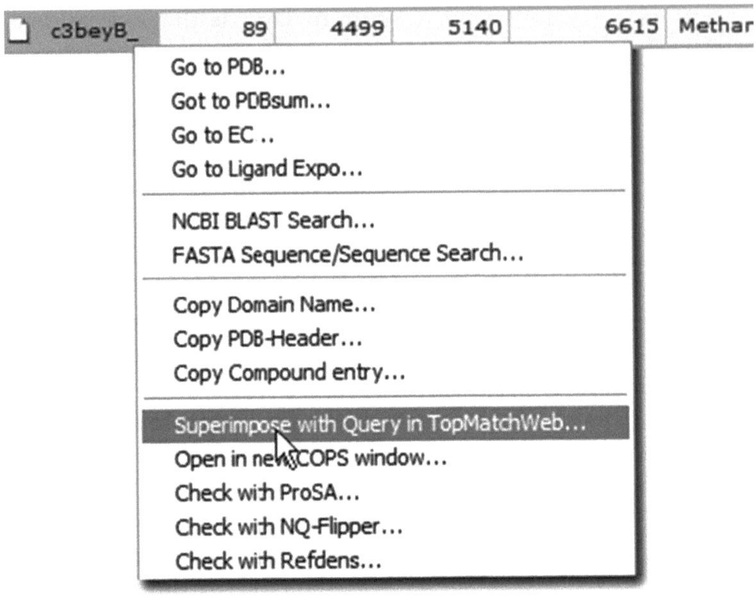

Fig. 3. The right-click context menu of the *Tree Result Table* is split into four sections. The first section contains entry-specific links to external resources such as PDB, PDBsum, Enzyme Classification (EC), Ligand Expo, and Pubmed (Primary Citation). The second section provides sequence search functionality and sequence data. Copy functionality is given in the third section, and the last section includes links to resources for structure comparison, structure search, and structure validation. For example, the first entry in the last section opens up a new window with the TopMatch (8) superimposition of the query and the selected target from the *Tree Result Table*. The second entry in the last section (*Open in new COPS window* ...) queries COPS with the selected target from the *Tree Result Table* in a new window.

selects the representative domain on the respective layer and all descendants of the representative are listed in the *Tree Result Table*. The name of the parent is shown within the tool tip that appears when the mouse pointer is placed over the respective layer icon. It is identical to the entry with the folder icon in the *Tree Result Table* (item 3b). The *Breadcrumb Navigation* is automatically updated if the selection in the *Tree Widget* or the *Selection Widget* is changed.

4. Superimposition Box (Fig. 2c)
   The *Superimposition Box* provides access to the TopMatch structure alignment server (8). *Query* and *Target* name for the structure alignment have to be provided in the correspondingly named text fields. Domain names can be entered directly into the text fields or, more conveniently, dragged and dropped into the respective text fields. Drag and drop is possible from any widget with domain names, particularly the *Selection Widget*, the *Tree Widget*, and the *Tree Result Table*. Once the *Query* and *Target* fields are filled in, a click on the *Superimpose*

button opens a new browser window where the detailed TopMatch structure alignment is displayed. The TopMatch superimpositions are always loaded into the same external window as long as the *New Window* check box besides the button is not selected.

5. Jmol Widget (Fig. 2f)
   The *Jmol Widget* contains Jmol (http://www.jmol.org/), an open-source Java viewer for chemical structures in 3D. Below the applet a small magnifier is located that can be used to maximize the 3D view. Additionally, the maximized view displays the ligands of the respective chain, too.

3. Application of COPS in Homology Modeling

The major goal in homology modeling is to obtain an accurate structural model for a given protein sequence with unknown structure. The first step on the way to the model is the identification of proper structural templates for the given sequence. This is an essential step, since the template structures form the basic framework upon which the model is constructed. Hence, the choice of the templates has a significant impact on the quality of the resulting model.

The first step in homology modeling is the identification of evolutionary-related proteins with known structure that can serve as suitable templates for a specific target sequence. There is a plethora of sequence-based homology detection methods available for this task (11) with distinct capabilities in detecting homologous sequences (12). In general, all methods return a hit list sorted by a similarity score indicating the relevance of the specific hits. Hits within a certain threshold are considered to be trustable results and those with available structure files are potential templates for protein core modeling.

Table 2 shows the hit list for CASP8 target T0408 (http://predictioncenter.org/casp8/target.cgi?id=23&view=all) obtained by the sequence-based HHsearch algorithm in a search against a nonredundant template data base (13). Recently, HHsearch outperformed other sequence-based algorithms in an analysis of sequence database search methods (12). Entries from the hit list within the trustable cutoff (Table 2) are our potential templates in the modeling process of T0408. At this point of the modeling procedure, nothing is known about the structural similarities between the template candidates, their domain organization and other structural characteristics that facilitate the selection of templates for subsequent model building.

In the process of homology modeling, COPS can be applied as soon as the first template candidates have been identified. These structures can then be analyzed in terms of structural relationships

Table 2
HHsearch results for CASP target T0408 retrieved from the HHsearch web server (13) using default parameters

No	Hit		Prob	*E* value	SeqId (%)
1	3d7i_A	Carboxymuconolactone de	100.0	7.2E-32	97
2	3bey_A	Conserved protein O2701	100.0	2.2E-28	20
3	1p8c_A	Conserved hypothetical	99.9	1.8E-24	19
4	2qeu_A	Putative carboxymuconol	99.9	3.1E-24	23
5	2af7_A	Gamma-carboxymuconolact	99.9	1E-24	20
6	1vke_A	Carboxymuconolactone de	99.9	2.6E-24	18
7	2cwq_A	Hypothetical protein TT	99.9	2E-22	23
8	2q0t_A	Putative gamma-carboxym	99.9	1.6E-21	20
9	2q0t_A	Putative gamma-carboxym	99.9	3.4E-21	21
10	2ouw_A	Alkylhydroperoxidase AH	99.7	3.1E-16	22
11	1gu9_A	Alkylhydroperoxidase D;	99.7	2.5E-16	13
12	3cl1_A	Putative antioxidant de	99.3	1.1E-10	10
13	2prr_A	Alkylhydroperoxidase AH	99.2	2.3E-10	13
14	2gmy_A	Hypothetical protein AT	99.2	1.2E-10	15
15	2o4d_A	Hypothetical protein PA	99.2	2E-10	14
16	3lvy_A	Carboxymuconolactone de	99.0	1E-09	8
17	2pfx_A	Uncharacterized peroxid	99.0	1.9E-09	6
18	2oyo_A	Uncharacterized peroxid	99.0	2.9E-09	9
19	1gu9_A	Alkylhydroperoxidase D	97.9	0.00015	12
20	3bjx_A	Halocarboxylic acid deh	97.6	5E-06	14
21	2pfx_A	Uncharacterized peroxid	96.7	0.003	15
22	3lvy_A	Carboxymuconolactone de	96.1	0.0088	21
23	2oyo_A	Uncharacterized peroxid	96.1	0.004	14
24	2gmy_A	Hypothetical protein AT	95.9	0.0095	8
25	2o4d_A	Hypothetical protein PA	95.9	0.0063	16

The hit list is sorted by the estimated probability (Prob) which is the most important criterion for homology. According to the HHsearch manual hits with a probability larger than 95% are nearly certainly homologous to the query sequence. Therefore, only hits above the 95% probability cutoff are included. Additionally, the *E* value and the sequence identity (SeqId) to the query sequence are shown. The structure of T0408 has been solved by X-ray crystallography and is available as PDB file 3d7i.

to other proteins in the PDB, as well as structural differences between the templates (see Subheading 3.1). Furthermore, the candidates can be characterized by features describing their biological context, like source organism or functional annotation (see Subheading 3.2). We exemplify the practical usage of COPS for homology modeling in the following two subsections using the templates from Table 2 and other examples.

3.1. How Diverse Are My Template Structures?

The protein structures in Table 2 are putative templates for our model. Hits with the highest score and *E* value are considered to be the best templates. However, nontrivial templates (query coverage ≤ 90% and sequence identity ≤ 90%) may have structural varieties that are not detectable from the initial template list, but that are essential for model building. Structure comparison of the templates is an indispensable step in the process of template selection and alignment correction. This is especially useful if the structural differences are visualized and the corresponding sequence alignments are available. Pairwise structural comparisons and their visualizations are cumbersome tasks, but COPS and TopMatch facilitate this process considerably.

The first hit in the template list (Table 2) is the solved structure of target T0408 as determined by X-ray crystallography and deposited in the PDB with the code 3d7i (14). Since this structure was not available during prediction season in CASP8, we perform a COPS search with the second hit, 3bey (15). After the search has been finished, all six structural domains of 3bey are listed in the *Selection Widget* (Fig. 2b), the first domain in the list (c3beyA) is selected and visualized in the *Jmol Widget*, and all domains of the respective *Equivalent* layer are displayed in the *Tree Result Table*. It is obvious from the COPS domain names that all six domains of 3bey are single chain domains, because no domain numbers are given but underscores. The found domains have at least 90% sequence identity indicated by identical *S30* and *S90* values. If we stain the domains by the *Structure* column entries it is easy to see that the domains are in different *Equivalent* layers except for c3beyC_ and c3beyF_, thus their relative structural similarities are less than 99%. The data from the *Selection Widget* addresses the internal organization and domain composition of a given protein structure. The data from the *Tree Result Table* explained in the following paragraphs deals with the structural similarities to other domains in the protein space.

The main goal of this section is to investigate the structural differences and similarities between our template candidates. Templates that cover the same regions of the target sequence are descendants of the same parent domain and can be found in the same layers of the *Tree Result Table*, presumed that they share the same structure. In this case, it is most straightforward to start with

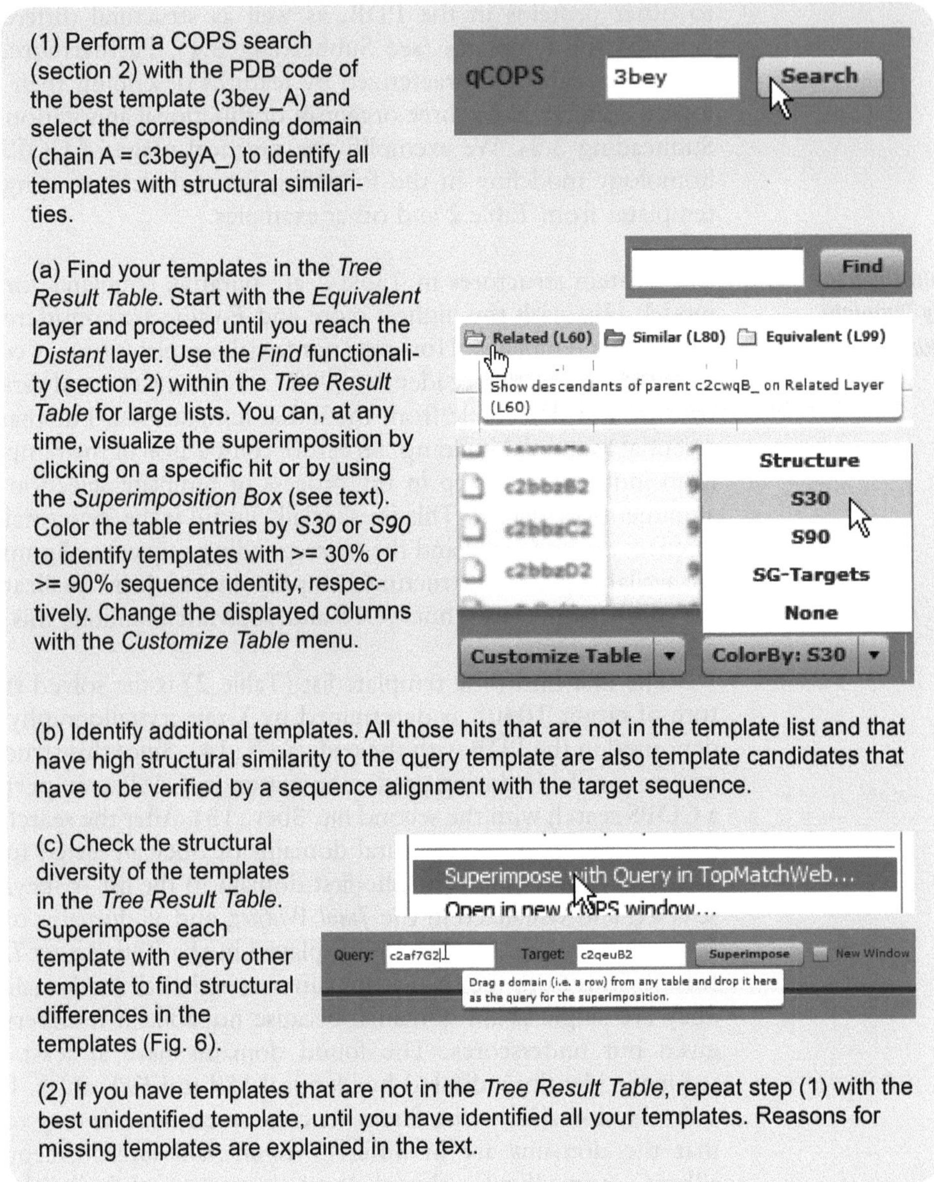

Fig. 4. Basic steps to investigate the structural diversity of a set of modeling templates. For details on the example used here, see Subheading 3.

the first template, browse through the hierarchical layers in COPS and identify the template structures from our template list from Table 2 For a condensed how-to manual of the following steps, refer to the box in Fig. 4.

The *Equivalent* layer of c3beyA_ contains one member and that is the domain itself. We switch to the next higher layer, the *Similar* layer, by clicking on the respective folder icon in the *Breadcrumb Navigation*. The parent c2cwqB_ on this *Similar* layer has nine descendants including itself. Six domains are from 3bey (i.e., chains A–F) and three domains are from PDB file 2cwq (i.e., chains A–C) (16). If we color the *Tree Result Table* by *S30*, we see that the domains of 3bey and 2cwq are in different *S30* sequence clusters that means the domains have less than 30% sequence identity. As a consequence, the domains of the two PDB files are in different *S90* clusters, too.

All three chains (A–C) of 2cwq are stored as single chain domains within COPS. More than 90% of the domain sequences are identical illustrated by equivalent *S90* ids. In the template list, 2cwq is represented by template seven (i.e., chain A or c2cwqA_ in COPS, respectively). Generally, not all domains (respectively chains) from the *Tree Result Table* have to be comprised in the template list, since similar templates are pooled by HHsearch. Within the *Tree Result Table*, it is straightforward to validate the pools by checking the sequence and structure layers. Moreover, additional data is available to select the appropriate template from a pool. Columns that contain essential information supporting template selection and validation include experimental method, resolution, and the ligand columns. We will cover specific COPS columns in more detail where applicable.

A mouse click on the row of c2cwqA_ in *Tree Result Table* displays the TopMatch superimposition of the two templates c2cwqA_ and c3beyA_ (in COPS called target and query, respectively) in the *Jmol Widget*. The visualization of the superimposition and the respective layer give a first clue about the structural differences and similarities between the two templates (see Fig. 5c). For a detailed investigation, it is advisable to switch to the TopMatch server using the *Superimposition Box* (see Subheading 2, item 4 for details). Instantly, the same TopMatch superimposition is opened in an additional browser window, together with the structure-based sequence alignment and all key values of the alignment. In the structure-based sequence alignment, the structurally equivalent regions are colored red and orange, respectively, and the conserved residues are accentuated with black vertical bars. The 3D position of any amino acid in the protein structure can be highlighted by moving the mouse over the corresponding entry in the alignment. Together with the visualization of the ligands, these structural alignments greatly assist the identification of the structural core of the templates, as well as the validation of multiple sequence alignments of the templates.

To identify more templates in the *Tree Result Table*, we switch to the next higher layer, the *Related* layer. The parent domain remains the same (c2cwqB_), but the number of descendants increases to 36, because the structural similarity cutoff on the *Related* layer shrinks to 60%. We use the *Find* button to identify remaining templates. In addition to the already identified template c2cwqA_ from the *Similar* layer, templates three to six (1p8c_A,

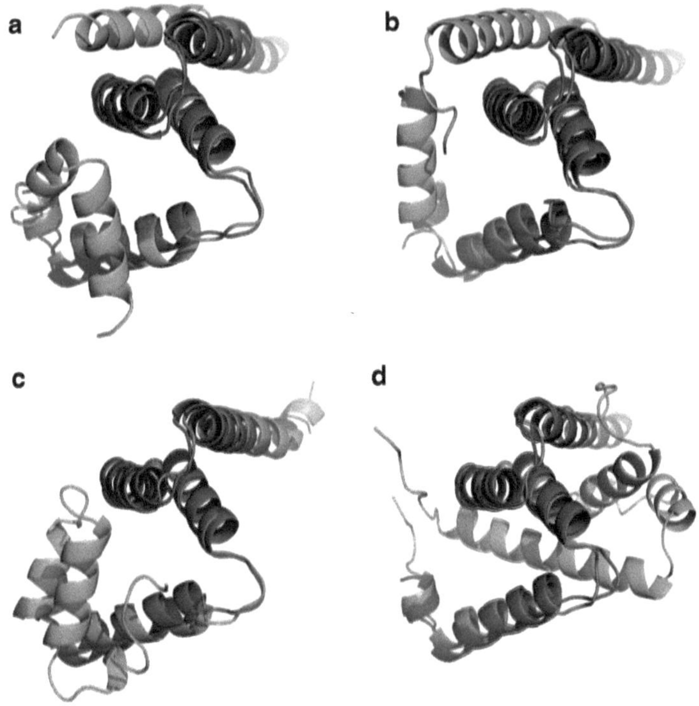

Fig. 5. Structural diversity among templates for CASP8 target T0408. The best hit (c3beyA_) from the HHsearch template list is superimposed with (**a**) c2af7A_, (**b**) c1vkeA_, (**c**) c2cwqA_, and (**d**) c2gmyA_. The first structure (query, here c3beyA_) is shown in *blue*, the second structure (target) in *green*, and the regions of similar structure are colored *red* (query) and orange (target).

2qeu_A, 2af7_A, and 1vke_A) are now present in the *Tree Result Table* of the *Related* layer. Again, we click on the rows of the respective templates to visually investigate the structural differences between the query (c3beyA_) and the other templates in the *Tree Result Table*. For example, structure 1p8c_A (17) is the second best template from the HHsearch template list (Table 2). Selecting the row of c1p8cA_ in the *Tree Result Table* displays the TopMatch superimposition of c1p8cA_ on c3beyA_. The superimposition in Fig. 6a reveals the structural similarity of c1p8cA_ and c3beyA_. c1p8cA_ covers 82% of c3beyA_ with an RMS of 1.8 Å, although the respective sequences have only 30% identical residues. Major structural differences are located at the carboxyl terminus (C terminus), where about half of the C-terminal α-helix of c3beyA_ is not superimposeable with c1p8cA_. This is the consequence of an almost 180° collapse in the α-helix of c1p8cA_, whereas the α-helix of c3beyA_ is elongated (see Fig. 6a). These unaligned regions are colored blue and green in the TopMatch alignment (Fig. 6a, b). One can easily determine the borders of the not superimposeable α-helices from the 3D view by moving the mouse over the sequences in the alignment. Here we have to decide if c1p8cA_ or c3beyA_ is

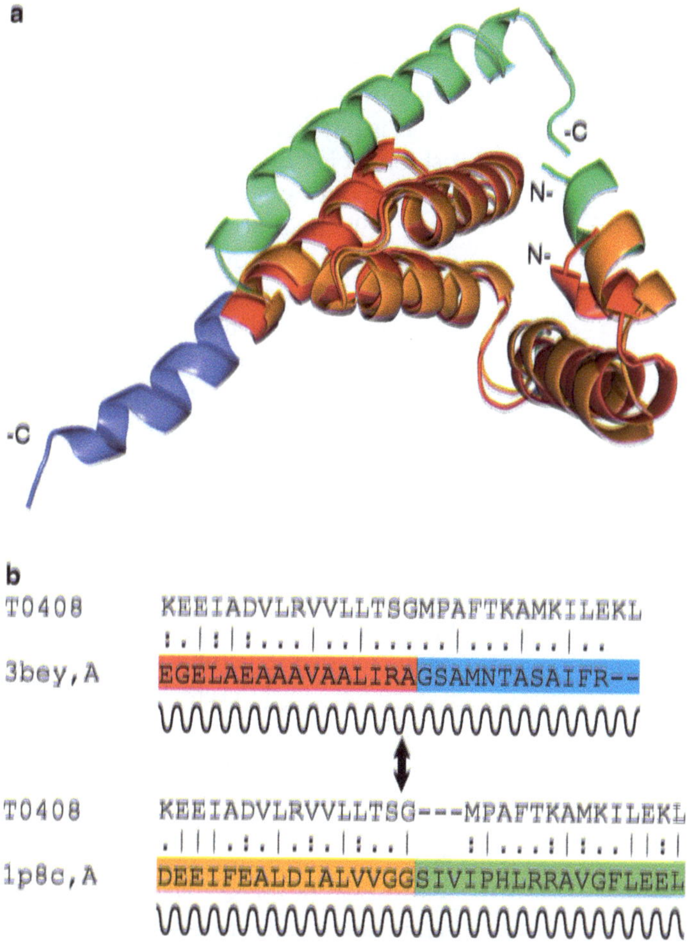

Fig. 6. Structural differences between the two best HHsearch templates for CASP target T0408 (Table 2). (**a**) TopMatch superimposition of first template 3bey,A (*blue* and *red*) with second template 1p8c,A (*green* and *orange*). *Red* and *orange parts* are structurally equivalent. The long C-terminal α-helix of 3bey,A cannot be superimposed on the corresponding α-helix of 1p8c,A over the full length of the helix. The reason is a considerable twist at residue GLY92 in 1p8c,A that involves an almost 180° collapse in the helix. (**b**) Pairwise sequence alignments of the C-terminal α-helices of the two templates with the target sequence (T0408). The *color coding* matches the TopMatch coloring from (**a**). The *black arrow* denotes the helix collapse. *Vertical bars mark* identical and *double dots* similar residues. Pairwise alignments were generated with EMBOSS (18).

the better template or if both structures are inadequate templates for this region. Best practice is to generate a pairwise sequence alignment of both templates with our target sequence (use the right-click menu explained in Fig. 3 to retrieve a specific protein sequence). Then the earlier defined borders of the respective α-helices from TopMatch can be identified in the pairwise sequence alignments (Fig. 6b). The target-template alignment shows higher sequence similarity at the collapsed α-helix of c1p8cA_ than at the elongated α-helix of c3beyA_. To play it safe, one would use both templates to generate different models and examine the modeled structures with appropriate validation tools (c.f. Note 1).

It is highly advisable to proceed the whole template list in this fashion, at least for the best templates that are considered for modeling. In our case, the next template candidate is chain A of protein 2qeu (19). By repeating the previous steps, we are able to identify this entry as c2qeuA2 in the *Tree Result Table* in the same *Related* layer we discussed earlier. The domain name specifies c2qeuA2 as domain two of chain A of 2qeu. Obviously our query template 3bey,A has a different domain configuration as 2qeu,A, which can easily be verified by the TopMatch superimposition of the two domains. Three α-helices are perfectly superimposeable, but c2qeuA2 lacks the twist in the C-terminal α-helix (cf. c1p8cA_) and, additionally, the N-terminal α-helix of c3beyA_. The N-terminal α-helix is part of the first domain (c2qeuA1) of 2qeu,A. The same domain configuration can be found in the fifth best template 2af7_A. Both domains of 2af7 (c2af7A1 and c2af7A2) have highly similar structures compared to the two domains of 2qeu (relative structural similarity >80%), although c2qeuA2 and c2af7A2 are in different *S30* layers.

All templates from the template list can be found at least on the next higher layer, the *Remote* layer, except for the template 3bjx_A on position 20. Even on the *Distant* layer, which is the highest COPS layer beneath the *Root*, where the descendants have only 30% relative structural similarity to the parent, this protein structure is missing. In some cases, it is possible that templates from the template list cannot be found in the layers of the *Tree Result Table*; for instance if the templates are matching on different parts of the target sequence. In this case, it is advisable to use the first unidentified template in the COPS search, just like we used chain A of 3bey in the previous example. Moreover, this is indicative of templates that match different domains of the target sequence.

Another reason for missing templates in the *Tree Result Table* is structural diversity among the templates. In the worst case, the result is a false positive, like 3bjx,A from the template list. The sequence similarity scores returned for this template are all considered to be significant, but pairwise structural comparisons to the other templates reveal no trustable structural equivalences (see Fig. 7). A single template with no significant structural similarity to other templates in the list should be regarded with caution. If the sequence similarity to the target is weak, too, and the template covers the same regions of the sequence as other, more trustable templates, it is save to skip this structure.

Further reasons for missing templates in the *Tree Result Table* include protein structures with similar sequences but different 3D structures. We report more on this phenomenon in Note 2.

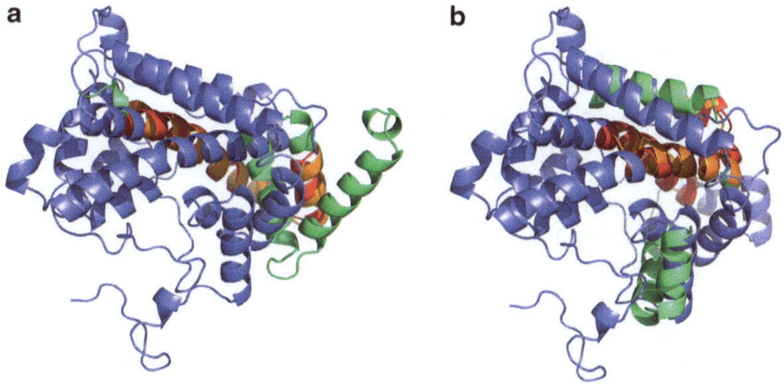

Fig. 7. Comparison of the potential template 3bjx_A (in *blue/red*) with (**a**) the best HHsearch template 3bey_A and (**b**) chain A of the released structure of CASP8 target T0408 (PDB code 3d7i). 3bjx_A is not a suitable template for T0408 although having significant scores (Table 2). More information about the characterization of potential false positives can be found in Subheading 3.1.

3.2. What Is the Biological Context of My Templates?

For many modeling targets, at least basic information is available about the biological context of the sequence, such as its source organism, its putative role in the cell or known binding partners. This information provides valuable clues for template selection in addition to sequence similarity and further data from experiments (e.g., chemical shifts, c.f. Note 3).

COPS domains shown in the *Selection Widget* or the *Tree Result Table* are annotated with several features that can be employed to narrow down the set of template candidates (see Fig. 8). For instance, the source organisms of the respective protein chains and their assignment to a taxonomic superkingdom can be compared across potential templates using the *Species* and *S-Kingdom* columns. Taking up our example above (T0408), we find that the target sequence was obtained from the archaeon *Methanocaldococcus jannaschii.* The HHsearch template list contains only two more proteins from archaea. The first is the highest ranking template 3bey_A and the second is structure 2af7_A at rank five; all other templates are from bacteria. In general, template structures from evolutionary-related organisms should be favored. Note, however, that a template from the same organism as the target sequence might have considerable changes in its fold, because proteins that result from the duplication of a gene (paralogs) are usually no longer subject to functional constraints (20–24).

The list of putative templates can also be characterized by functional aspects of the respective proteins. According to the *PDB-Header* column in COPS, the template list contains ten proteins with unknown function, eight oxidoreductases, and five lyases. Together with the more detailed *Compound* data this information can be used to find templates that match descriptions of function available for the target sequence.

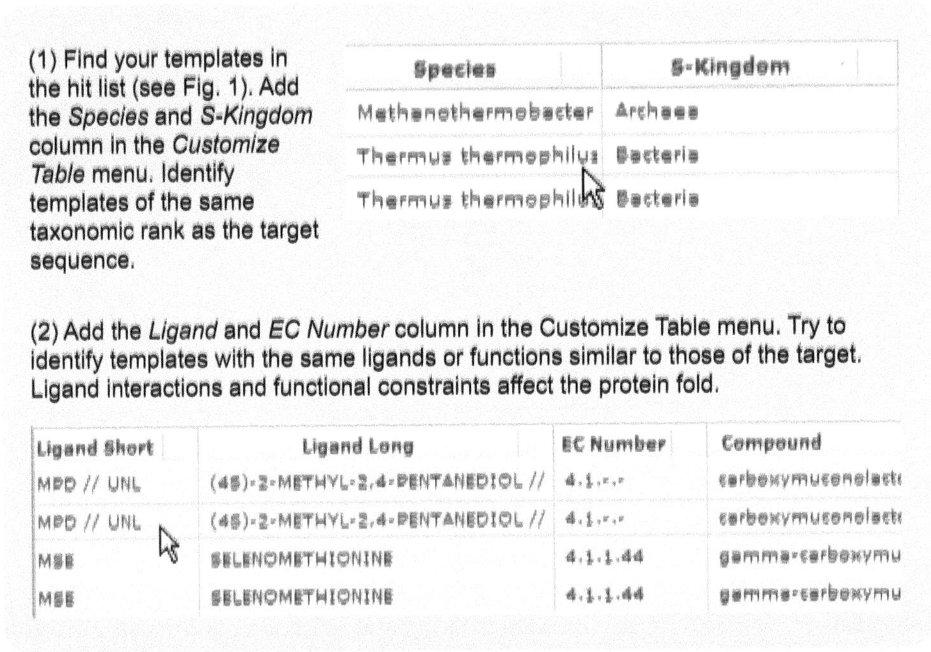

Fig. 8. Basic steps to investigate the biological context of putative template structures in COPS.

Ligands are another important source for clues on the biochemical function of proteins. They often affect the 3D structure of proteins resulting in considerable differences between the plain and the ligand bound conformations. Interfaces where ligands are bound depend on specific residues that interact with the ligand. Frequently, these residues are conserved across species. For example, the apoptotic protease-activating factor 1 (Apaf-1, PDB code 1z6t (10)) from *Homo sapiens* comprises five distinct domains in its chain A: (1) CARD, (2) an α/β fold, (3) helical domain I, (4) a winged-helix domain, and (5) helical domain II. Apaf-1 is bound to the ligand ADP. Three domains of Apaf-1 (the α/β fold, helical domain I, and the winged-helix domain) have equivalent domains in chain C of the apoptosis regulator *CED*-4-*CED*-9 (PDB code 2a5y (25)) from *Caenorhabditis elegans.* If superimposed pairwise, the equivalent domains have high structural similarities but sequence similarities below 30% (1). On chain level only the CARD domain and the α/β-fold can be superimposed simultaneously. This means that the arrangement of the domains in the protein chains is different for the ATP-bound 2a5y and the ADP-bound 1z6t. Both conformations are a consequence of the bound ligands. In particular, ADP locks Apaf-1 in the inactive conformation because it promotes the interactions between the domains of 1z6t (10). This is a clear example of how ligand binding can alter the structure of a protein. Even so, five residues of the eight residues that bind ADP and ATP, respectively, are conserved and structurally equivalent.

Regions of proteins that lack a well-defined three-dimensional structure may switch to an ordered state upon interaction with a ligand (26). Automated methods may confusingly predict such regions as having a specific secondary structure as well as being disordered (27). If a template aligns to a region predicted to be disordered in the target, the ligand information given in COPS and the 3D visualization of their location in Jmol assist in the identification and validation of these regions.

To gather information on ligands in COPS and compare it across the templates, enable the *Ligand Short/Ligand Long* columns in the *Tree Result Table.* Additionally, the location of the ligands in the 3D structure can be visualized in the maximized *Jmol Widget* (Fig. 2f) and the external TopMatch window. The *Ligand* columns display all ligands associated with the respective PDB chain, separated by two slashes. In *Ligand Short*, ligands are represented by their shortcuts as defined by PDB. The entry *Go to Ligand Expo* in the context menu of the hit list links to the corresponding Ligand Expo page of PDB. This page offers 3D visualization of the selected ligand as well as detailed chemical and structural information. Enzymes in the *Tree Result Table* are further characterized by the entries in the *EC Number* column. This column contains the Enzyme Classification numbers as provided by the IUBMB (http://www.chem.qmul.ac.uk/iubmb/enzyme/). The detailed description of each enzymatic reaction can be opened with the *Go to EC* entry in the context menu of the *Tree Result Table.*

4. Notes

1. Final model quality is affected by a multitude of factors. Since each step in homology modeling implies its own pitfalls and error sources, it is vital to continuously check potential model structures for inaccuracies introduced by the modeling pipeline. In particular, care should be taken in template selection by choosing templates with high quality. Various parameters that can be used to winnow template structures in terms of quality directly originate from experimental structure determination, like crystallographic resolution or *R*-factor (28). In the *Tree Result Table* of COPS, the *Method* and *Resolution* columns can be consulted to get first clues on template quality. In addition, several tools directly linked from COPS provide independent quality estimates of potential template structures as well as the resulting models. ProSA (29, 30) employs knowledge-based potentials to recognize erroneous coordinates of protein structures. Besides a global quality measure, ProSA yields quality scores on residue level which allows to identify problematic parts of the template. Following a related approach, NQ-Flipper (31) recognizes unfavorable rotamers of asparagine and glutamine residues and provides means to download a corrected model. Side-chain correctness, in general, may be

analyzed by using a different approach (32) which compares local electron density distributions to their expected analogs. Using this method, it is possible to detect a wide variety of problems including unrealistic atomic contacts, unusual rotamers, and incorrect atom naming. Further computational tools widely used for model validation include Procheck (33), MolProbity (34), and WHAT_CHECK (35).

2. Currently only a few cases of pairs of proteins with high sequence similarity and different conformations are known, but this phenomenon may be more common than previously thought (36, 37). Designed proteins with these properties have been reported (38, 39), and there are also examples of naturally occurring proteins of this kind. Roessler et al. (40) found two members of the Cro repressor family having sequence identities as high as 40%, although half of their structures have switched from helices to strands. Moreover, some proteins have the ability to switch between several stable conformations (41–43). For instance, the chemokine lymphotactin adopts two distinct folds at equilibrium under physiological conditions (44). In the CASP6 experiment, the experimentally solved structure of one of the targets showed a conformation considerably different to that of the best template although having the same sequence (45). In a large-scale analysis with 13,000 protein chains (46), sequence alignment-based structural superpositions and geometry-based structural alignments for protein pairs were carried out to determine the extent to which sequence similarity ensures structural similarity. There were many examples where two proteins that are similar in sequence have structures that differ significantly. Some homology detection tools are searching against a nonredundant set of templates defined by sequence similarity. Important structure information for the modeling process can be lost if a nonredundant set of structures is constructed based merely on sequence similarity. TopMatch provides the possibility to perform both sequence-based superpositions and structure-based superpositions for a detailed investigation of such cases.

3. Chemical shifts are the "mileposts" of NMR spectroscopy (47). They are used for direct refinement of protein structures (48), prediction of protein secondary structure (49, 50), inference of protein backbone angles (51, 52), structure validation (53), and detection of structural similarities in proteins (54). Supplementing modeling by chemical shift information has gained interest (again) over the past years. In 2008, the CS23D Server (51) was presented which rapidly generates structures from both chemical shift and sequence information. In the beginning of 2009, Shen ea. (52) published a modified version of the structure prediction tool Rosetta which applies a chemical shift filter to improve the quality of the fragments used for

model generation. Finally, Ginzinger and Coles (55) published work on a fast structure database search which uses the chemical shifts of the target protein to reliably identify structural templates even in cases of low amino acid sequence similarity.

Acknowledgments

This work was supported by FWF Austria grant number P21294-B12.

References

1. Suhrer SJ, Wiederstein M, Gruber M, et al. (2009) COPS-a novel workbench for explorations in fold space. Nucleic Acids Res 37:W539–W544
2. Suhrer SJ, Wiederstein M, Sippl MJ (2007) QSCOP – SCOP quantified by structural relationships. Bioinformatics 23:513–514
3. Suhrer SJ, Gruber M, Sippl MJ (2007) QSCOP-BLAST–fast retrieval of quantified structural information for protein sequences of unknown structure. Nucleic Acids Res 35:W411–W415
4. Choi WS, Jeong BC, Joo YJ, et al. (2010) Structural basis for the recognition of N-end rule substrates by the UBR box of ubiquitin ligases. Nat Struct Mol Biol 17:1175–1181
5. Norambuena T, Melo F (2010) The Protein-DNA Interface database. BMC Bioinformatics 11:262
6. Berman HM, Westbrook J, Feng Z, et al. (2000) The Protein Data Bank. Nucleic Acids Res 28:235–242
7. Chothia C, Lesk AM (1986) The relation between the divergence of sequence and structure in proteins. EMBO J 5:823–826
8. Sippl MJ, Wiederstein M (2008) A note on difficult structure alignment problems. Bioinformatics 24:426–427
9. Sippl MJ, Suhrer SJ, Gruber M, et al. (2008) A discrete view on fold space. Bioinformatics 24:870–871
10. Riedl SJ, Li W, Chao Y, et al. (2005) Structure of the apoptotic protease-activating factor 1 bound to ADP. Nature 434:926–933
11. Cozzetto D, Kryshtafovych A, Fidelis K, et al. (2009) Evaluation of template-based models in CASP8 with standard measures. Proteins 77 Suppl 9:18–28
12. Frank K, Gruber M, Sippl MJ (2010) COPS Benchmark: interactive analysis of database search methods. Bioinformatics 26:574–575
13. Söding J (2005) Protein homology detection by HMM-HMM comparison. Bioinformatics 21:951–960
14. JCSG (2008) Crystal structure of carboxymuconolactone decarboxylase family protein possibly involved in oxygen detoxification (1591455) from *Methanococcus jannaschii* at 1.75Å resolution. To be published
15. Kuzin A, Xu JGX, Neely H, et al. (2007) Crystal structure of the protein O27018 from *Methanobacterium thermoautotrophicum*. To be published
16. Ito K, Arai R, Fusatomi E, et al. (2006) Crystal structure of the conserved protein TTHA0727 from Thermus thermophilus HB8 at 1.9 A resolution: A CMD family member distinct from carboxymuconolactone decarboxylase (CMD) and AhpD. Protein Sci 15:1187–1192
17. Kim Y, Joachimiak A, Brunzelle J, et al. (2003) Crystal Structure Analysis of *Thermotoga maritima* protein TM1620 (APC4843). To be Published
18. Rice P, Longden I, Bleasby A (2000) EMBOSS: the European Molecular Biology Open Software Suite. Trends Genet 16:276–277
19. JCSG (2007) Crystal structure of Putative carboxymuconolactone decarboxylase (YP-555818.1) from *Burkholderia xenovorans* LB400 at 1.65Å resolution
20. Koonin EV (2005) Orthologs, paralogs, and evolutionary genomics. Annu Rev Genet 39:309–338
21. Pál C, Papp B, Lercher MJ (2006) An integrated view of protein evolution. Nat Rev Genet 7:337–348
22. Andreeva A, Murzin AG (2006) Evolution of protein fold in the presence of functional constraints. Curr Opin Struct Biol 16:399–408
23. Chothia C, Gough J (2009) Genomic and structural aspects of protein evolution. Biochem J 419:15–28

24. Worth CL, Gong S, Blundell TL (2009) Structural and functional constraints in the evolution of protein families. Nat Rev Mol Cell Biol 10:709–720
25. Yan N, Chai J, Lee ES, et al. (2005) Structure of the CED-4-CED-9 complex provides insights into programmed cell death in *Caenorhabditis elegans*. Nature 437:831–837
26. Dyson HJ, Wright PE (2005) Intrinsically unstructured proteins and their functions. Nat Rev Mol Cell Biol 6:197–208
27. Bordoli L, Kiefer F, Arnold K, et al. (2009) Protein structure homology modeling using SWISS-MODEL workspace. Nat Protoc 4:1–13
28. Wlodawer A, Minor W, Dauter Z, et al. (2008) Protein crystallography for non-crystallographers, or how to get the best (but not more) from published macromolecular structures. FEBS J 275:1–21
29. Sippl MJ (1993) Recognition of errors in three-dimensional structures of proteins. Proteins 17:355–362
30. Wiederstein M, Sippl MJ (2007) ProSA-web: interactive web service for the recognition of errors in three-dimensional structures of proteins. Nucleic Acids Res 35:W407–W410
31. Weichenberger CX, Byzia P, Sippl MJ (2008) Visualization of unfavorable interactions in protein folds. Bioinformatics 24:1206–1207
32. Ginzinger SW, Weichenberger CX, Sippl MJ (2010) Detection of unrealistic molecular environments in protein structures based on expected electron densities. J Biomol NMR 47:33–40
33. Laskowski RA, MacArthur MW, Moss DS, et al. (1993) PROCHECK: a program to check the stereochemical quality of protein structures. J Appl Crystallogr 26:283–291
34. Chen VB, Arendall WB, Headd JJ, et al. (2010) MolProbity: all-atom structure validation for macromolecular crystallography. Acta Crystallogr D Biol Crystallogr 66:12–21
35. Hooft RW, Vriend G, Sander C, et al. (1996) Errors in protein structures. Nature 381:272
36. Davidson AR (2008) A folding space odyssey. Proc Natl Acad Sci U S A 105:2759–2760
37. Sippl MJ (2009) Fold space unlimited. Curr Opin Struct Biol 19:312–320
38. Dalal S, Balasubramanian S, Regan L (1997) Protein alchemy: changing beta-sheet into alpha-helix. Nat Struct Biol 4:548–552
39. He Y, Chen Y, Alexander P, et al. (2008) NMR structures of two designed proteins with high sequence identity but different fold and function. Proc Natl Acad Sci U S A 105:14412–14417
40. Roessler CG, Hall BM, Anderson WJ, et al. (2008) Transitive homology-guided structural studies lead to discovery of Cro proteins with 40% sequence identity but different folds. Proc Natl Acad Sci U S A 105:2343–2348
41. Murzin AG (2008) Metamorphic Proteins. Science 320:1725–1726
42. Gambin Y, Schug A, Lemke EA, et al. (2009) Direct single-molecule observation of a protein living in two opposed native structures. Proc Natl Acad Sci U S A 106:10153–10158
43. Bryan PN, Orban J (2010) Proteins that switch folds. Curr Opin Struct Biol 20:482–488
44. Tuinstra RL, Peterson FC, Kutlesa S, et al. (2008) Interconversion between two unrelated protein folds in the lymphotactin native state. Proc Natl Acad Sci U S A 105:5057–5062
45. Ginalski K (2006) Comparative modeling for protein structure prediction. Curr Opin Struct Biol 16:172–177
46. Kosloff M, Kolodny R (2008) Sequence-similar, structure-dissimilar protein pairs in the PDB. Proteins 71:891–902
47. Zhang H, Neal S, Wishart DS (2003) RefDB: a database of uniformly referenced protein chemical shifts. J Biomol NMR 25:173–195
48. Schwieters CD, Kuszewski JJ, Tjandra N, et al. (2003) The Xplor-NIH NMR molecular structure determination package. J Magn Reson 160:65–73
49. Wishart DS, Sykes BD, Richards FM (1992) The chemical shift index: a fast and simple method for the assignment of protein secondary structure through NMR spectroscopy. Biochemistry 31:1647–1651
50. Wang Y, Jardetzky O (2002) Probability-based protein secondary structure identification using combined NMR chemical-shift data. Protein Sci 11:852–861
51. Berjanskii MV, Neal S, Wishart DS (2006) PREDITOR: a web server for predicting protein torsion angle restraints. Nucleic Acids Res 34:W63–W69
52. Shen Y, Delaglio F, Cornilescu G, et al. (2009) TALOS+: a hybrid method for predicting protein backbone torsion angles from NMR chemical shifts. J Biomol NMR 44:213–223
53. Oldfield E (1995) Chemical shifts and three-dimensional protein structures. J Biomol NMR 5:217–225
54. Ginzinger SW, Fischer J (2006) SimShift: identifying structural similarities from NMR chemical shifts. Bioinformatics 22:460–465
55. Ginzinger SW, Coles M (2009) SimShiftDB; local conformational restraints derived from chemical shift similarity searches on a large synthetic database. J Biomol NMR 43:179–185

Chapter 3

Methods for Sequence–Structure Alignment

Česlovas Venclovas

Abstract

Homology modeling is based on the observation that related protein sequences adopt similar three-dimensional structures. Hence, a homology model of a protein can be derived using related protein structure(s) as modeling template(s). A key step in this approach is the establishment of correspondence between residues of the protein to be modeled and those of modeling template(s). This step, often referred to as sequence–structure alignment, is one of the major determinants of the accuracy of a homology model.

This chapter gives an overview of methods for deriving sequence–structure alignments and discusses recent methodological developments leading to improved performance. However, no method is perfect. How to find alignment regions that may have errors and how to make improvements? This is another focus of this chapter. Finally, the chapter provides a practical guidance of how to get the most of the available tools in maximizing the accuracy of sequence–structure alignments.

Key words: Homology modeling, Protein structure, Sequence profiles, Hidden Markov models, Alignment accuracy, Model quality

1. Introduction

At present, homology or comparative modeling is the most accurate and therefore the most widely used protein structure prediction approach. Homology modeling is based on the empirical observation that evolutionary-related proteins (to be more precise—evolutionary-related protein domains) tend to have similar three-dimensional (3D) structures. Moreover, protein structural features often remain preserved long after the sequence signal is lost to mutations, insertions, and deletions. Therefore, 3D structure is considered to be the most robustly conserved feature of homologous proteins, certainly more conserved than the sequence or molecular function. Although there are some convincing exceptions to this rule (1), it still holds for the absolute majority of cases.

Andrew J.W. Orry and Ruben Abagyan (eds.), *Homology Modeling: Methods and Protocols*, Methods in Molecular Biology, vol. 857, DOI 10.1007/978-1-61779-588-6_3, © Springer Science+Business Media, LLC 2012

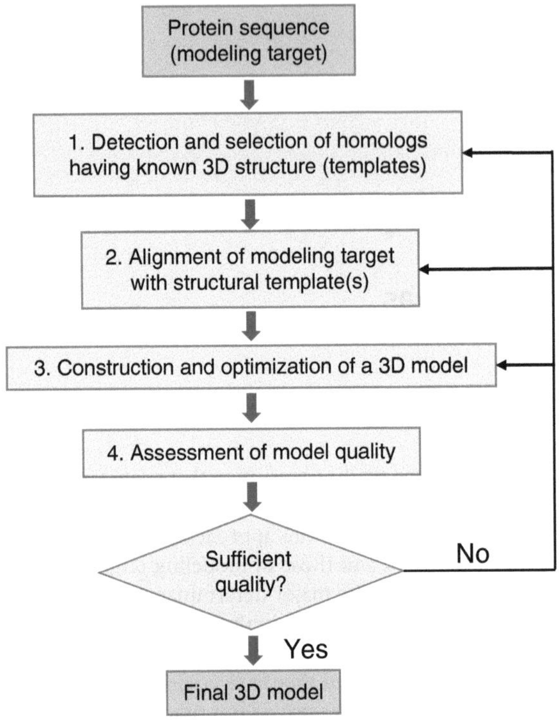

Fig. 1. Homology modeling flowchart.

Homology modeling is used to build a 3D structural model of a protein (modeling target) on the basis of the alignment of its amino acid sequence with a related protein of known structure (template). Any homology modeling approach consists of four main steps: (1) identification of related proteins that have experimentally determined structures and therefore can be used as structural templates for modeling, (2) mapping corresponding residues between the target sequence and template structure, the process often referred to as sequence–structure alignment, (3) generating a 3D model of a target protein on the basis of the sequence–structure alignment, and (4) estimating the correctness of the resulting model. The whole process may be iterated (restarting at any of the steps) until the satisfactory estimated quality is obtained or until the model can no longer be improved (Fig. 1).

This chapter focuses on the second step in the homology modeling process—producing sequence–structure alignment—and will only touch upon other steps as necessary.

2. Sequence–Structure Alignment Problem

Once a suitable structural homolog (template) is identified, the accurate mapping of target sequence onto template structure becomes a major determinant of the resulting model quality.

What does it mean to produce an accurate sequence–structure mapping/alignment? Let us suppose that we know 3D structures of both the template and the target. If we superimpose those two structures, we will find out that for structurally similar regions of both proteins we can derive an unequivocal correspondence between residues. The sequence–structure alignment step in homology modeling aims to reproduce this correspondence as accurately as possible, but without the benefit of knowing the "real" (experimental) structure of the modeling target. Obviously, unless target and template are very closely related, there may be regions displaying significant structural differences between the two. These structurally dissimilar regions most often result from insertions, deletions, or extensive changes in the amino acid sequence. Therefore, in such regions, the assignment of residue correspondence is not always straightforward and sometimes plainly meaningless. In other words, an accurate sequence–structure alignment should include all the structurally and evolutionary equivalent residue pairs, at the same time leaving out structurally different regions. As the number of experimentally determined structures continues to grow steadily, in many cases a modeling target can be aligned not only to a single but also to a number (sometimes very large) of available structural templates. Often, an accurate alignment over the entire target length cannot be achieved with the same template; instead, different target regions (sometimes quite short) can be aligned to different templates. This provides opportunity for the model improvement but at the same time introduces additional complexity into the modeling procedure.

The sequence–structure alignment problem can be subdivided into the three subproblems: (1) generating initial sequence–structure alignment, (2) finding out which alignment regions may need adjustment, and (3) improving the alignment.

3. Sequence-Based Methods for Sequence–Structure Alignment

Usually, the construction of initial sequence alignment between the target and the template coincides with the first step in homology modeling (Fig. 1), template identification. Therefore, template identification will be discussed along with the sequence–structure alignment. Since for the modeling target only amino acid sequence is known to start with, sequence comparison is the primary means to detect related protein(s) having known experimental 3D structure. If aligned sequences share a statistically significant sequence similarity (the similarity which could not be expected by chance), it is considered that the sequences share common evolutionary origin. It further means that their 3D structures can also be expected to be similar.

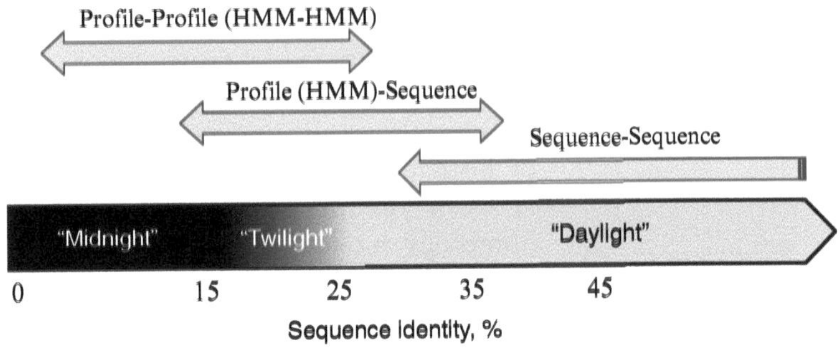

Fig. 2. Different types of homology detection and alignment methods are most effective for different sequence similarity ranges. Sequence similarity is partitioned into three approximate intervals corresponding to the decreasing difficulty of identifying homology from sequence: the "midnight" zone (<15% sequence identity), the "twilight" zone (~15–25%), and the "daylight" zone (>25%).

Depending on the evolutionary distance between proteins, sequence-based methods of different complexity may be required to detect their relationship (Fig. 2). These methods can be grouped on the basis of the increasingly complex sequence information they use:

1. Alignment of a pair of sequences
2. Profile–sequence and hidden Markov model (HMM)–sequence alignments
3. Profile–profile and HMM–HMM alignments.

3.1. Pairwise Sequence Alignment Methods

Methods that detect homology through the alignment of a pair of sequences (pairwise alignment) have emerged earliest and are conceptually the simplest. They use only amino acid sequences of two proteins, a scoring table for residue substitutions and an algorithm to produce an alignment. Usually, pairwise alignment methods report the statistical significance of the resulting alignments, allowing to use them for sequence database searches. Undoubtedly, the most popular database search tool based on pairwise alignment is BLAST (2, 3). It is very fast and has a solid statistical foundation for homology inference, provided by the incorporation of the Karlin–Altschul extreme value statistics (4). The integration of BLAST suite of programs together with major sequence databases at the National Center for Biotechnology Information (NCBI; http://www.ncbi.nlm.nih.gov/) is another important factor contributing to the popularity of BLAST. FASTA (5) and Ssearch (6, 7) are two other widely used pairwise alignment and database search methods. Pairwise sequence comparison programs can provide a fast initial estimate of the difficulty level of homology modeling. They can be adequate for detecting evolutionary-related proteins that share over 25–30% identical residues, the range of sequence similarity that may be called a "daylight" zone (Fig. 2). However, in many cases, corresponding alignments need improvements. Only if aligned sequences are over 40–50% identical to each other and have few or no gaps, it can be expected that alignments may be accurate in a structural sense.

Despite the limited and ever decreasing use of pairwise sequence comparison to obtain sequence–structure alignments for direct use in modeling, this is the initial step essentially in all of the more sophisticated sequence comparison techniques that utilize information from multiple related sequences. Therefore, the improvements in the initial pairwise comparison step may have a profound effect on the final results. Recently, a significant step forward was made by the development of the context-specific BLAST (CS-BLAST) (8). Unlike the original BLAST, which treats sequence positions independently of each other, CS-BLAST considers the substitution probability at a particular position to depend on the neighboring residues (sequence context). This methodological innovation led not only to a higher sensitivity in homology detection but also to a significant improvement of the alignment quality (8). CS-BLAST may be especially promising for application to *singleton sequences* (sequences without detectable homologs), because the lack of related sequences precludes the use of methods based on profile–sequence or profile–profile alignments that are discussed next.

3.2. Profile–Sequence and Hidden Markov Model–Sequence Alignment Methods

When the evolutionary relationship is more distant (sequence similarity is fading into the "twilight" zone; Fig. 2), the pairwise sequence comparison may not be sufficient to reliably identify homology and to produce an accurate alignment. In such cases, methods that use information from aligned multiple sequences represented by either sequence profiles (9) or HMMs (10) can be much more effective. The power of profiles and HMMs stems from a comprehensive statistical model generated for the aligned group of related sequences. This model indicates which positions are conserved and which are variable and where insertions or deletions are most likely to occur. Therefore, a comparison of a profile with database sequences can both provide more sensitive detection of homologs and generate more accurate alignments. Currently, the most widely used profile–sequence comparison method is position-specific iterated BLAST (PSI-BLAST) (3). PSI-BLAST uses a multiple alignment of the highest-scoring matches returned in an initial BLAST search to construct a position-specific scoring matrix (PSSM). The constructed PSSM replaces the generic substitution matrix (e.g., BLOSUM or PAM series) in a subsequent round of the BLAST search. This process can be repeated a number of times. Every time, new sequences detected above the predefined threshold are used to adjust the profile. Thus, with each iteration more and more distantly related sequences are included making the profile more inclusive yet still specific for the sequence family.

This makes PSI-BLAST a very powerful sequence search and comparison tool that can often detect and align homologs having sequence identities of 15% or even lower (both "twilight" and "midnight" zones of sequence similarity). Since the elementary step in PSI-BLAST is based on BLAST, it also treats positions as being independent from each other. Just like CS-BLAST, context-specific iterated BLAST (CSI-BLAST) (8) has been shown to outperform PSI-BLAST, suggesting that the incorporation of sequence context into sequence or profile comparisons is a promising avenue for improvements.

HMMER (11) and sequence alignment and modeling (SAM) (12) tool suites are the best known HMM–sequence comparison methods. HMMs are similar to sequence profiles, but they use probability theory to guide how all the scoring parameters should be set. HMMs also have additional probabilities for insertions and deletions at each position of the profile. The latter feature of HMMs is important in trying to better represent properties of protein sequence evolution. It is obvious that the probability of insertions and deletions within the protein sequence is very much position-dependent because of varying structural and/or functional constraints. While insertions/deletions may be detrimental within the structural core, they are more likely to be tolerated within solvent-exposed structurally variable regions such as loops. HMMs, however, have important limitations too. Just like sequence profiles (PSSMs), HMMs treat a particular position independent of all the other positions, and thus are not able to capture any higher-order correlations that may exist (and we know that they do!) in protein sequences. Despite seeming methodological advantages, HMM–sequence-based methods have not been used as widely as PSI-BLAST. Why so? For one, so far HMM–sequence comparison methods have been much slower than PSI-BLAST. Besides, it has been difficult to devise an iteration procedure for HMMs that would work as smoothly and seamlessly as in PSI-BLAST. However, the HMM field has made significant advances. For example, SAM-T08 (13), the latest protein structure prediction method based on SAM tool suite, features several iterative procedures. The use of heuristics has also recently helped to achieve a significant speedup and to introduce an iterative search protocol for HMMER (14). Reportedly, HMMER is now roughly on a par with BLAST according to the speed of database search, and its iterative search procedure (jackhmmer) rivals PSI-BLAST in sensitivity and alignment accuracy.

3.3. Profile–Profile and HMM–HMM Alignment Methods

Evolutionary relationships that are too distant to be detected either by pairwise sequence or by profile–sequence (HMM–sequence) comparisons ("midnight" zone; Fig. 2) may still be identified by methods that are based on profile–profile or HMM–HMM alignments. These methods add another level of complexity by comparing two sequence profiles (HMMs) instead of a profile (HMM) with a single sequence. In other words, instead of asking the question of whether a sequence belongs to the family, these methods are asking the question of whether two sequence families are evolutionary related. This generalization brought about a previously unseen sensitivity of homology detection and, albeit less dramatic, an improvement in the alignment accuracy (15–20). Although in sensitivity and alignment accuracy they still lag behind the methods based on 3D structure comparison such as DALI (21), it is possible to see examples of the opposite (17). Some of the best performers among methods based on HMM–HMM comparison include HHsearch (16) and PRC (19), while COMPASS (15), COMA (17), and PROCAIN (22) represent those based on profile–profile comparison. At present, both methodologies (profile and HMM-based) are being actively developed, and it is not clear whether one of the two will be dominating in the future. There are pros and cons on both sides. Traditionally, sequence profile–profile alignments have been using fixed gap penalties, while the HMM framework naturally accommodates more biologically relevant position-dependent gap penalties. Nonetheless, position-dependent gap penalties can be successfully implemented in profile–profile methods, as recently has been demonstrated in COMA (17). The Karlin–Altschul statistics introduced in BLAST and PSI-BLAST can be more easily extended for profile–profile than for the HMM–HMM comparison. On the other hand, recently a probabilistic model of local sequence alignment amenable to the Karlin–Altschul statistics has been introduced in HMMER. This has significantly reduced the computational cost for statistical significance estimation without sacrificing the accuracy (23). Both profile–profile and HMM–HMM methods consider sequence positions to be independent of each other, but as demonstrated by the success of CS/CSI-BLAST (8), this is clearly a non-optimal representation of protein sequence information. Indirectly, the importance of positional context in the profile–profile (HMM–HMM) comparison has been demonstrated by a boost in performance with the incorporation of additional information (16, 22). The largest impact has been observed by the inclusion of the secondary structure (SS) information, which may be considered as a particular representation of context dependency. Thus, a further improvement of the context-specific scoring may be a promising direction for increasing homology detection sensitivity and alignment accuracy.

A brief summary of different types of alignment methods is provided in Table 1.

3.4. Multiple Sequence Alignment Methods

Multiple sequence alignment (MSA) methods represent a distinct case as they are not designed to detect homologous sequences. Instead, they align a set of homologous sequences already identified by other methods, such as those discussed above. MSA methods may be useful in at least two different ways. First, these methods

Table 1
Sequence-based methods for homology detection and sequence–structure alignment construction

Method	Type	Address
BLAST	Sequence–Sequence	http://blast.ncbi.nlm.nih.gov/
FASTA/Ssearch	Sequence–Sequence	http://fasta.bioch.virginia.edu/ http://www.ebi.ac.uk/Tools/sss/fasta/
CS-BLAST	Sequence (profile)–Sequence	http://toolkit.lmb.uni-muenchen.de/cs_blast/
PSI-BLAST	Profile–Sequence	http://blast.ncbi.nlm.nih.gov/
CSI-BLAST	Profile–Sequence	http://toolkit.lmb.uni-muenchen.de/cs_blast/
HMMER	HMM–Sequence	http://hmmer.org/
SAM	HMM–Sequence	http://compbio.soe.ucsc.edu/HMM-apps/
COMPASS	Profile–Profile	http://prodata.swmed.edu/compass/
PROCAIN	Profile–Profile + additional sequence features + SS[a]	http://prodata.swmed.edu/procain/
COMA	Profile–Profile	http://www.ibt.lt/bioinformatics/coma/
HHsearch	HMM–HMM + SS[a]	http://toolkit.lmb.uni-muenchen.de/hhpred/
PRC	HMM–HMM	http://supfam.org/PRC http://www.ibi.vu.nl/programs/prcwww/

[a]Secondary structure

may be used to improve the quality of MSAs, from which profiles (HMMs) for homology search and alignment are constructed. Second, if both target and template are in the set of sequences to be aligned, target-template alignment can be directly obtained in the context of resulting MSA.

Given a set of sequences, MSA methods aim to construct an alignment in which columns represent evolutionary (structurally) equivalent residues. Although in theory dynamic programming algorithms for pairwise alignment can be extended for computing an optimal alignment of multiple sequences, they are too computationally demanding to be practically useful. As a result, most current techniques use various approximations and heuristics. These methods are not guaranteed to derive an optimal MSA, but in practice they can often produce good alignments using modest computational resources. Most of the modern MSA tools use heuristics known as *progressive alignment*. In this strategy, an approximate alignment guide tree is first constructed based on pairwise sequence similarities. Using this guide tree, the most closely related sequences are aligned first. Next, these subalignments are aligned to each other until all sequences are incorporated into MSA.

Thus, the progressive alignment substitutes the task of MSA into a series of pairwise alignments. ClustalW (24), one of the earliest programs and still a very popular choice, is a representative of progressive alignment methods. The main drawback of the progressive alignment strategy is that errors made early on in the construction of guide trees or pairwise alignments (especially in the initial stages) cannot be corrected and tend to propagate in the entire alignment. Thus, ClustalW can produce good alignments for closely related sequences, but alignments for divergent sequence sets may be poor. Therefore, a number of approaches have been devised to avoid the problems associated with an application of progressive alignment. For more details on recent methodological and algorithmic improvements, the reader is referred to recent reviews (25–27). Here, only several methods that had been reported to perform well in various benchmarks are briefly discussed.

One of the strategies to deal with errors in progressive alignments is to perform an iterative refinement. MAFFT (28) and MUSCLE (29) are two representative MSA methods that use such an iterative refinement strategy. Both are very fast and flexible: depending on the number of sequences the balance between the accuracy and speed can be easily adjusted.

Another strategy to improve initial progressive alignments is to use consistency information. The consistency concept is very simple. Let us suppose that we have three sequences (A, B, and C) and the corresponding pairwise alignments. If residue A_i is aligned to residue B_j and residue B_j is aligned to residue C_k, this implies that in A-C alignment A_i should be aligned with C_k. In other words, pairwise alignments induced by multiple alignments should be consistent. This transitivity condition is taken into account in scoring the alignment of two sequences (or group of sequences) by considering the information of their alignment to other sequences not involved in pairwise merge. T-coffee (30) and ProbCons (31) are examples of methods that make use of consistency-based scoring. In general, consistency-based methods are more accurate than those based on iterative refinement, but are more computationally demanding. However, in some cases, such as in recent versions of MAFFT (32), a simpler version of consistency measure has helped to keep the program fast. While being much faster, MAFFT now rivals the accuracy of both T-coffee and ProbCons (33).

Other strategies to improve the alignment accuracy include combination of several methods, as in M-coffee (34), or the incorporation of additional information. The additional information may be evolutionary (e.g., additional homologous sequences) or structural, since a 3D structure evolves more slowly than a sequence. For example, the MAFFT package has an option to add close homologs (35) detected using a BLAST search to improve the alignment accuracy of the initially submitted set of multiple sequences. One of the recently developed programs, PROMALS (36), uses a number of sources for additional information. First, it detects

Table 2
Multiple sequence alignment methods

Method	Type of information used	Address
ClustalW	Sequence	http://www.clustal.org/
MAFFT MAFFT-homologs	Sequence Sequence + homologs	http://mafft.cbrc.jp/alignment/software/
MUSCLE	Sequence	http://www.drive5.com/muscle/, http://www.ebi.ac.uk/Tools/muscle/index.html
ProbCons	Sequence	http://probcons.stanford.edu/
PROMALS	Sequence + homologs + SS[a]	http://prodata.swemd.edu/promals/
PROMALS3D	Sequence + homologs + SS[a] + 3D[b]	http://prodata.swemd.edu/promals3d/
T-coffee M-coffee 3DCoffee/Expresso	Sequence Consensus Sequence + 3D[b]	http://www.tcoffee.org/

[a]Secondary structure
[b]Three-dimensional structure

sequence homologs with PSI-BLAST and uses the obtained profiles to predict secondary structure. Next, profile–profile comparisons enhanced with predicted secondary structures are used in the alignment processes. If the 3D structural information is available, it can also be combined with sequence data within the consistency framework to improve accuracy of MSAs. The automatic incorporation of the available 3D structural information has been implemented in programs such as PROMALS3D (37), a successor of PROMALS, and 3DCoffee/Expresso (38, 39).

The MSA methods discussed here are summarized in Table 2. It should be emphasized that, depending on the situation, different MSA methods may be optimal. In general, when sequences to be aligned are fairly similar (over 35% sequence identity; the "daylight" zone), any method is likely to produce an accurate alignment. The alignment accuracy starts deteriorating when sequence similarity falls into the "twilight" zone (<25%) and/or the number of sequences is small. In such cases, despite being slower, methods that use additional sequence and/or structure information may be more suitable.

4. Hybrid Methods, Fully Integrated Automatic Servers and Meta-servers

A growing number of contemporary modeling methods derive sequence–structure mapping (alignment) by combining multiple sequence and structure features. Moreover, often a number of alignments with multiple templates or their fragments are considered simultaneously in deriving protein models based on homology. Even the concept of sequence–structure alignment sometimes becomes blurred because the derived final model cannot be easily attributed to one or more explicit sequence–structure alignments. Another popular trend is the use of meta-approaches. By combining results of different algorithms, these approaches attempt to identify the closest structural templates and the most accurate sequence–structure alignments. It would be impossible to provide an in-depth description for each of the multitude of methods presently available. Therefore, here only several popular methods that performed well in recent international blind trials of protein structure prediction known as CASP (40), and at the time of writing were accessible as public Web servers on the Internet (Table 3), are briefly discussed.

I-TASSER (41), one of the top hybrid protein structure modeling methods, uses combined results from multiple profile–profile comparison algorithms to detect suitable structural templates and to generate sequence–structure alignments. During next steps, the continuous fragments of initial alignments are reassembled into full-length models using iterative rounds of structure construction, model assessment, and refinement. In a sense, I-TASSER represents a meta-server for distant homology detection combined with techniques for structure simulation and evaluation. A similar approach is used in pro-Sp3-TASSER (42) with the difference being mostly in the methods used for the construction of initial sequence–structure alignments and model evaluation. The SAM-T08 server (13) uses the HMM-based sequence comparison

Table 3
Hybrid methods, fully integrated protein modeling servers and meta-servers

Method	Type	Address
I-TASSER	Server	http://zhanglab.ccmb.med.umich.edu/I-TASSER/
Pro-sp3-TASSER	Server	http://cssb.biology.gatech.edu/skolnick/webservice/pro-sp3-TASSER/
Robbeta	Server	http://robetta.bakerlab.org/
Phyre	Server	http://www.sbg.bio.ic.ac.uk/~phyre/
MULTICOM	Server	http://casp.rnet.missouri.edu/multicom_3d.html
SAM-T08	Server	http://compbio.soe.ucsc.edu/SAM_T08/T08-query.html
pGenTHREADER	Server	http://bioinf.cs.ucl.ac.uk/psipred/
GeneSilico	Meta-server	http://genesilico.pl/meta2/
Pcons.net	Meta-server	http://pcons.net/

enriched with predicted local structural features to detect templates and to generate several alignments with each of them. Models are then assembled using the templates, the local structure predictions, the distance constraints, and the contact predictions. Robetta (43) in the homology modeling regime uses profile-based methods to detect templates. Next, an ensemble of sequence–structure alignments is generated, followed by structure simulation and refinement. Perhaps the most important difference between Robetta and other methods discussed here is that in structure simulation it uses extensive conformational sampling coupled with physics-based all atom refinement. However, this means that much larger computational resources are needed. Phyre (44) is based on an ensemble of algorithmic variants for remote homology detection (essentially an in-house meta-server) combined with model construction and selection. MULTICOM (45) implements a combination of data at multiple modeling levels including templates, alignments, and models. pGenTHREADER (46), the latest implementation of GenTHREADER (47), the classical threading method, uses a linear combination of profile–profile alignments with secondary-structure-specific gap-penalties and classic pair- and solvation potentials.

There are also a number of meta-servers that apply a consensus approach either to select a best model or to construct a consensus model using the results obtained from different methods. GeneSilico (48) and Pcons.net (49) are among those meta-servers that are being continuously developed and updated.

Although now there are a large number of fully automated methods for homology modeling, one should keep in mind that the use of a more sophisticated procedure does not necessarily guarantee a better quality of the final model. It has been observed over and over again that no matter which template-based techniques are used to arrive at the final model, the largest contribution to its quality comes from the optimal template selection and the improvement of sequence–structure alignment (50). Therefore, a method that generates accurate alignments may sometimes outperform those with multiple layers of complexity. A vivid example of that was provided in CASP8 (51) by HHpred (52), a server implementation of the HHsearch method (16). HHpred was ranked among top servers despite the fact that it was neither exploring alternative alignments, nor reassembling structures from fragments, nor using additional structural features and optimization procedures. At the same time, HHpred was orders of magnitude faster than any other of the top servers. When just single domain targets were considered, it was second to only I-TASSER (52). This example clearly shows that the optimal selection of template(s) and especially the accuracy of the sequence–structure alignment are of paramount importance.

5. Accuracy of the Sequence–Structure Mapping

The construction of the initial sequence–structure alignment either through database searching or by using MSA methods on a predefined set of sequences is usually straightforward. However, unless the alignment between the modeling target and the structural template(s) is trivial (sequence identity over 40–50% and no or only few gaps), its reliability should be carefully evaluated.

5.1. Non-trivial Relationship Between Sequence Similarity, Statistical Significance, and Alignment Accuracy

In general, with the increase of evolutionary distance, both structures and sequences of homologous proteins become less similar, making homology detection more challenging. Intuition suggests that a lower sequence similarity might also be expected to result in the decreased accuracy of sequence–structure mapping. However, it turns out that the relationship between sequence similarity, statistical significance of the alignment, and its accuracy is not simple. In distant homology cases, sequence similarity between the target and template by itself is a poor predictor of alignment accuracy, because most commonly, the target-template pairwise alignment is derived in the context of multiple aligned sequences (sequence profiles, HMMs, or explicitly derived MSAs). Therefore, the number and the similarity distribution of additional homologous sequences seem to play a major role in determining both the sensitivity of homology detection and the overall alignment accuracy. As in crossing a river by hopping from one stone to the next, intermediate homologs may serve as "bridging stones" helping to link the target and the template (53). It is apparent that the more intermediate sequences are available and the smoother is their similarity transition, the more accurate alignment may be expected. A higher statistical significance of an alignment usually means a higher alignment accuracy. However, in distant homology cases, it would be a big mistake to think that highly statistically significant alignments are always highly accurate. This is illustrated in Fig. 3 with a distantly homologous pair of DNA sliding clamps. While BLAST is not able to detect this relationship at all, PSI-BLAST, HMMER, COMA, and HHpred, representing both profile- and HMM-based methods, detect it with a very high confidence. However, all of the corresponding alignments show significant discrepancies with the "gold standard" alignment derived from structure comparison with DaliLite (54). In other words, there is no strict dependency between alignment accuracy and homology detection ability. At the same time, this example seems to support observations (e.g., refs. 17, 55) that profile–profile alignments are in general more accurate than profile–sequence alignments. Alignment accuracy may also depend on inherent properties of a protein family. In particular, it has been observed that families with a high diversity of confident homologs tend to produce lower quality profile–profile alignments

a

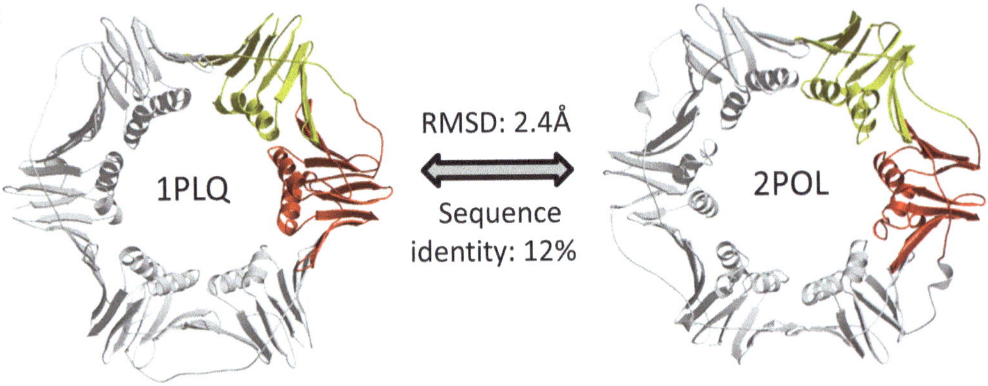

b

```
             EEEEE  HHHHHHHHHHH     EEEEEEE  EEEEEEE     EEEEEEEEEHHH  EE     EEEEE
1plq        MLEAKFEEASLFKRIIDGFKDCVQLVNFQCKEDGIIAQAVDDSRVLLVSLEIGVEAFQEYRCDHPVTLGM
2pol_DALI   EVEFTL-PQATMKRLIEATQFSMNGMLFETEGEELRTVATD--GHRLAVCSMPIG----QSLP-SHSVIV
2pol_PSI    ----RLIEATQFSMAHQDVRYYLNGMLFETEGEELRTVATDGHRLAVCSMPIGQSLPSH------SVIV
2pol_HMMER  ----------------------NGMLFETEGEELRTVATDGHRLAVCSMPIGQSLPS-------HSVIV
2pol_COMA   -F--TL-PQATMKRLIQDVRYYLNGMLFETEGEELRTVATDGHRLAVCSMPIGQS-------LPSHSVIV
2pol_HHPRED ------------------------MLFETEGEELRTVATDGHRLAVCSMPIGQS-------LPSHSVIV

            EHHHHHHH        EEEEEE     EEEEEEE      EEEEEEE                 EEEEEE
1plq        DLTSLSKILRCGNNTDTLTLIADNTPDSIILLFEDTKKDRIAEYSLKLMDIDADFLKIEELQYDSTLSLP
2pol_DALI   PRKGVIELMRMLDGGNPLRVQIG--SNNIRAHVG------DFIFTSKLVDGRFPDYRVLPKNPDKHLEAG
2pol_PSI    PRKGVIELMRMLDGDNPLRVQIGSNNIRAHVGDF--------IFTSKLVDGRFPDYRVLPKNPDKHLEAG
2pol_HMMER  PRKGVIELMRMLDGDNPLRVQIGS--NNIRAHVGD------FIFTSKLVDGRFPDYRVLPKNPDKHLEAG
2pol_COMA   PRKGVIELMRMLDGGDPLRVQI--GSNNIRAHVGD------FIFTSKLVDGRFPDYRVLPKNPDKHLEAG
2pol_HHPRED PRKGVIELMRMLDGGDPLRVQIGSNN---IRAHVGDF----IFTSKLVDGRFPDYRLPK-NPDKHLEAG

            HHHHHHHHHHHHH   EEEEEEE  EEEEEEE    EEEEEE        HHH  EEEE   EEEEEEEHH
1plq        SSEFSKIVRDLSQLSDSINIMITKETIKFVADGDIGSGSVIIKPFVDMEHPETSIKLEMDQPVDLTFGAK
2pol_DALI   CDLLKQAFARAAILSRGVRLYVSENQLKITANNPQEEAEEILD-----------VTYSG-AEMEIGFNVS
2pol_PSI    CDLLKQAFARAAILSNEVRLYVSENQLKITAN----------NPEQEEAEEILDVTYSGAEMEIGFNVS
2pol_HMMER  CDLLKQAFARAAILSNEVRLYVSENQLKITANNPEQEAEEILD------------VTYSGAEMEIGFNVS
2pol_COMA   CDLLKQAFARAAAASNEVRLYVSENQLKITANNEQEEA------------EEILDVTYSGAEMEIGFNVS
2pol_HHPRED CDLLKQAFARAAILSNGVRLYVSENQLKITANPEQEEAEEILDVTYS------------GAEMEIGFNVS

            HHHHHHHHHH   EEEEEE     EEEEEEE  EEEEEEE
1plq        YLLDIIKGSSLSDRVGIRLSSEAPALFQFDLKSGFLQFFLAPKFNDEE
2pol_DALI   YVLDVLNALK-CENVRMMLTDVSSVQIED-AASQSAAYVVMPMRL---
2pol_PSI    YVLDVLNALK-CENVRMMLTDSVSSVQIEDAASQSAAYVVMPM-----
2pol_HMMER  YVLDVLNALK-CENVRMMLTDVSSVQIE--DAAQSAAYVVMPM-----
2pol_COMA   YVLDVLNALK-CENVRMMLDSVSSVQIED-AASQSAAYVVMP------
2pol_HHPRED YVLDVLNALKC-ENVRMMLTDSSSVQIE-DAASQSAAYVVMP------
```

Fig. 3. Structure and sequence comparison of distantly homologous DNA sliding clamps from yeast (PDB code: 1plq) and *E. coli* (2pol). (**a**) Their 3D structures are similar despite sharing only 12% identical residues. (**b**) Comparison of DaliLite (DALI) structure-based alignment between 1plq and 2pol with the alignments produced by PSI-BLAST (PSI; *E* value = 3e–30), HHMER (*E* value = 2e–32), COMA (*E* value = 3e–13), and HHpred (probability = 99%). Alignments were obtained by searching PDB with 1plq sequence profiles (HMMs) that were obtained by running up to five iterations of PSI-BLAST (jackhmmer in the case of HMMER) with the 1plq sequence as a query against the filtered "nr" database. For easier comparison, columns corresponding to gaps in 1plq sequence were removed from all the alignments. Alignment positions showing discrepancies between DaliLite and each of the methods are shaded. Only positions corresponding to secondary structure elements ("H," helix, "E," strand) in 1plq were considered. The best agreement with the DaliLite alignment is shown by COMA, followed by HMMER, HHsearch, and PSI-BLAST.

with their remote relatives (56). However, this lower alignment accuracy cannot be improved when the most distant members of these families are excluded from their profiles. On the contrary, the presence of more diverse members has been found to result in more accurate alignments. This implies that the growth of the sequence databases should automatically result in more accurate alignments for the same level of sequence identities. However, this conclusion appears to hold only for confident high-quality homologous sequences. The inclusion of spurious contaminating sequences or even low-quality metagenomic sequences may negatively impact the target-template alignment accuracy (57).

5.2. Estimation of the Region-Specific Alignment Reliability

Sequence–structure alignment by itself does not tell which regions are aligned reliably (provide the correct residue mapping) and which ones may require adjustment. Therefore, to improve an alignment, the first task is to identify those alignment regions that can be trusted. Once the reliable regions are identified, the remaining alignment stretches can be either subjected to refinement or (if a significant conformational change is anticipated) rebuilding using different templates or template fragments.

The earliest methods for identification of reliable alignment regions (58–60) were focusing on pairwise sequence alignments that are largely irrelevant for the present day comparative modeling approaches. For target-template alignments constructed in the context of sequence profile- (or HMM)-based methods, several approaches were shown to be useful. Perhaps the simplest approach is based on the scores of individual positions within the profile–profile alignment. It was shown that the regions containing high scoring positions correlate well with the correctness of their alignment (61). More commonly, the positional reliability of sequence–structure alignments is estimated by assessing the region-specific alignment stability. There are two general strategies to generate sufficient alignment variability from which stable alignment regions can then be identified. The first strategy relies on a single method to generate alignment variability. This has been done either by using suboptimal alignments derived from the same sequence data (62, 63) or by diversifying alignments through the sampling of the available sequence space of homologs as in PSI-BLAST-ISS (64). The second strategy is based on the use of multiple methods to generate corresponding alignments followed by the analysis of alignment regions that do or do not agree between these different methods (65). Independently of which strategy is used, a strong consensus is considered to indicate reliably aligned regions. The lack of consensus may be caused by different reasons such as weak sequence conservation, insertions/deletions, or a significant conformational change. Figures 4 and 5 illustrate two typical situations resulting in unreliable alignment regions delineated with PSI-BLAST-ISS (64). In Fig. 4, the region of unreliable alignment coincides with a significant difference in orientation of corresponding α-helices.

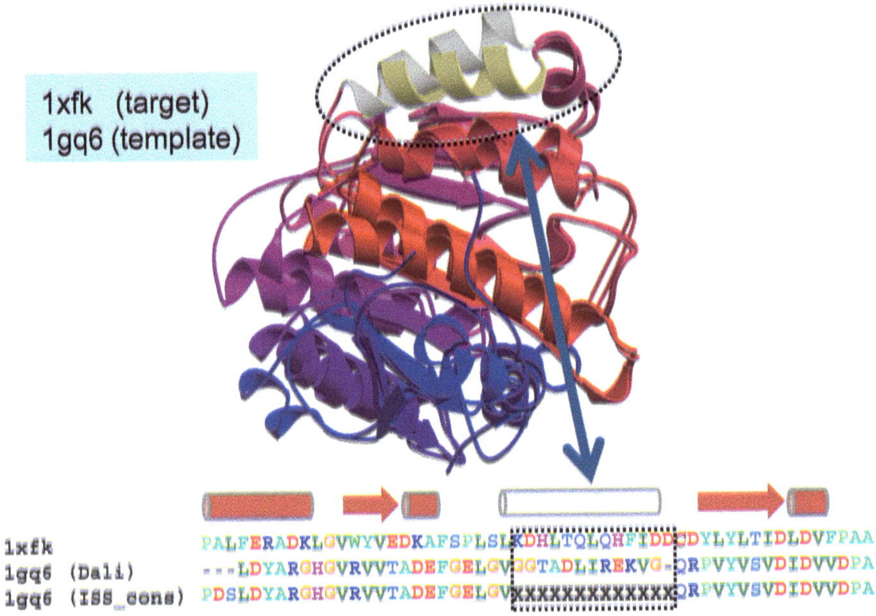

Fig. 4. Example of an unreliable alignment region corresponding to a structurally divergent motif. This motif is represented by an α-helix shown in *light colors* (enclosed in an *ellipse*) in superimposed structures of the modeling target (PDB code: 1xfk) and the template (1gq6). Below, the 1xfk is aligned with 1gq6 according to both structural correspondence (Dali) and a consensus alignment produced by PSI-BLAST-ISS (ISS_cons). "X" denotes positions lacking the consensus. The secondary structure of the 1xft is shown above the alignment. Figure adopted from ref. 64.

The unreliable region in Fig. 5 corresponds to a structurally conserved α-helix, which, however, has an insertion at one end and a deletion at the other end. Aligning this region correctly for sequence-based methods is difficult because of their tendency to cancel out the insertion and the deletion adjacent to the α-helix by shifting (incorrectly) its sequence. Yet, among individual alignment variants suggested by PSI-BLAST-ISS, there is one that corresponds to the structurally accurate alignment.

5.3. Improvement of Sequence–Structure Alignments

Although it is useful to know which regions in the model may be misaligned, the desirable goal is to achieve the highest possible sequence–structure alignment accuracy. Since sequence features alone are of little help in resolving alignment ambiguities, the often used recipe is to apply the assessment of alternative alignments in the context of a corresponding 3D model. To do this, one needs some sort of diagnostic tool for evaluating model quality in a region-specific way. Until recently, there were only few such tools available for performing the task. For quite some time, classical methods, ProSA (66) and Verify3D (67), have been popular choices for both the overall (global) and the position-specific (local) protein structure quality assessment. An important stimulus for development of new methods has appeared a few years back with the introduction

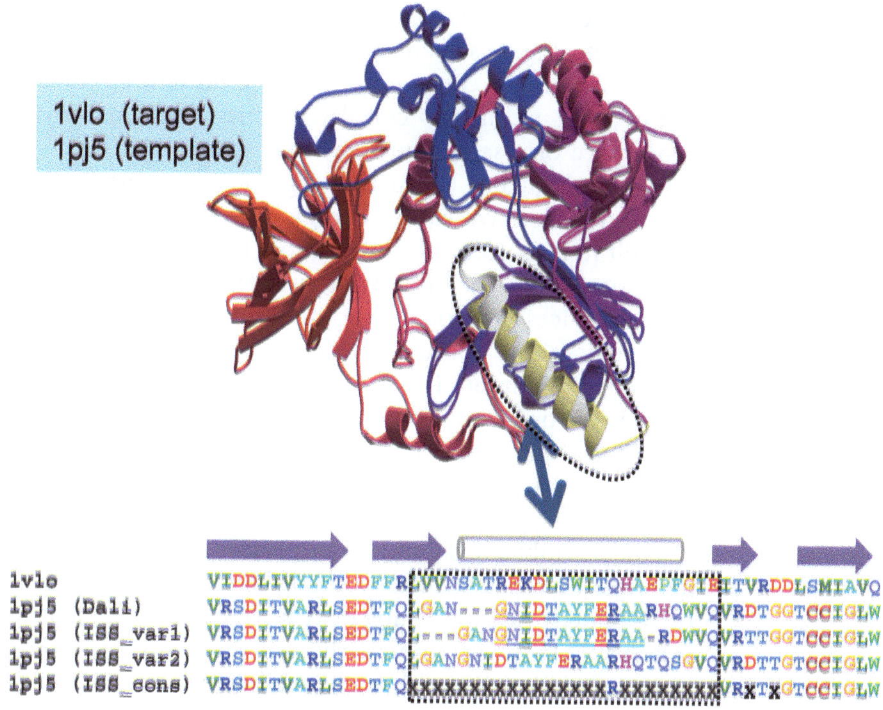

Fig. 5. Example of an unreliable alignment region corresponding to a structurally conserved motif surrounded with variable adjacent regions. The motif includes a structurally conserved α-helix (shown in *light color* and marked by an *ellipse*) in superimposed structures of the modeling target (PDB code: 1vlo) and the template (1pj5). However, one of the adjacent loops has an insertion and the other one has a deletion. The alignment shows structural correspondence (Dali), the PSI-BLAST-ISS consensus alignment (cons), and two individual variants (var1 and var2). "X" denotes positions lacking the consensus. One of the variants (var1) reproduces most of the structure-based mapping for the conserved α-helix (sequence underlined). Figure adopted from ref. 64.

of the model quality assessment category in CASP experiments (68). Quite a few approaches for estimating both the global and the local quality of a protein model have been developed since. Clustering- or consensus-based methods currently are the most accurate and the best such methods show a respectable accuracy in predicting global model quality (69). However, to work well, they require a large ensemble of models generated by different methods. Unfortunately, while this setting is natural for CASP, it has little to do with real modeling projects. In addition, even clustering-based methods perform significantly worse in the local model quality assessment mode, which is critical for the alignment improvement task. Nevertheless, promising new methods such as QMEAN (70, 71) that are capable of assessing position-specific quality of individual models have also emerged.

CASP results revealed that the systematic identification of correct alignment variants in unreliable regions is still difficult. Analysis of common alignment failures showed that the error-prone regions often share similar traits (72, 73). These regions often correspond to peripheral secondary structure elements (β-strands at the edge of β-sheets, highly solvent-exposed α-helices) that are under lesser structural/energy constraints than the structural core. Another feature that frequently correlates with alignment errors is the appearance or disappearance of small structural "defects" such as β-bulges. Arguably, alternative alignment variants in such error-prone regions have subtle energy differences and therefore are difficult to rank correctly. In addition, template structure is just an approximation of the native structure of modeling target. Inevitably, this introduces additional error during the evaluation of alternative alignments, and because of that even an effective assessment technique might fail. It is intuitively apparent that the more accurately is the protein main chain modeled, the easier it should be to distinguish the correct residue mapping from the erroneous one. In other words, perhaps the most effective, although computationally expensive, way to identify the native alignment would be to test an ensemble of alignments by performing simultaneous refinement for each of the corresponding models. In fact, the sampling of alignment variants coupled with all-atom refinement has been tested at CASP, with impressive results for some modeling targets (74). Less successful results were attributed to insufficient sampling and imperfect energy estimation (74).

Thus, the accurate mapping of sequence onto structure remains one of the important bottlenecks in homology modeling. Although there are signs of improvement, a lot more will have to be done in developing more effective approaches for sampling alignments and conformations, together with better methods for the local model quality estimation.

6. Practical Guide for Sequence–Structure Alignment

The following is a brief description of practical steps for aligning a sequence to known structure(s), estimating the reliability of alignment regions and selecting the best alignment. To a large degree, this rough guide is based on an updated protocol (73) used to achieve the top-ranked results in the homology (template-based) modeling category during the CASP8 experiment (75). The flowchart depicting main steps in sequence–structure alignment is presented in Fig. 6.

6.1. Searching for Structural Templates and Constructing Initial Alignments

First, it is useful to find out what is the level of difficulty for generating accurate sequence–structure alignment. The initial estimate can be made, once it is known if there are closely related experimental 3D structures available. If so, how similar their sequences are to the protein of interest? How many structures are available? How many additional homologs can be detected in sequence databases and how closely they are related to the target?

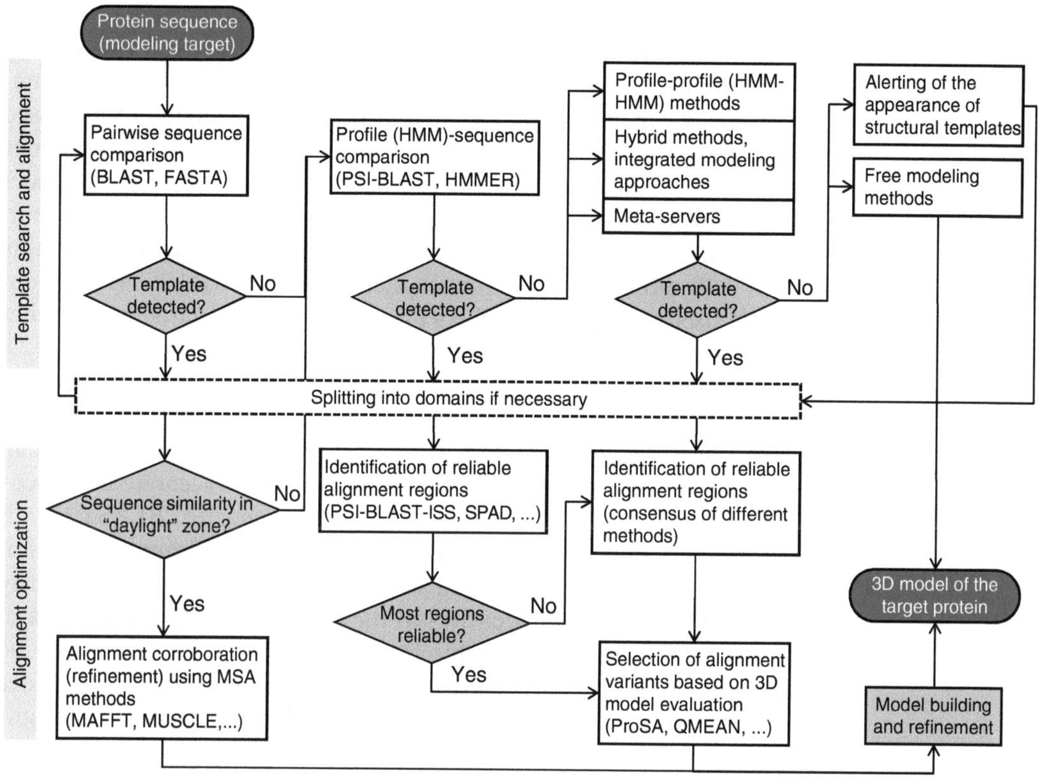

Fig. 6. Flowchart of major steps in sequence to structure alignment.

The best idea is to start with a simple sequence search using BLAST (3). It is useful to have the BLAST suite of programs including both BLAST and PSI-BLAST as well as protein sequence databases installed locally. This provides an increased flexibility in using these programs. The BLAST program suite and sequence databases can be obtained from the NCBI FTP site at ftp://ftp.ncbi.nlm.nih.gov/blast/. Sequence databases at NCBI are updated daily and can be retrieved automatically using the *update_blastdb.pl* script, which is provided freely as part of the BLAST documentation at NCBI. For the local installation, it is important to have at least two protein sequence databases: nonredundant sequence database (nr) containing all nonredundant protein sequences (except those from metagenomic projects) and the PDB sequence database (pdbaa), which contains protein sequences of known 3D structures. The latter sequences are also available for downloading directly from PDB (http://www.pdb.org). Since the nonredundant (nr) sequence database is huge and continues to grow fast, it is advisable to have several smaller versions of this database with very similar sequences removed. It is a common practice to remove sequences up to 90, 80, and 70% identical to each other. This helps to reduce the database size significantly without negatively affecting homology search results. The filtering of sequence databases can be done with clustering tools such as CD-HIT (76). If the filtering of the locally installed "nr" database turns out to be too computationally expensive, the user may choose to download preprocessed UniRef sequence databases with the reduced levels of redundancy from UniProt (http://www.uniprot.org/). These sequence databases are also aiming at a complete coverage of sequence space. At present, UniRef100, UniRef90, and UniRef50 filtered correspondingly at 100, 90, and 50% sequence identity, are available. Alternatively, the user can run both BLAST and PSI-BLAST sequence searches using web servers either at NCBI (http://blast.ncbi.nlm.nih.gov/), EBI (http://www.ebi.ac.uk/Tools/sss/), or at many other locations on the Internet.

The results of BLAST search against PDB sequences give an approximate estimate of the difficulty to derive an accurate sequence–structure alignment. During the simplest scenario, BLAST search detects a PDB sequence with a statistically significant expectation value (E value < 0.001) and a relatively high sequence similarity (over 40% sequence identity) to the modeling target. In such case, the homologous relationship is obvious and the alignment may be structurally optimal. However, even if such pairwise alignment does not have any gaps, it is still recommended to substantiate the alignment with methods that rely on information derived from multiple sequences. This can be done by collecting additional close sequence homologs with BLAST, pooling them together with target and template sequences and aligning with one of the fast MSA methods such as MAFFT (28) or MUSCLE (29). If sequence identity is lower than 40% and there are gaps, the alignment almost certainly will need some adjustments such as the placement of the gaps or their boundaries. In such case, an MSA might also help to refine the target-template alignment. However, if the sequence similarity enters the "twilight" zone, MSA methods that use additional information (predicted secondary structure, 3D structural information) such as PROMALS/PROMALS3D (36, 37) and 3DCoffee/Expresso (38, 39) might be more appropriate. The use of PSI-BLAST and other profile (HMM)-based methods is also recommended in more distant homology cases (see below).

If no PDB sequences with statistically significant E values are detected with BLAST, more sensitive methods such as PSI-BLAST should be used next. The power of PSI-BLAST is in rich sequence profiles generated from aligned multiple homologous sequences. The PDB sequence database is too small to perform the iterative PSI-BLAST searches against it directly. Usually, potential structural templates are detected and aligned with the target sequence using the so-called PDB-BLAST procedure. It involves performing several iterations of PSI-BLAST search against a large sequence database (e.g., "nr" or its derivatives) and then using the constructed profile to run the last iteration against the PDB sequence database.

It is worthwhile to make several PDB-BLAST runs, every time generating a more inclusive profile by increasing the number of iterations against the "nr" database or its derivatives. The change in the number of detected PDB sequences and the corresponding *E* values will give an approximate estimate of evolutionary distance between the target sequence and the confidently (*E* value < 0.001) detected structures. If PSI-BLAST and sequence databases are not installed locally, it is still possible to perform PDB-BLAST-like searches using the NCBI BLAST server through several manual steps. Automatic PDB-BLAST searches can be performed both locally and remotely (at NCBI) using Re-searcher (77). Note that PSI-BLAST is not the only available option. Recently, an iterative procedure similar to that in PSI-BLAST was implemented in HMMER (http://hmmer.org/). With the reported high speed and sensitivity, the iterative HMMER3 procedure (jackhmmer) is at least as good as PSI-BLAST.

If sequence searches with profiles (PSI-BLAST) or HMMs (e.g., HMMER) do not reveal any obvious structural homologs, it does not necessarily mean that they are absent from the PDB. It may be that the evolutionary relationship is too distant to be detected by profile (HMM)–sequence comparisons. In such case the obvious next step is to turn to the even more sensitive profile–profile, HMM–HMM, or hybrid sequence–structure methods. There are now a large number of such methods available and only a small fraction is listed in Tables 2 and 3. One of the best choices to start with is HHsearch (16), a very fast and one of the most sensitive homology detection methods. Based on HMM–HMM comparison, HHsearch is available both as a standalone toolkit and as part of the HHpred web server (78). Other sensitive alternatives to HHsearch include PRC (19, 79), COMA (17, 80), COMPASS (15, 81), and PROCAIN (22, 82). Both HHpred and COMA servers also have a useful option to produce 3D models based on the reported sequence–structure alignments. Among the fully integrated modeling approaches I-TASSER (41) at present is clearly the best choice. As many other integrated hybrid modeling methods it will return the final 3D model, which may not necessarily correspond to any of the initial sequence–structure alignments used. Meta-servers such as Genesilico (48) or Pcons.net (49) may also be useful, since they provide results from several methods simultaneously. In general, many new methods are continuously reported, making it difficult to select the best methods at a given time. It may be instructive to check the server results during latest CASP experiments (http://www.predictioncenter.org/). However, not always well-performing methods at CASP are available as public servers and not all well-performing methods take part in CASP. Independently of which servers you use, check when the databases were last updated; even the best methods will likely perform poorly on old sequence and structure databases.

Initial template search results usually reveal the domain composition of the modeling target. If it is a multidomain protein, it may be beneficial or even necessary to partition the sequence into chunks corresponding to individual domains. First, individual protein domains may have a closer relationship with different structural templates. In such case, treating domains individually may improve the selection of templates and/or the accuracy of sequence–structure alignments. Second, the partition of the sequence into domains may help to avoid homologous over-extension (HOE), an important source of errors in iterative profile-based searches (83). This error occurs when the alignment initially covering only homologous domains over the course of iterations is extended into nonhomologous regions.

6.2. Estimation of Position-Dependent Alignment Reliability

Typically, sequence–structure alignments produced within the "twilight" or "midnight" zones of sequence similarity will have inaccuracies. However, a visual inspection at this level of sequence similarity is virtually useless in spotting them. How then to distinguish alignment regions that are reliable from those that may be incorrect and will likely require refinement? One of the options is to use alignment stability as an indicator of reliability. One of the available tools that use this idea is PSI-BLAST-ISS (64). It is based on multiple PSI-BLAST searches with different yet related queries. PSI-BLAST-ISS results simultaneously provide several types of information: (1) automatically detected structural templates and corresponding alignments, (2) data suggesting which one of the templates may be the closest to the target, and (3) the region-specific alignment reliability indication for each of the templates. The drawback of PSI-BLAST-ISS is that it takes time to run all the PSI-BLAST searches (typically 50–100) and that parameter settings may need adjustment depending on the target. PSI-BLAST-ISS is also useless in cases of very distant homology, when PSI-BLAST is not sensitive enough to detect templates. In such cases, perhaps the simplest way to estimate regional alignment reliability is to use the agreement between the sequence–structure alignments produced by different methods. However, different methods may provide alignments or build models using different templates. To cope with this potential heterogeneity of results, it is useful to convert all the outputs into a common format such as 3D structure. Nowadays, many methods generate 3D models as the final output or at least provide an option to construct models using the resulting alignments. However, if models are unavailable, they can be easily constructed from sequence–structure alignments using one of the modeling tools such as MODELLER (84), Nest (85), and Swiss-PdbViewer (86). There are also web servers for converting sequence–structure alignments to structural models. For example, "alignment mode" of SwissModel (86), one of the popular modeling servers, can be used for this purpose. Comparison of the resulting models with one of the representative templates provides the underlying sequence–structure mappings. After that, all the pairwise alignments can be merged into a single PSI-BLAST-ISS-like alignment, in which a template is aligned to the target sequence variants corresponding to different models. Both pairwise structure comparisons and merging of the corresponding alignments can be easily performed in one step using the *dali_sp.pl* wrapper (http://www.ibt.lt/bioinformatics/software/) for DaliLite (54). Just like in the case of PSI-BLAST-ISS, the agreement between different methods tends to indicate reliable regions of the alignment, while the lack of consistency points to the need of further analysis.

6.3. Improving Alignments

If the sequence of the modeling target is aligned reliably with all the structurally conserved regions of the template(s) the sequence–structure mapping is done. In such case, the final quality of the homology model will be determined by other steps such as the ability to accurately model variable regions and to drive the model structure closer to the native one. The tricky part begins with the regions that are not reliably aligned, because first it is important to understand whether the uncertainty is caused by the conformational change or simply by the lack of sequence conservation. Only if there are hints from available template(s) that the region is structurally conserved, there is a good chance to identify structurally/evolutionary meaningful alignment for this region without modifying the template backbone. In that case, the assessment of sequence–structure mapping within the context of 3D structure (i.e., assessing a structural model based on a particular sequence–structure alignment) perhaps is the most promising. Structure quality evaluation methods such as ProSA (66, 87) or QMEAN (70, 71) can help identify the correct alignment by estimating both the overall and region-specific model quality. Often, the problem with the evaluation of models based on alternative alignment variants is the noisiness of the results. More often than not, the evaluation results do not show a clear preference towards a particular alignment variant. One way to deal with the noisy signal is to include additional homologs of the target sequence into the analysis. The homologs should be selected such that their alignment with the target sequence would be unambiguous. The consensus of evaluation results of models based on alternative sequence–structure alignments for multiple family members may help rank the alignment variants more effectively. However, the consistent improvement of the sequence–structure mapping based on model evaluation is still an unresolved problem.

6.4. What Can Be Done If No Template Is Detected Reliably?

If none of the most sensitive profile (HMM)-based methods can reliably detect any structural template it may mean that indeed there is no related template in the PDB. Alternatively, the relationship might be too distant, beyond the sensitivity limits of current methods. In both cases, there are at least two ways to approach the problem.

If obtaining the 3D model is not the most urgent task, the first option is to use alerting systems such as Re-searcher (77) or PDBalert (88) for performing automatic recurrent searches of homologous structures in PDB. Re-searcher uses PSI-BLAST as the search engine, and PDBalert is based on even more sensitive method, HHsearch. Usually the confident detection of a modeling template is the result of new homologous structure being deposited into PDB. However, in some cases, merely an increase of the number of sequence homologs may be sufficient to reliably detect templates that have already been present in PDB. This may happen because additional sequences help to build more representative sequence profiles (or HMMs). The serious drawback of this option is the unpredictability of the time frame when the suitable template will be detected. It may happen within days, but it may also happen years later, when the structure of a homolog is solved and deposited into PDB.

The second option is to use free modeling (FM) methods that do not have to rely on explicit templates and sequence–structure alignments to construct 3D models. Currently, there are a number of methods that would automatically shift to the free modeling mode if no suitable templates could be detected. Some of the most effective such methods include Robetta (43), an automatic server based on Rosetta, a highly successful fragment-based approach (89), I-TASSER (41, 90) and its relative Pro-sp3-TASSER (42, 91), SAM-T08 (13), MULTICOM (45). As it has been observed in CASP trials, these approaches can produce models of reasonable quality for small proteins (up to ~100 residues) having simple topology. However, at present, it would be too optimistic to expect consistently good models from FM approaches. Therefore, the confident detection of even remotely homologous structural template may help to improve modeling results considerably.

7. Conclusions

A steady growth of experimentally determined protein structures coupled with a dramatic increase of sequence data has made homology modeling both widely applicable and practically useful. In recent years, there have also been significant advances in distant homology detection and sequence alignment. The largest progress has been made mainly due to the application of sequence profiles and HMMs. At the same time, there are a number of remaining issues. In particular, there is a great need for improvement of the sequence–structure alignment accuracy, which is a key factor determining the quality of a homology model. This issue is tightly linked with the ability to accurately estimate local errors in protein models. As indicated by CASP blind trials this is a notoriously difficult problem. However, with the recent emphasis within the modeler community on the accurate model quality estimates there is hope for significant breakthroughs in this area. On the other hand, even currently available tools provide users with a lot of possibilities to construct, assess, and improve sequence–structure alignments for homology modeling.

Acknowledgments

Ana Vencloviené and members of Venclovas' lab are gratefully acknowledged for useful comments and suggestions.

References

1. Grishin, N. V. (2001) Fold change in evolution of protein structures, *J Struct Biol* ***134***, 167–185.
2. Altschul, S. F., Gish, W., Miller, W., Myers, E. W., and Lipman, D. J. (1990) Basic local alignment search tool, *J Mol Biol* ***215***, 403–410.
3. Altschul, S. F., Madden, T. L., Schaffer, A. A., Zhang, J., Zhang, Z., Miller, W., and Lipman, D. J. (1997) Gapped BLAST and PSI-BLAST: a new generation of protein database search programs, *Nucleic Acids Res* ***25***, 3389–3402.
4. Karlin, S., and Altschul, S. F. (1990) Methods for assessing the statistical significance of molecular sequence features by using general scoring schemes, *Proc Natl Acad Sci U S A* ***87***, 2264–2268.
5. Pearson, W. R., and Lipman, D. J. (1988) Improved tools for biological sequence comparison, *Proc Natl Acad Sci U S A* ***85***, 2444–2448.
6. Smith, T. F., and Waterman, M. S. (1981) Identification of common molecular subsequences, *J Mol Biol* ***147***, 195–197.
7. Pearson, W. R. (1991) Searching protein sequence libraries: comparison of the sensitivity and selectivity of the Smith-Waterman and FASTA algorithms, *Genomics* ***11***, 635–650.
8. Biegert, A., and Söding, J. (2009) Sequence context-specific profiles for homology searching, *Proc Natl Acad Sci U S A* ***106***, 3770–3775.
9. Gribskov, M., McLachlan, A. D., and Eisenberg, D. (1987) Profile analysis: detection of distantly related proteins, *Proc Natl Acad Sci U S A* ***84***, 4355–4358.
10. Durbin, R., Eddy, S., Krogh, A., and Mitchison, G. (1999) *Biological Sequence Analysis: Probabilistic Models of Proteins and Nucleic Acids*, Cambridge University Press.
11. Eddy, S. R. (1998) Profile hidden Markov models, *Bioinformatics* ***14***, 755–763.
12. Hughey, R., and Krogh, A. (1996) Hidden Markov models for sequence analysis: extension and analysis of the basic method, *Comput Appl Biosci* ***12***, 95–107.
13. Karplus, K. (2009) SAM-T08, HMM-based protein structure prediction, *Nucleic Acids Res* ***37***, W492–497.
14. Johnson, L. S., Eddy, S. R., and Portugaly, E. (2010) Hidden Markov model speed heuristic and iterative HMM search procedure, *BMC Bioinformatics* ***11***, 431.
15. Sadreyev, R., and Grishin, N. (2003) COMPASS: a tool for comparison of multiple protein alignments with assessment of statistical significance, *J Mol Biol* ***326***, 317–336.
16. Söding, J. (2005) Protein homology detection by HMM-HMM comparison, *Bioinformatics* ***21***, 951–960.
17. Margelevičius, M., and Venclovas, Č. (2010) Detection of distant evolutionary relationships between protein families using theory of sequence profile-profile comparison, *BMC Bioinformatics* ***11***, 89.
18. Yona, G., and Levitt, M. (2002) Within the twilight zone: a sensitive profile-profile comparison tool based on information theory, *J Mol Biol* ***315***, 1257–1275.
19. Madera, M. (2008) Profile Comparer: a program for scoring and aligning profile hidden Markov models, *Bioinformatics* ***24***, 2630–2631.
20. Rychlewski, L., Jaroszewski, L., Li, W., and Godzik, A. (2000) Comparison of sequence profiles. Strategies for structural predictions using sequence information, *Protein Sci* ***9***, 232–241.

21. Holm, L., and Sander, C. (1993) Protein structure comparison by alignment of distance matrices, *J Mol Biol 233*, 123–138.
22. Wang, Y., Sadreyev, R. I., and Grishin, N. V. (2009) PROCAIN: protein profile comparison with assisting information, *Nucleic Acids Res 37*, 3522–3530.
23. Eddy, S. R. (2008) A probabilistic model of local sequence alignment that simplifies statistical significance estimation, *PLoS Comput Biol 4*, e1000069.
24. Thompson, J. D., Higgins, D. G., and Gibson, T. J. (1994) CLUSTAL W: improving the sensitivity of progressive multiple sequence alignment through sequence weighting, position-specific gap penalties and weight matrix choice, *Nucleic Acids Res 22*, 4673–4680.
25. Do, C. B., and Katoh, K. (2008) Protein multiple sequence alignment, *Methods Mol Biol 484*, 379–413.
26. Pei, J. (2008) Multiple protein sequence alignment, *Curr Opin Struct Biol 18*, 382–386.
27. Kemena, C., and Notredame, C. (2009) Upcoming challenges for multiple sequence alignment methods in the high-throughput era, *Bioinformatics 25*, 2455–2465.
28. Katoh, K., Misawa, K., Kuma, K., and Miyata, T. (2002) MAFFT: a novel method for rapid multiple sequence alignment based on fast Fourier transform, *Nucleic Acids Res 30*, 3059–3066.
29. Edgar, R. C. (2004) MUSCLE: multiple sequence alignment with high accuracy and high throughput, *Nucleic Acids Res 32*, 1792–1797.
30. Notredame, C., Higgins, D. G., and Heringa, J. (2000) T-Coffee: A novel method for fast and accurate multiple sequence alignment, *J Mol Biol 302*, 205–217.
31. Do, C. B., Mahabhashyam, M. S., Brudno, M., and Batzoglou, S. (2005) ProbCons: Probabilistic consistency-based multiple sequence alignment, *Genome Res 15*, 330–340.
32. Katoh, K., Kuma, K., Toh, H., and Miyata, T. (2005) MAFFT version 5: improvement in accuracy of multiple sequence alignment, *Nucleic Acids Res 33*, 511–518.
33. Edgar, R. C., and Batzoglou, S. (2006) Multiple sequence alignment, *Curr Opin Struct Biol 16*, 368–373.
34. Wallace, I. M., O'Sullivan, O., Higgins, D. G., and Notredame, C. (2006) M-Coffee: combining multiple sequence alignment methods with T-Coffee, *Nucleic Acids Res 34*, 1692–1699.
35. Katoh, K., Kuma, K., Miyata, T., and Toh, H. (2005) Improvement in the accuracy of multiple sequence alignment program MAFFT, *Genome Inform 16*, 22–33.
36. Pei, J., and Grishin, N. V. (2007) PROMALS: towards accurate multiple sequence alignments of distantly related proteins, *Bioinformatics 23*, 802–808.
37. Pei, J., Kim, B. H., and Grishin, N. V. (2008) PROMALS3D: a tool for multiple protein sequence and structure alignments, *Nucleic Acids Res 36*, 2295–2300.
38. O'Sullivan, O., Suhre, K., Abergel, C., Higgins, D. G., and Notredame, C. (2004) 3DCoffee: combining protein sequences and structures within multiple sequence alignments, *J Mol Biol 340*, 385–395.
39. Armougom, F., Moretti, S., Poirot, O., Audic, S., Dumas, P., Schaeli, B., Keduas, V., and Notredame, C. (2006) Expresso: automatic incorporation of structural information in multiple sequence alignments using 3D-Coffee, *Nucleic Acids Res 34*, W604–608.
40. Moult, J. (2005) A decade of CASP: progress, bottlenecks and prognosis in protein structure prediction, *Curr Opin Struct Biol 15*, 285–289.
41. Roy, A., Kucukural, A., and Zhang, Y. (2010) I-TASSER: a unified platform for automated protein structure and function prediction, *Nat Protoc 5*, 725–738.
42. Zhou, H., and Skolnick, J. (2009) Protein structure prediction by pro-Sp3-TASSER, *Biophys J 96*, 2119–2127.
43. Kim, D. E., Chivian, D., and Baker, D. (2004) Protein structure prediction and analysis using the Robetta server, *Nucleic Acids Res 32*, W526–531.
44. Kelley, L. A., and Sternberg, M. J. (2009) Protein structure prediction on the Web: a case study using the Phyre server, *Nat Protoc 4*, 363–371.
45. Wang, Z., Eickholt, J., and Cheng, J. (2010) MULTICOM: a multi-level combination approach to protein structure prediction and its assessments in CASP8, *Bioinformatics 26*, 882–888.
46. Lobley, A., Sadowski, M. I., and Jones, D. T. (2009) pGenTHREADER and pDomTHREADER: new methods for improved protein fold recognition and superfamily discrimination, *Bioinformatics 25*, 1761–1767.
47. Jones, D. T. (1999) GenTHREADER: an efficient and reliable protein fold recognition method for genomic sequences, *J Mol Biol 287*, 797–815.
48. Kurowski, M. A., and Bujnicki, J. M. (2003) GeneSilico protein structure prediction meta-server, *Nucleic Acids Res 31*, 3305–3307.
49. Wallner, B., Larsson, P., and Elofsson, A. (2007) Pcons.net: protein structure prediction meta server, *Nucleic Acids Res 35*, W369–374.

50. Ginalski, K. (2006) Comparative modeling for protein structure prediction, *Curr Opin Struct Biol* *16*, 172–177.
51. Moult, J., Fidelis, K., Kryshtafovych, A., Rost, B., and Tramontano, A. (2009) Critical assessment of methods of protein structure prediction - Round VIII, *Proteins* 77 Suppl *9*, 1–4.
52. Hildebrand, A., Remmert, M., Biegert, A., and Söding, J. (2009) Fast and accurate automatic structure prediction with HHpred, *Proteins* 77 Suppl *9*, 128–132.
53. Cozzetto, D., and Tramontano, A. (2005) Relationship between multiple sequence alignments and quality of protein comparative models, *Proteins* *58*, 151–157.
54. Holm, L., Kaariainen, S., Rosenstrom, P., and Schenkel, A. (2008) Searching protein structure databases with DaliLite v.3, *Bioinformatics* *24*, 2780–2781.
55. Qi, Y., Sadreyev, R. I., Wang, Y., Kim, B. H., and Grishin, N. V. (2007) A comprehensive system for evaluation of remote sequence similarity detection, *BMC Bioinformatics* *8*, 314.
56. Sadreyev, R. I., and Grishin, N. V. (2004) Quality of alignment comparison by COMPASS improves with inclusion of diverse confident homologs, *Bioinformatics* *20*, 818–828.
57. Tress, M. L., Cozzetto, D., Tramontano, A., and Valencia, A. (2006) An analysis of the Sargasso Sea resource and the consequences for database composition, *BMC Bioinformatics* *7*, 213.
58. Chao, K. M., Hardison, R. C., and Miller, W. (1993) Locating well-conserved regions within a pairwise alignment, *Comput Appl Biosci* *9*, 387–396.
59. Vingron, M., and Argos, P. (1990) Determination of reliable regions in protein sequence alignments, *Protein Eng* *3*, 565–569.
60. Mevissen, H. T., and Vingron, M. (1996) Quantifying the local reliability of a sequence alignment, *Protein Eng* *9*, 127–132.
61. Tress, M. L., Jones, D., and Valencia, A. (2003) Predicting reliable regions in protein alignments from sequence profiles, *J Mol Biol* *330*, 705–718.
62. Cline, M., Hughey, R., and Karplus, K. (2002) Predicting reliable regions in protein sequence alignments, *Bioinformatics* *18*, 306–314.
63. Chen, H., and Kihara, D. (2008) Estimating quality of template-based protein models by alignment stability, *Proteins* *71*, 1255–1274.
64. Margelevičius, M., and Venclovas, Č. (2005) PSI-BLAST-ISS: an intermediate sequence search tool for estimation of the position-specific alignment reliability, *BMC Bioinformatics* *6*, 185.
65. Prasad, J. C., Comeau, S. R., Vajda, S., and Camacho, C. J. (2003) Consensus alignment for reliable framework prediction in homology modeling, *Bioinformatics* *19*, 1682–1691.
66. Sippl, M. J. (1993) Recognition of errors in three-dimensional structures of proteins, *Proteins* *17*, 355–362.
67. Eisenberg, D., Luthy, R., and Bowie, J. U. (1997) VERIFY3D: assessment of protein models with three-dimensional profiles, *Methods Enzymol* *277*, 396–404.
68. Cozzetto, D., Kryshtafovych, A., Ceriani, M., and Tramontano, A. (2007) Assessment of predictions in the model quality assessment category, *Proteins* 69 Suppl *8*, 175–183.
69. Cozzetto, D., Kryshtafovych, A., and Tramontano, A. (2009) Evaluation of CASP8 model quality predictions, *Proteins* 77 Suppl *9*, 157–166.
70. Benkert, P., Kunzli, M., and Schwede, T. (2009) QMEAN server for protein model quality estimation, *Nucleic Acids Res* *37*, W510–514.
71. Benkert, P., Tosatto, S. C., and Schomburg, D. (2008) QMEAN: A comprehensive scoring function for model quality assessment, *Proteins* *71*, 261–277.
72. Venclovas, Č. (2003) Comparative modeling in CASP5: progress is evident, but alignment errors remain a significant hindrance, *Proteins* 53 Suppl *6*, 380–388.
73. Venclovas, Č., and Margelevičius, M. (2009) The use of automatic tools and human expertise in template-based modeling of CASP8 target proteins, *Proteins* 77 Suppl *9*, 81–88.
74. Raman, S., Vernon, R., Thompson, J., Tyka, M., Sadreyev, R., Pei, J., Kim, D., Kellogg, E., DiMaio, F., Lange, O., Kinch, L., Sheffler, W., Kim, B. H., Das, R., Grishin, N. V., and Baker, D. (2009) Structure prediction for CASP8 with all-atom refinement using Rosetta, *Proteins* 77 Suppl *9*, 89–99.
75. Cozzetto, D., Kryshtafovych, A., Fidelis, K., Moult, J., Rost, B., and Tramontano, A. (2009) Evaluation of template-based models in CASP8 with standard measures, *Proteins* 77 Suppl *9*, 18–28.
76. Li, W., and Godzik, A. (2006) Cd-hit: a fast program for clustering and comparing large sets of protein or nucleotide sequences, *Bioinformatics* *22*, 1658–1659.
77. Repšys, V., Margelevičius, M., and Venclovas, Č. (2008) Re-searcher: a system for recurrent detection of homologous protein sequences, *BMC Bioinformatics* *9*, 296.
78. Söding, J., Biegert, A., and Lupas, A. N. (2005) The HHpred interactive server for protein homology detection and structure prediction, *Nucleic Acids Res* *33*, W244–248.
79. Brandt, B. W., and Heringa, J. (2009) web-PRC: the Profile Comparer for alignment-based

searching of public domain databases, *Nucleic Acids Res* *37*, W48–52.

80. Margelevičius, M., Laganeckas, M., and Venclovas, Č. (2010) COMA server for protein distant homology search, *Bioinformatics* *26*, 1905–1906.
81. Sadreyev, R. I., Tang, M., Kim, B. H., and Grishin, N. V. (2007) COMPASS server for remote homology inference, *Nucleic Acids Res* *35*, W653–658.
82. Wang, Y., Sadreyev, R. I., and Grishin, N. V. (2009) PROCAIN server for remote protein sequence similarity search, *Bioinformatics* *25*, 2076–2077.
83. Gonzalez, M. W., and Pearson, W. R. (2010) Homologous over-extension: a challenge for iterative similarity searches, *Nucleic Acids Res* *38*, 2177–2189.
84. Sali, A., and Blundell, T. L. (1993) Comparative protein modelling by satisfaction of spatial restraints, *J Mol Biol* *234*, 779–815.
85. Petrey, D., Xiang, Z., Tang, C. L., Xie, L., Gimpelev, M., Mitros, T., Soto, C. S., Goldsmith-Fischman, S., Kernytsky, A., Schlessinger, A., Koh, I. Y., Alexov, E., and Honig, B. (2003) Using multiple structure alignments, fast model building, and energetic analysis in fold recognition and homology modeling, *Proteins* 53 Suppl *6*, 430–435.
86. Guex, N., Peitsch, M. C., and Schwede, T. (2009) Automated comparative protein structure modeling with SWISS-MODEL and Swiss-PdbViewer: a historical perspective, *Electrophoresis* 30 Suppl *1*, S162–173.
87. Wiederstein, M., and Sippl, M. J. (2007) ProSA-web: interactive web service for the recognition of errors in three-dimensional structures of proteins, *Nucleic Acids Res* *35*, W407–410.
88. Agarwal, V., Remmert, M., Biegert, A., and Söding, J. (2008) PDBalert: automatic, recurrent remote homology tracking and protein structure prediction, *BMC Struct Biol* *8*, 51.
89. Bradley, P., Malmstrom, L., Qian, B., Schonbrun, J., Chivian, D., Kim, D. E., Meiler, J., Misura, K. M., and Baker, D. (2005) Free modeling with Rosetta in CASP6, *Proteins* *61* Suppl *7*, 128–134.
90. Zhang, Y. (2009) I-TASSER: fully automated protein structure prediction in CASP8, *Proteins* 77 Suppl *9*, 100–113.
91. Zhou, H., Pandit, S. B., and Skolnick, J. (2009) Performance of the Pro-sp3-TASSER server in CASP8, *Proteins* 77 Suppl *9*, 123–127.

Chapter 4

Force Fields for Homology Modeling

Andrew J. Bordner

Abstract

Accurate all-atom energy functions are crucial for successful high-resolution protein structure prediction. In this chapter, we review both physics-based force fields and knowledge-based potentials used in protein modeling. Because it is important to calculate the energy as accurately as possible given the limitations imposed by sampling convergence, different components of the energy, and force fields representing them to varying degrees of detail and complexity are discussed. Force fields using Cartesian as well as torsion angle representations of protein geometry are covered. Since solvent is important for protein energetics, different aqueous and membrane solvation models for protein simulations are also described. Finally, we summarize recent progress in protein structure refinement using new force fields.

Key words: Force field, Knowledge-based potential, Homology modeling, Implicit solvation, Protein structure refinement

1. Introduction

Much of computational protein modeling, including homology modeling, is based on Anfinsen's thermodynamic hypothesis, that a protein's native structure is uniquely determined by its amino acid sequence and that the native structure is the conformation with the lowest free energy (1). This offers a conceptually simple approach to protein structure prediction: find the minimum energy structure. In practice, however, this is extremely difficult due to the two primary challenges of computational protein structure prediction: (1) accurate calculation of the free energy for any protein conformation including the effects of aqueous or membrane solvation and (2) global optimization of a free energy function that is computationally intensive to calculate and is rough, i.e., has many local minima in conformational space. Homology modeling

Andrew J.W. Orry and Ruben Abagyan (eds.), *Homology Modeling: Methods and Protocols*, Methods in Molecular Biology, vol. 857, DOI 10.1007/978-1-61779-588-6_4, © Springer Science+Business Media, LLC 2012

approaches challenge 2 by starting with approximate initial structures based on existing experimental protein structures with recognizable sequence similarity, and thus presumably possessing similar structures (2–4). An accurate energy function is required to generate initial models with near-native geometry and also to further refine these structures so that challenge 1 remains important for homology modeling. These energy functions used in homology modeling methods are the subject of this chapter. Because it is impossible to provide a single detailed yet universal protocol for employing force fields in homology modeling that is applicable to the many commonly used methods and associated computer programs, we instead provide an introductory overview that aims to be a guide in choosing appropriate energy functions for each homology modeling task, in understanding the approximations implicit in each energy function, and in interpreting the homology modeling results in terms of these energy functions. Furthermore, both the modeling program (see Note 1) and available computer resources (see Note 2) dictate which force fields can be used for a particular homology modeling task.

Energy functions are used in both comparative and ab initio protein homology modeling for a number of different tasks that include (1) enforcing the correct covalent geometry, (2) avoiding steric clashes or atomic overlap, (3) selecting the near-native structure from among a set of potential model structures, and (4) assessing final model quality. Conformational sampling is achieved either by molecular dynamics (MD), in which the motion of the protein and possibly surrounding solvent are calculated using Newtonian mechanics, or by molecular mechanics (MM), in which sophisticated optimization techniques are used to find the global minimum of the energy function.

The energy functions employed in homology modeling, and indeed in any protein modeling task, can be divided into three basic types: physics-based force fields, knowledge-based potentials, and hybrid potentials that are a combination of the first two types. Physics-based force fields attempt to accurately approximate the actual physical energy of a protein conformation. On the other hand, knowledge-based potentials, also called statistical potentials, are derived based on the observed distribution of protein conformational variables, such as atomic separations, in a set of known experimental structures. Usually a Boltzmann distribution is assumed, insuring that commonly occurring conformations have a favorable (lower) energy than less common ones. The conversion from conformational frequencies to a physical energy scale in knowledge-based potentials also allows both types of energy functions, physics-based and knowledge-based, to be combined into a hybrid potential in which the interaction terms are a mixture of these two types.

In this chapter, we only discuss all-atom protein force fields. There are many coarse-grained force fields, in which the protein molecule is represented in a simplified manner by considering neighboring atoms in groups. One example is representing the position of a residue side chain by only its centroid and deriving interaction parameters based on this simplified representation. While such force fields have proven invaluable in protein design, generating initial near-native structures for protein structure prediction, and scoring potential structure solutions (near-native/decoy discrimination), we instead focus here on the all-atom energy functions needed for predicting protein structures with atomic level accuracy.

2. Physics-Based Force Fields

Physics-based force fields are a direct approximation of the physical energy for a collection of biomolecules in a particular conformation. Although many force fields have also been parameterized for a wide variety of other biomolecules and drug compounds, here we will only consider proteins and water molecules as the molecules most directly relevant to homology modeling (see Note 3). Physics-based force fields generally fall into two categories: (1) Cartesian force fields that account for all $3N$ degrees of freedom for N atoms and (2) torsion angle or internal coordinate force fields in which the stiff degrees of freedom, namely bond lengths and angles, are kept fixed. As a general rule, molecular dynamics simulations usually employ Cartesian force fields while molecular mechanics stimulation use torsion angle force fields.

Some of the most widely used Cartesian force fields are CHARMM22 (5, 6), AMBER (ff94 (7), ff99 (8), and ff03 (9) versions), GROMOS (10), and OPLS-AA (11). These and other force fields are under continuous development so that usually the latest available version, which is presumably the most accurate one, should be used if possible. There are also CHARMM (12), AMBER (13), and GROMOS (14) molecular mechanics programs that implement their respective force fields. Other commonly used molecular dynamics programs suited for protein simulations implement these force fields including NAMD (15) (CHARMM, AMBER, OPLS), GROMACS (16) (AMBER, CHARMM, GROMOS, OPLS), Desmond (17) (CHARMM, AMBER, OPLS), and TINKER (18) (CHARMM, AMBER, OPLS). In addition, the MODELLER (19, 20) homology modeling program and the SWISS-MODEL (21) server utilize the CHARMM and GROMOS force fields in their respective modeling procedures.

The parameters of physics-based force fields are determined by fitting to ab initio quantum mechanical energies and electrostatic potentials and experimental data such as neat liquid properties, crystal geometries and thermodynamic properties, solvation free energies, and vibrational spectra. To keep the fitting procedure tractable, the parameters are derived to fit properties of small compounds, such as small side chain analog compounds, terminal-blocked amino acids, or short peptides, with the assumption that the derived parameters will be transferable to proteins. Some force fields, including the four mentioned above, also have parameters for other biologically important molecules, including lipids, nucleic acids, and carbohydrates.

In physics-based force fields, the total energy is decomposed into a sum of contributions from different components. Furthermore, the energy components can be grouped into bonded interactions between atoms separated by one (1–2), two (1–3), or three (1–4) covalent bonds and nonbonded interactions. Nonbonded interactions generally include intramolecular interactions between atoms separated by ≥3 bonds in addition to intermolecular interactions. In other words, the total energy E for a conformation can be expressed as $E = E_{\text{bonded}} + E_{\text{nonbonded}}$.

Each atom in the protein is assigned a type and the force field terms used to compute the total energy depend on the particular atom types involved. The atom types generally differ between force fields and reflect the atom's characteristic chemical properties, such as element, charge, hybridization (e.g., sp^2 or sp^3), and aromaticity. All force field parameters depend on the atom types of the atoms involved. Next, we separately examine the individual bonded and nonbonded terms in a typical basic, or so-called class I, force field.

2.1. Bonded Interactions

The bonded component of the total conformational energy may be expressed as

$$E_{\text{bonded}} = \sum_{\text{bonds}} C_b \left(b - b_0\right)^2 + \sum_{\text{angles}} C_\theta \left(\theta - \theta_0\right)^2 + \sum_{\text{dihedrals}} C_\phi \left(1 + \cos(n\phi + \delta)\right) + \sum_{\text{impropers}} C_\alpha \left(\alpha - \alpha_0\right)^2. \quad (1)$$

The first term represents the energy of stretching a bond from its equilibrium length, b_0 to b. Its quadratic form is the same as Hooke's law for a spring. The second component accounts for the energy of changing the angle between two adjacent bonds from its equilibrium value, θ_0 to θ. The dihedral component in the third term is the energy of rotating about a dihedral, or torsion, angle ϕ defined by three consecutive bonds. Each term in the sum is necessarily periodic and has n minima. For four consecutive bonded atoms i, j, k, and l, the dihedral angle about the j–k bond, ϕ is the angle between the plane containing the atoms i, j, and k and the

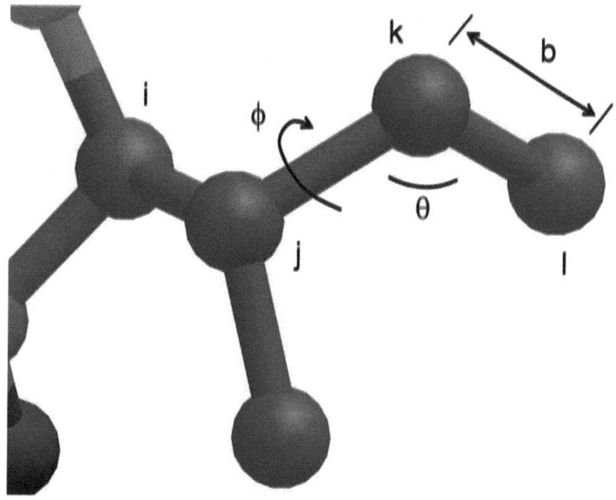

Fig. 1. An illustration of bonded interaction variables for the bond length (b), bond angle (θ), and dihedral angle (ϕ). Typical energy terms for these variables are given in Eq. 1.

plane containing the atoms *j*, *k*, and *l* (see Fig. 1). An accurate representation of the dihedral energy dependence is crucial for predicting correct side chain and loop backbone conformations, which are primary modeling tasks for homology model refinement. The dihedral parameters are usually some of the last parameters to be fit during force field development and so effectively contain whatever interactions are not accounted for by the other bonded and nonbonded terms. Because the division of intermolecular interactions between bonded and nonbonded components is to some extent arbitrary, since only the total energy is relevant, force fields can have different dihedral potentials depending on how they handle 1–4 bonded interactions (see below). This also highlights the fact that mixing parameter between different force fields is not a good idea and that improvements to a subset of parameters often necessitates refitting of the remaining force field parameters to maintain accuracy.

Many force fields also have an improper torsion term, the last term in Eq. 1, to enforce the geometry of certain chemical groups formed by three atoms bonded to a central atom. This includes the approximate planarity of a group with a central sp^2 hybridized atom or the chirality of tetrahedrally arranged atoms about a central sp^3 atom. For example, this term can be used to maintain the planarity of peptide bonds and aromatic rings in protein structures. For an arrangement of three atoms *j*, *k*, *l* bonded to the central atom *i*, the improper torsion angle α is defined to be the angle between the plane containing atoms *i*, *j*, and *k* and the one containing atoms *j*, *k*, and *l*. Thus, it involves the same calculation as for a usual dihedral angle, except for a different connectivity of the four atoms involved.

2.2. Nonbonded Interactions

A typical minimal expression for the nonbonded energy component is

$$E_{\text{nonbonded}} = \sum_{\text{nonbonded}} \varepsilon_{ij} \left[\left(\frac{r_{ij}}{r_{ij}^{\text{min}}} \right)^{-12} - 2 \left(\frac{r_{ij}}{r_{ij}^{\text{min}}} \right)^{-6} \right] + \frac{q_i q_j}{\varepsilon r_{ij}}. \quad (2)$$

Nonbonded interactions are more computationally intensive than bonded interactions because they are longer range and so involve more terms. Because of this, they are usually limited to only pairwise interactions between atoms. Interactions between atoms separated by >3 bonds are usually included in nonbonded interactions. Nonbonded interaction terms for atoms separated by two bonds (1–4 interactions) are also often included and are multiplied by a reduction factor in some force fields. This is done to better reproduce the torsion angle energy profile, which is a sum of the (scaled) nonbonded interactions and the bonded dihedral energy component.

The first term in Eq. 2 is the van der Waals energy. This component actually account for two different physical forces. One is the weak attractive dispersion force due to dipole-induced dipole interactions caused by transient charge fluctuations described by quantum mechanics. This force acts between all atoms and molecules and falls off to zero as r^{-6} at large distances, as does this 6-12 Lennard-Jones form of the potential. The other force is the so-called steric exclusion force that causes atoms to repel each other at small separation distances. This is due to another quantum mechanical effect, namely the Pauli exclusion principle that, roughly speaking, opposes significant overlap of the two atoms' electron clouds. As

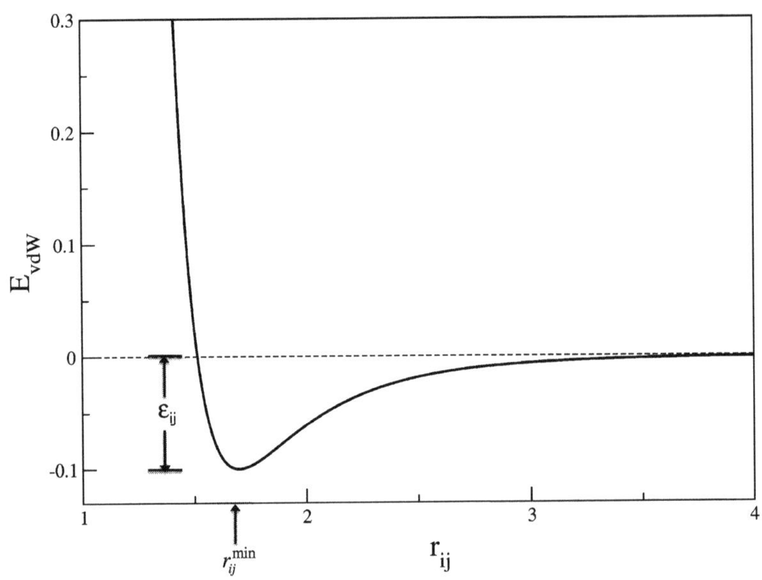

Fig. 2. An example of the Lennard-Jones form of the van der Waals potential between two atoms included in Eq. 2.

shown in Fig. 2, the van der Waals energy is high at short distances in which the atoms have significant steric overlap, reaches a minimum due to the weak dispersion force, and then rapidly approaches zero at large separation distances. The functional form of the Lennard-Jones potential is chosen for computational efficiency since r^{-12} may be simply calculated as the square of r^{-6}. The alternative Buckingham (22), or Exp-6, van der Waals potential function retains the r^{-6} attractive term of Eq. 2 but instead has an exponential repulsive term, $A\exp(-Br)$. This repulsive term is more physically realistic than the r^{-12} Lennard-Jones repulsive term, however, the Buckingham potential becomes unphysically attractive at small distances and is slower to calculate.

The van der Waals parameters, ε_{ij} and r_{ij}, for the interaction term between two atoms are determined from respective atomic parameters, (ε_i, r_i) and (ε_j, r_j), through the use of so-called combination rules. Because there is no theoretical basis for such rules, they tend to vary between different force fields, with either arithmetic or geometric averages as common choices.

The divergence of the van der Waals potential as the separation distance approaches zero is problematic for protein structure optimization. The extreme sensitivity of the potential to small conformational changes, on the order of a fraction of an Ångstrom, can cause the native conformation to have unfavorable high energy due to inaccuracies in the force field. It also leads to a rough energy surface rendering global optimization difficult and also can cause numerical instabilities in local optimization routines. One solution that is often implemented in molecular mechanics programs is to remove the van der Waals potential divergence by modifying it so that it smoothly approaches a finite value at zero separation. This simple prescription can speed up energy optimization and yield a more accurate final structure (see Note 4).

The last term in Eq. 2 represents the electrostatic energy of the conformation. This component accounts for the interaction energy of the electrostatic charge distribution of the electrons and nuclei. For computational efficiency the molecular charge distribution is usually approximated by partial point charges, q_i, at atomic centers. The sum of atomic charges for a molecule is required to equal its total formal charge. The dielectric constant, ε, has the value 1 in vacuum, as is the case of protein simulations with explicit solvent. If an implicit solvation model is employed, the electrostatic energy contribution must be further modified to account for solvent polarization or charge screening, which reduces the interaction strength. These models will be discussed below.

2.3. Other Energy Terms

2.3.1. Hydrogen Bond

Hydrogen bond interactions make a significant contribution to the protein and solvent energy and are a major factor in determining protein structure since the interaction is relatively strong (~5–6 kcal/mol for isolated bonds (23–25)), local, and directional. However, these interactions are incorporated into different force fields in diverse ways. Some force fields, such as CHARMM and AMBER, that include hydrogen atoms do not have an explicit hydrogen bond term but instead account for the interaction via the electrostatic and van der Waals terms. In this case, the favorable hydrogen bond energy is largely due to the interaction between a dipole formed by the donor proton and bound electronegative atom on one side of the hydrogen bond and an aligned dipole formed by the electronegative acceptor and bound atom on the other side. Although this scheme simplifies the force field additional charge centers or multipoles can more accurately reproduce hydrogen bond directionality at, for example, donor atoms with lone pair electrons, but at the expense of introducing more parameters (26–29).

2.3.2. Additional Terms

Additional terms beyond the basic ones outlined above may be included to improve accuracy. These include cross-terms, higher order polynomial terms, and Urey–Bradley terms. Such terms may be added to better reproduce experimental data, such as vibrational spectra. Their added complexity results in increased time to evaluate the energy. The CHARMM22 force field includes a Urey–Bradley term, which is a harmonic term between some atoms separated by two bonds. One force field that makes extensive use of such additional terms is CFF91, a member of the consistent family of force fields parameterized for a wide range of compounds in addition to proteins (30, 31). This force field includes higher order (quartic) polynomials for bond stretching and bending as well as cross-terms between bond stretching, bond bending, and dihedral terms. CFF91 and the newer CFF cover a wide range of compounds beyond proteins and as such have been mainly applied to smaller molecules rather than proteins. The CFF force field is implemented in the Cerius^2 modeling program (Accelrys, Inc.).

Most of the widely used force fields are periodically updated so that usually the latest version is preferred. In particular, the revision of the AMBER ff94 force field to the ff99 version (8) was largely to correct the α-helical preference of the ff94 backbone torsion potential parameters. Likewise, the CHARMM22 backbone torsion potential was modified to improve the agreement of backbone torsion angles in α-helical and β-sheet regions of proteins (6). Rather than refitting dihedral parameters, this was accomplished by adding a grid-based correction term (CMAP) depending on two neighboring dihedrals.

3. Knowledge-Based Potentials

The basic premise of knowledge-based potentials is that the observed distribution of conformational variables in experimental protein structures follows a Boltzmann distribution so that the energy can be derived from the estimated distributions of conformational variables, x_i, in the native state, $p_{\text{native}}(.)$, and in a reference state, $p_{\text{ref}}(.)$, as

$$E = -kT\log\left(\frac{p_{\text{native}}(x_1, x_2, \ldots, x_N)}{p_{\text{ref}}(x_1, x_2, \ldots, x_N)}\right)$$

$$= -kT\sum_i -kT\log\left(\frac{p^{(i)}_{\text{native}}(x_i)}{p^{(i)}_{\text{ref}}(x_i)}\right) \equiv \sum_i S_i(x_i) \qquad (3)$$

in which kT is the Boltzmann constant times the temperature. Furthermore, the conformational variables are assumed to be independent so that the total potential is a sum over terms, or scores $S_i(x_i)$, for each variable. As in physics-based force fields, atom types are defined and the parameters (scores) depend on them. Although the assumption of a Boltzmann distribution is not strictly justified (32), the temperature is an overall multiplicative factor and so does not affect relative energies, unless the knowledge-based potential is combined with a physics-based force field. This fact allows an alternative Bayesian statistical interpretation of knowledge-based potentials (33, 34). Regardless of their interpretation, knowledge-based potentials perform well in many protein modeling tasks and have been used successfully for homology model structure refinement and scoring.

One type of knowledge-based potential depends on the separation distances between pairs of atoms in a protein. Distance-dependent atom pair potentials are calculated as a sum over all atoms in different residues

$$E = \sum_{i>j} f_{ij}(r_{ij}), \qquad (4)$$

in which $f_{ij}(r_{ij})$ is the interaction potential for atom types i and j and r_{ij} is their separation distance. One example is the DFIRE potential (35, 36), whose key feature is the use of a finite ideal gas reference state in deriving the atom pair potentials. Another distance-dependent atom pair potential, DOPE, also accounts for the finite size in the reference state (37). The DOPE potential is currently used in the MODELLER homology modeling program. Both potentials have been employed for scoring alternative homology models to select the best structure.

SCWRL is a useful program for predicting side chain conformations in proteins and can be used for side chain placement in homology models (38). The latest version of this program, SCWRL4, relies on a knowledge-based side chain-dependent rotamer potential combined with a smoothed van der Waals potential and orientation-dependent hydrogen bond term. Optimization is accomplished via a fast graph-based algorithm.

4. Torsion Angle Force Fields

Protein bond lengths and bond angles fluctuate relatively little about their equilibrium values. This allows the approximation of representing the protein covalent geometry in torsion angle space (also called dihedral angle space or internal coordinate space) in which these stiff degrees of freedom are fixed and only the remaining torsion angles are sampled. The torsion angle representation greatly speeds up conformational sampling since the number of sampling steps necessary to find the global optimal structure scales exponentially with the number of degrees of freedom, which is reduced by about a factor of 5–10. The radius of convergence for structure optimization, an important consideration for homology model refinement, is also higher than for a Cartesian representation (39). One potential disadvantage of torsion angle force fields is that they may result in too high energies for some conformations and conformational energy barriers.

Two torsion angle force fields that are widely used for protein molecular mechanics are the ECEPP and Rosetta all-atom force fields. Their main difference is that ECEPP is a physics-based force field, while the Rosetta force field is primarily knowledge-based.

4.1. Physics-Based Torsion Angle Force Fields

The ECEPP force fields were continually developed over a number of years by the Scheraga group (40–42) and are implemented in their molecular mechanics program of the same name (also released as ECEPPAK). ECEPP/3 is also implemented in the ICM program (Molsoft LLC) (39). Special features of the ECEPP/3 force field include a 10-12 Lennard-Jones potential for atom pairs forming hydrogen bonds and scaling of the repulsive r^{-12} term in the Lennard-Jones van der Waals term (see Eq. 2) for atoms separated by three bonds by a factor of ½. The latest version, ECEPP-05, exploits the increased quantity of experimental and ab initio quantum mechanical data available for parameter fitting to update the force field (43). Major changes over ECEPP/3 include no 1–4 van der Waals scaling, no special hydrogen bonding terms (so that it is now included in electrostatics and van der Waals terms), and a different Buckingham potential for the van der Waals potential. This new version is not yet implemented in available modeling programs. As with other physics-based force fields, the ECEPP parameters were fit to both experimental data and energies calculated using ab initio quantum mechanics. To accurately reproduce torsional energy barriers, the torsion representation potentials were fit to ab initio energies calculated using an adiabatic approximation in which the torsion angle is fixed and the remaining degrees of freedom are relaxed by energy optimization.

The recently developed ICMFF force field (44) is based on earlier ECEPP force fields and optimized for loop modeling, an important task in homology modeling. New features include (1) parameterization using a dielectric constant, $\varepsilon=2$ that is relevant to the condensed state (see discussion below), (2) an improved description of hydrogen bond interactions that utilizes an additional set of van der Waals parameters for interactions between heavy (non-hydrogen) and hydrogen atoms, and (3) more accurate backbone torsion angle potentials that include corrections to the basic potential function in Eq. 1.

4.2. Rosetta All-Atom Force Field

Two energy functions are implemented in the Rosetta molecular mechanics program. One is a coarse-grained potential in which each residue side chain is represented by a single centroid. This is employed in the early stages of ab initio protein structure prediction. The other is an all-atom energy function that is used for refinement and scoring of protein structures from the initial ab initio structure search or from comparative modeling.

The Rosetta all-atom energy function is a sum of knowledge-based terms and one physics-based term that are each multiplied by (optimized) constant weight factors. The physics-based contribution is a van der Waals potential using CHARMM19 parameters with an optional damping via a linear approach to a finite value at zero separation. The remaining knowledge-based components include backbone torsion potential, backbone-dependent rotamer energy, a four-dimensional orientation-dependent hydrogen bond potential, residue pair interactions, and the EEF1 implicit solvation model (45). The Rosetta hydrogen bond potential is of particular interest as it was shown to better reproduce the angular dependence of high-level ab initio quantum mechanical energies for hydrogen-bonded side chain analogs than traditional physics-based force fields without explicit hydrogen bond terms (46). The optimized hydrogen bond geometry for the physics-based force fields were approximately linear, presumably due to a favorable linear geometry for the dipole–dipole interaction of the donor and acceptor groups rather than the correct angle at the acceptor group near 120°.

5. Polarization

Polarization is the redistribution of the molecular charge density in response to the electric field generated by surrounding atoms. The induced charge difference in turn contributes to the total electrostatic energy of the system. The standard fixed-charge force fields discussed so far account for polarization only in an average, or mean field, sense. This has been accomplished by, for example, fitting atomic charges using quantum mechanics derived potentials (from, e.g., HF/6-31G*) that systematically overestimate bond dipoles to mimic solvent-induced solute polarization, fitting to potentials using quantum mechanics potentials calculated with a continuum solvent model (9), and/or adjusting fit charges to obtain larger dipole moments (5). Despite the importance of polarization in accurate protein and solvent energetics, there is good reason to employ a fixed charge approximation since incorporating polarization requires many additional force field parameters to be fit, which significantly increases the computational cost of evaluating the conformational energy. However, the rapid increase in computer speed is expected to make polarizable force fields more attractive for protein simulations in the future (see Note 5). Several polarizable force fields for proteins have already been developed including AMBER ff02 (47), AMOEBA (48), PFF (derived from OPLS-AA) (49), and CHARMM fluctuating charge (CHEQ) (50, 51) and Drude oscillator models (52, 53). AMBER ff02 and AMOEBA are available in the AMBER molecular dynamics program, while the two polarizable CHARMM force fields are available in the CHARMM program. Because development continues for these force fields, they have not yet been extensively tested in protein simulations.

6. Solvation

Under physiological conditions, proteins exist in solution with water and usually also dissolved ions. Indeed, solvation is responsible for many of the forces that drive protein folding, especially the burial of hydrophobic residues in the protein interior (54–56). Because proteins only assume their native structure in solution it is crucial to account for solvation effect in the energy function. Solvation may be either explicit, through the inclusion of water molecules in the simulation used for structure optimization, or implicit, in which the effects of the solvent are accounted for in an average manner. Implicit solvation models are more approximate than explicit solvation but offer the advantages of a significant reduction in the computational cost and faster sampling of protein conformations in molecular dynamics simulations due to the absence of solvent viscosity.

6.1. Explicit Solvation

Explicit solvation is simply the inclusion of water molecules in the protein simulation. Explicit solvent is usually employed in molecular dynamics simulations but not in molecular mechanics simulations. This is because their effects on the protein conformation should be averaged whereas a molecular mechanics simulation would only find a single lowest energy conformation. One exception is when modeling specifically bound water molecules, often observed in high-resolution X-ray crystal structures, that are important for maintaining the correct structure and stability of a protein or protein complex.

Numerous parameters have been developed for water models (as reviewed in ref. 57). Commonly employed water models include SPC/E (58), TIP3P (59), and TIP4P (60). More detailed models incorporate electrostatic polarizability (61) and bond flexibility (62, 63). However, because a large proportion of the atoms in an explicit solvent protein simulation are for water and the computational cost for an N-site water model increases as N^2, such models come at a considerably higher computational expense, and so are less widely used. One consideration regarding the use of molecular dynamics simulations in explicit water is that a protein force field may be parameterized using a particular water model. For example, the CHARMM22 force field parameters were derived using a modified TIP3P water model (5, 6). Because of this implicit dependence on the water model, protein simulations using a different water model may yield less accurate results.

6.2. Implicit Solvation

The solvent contribution to the energy of a solvated protein can be divided into polar, or electrostatic, and nonpolar, or hydrophobic, contributions. The electrostatic contribution is modeled by considering water as a polarizable continuous medium with a uniform dielectric constant of approximately 80. The protein interior is also often assumed to have a dielectric constant of ~2–4 to account for its polarizability. Various values have been used for different modeling tasks and there has been some discussion about what values are appropriate (64, 65). This can be attributed to the fact that the protein interior is a highly heterogeneous environment, the effects of water penetration, and uncertainty on which polarization effects are implicitly included in the dielectric model. Next, we describe common polar implicit solvation models in decreasing order of accuracy and increasing order of speed.

6.2.1. Implicit Polar (Electrostatic) Solvation Models

Numerical solution of the Poisson–Boltzmann (PB) equation provides the most detailed and accurate implicit polar solvation model. Again, the protein interior is considered a dielectric continuum with a low dielectric constant and partial charges at atom centers while the exterior solvent region is assigned a high dielectric constant. This model also approximates the effects of ionic screening, which is significant for proteins in physiological ion concentrations of ~0.1 M. Many computer programs are available that use various numerical techniques to solve the PB equation, such as finite difference (DelPhi (66, 67) and Zap (68, 69)), multigrid finite element (APBS (70, 71)), and boundary element (ICM (72)) methods.

Although PB solvers are well suited for accurate energy calculations on individual structures to evaluate alternative homology models, they are not generally used for molecular dynamics simulations or structure optimization of proteins because of their slow speed. Generalized Born (GB) models (73, 74) using a pairwise descreening approximation (75–77) offer an efficient approximation to PB electrostatics that addresses this problem. GB models have been implemented in many molecular dynamics and molecular mechanics packages.

The most approximate but simplest polar solvation model is to use Coulomb electrostatics, as in Eq. 2, but with a dielectric constant ε that linearly increases with distance r, i.e., $\varepsilon = cr$, with c a constant. This roughly approximates the solvent screening of atomic charges by decreasing electrostatic interactions at large distances.

6.2.2. Implicit Nonpolar (Hydrophobic) Solvation Models

The most widely used nonpolar solvation model is a surface tension model in which the energy is proportional to the total protein solvent accessible surface area (SASA). The constant of proportionality is typically in the range of 20–30 cal/(mol Å²), in accordance with experimentally determined values (78, 79). When combined with the PB or GB polar solvation models, the resulting implicit solvation models are called PBSA or GBSA, respectively. Analytical derivatives of SASA are available for MM local optimization and MD (80, 81) but are complicated to calculate.

6.2.3. Other Implicit Solvation Models

Another approach to implicit solvation is to estimate the solvation energy as a sum of contributions from each protein atom, each of which is proportional to its respective SASA. In other words, the total solvation energy, E_{ASP}, is calculated as

$$E_{\mathrm{ASP}} = \sum_i \sigma_i A_i, \tag{5}$$

in which A_i are the SASAs, σ_i are the atomic solvation parameters (ASPs), and the sum is over all non-hydrogen atoms. Aqueous solvation parameters for a reduced set of five atom types were derived in an early paper by Wesson and Eisenberg (82) and designed to include both the hydrophobic and electrostatic components of solvation. This model is available in the CHARMM and ICM programs. In addition, ASPs for use with the new ICMFF force field implemented in ICM have been optimized for protein loop modeling (44). Another ASP model with only two parameters is also implemented in CHARMM and is designed to be used in conjunction with a simplified electrostatics model (83).

The EEF1 model of Lazaridis and Karplus is another computationally efficient approach to implicit solvation (45). This model has been implemented in the CHARMM and Rosetta programs. In this model, the electrostatic contribution to the solvation free energy is calculated using a distance-dependent dielectric constant, $\varepsilon = r$, to approximately account for charge screening and also ionic side chains are neutralized. The remaining solvation free energy is then calculated as a sum over contributions for atom i

$$\Delta G_i^{\mathrm{EEF1}} = \Delta G_i^{\mathrm{ref}} - \alpha_i \sum_{j \neq i} \exp\left[-\left(\frac{r_{ij} - R_i}{\lambda_i} \right)^2 \right] V_j, \qquad (6)$$

in which r_{ij} is the separation distance between atoms i and j, V_j is an effective volume, and $\Delta G_i^{\mathrm{ref}}$, α_i, and λ_i are parameters depending on the atom type. The sum over all atoms accounts for solvent exclusion. This model is roughly comparable to the ASP model in terms of both accuracy and computational efficiency, being only about 50% slower than a vacuum simulation without solvation.

6.2.4. Membrane Implicit Solvation Models

Membrane proteins constitute a significant fraction of the proteome in sequenced organisms (84) and also are the targets of about one half of all current drugs on the market (85, 86). However, despite their prevalence and biomedical importance, relatively few experimental X-ray crystallographic structures are available due to technical challenges (87). This provides motivation for the growing interest in predicting membrane protein structures (88, 89), particularly as new template structures become available for comparative modeling (90).

Implicit solvation models that account for the membrane environment as well as surrounding solvent can be used for membrane protein structure prediction and refinement at a greatly reduced computational cost compared with explicit membrane simulations. An actual biological membrane is generally composed of diverse mixtures of component lipids that depend on its cellular origin. Also because the lipids are ordered with their hydrophilic, and possibly charged, head groups at the interface and their hydrophobic hydrocarbon tails in the membrane interior, the average physiochemical environment of the membrane protein varies continuously with depth. For simplicity, and consequently computational efficiency, most commonly used models are parameterized for a single membrane environment that is characterized by two regions, the hydrophobic membrane core and the solvent, possibly with a smooth transition of the solvation energy between them.

Implicit solvation models contribute to two components of membrane structure prediction: (1) ensuring the correct degree of surface exposure of residues within the membrane and (2) helping stabilize the conformation with the correct position and tilt angle of transmembrane segments by minimizing any hydrophobic mismatch. While component (1) is analogous to the corresponding partitioning of surface and buried residues in non-membrane proteins and (2) is unique to membrane proteins. Implicit membrane solvation models have only been implemented in a few molecular modeling packages with two available models: generalized Born/solvent accessibility (GBSA) and IMM1. A modification of the GBSA model for membranes was introduced by Spassov et al. (91) and implemented in CHARMM. In this model, the membrane was represented as an infinite slab with the same low dielectric constant as the protein interior (~1–2), while the solvent region has a high dielectric constant (80). Also the nonpolar SASA solvation term is only active in the aqueous solvent region. The IMM1 model is a modification of EEF1 that includes a smooth transition as a function of the transverse membrane coordinate from water to membrane parameters (92) and is available both in CHARMM and Rosetta. Finally, coarse-grained lipid models, such as those available in the GROMACS program, provide a more detailed representation of the membrane at a higher but still reasonable computational cost for structure refinement.

6.3. pH and Ion Concentration Dependence of the Electrostatic Energy

The effects of pH and solvent ion concentration on the overall electrostatic energy of a protein, and hence its native conformation are often neglected in homology modeling. Instead, a lowest-order approximation is assumed, with ionizable residues and terminal groups in their unperturbed charge state at neutral pH and ionic screening is either neglected or roughly accounted for by a distance-dependent dielectric constant. Although most ionizable buried residues appear to remain charged due to compensating salt bridge and hydrogen bond interactions (93), so that this prescription is correct for the majority of residues, even a few misassigned charges can have a large effect on the total energy. The charge on a histidine residue is particular difficult to determine due to the fact that its intrinsic p*K*a, when fully solvated and without the influence of surrounding residues, of ~6.5 is near physiological pH values. While detailed p*K*a calculation during the conformational search is likely impractical, it is worthwhile to check charge states in the final structure using one of the available p*K*a web servers (e.g., H++ (http://biiophysics.cs.vt.edu/H++/) (94) or PROPKA (http://propka.ki.ku.dk) (95)) and to adjust charges and structure if necessary. Ionic screening of charges can be accounted for in explicit solvent by including ions in the simulation or in implicit solvent by using Poisson–Boltzmann electrostatics with a non-zero ionic strength. In any case, ions must be added to neutralize the protein charge in MD simulations and so yield a neutral system as required by Ewald summation methods (96) used to calculate electrostatic interactions with periodic boundary conditions. The GB electrostatics method has also been modified to account for ionic screening (97) and is implemented in the AMBER MD program.

7. Force Fields in Structure Refinement and Loop Modeling

One important and challenging application of energy functions is in the refinement, or optimization, of initial homology model structures. The goal of refinement is to improve an approximately correct model structure by moving it closer to the correct native structure. A more easily obtainable, but still important, goal is to simply make limited improvements to the model, for example remove steric clashes, adjust side chain conformations, or shift secondary structure elements, that lead to a better ranking of alternative models by the energy function.

The general view a decade ago, expressed in a published assessment of CASP3 results (98), was that energy optimization with molecular mechanics or molecular dynamics generally moved initial homology models farther from the native structure. More recently, a number of studies have demonstrated successful refinement of near-native models using molecular mechanics or molecular dynamics optimization with all-atom force fields, although structure refinement remains a challenging problem. Progress can be attributed to continuous improvements in force fields and solvation models as well as to new refinement protocols, particularly the judicious use of structural restraints in simulations. Restrained molecular dynamics simulations using the GROMACS force field with explicit solvent (99) and, more recently the CHARMM/CMAP force field with GBSA implicit solvent (100) improved model structures. There have also been a number of reports of success in loop modeling, an important part of structure refinement. One pair of studies employed molecular mechanics with the OPLS-AA force field and implicit solvation with GB electrostatics and a novel nonpolar solvation model (101, 102). Another study employed molecular dynamics using the AMBER ff03 force field with explicit solvent (103). Also, the ICMFF force field, implemented in ICM, has been optimized for loop modeling and achieved accuracies at least as good as any previous method on a benchmark set of protein loop structures (44). Knowledge-based potentials have also been used to demonstrate model improvement including an atom pair potential (104) and the Rosetta all-atom potential (105). One interesting approach is to optimize a force field so that it moves initial models closer to rather than away from the native structure (106–108). The significant improvements in all-atom refinement of homology models since CASP3 are reflected in a report on four different modeling algorithms that performed well in optimizing atomic structures in the recent CASP8 experiment (109).

8. Notes

1. Each molecular mechanics or molecular dynamics program only implements a limited set of force fields and solvation methods. This means that the choice of simulation method must necessarily be considered along with the force field. It is useful to examine the complete set of options for a program before choosing the best ones for the modeling task at hand

since the default settings may not always be appropriate. Most commonly used force fields are periodically updated to improve accuracy and are implemented in the latest version of the simulation program. Previously published applications of a program to homology modeling provide a useful starting point for choosing an appropriate energy model and also give an indication of what accuracy to expect.

2. There is usually a tradeoff between speed and accuracy so that a general rule is to use the most detailed force field and solvent representation for which the simulations will converge within a reasonable amount of time (depending on available computer resources). All-atom molecular mechanics with implicit solvation works well for initial prediction of loop regions and side chain conformations. Confidently assigned backbone regions, with an accurate sequence alignment and an ordered secondary structure in the protein core, should be constrained during the simulations. This can be accomplished using quadratic restraints on atom positions or simply not sampling the conformations of residues distant from the region of interest. Multiple (~5) independent simulations can be used to monitor convergence by verifying that the final energies approach a common value. More computationally expensive molecular dynamics simulations with explicit solvent can be used to further refine the initial predicted structures. Again, including some type of constraints on atomic positions are often necessary to prevent the conformations from moving too far away from the initial model structure. Also ions must be included in the molecular dynamics simulations to neutralize the system and to reproduce a physiologically relevant ion strength that properly screens electrostatic interactions.

3. Force fields specifically developed for proteins should be used for homology modeling. These include the ECEPP, ICMFF, and Rosetta torsion angle force fields for molecular mechanics as well as the CHARMM, AMBER, GROMOS, and OPLS-AA Cartesian force fields for molecular dynamics simulations discussed above. Other force fields, such as CFF, MMFF94 (110–114), and MM2-4 (115–118), were originally optimized for more chemically diverse small molecules and so are not appropriate for protein modeling.

4. In general, knowledge-based potentials are less sensitive to small conformational deviations than physics-based potentials. This is mainly due to the steep increase in the physical van der Waals potential at small atomic separation distances. This makes knowledge-based potentials a good choice for selecting near-native structures from among a set of incorrect, or decoy, structures in ab initio modeling or for assessing the quality of homology model structures. Physics-based force fields in which

the van der Waals potential is modified so that it approaches a finite value at small separations can also be use for these tasks. Such truncated van der Waals potentials are also recommended for use in molecular mechanics refinement of initial homology model structures to speed up convergence and avoid numerical instabilities.

5. Polarizable force fields offer a potentially more accurate representation of electrostatic interactions but at a significantly higher computational cost and so are less widely used than traditional nonpolarizable force fields. They are still under active development and have not yet been extensively tested for homology model refinement and so are not currently recommended for routine modeling projects.

Acknowledgments

This work was funded by the Mayo Clinic.

References

1. Anfinsen, C. B. (1973) Principles that govern the folding of protein chains, *Science 181*, 223–230.
2. Chothia, C., and Lesk, A. M. (1986) The relation between the divergence of sequence and structure in proteins, *EMBO J 5*, 823–826.
3. Levitt, M., and Gerstein, M. (1998) A unified statistical framework for sequence comparison and structure comparison, *Proc Natl Acad Sci U S A 95*, 5913–5920.
4. Russell, R. B., Saqi, M. A., Sayle, R. A., Bates, P. A., and Sternberg, M. J. (1997) Recognition of analogous and homologous protein folds: analysis of sequence and structure conservation, *J Mol Biol 269*, 423–439.
5. MacKerell Jr., A. D., Bashford, D., Bellott, M., Dunbrack Jr., R. L., Evanseck, J. D., Field, M. J., Fischer, S., Gao, J., Guo, H., Ha, S., Joseph-McCarthy, D., Kuchnir, L., Kuczera, K., Lau, F. T. K., Mattos, C., Michnick, S., Ngo, T., Nguyen, D. T., Prodhom, B., Reiher III, W. E., Roux, B., Schlenkrich, M., Smith, J. C., Stote, R., Straub, J., Watanabe, M., Wlorkiewicz-Kuczera, J., Yin, D., and Karplus, M. (1998) All-atom empirical potential for molecular modeling and dynamics studies of proteins, *J Phys Chem B 102*, 3586–3616.
6. Mackerell, A. D., Jr., Feig, M., and Brooks, C. L., 3rd. (2004) Extending the treatment of backbone energetics in protein force fields: limitations of gas-phase quantum mechanics in reproducing protein conformational distributions in molecular dynamics simulations, *J Comput Chem 25*, 1400–1415.
7. Cornell, W. D., P., C., Bayley, C. I., Gould, I. R., Merz Jr., K. M., Ferguson, D. M., Spellmeyer, D. C., Fox, T., Caldwell, J. W., and Kollman, P. A. (1995) A second generation force field for the simulation of proteins, nucleic acids, and organic molecules, *J Am Chem Soc 117*, 5179–5197.
8. Wang, J., Cieplak, P., and Kollman, P. A. (2000) How well does a restrained electrostatic potential (RESP) model perform in calculating conformation energies of organic and biological molecules?, *J Comput Chem 21*, 1049–1074.
9. Duan, Y., Wu, C., Chowdhury, S., Lee, M. C., Xiong, G., Zhang, W., Yang, R., Cieplak, P., Luo, R., Lee, T., Caldwell, J., Wang, J., and Kollman, P. (2003) A point-charge force field for molecular mechanics simulations of proteins based on condensed-phase quantum mechanical calculations, *J Comput Chem 24*, 1999–2012.
10. Oostenbrink, C., Villa, A., Mark, A. E., and van Gunsteren, W. F. (2004) A biomolecular force field based on the free enthalpy of hydration and solvation: the GROMOS force-field parameter sets 53A5 and 53A6, *J Comput Chem 25*, 1656–1676.
11. Jorgensen, W. L., Maxwell, D. S., and Tirado-Rives, J. (1996) Development and testing of the

OPLS all-atom force field on conformational energetics and properties of organic liquids, *J Am Chem Soc 118*, 11225–11236.

12. Brooks, B. R., Brooks, C. L., 3rd, Mackerell, A. D., Jr., Nilsson, L., Petrella, R. J., Roux, B., Won, Y., Archontis, G., Bartels, C., Boresch, S., Caflisch, A., Caves, L., Cui, Q., Dinner, A. R., Feig, M., Fischer, S., Gao, J., Hodoscek, M., Im, W., Kuczera, K., Lazaridis, T., Ma, J., Ovchinnikov, V., Paci, E., Pastor, R. W., Post, C. B., Pu, J. Z., Schaefer, M., Tidor, B., Venable, R. M., Woodcock, H. L., Wu, X., Yang, W., York, D. M., and Karplus, M. (2009) CHARMM: the biomolecular simulation program, *J Comput Chem 30*, 1545–1614.
13. Case, D. A., Cheatham, T. E., 3rd, Darden, T., Gohlke, H., Luo, R., Merz, K. M., Jr., Onufriev, A., Simmerling, C., Wang, B., and Woods, R. J. (2005) The Amber biomolecular simulation programs, *J Comput Chem 26*, 1668–1688.
14. Christen, M., Hunenberger, P. H., Bakowies, D., Baron, R., Burgi, R., Geerke, D. P., Heinz, T. N., Kastenholz, M. A., Krautler, V., Oostenbrink, C., Peter, C., Trzesniak, D., and van Gunsteren, W. F. (2005) The GROMOS software for biomolecular simulation: GROMOS05, *J Comput Chem 26*, 1719–1751.
15. Phillips, J. C., Braun, R., Wang, W., Gumbart, J., Tajkhorshid, E., Villa, E., Chipot, C., Skeel, R. D., Kale, L., and Schulten, K. (2005) Scalable molecular dynamics with NAMD, *J Comput Chem 26*, 1781–1802.
16. Hess, B., Kutzner, C., van der Spoel, D., and Lindahl, E. (2008) GROMACS 4: Algorithms or highly efficient, load-balanced, and scalable molecular simulation, *J Chem Theory Comput 4*, 435–447.
17. Bowers, K. J., Chow, E., Xu, H., Dror, R. O., Eastwood, M. P., Gregersen, B. A., Klepeis, J. L., Kolossvary, I., Moraes, M. A., Sacerdoti, F. D., Salmon, J. K., Shan, Y., and Shaw, D. E. (2006) Scalable algorithms for molecular dynamics simulations on commodity clusters, in *ACM/IEEE Conference on Supercomputing (SC06)*, ACM, Tampa, FL.
18. Ponder J. (2011) TINKER Molecular Modeling Package, http://dasher.wustl.edu/ffe/.
19. Sali, A., and Blundell, T. L. (1993) Comparative protein modelling by satisfaction of spatial restraints, *J Mol Biol 234*, 779–815.
20. Eswar, N., Eramian, D., Webb, B., Shen, M. Y., and Sali, A. (2008) Protein structure modeling with MODELLER, *Methods Mol Biol 426*, 145–159.
21. Schwede, T., Kopp, J., Guex, N., and Peitsch, M. C. (2003) SWISS-MODEL: An automated protein homology-modeling server, *Nucleic Acids Res 31*, 3381–3385.
22. Buckingham, R. A. (1938) The classical equation of state of gaseous helium, neon, and argon, *Proc R Soc Lond. A 168*, 264–283.
23. Avbelj, F., Luo, P., and Baldwin, R. L. (2000) Energetics of the interaction between water and the helical peptide group and its role in determining helix propensities, *Proc Natl Acad Sci U S A 97*, 10786–10791.
24. Ben-Tal, N., Sitkoff, D., Topol, I. A., Yang, A. S., Burt, S. K., and Honig, B. (1997) Free energy of amide hydrogen bond formation in vacuum, in water, and in liquid alkane solution, *J Phys Chem B 101*, 450–457.
25. Sheu, S. Y., Yang, D. Y., Selzle, H. L., and Schlag, E. W. (2003) Energetics of hydrogen bonds in peptides, *Proc Natl Acad Sci U S A 100*, 12683–12687.
26. Mitchell, J. B. O., and Price, S. L. (1989) On the electrostatic directionality of N-H...O=C hydrogen bonding, *Chem Phys Lett 154*, 267–272.
27. Zhao, D. X., Liu, C., Wang, F. F., Yu, C. Y., Gong, L. D., Liu, S. B., and Yang, Z. Z. (2010) Development of a polarizable force field using multiple fluctuating charges per atom, *J Chem Theory Comput 6*, 795–804.
28. Allinger, N. L., and Chung, D. Y. (1976) Conformational analysis. 118. Application of the molecular-mechanics method to alcohols and ethers, *J Am Chem Soc 98*, 6798–6803.
29. Dixon, R. W., and Kollman, P. A. (1997) Advancing beyond the atom-centered model in additive and nonadditive molecular mechanics, *J Comput Chem 18*, 1632–1646.
30. Maple, J. R., Dinur, U., and Hagler, A. T. (1988) Derivation of force fields for molecular mechanics and dynamics from ab initio energy surfaces, *Proc Natl Acad Sci U S A 85*, 5350–5354.
31. Maple, J. R., Hwang, M. J., Stockfisch, T. P., Dinur, U., Waldman, M., Ewig, C. S., and Hagler, A. T. (1994) Derivation of class II force fields. 1. Methodology and quantum force field for the alkyl functional group and alkane molecules, *J Comput Chem 15*, 162–182.
32. Thomas, P. D., and Dill, K. A. (1996) Statistical potentials extracted from protein structures: how accurate are they?, *J Mol Biol 257*, 457–469.
33. Simons, K. T., Kooperberg, C., Huang, E., and Baker, D. (1997) Assembly of protein tertiary structures from fragments with similar local sequences using simulated annealing and Bayesian scoring functions, *J Mol Biol 268*, 209–225.

34. Bordner, A. J. (2010) Orientation-dependent backbone-only residue pair scoring functions for fixed backbone protein design, *Bmc Bioinformatics 11*, 192.
35. Zhou, H., and Zhou, Y. (2002) Distance-scaled, finite ideal-gas reference state improves structure-derived potentials of mean force for structure selection and stability prediction, *Protein Sci 11*, 2714–2726.
36. Yang, Y., and Zhou, Y. (2008) Ab initio folding of terminal segments with secondary structures reveals the fine difference between two closely related all-atom statistical energy functions, *Protein Sci 17*, 1212–1219.
37. Shen, M. Y., and Sali, A. (2006) Statistical potential for assessment and prediction of protein structures, *Protein Sci 15*, 2507–2524.
38. Krivov, G. G., Shapovalov, M. V., and Dunbrack, R. L., Jr. (2009) Improved prediction of protein side-chain conformations with SCWRL4, *Proteins 77*, 778–795.
39. Abagyan, R., Totrov, M., and Kuznetsov, D. (1994) ICM - A new method for protein modeling and design: Applications to docking and structure prediction from the distorted native conformation, *J Comput Chem 15*, 488–506.
40. Momany, F. A., McGuire, R. F., Burgess, A. W., and Scheraga, H. A. (1975) Energy parameters in polypeptides. VII. Geometric parameters, partial atomic charges, nonbonded interactions, hydrogen bond interactions, and intrinsic torsional potentials or the naturally occurring amino acids, *J Phys Chem 79*, 2361–2381.
41. Nemethy, G., Pottle, M. S., and Scheraga, H. A. (1983) Energy parameters in polypeptides. 9. Updating of geometric parameters, nonbonded interactions and hydrogen bond interactions for the naturally occurring amino acids, *J Phys Chem 87*, 1883–1887.
42. Nemethy, G., Gibson, K. D., Palmer, K. A., Yoon, C. N., Paterlini, G., Zagari, A., Rumsey, S., and Scheraga, H. A. (1992) Energy parameters in polypeptides. 10. Improved geometric parameters and nonbonded interactions for use in the ECEPP/3 algorithm, with application to proline-containing peptides, *J Phys Chem 96*, 6472–6484.
43. Arnautova, Y. A., Jagielska, A., and Scheraga, H. A. (2006) A new force field (ECEPP-05) for peptides, proteins, and organic molecules, *J Phys Chem B 110*, 5025–5044.
44. Arnautova, Y. A., Abagyan, R. A., and Totrov, M. (2011) Development of a new physics-based internal coordinate mechanics force field and its application to protein loop modeling, *Proteins 79*, 477–498.
45. Lazaridis, T., and Karplus, M. (1999) Effective energy function for proteins in solution, *Proteins 35*, 133–152.
46. Morozov, A. V., Kortemme, T., Tsemekhman, K., and Baker, D. (2004) Close agreement between the orientation dependence of hydrogen bonds observed in protein structures and quantum mechanical calculations, *Proc Natl Acad Sci U S A 101*, 6946–6951.
47. Cieplak, P., Caldwell, J., and Kollman, P. (2001) Molecular mechanical models for organic and biological systems going beyond the atom centered two body additive approximation: aqueous solution free energies of methanol and N-methyl acetamide, nucleic acid base, and amide hydrogen bonding and chloroform/water partition coefficients of the nucleic acid bases, *J Comput Chem 22*, 1048–1057.
48. Ponder, J. W., Wu, C., Ren, P., Pande, V. S., Chodera, J. D., Schnieders, M. J., Haque, I., Mobley, D. L., Lambrecht, D. S., DiStasio, R. A., Jr., Head-Gordon, M., Clark, G. N., Johnson, M. E., and Head-Gordon, T. Current status of the AMOEBA polarizable force field, *J Phys Chem B 114*, 2549–2564.
49. Kaminski, G. A., Stern, H. A., Berne, B. J., Friesner, R. A., Cao, Y. X., Murphy, R. B., Zhou, R., and Halgren, T. A. (2002) Development of a polarizable force field for proteins via ab initio quantum chemistry: First generation model and gas phase tests, *J Comput Chem 23*, 1515–1531.
50. Patel, S., and Brooks, C. L., 3rd. (2004) CHARMM fluctuating charge force field for proteins: I parameterization and application to bulk organic liquid simulations, *J Comput Chem 25*, 1–15.
51. Patel, S., Mackerell, A. D., Jr., and Brooks, C. L., 3 rd. (2004) CHARMM fluctuating charge force field for proteins: II protein/solvent properties from molecular dynamics simulations using a nonadditive electrostatic model, *J Comput Chem 25*, 1504–1514.
52. Lamoureux, G., and Roux, B. (2003) Modeling induced with classical Drude Oscillators: Theory and molecular dynamics simulation algorithm, *J Chem Phys 119*, 245–249.
53. Lamoureux, G., Harder, E., Vorobyov, I. V., Roux, B., and MacKerell, A. D. (2006) A polarizable model of water for molecular dynamics simulations of biomolecules, *Chem Phys Lett 418*, 245–249.
54. Chothia, C. (1976) The nature of the accessible and buried surfaces in proteins, *J Mol Biol 105*, 1–12.
55. Tanford, C. (1978) The hydrophobic effect and the organization of living matter, *Science 200*, 1012–1018.

56. Wolfenden, R. (1983) Waterlogged molecules, *Science 222*, 1087–1093.
57. Guillot, B. (2002) A reappraisal of what we have learnt during three decades of computer simulations on water, *J Mol Liq 101*, 219–260.
58. Berendsen, H. J. C., Grigera, J. R., and Straatsma, T. P. (1987) The missing term in effective pair potentials, *J Phys Chem 91*, 6269–6271.
59. Jorgensen, W. L., Chandrasekhar, J., Madura, J. D., Impey, R. W., and Klein, M. L. (1983) Comparison of simple potential functions for simulating liquid water, *J Chem Phys 79*, 926–935.
60. Jorgensen, W. L., and Madura, J. D. (1985) Temperature and size dependence for Monte Carlo simulations of TIP4P water, *Mol Phys 56*, 1381–1380.
61. Rick, S. W. (2001) Simulations of ice and liquid water over a range of temperatures using the fluctuating charge model, *J Chem Phys 114*, 2276–2283.
62. Anderson, J., Ullo, J. J., and S., Y. (1987) Molecular dynamics simulation of dielectric properties of water, *J Chem Phys 87*, 1726–1732.
63. Toukan, K., and Rahman, A. (1985) Molecular-dynamics study of atomic motions in water, *Phys Rev B 31*, 2643–2648.
64. Schutz, C. N., and Warshel, A. (2001) What are the dielectric "constants" of proteins and how to validate electrostatic models?, *Proteins 44*, 400–417.
65. Simonson, T., and Brooks III, C. D. (1996) Charge screening and the dielectric constant of proteins: Insights from molecular mechanics, *J Am Chem Soc 118*, 8452–8458.
66. Rocchia, W., Sridharan, S., Nicholls, A., Alexov, E., Chiabrera, A., and Honig, B. (2002) Rapid grid-based construction of the molecular surface and the use of induced surface charge to calculate reaction field energies: applications to the molecular systems and geometric objects, *J Comput Chem 23*, 128–137.
67. Honig, B. (2010) Software: DelPhi, A finite difference Poisson-Boltzmann solver.
68. Grant, J. A., Pickup, B. T., and Nicholls, A. (2001) A smooth permittivity function for Poisson-Boltzmann solvation methods, *J Comput Chem 22*, 608–640.
69. OpenEye Scientific Software (2011) Modeling Toolkits: Programming Libraries for Molecular Modeling, http://www.eyesopen.com/products/toolkits/modeling-toolkits.html
70. Baker, N. A., Sept, D., Joseph, S., Holst, M. J., and McCammon, J. A. (2001) Electrostatics of nanosystems: application to microtubules and the ribosome, *Proc Natl Acad Sci U S A 98*, 10037–10041.
71. Baker, N. (2010) Adaptive Poisson-Boltzmann Solver (APBS) – Software for evaluating the elecrostatic properties of nanoscale biomolecular systems, http://www.poissonboltzmann.org/apbs/
72. Totrov, M., and Abagyan, R. (2001) Rapid boundary element solvation electrostatics calculations in folding simulations: successful folding of a 23-residue peptide, *Biopolymers 60*, 124–133.
73. Still, W. C., Tempczyk, A., Hawley, R. C., and Hendrickson, T. (1990) Semianalytical treatment of solvation for molecular mechanics and dynamics, *J Am Chem Soc 112*, 6127–6129.
74. Bashford, D., and Case, D. A. (2000) Generalized born models of macromolecular solvation effects, *Annu Rev Phys Chem 51*, 129–152.
75. Hawkins, G. D., Cramer, C. J., and Truhlar, D. G. (1995) Pairwise Solute Descreening of Solute Charges from a Dielectric Medium, *Chemical Physics Letters 246*, 122–129.
76. Hawkins, G. D., Cramer, C. J., and Truhlar, D. G. (1996) Parameterized models of aqueous free energies of solvation based on pairwise descreening of solute atomic charges from a dielectric medium, *J Phys Chem 100*, 19824–19839.
77. Qiu, D., Shenkin, P. S., Hollinger, F. P., and Still, W. C. (1997) The GB/SA continuum model for solvation. A fast analytical method for the calculation of approximate Born radii, *Journal of Physical Chemistry A 101*, 3005–3014.
78. Chothia, C. (1974) Hydrophobic bonding and accessible surface area in proteins, *Nature 248*, 338–339.
79. Richards, F. M. (1977) Areas, volumes, packing and protein structure, *Annu Rev Biophys Bioeng 6*, 151–176.
80. Sridharan, S., Nicholls, A., and Sharp, K. A. (2004) A rapid method for calculating derivatives of solvent accessible surface areas of molecules, *J Comput Chem 16*, 1038–1044.
81. Richmond, T. J. (1984) Solvent accessible surface area and excluded volume in proteins. Analytical equations for overlapping spheres and implications for the hydrophobic effect, *J Mol Biol 178*, 63–89.
82. Wesson, L., and Eisenberg, D. (1992) Atomic solvation parameters applied to molecular dynamics of proteins in solution, *Protein Sci 1*, 227–235.
83. Ferrara, P., Apostolakis, J., and Caflisch, A. (2002) Evaluation of a fast implicit solvent

model for molecular dynamics simulations, *Proteins 46*, 24–33.

84. Wallin, E., and von Heijne, G. (1998) Genome-wide analysis of integral membrane proteins from eubacterial, archaean, and eukaryotic organisms, *Protein Sci 7*, 1029–1038.
85. Bakheet, T. M., and Doig, A. J. (2009) Properties and identification of human protein drug targets, *Bioinformatics 25*, 451–457.
86. Yildirim, M. A., Goh, K. I., Cusick, M. E., Barabasi, A. L., and Vidal, M. (2007) Drug-target network, *Nat Biotechnol 25*, 1119–1126.
87. Lacapere, J. J., Pebay-Peyroula, E., Neumann, J. M., and Etchebest, C. (2007) Determining membrane protein structures: still a challenge!, *Trends Biochem Sci 32*, 259–270.
88. O'Mara, M. L., and Tieleman, D. P. (2007) P-glycoprotein models of the apo and ATP-bound states based on homology with Sav1866 and MalK, *FEBS Lett 581*, 4217–4222.
89. Yarnitzky, T., Levit, A., and Niv, M. Y. (2010) Homology modeling of G-protein-coupled receptors with X-ray structures on the rise, *Curr Opin Drug Discov Devel 13*, 317–325.
90. Yarnitzky, T., Levit, A., and Niv, M. Y. Homology modeling of G-protein-coupled receptors with X-ray structures on the rise, *Curr Opin Drug Discov Devel 13*, 317–325.
91. Spassov, V. Z., Yan, L., and Szalma, S. (2002) Introducing an implicit membrane in generalized Born/solvent accessibility continuum solvent models, *J Phys Chem B 106*, 8726–8738.
92. Lazaridis, T. (2003) Effective energy function for proteins in lipid membranes, *Proteins 52*, 176–192.
93. Kim, J., Mao, J., and Gunner, M. R. (2005) Are acidic and basic groups in buried proteins predicted to be ionized?, *J Mol Biol 348*, 1283–1298.
94. Gordon, J. C., Myers, J. B., Folta, T., Shoja, V., Heath, L. S., and Onufriev, A. (2005) H++: a server for estimating pKas and adding missing hydrogens to macromolecules, *Nucleic Acids Res 33*, W368–371.
95. Li, H., Robertson, A. D., and Jensen, J. H. (2005) Very fast empirical prediction and rationalization of protein pKa values, *Proteins 61*, 704–721.
96. Darden, T., York, D., and Pedersen, L. (1993) Particle mesh Ewald: a N.log(N) method for Ewald sums in large systems, *J Chem Phys 98*, 10089–10092.
97. Srinivasan, J., Trevathan, M. W., Beroza, P., and Case, D. A. (1999) Application of a pairwise generalized Born model to proteins and nucleic acids: inclusion of salt effects, *Theoretical Chemistry Accounts 101*, 426–434.
98. Koehl, P., and Levitt, M. (1999) A brighter future for protein structure prediction, *Nat Struct Biol 6*, 108–111.
99. Flohil, J. A., Vriend, G., and Berendsen, H. J. (2002) Completion and refinement of 3-D homology models with restricted molecular dynamics: application to targets 47, 58, and 111 in the CASP modeling competition and posterior analysis, *Proteins 48*, 593–604.
100. Chen, J., and Brooks, C. L., 3rd. (2007) Can molecular dynamics simulations provide high-resolution refinement of protein structure?, *Proteins 67*, 922–930.
101. Sellers, B. D., Zhu, K., Zhao, S., Friesner, R. A., and Jacobson, M. P. (2008) Toward better refinement of comparative models: predicting loops in inexact environments, *Proteins 72*, 959–971.
102. Sellers, B. D., Nilmeier, J. P., and Jacobson, M. P. (2010) Antibodies as a model system for comparative model refinement, *Proteins 78*, 2490–2505.
103. Kannan, S., and Zacharias, M. (2010) Application of biasing-potential replica-exchange simulations for loop modeling and refinement of proteins in explicit solvent, *Proteins 78*, 2809–2819.
104. Chopra, G., Kalisman, N., and Levitt, M. (2010) Consistent refinement of submitted models at CASP using a knowledge-based potential, *Proteins, 78*, 2668–2678.
105. Misura, K. M., Chivian, D., Rohl, C. A., Kim, D. E., and Baker, D. (2006) Physically realistic homology models built with ROSETTA can be more accurate than their templates, *Proc Natl Acad Sci U S A 103*, 5361–5366.
106. Krieger, E., Koraimann, G., and Vriend, G. (2002) Increasing the precision of comparative models with YASARA NOVA – a self-parameterizing force field, *Proteins 47*, 393–402.
107. Krieger, E., Darden, T., Nabuurs, S. B., Finkelstein, A., and Vriend, G. (2004) Making optimal use of empirical energy functions: force-field parameterization in crystal space, *Proteins 57*, 678–683.
108. Jagielska, A., Wroblewska, L., and Skolnick, J. (2008) Protein model refinement using an optimized physics-based all-atom force field, *Proc Natl Acad Sci U S A 105*, 8268–8273.
109. Krieger, E., Joo, K., Lee, J., Raman, S., Thompson, J., Tyka, M., Baker, D., and Karplus, K. (2009) Improving physical realism, stereochemistry, and side-chain accuracy in homology modeling: Four approaches that performed well in CASP8, *Proteins 77 Suppl 9*, 114–122.

110. Halgren, T. A. (1996) Merck molecular force field. I. Basis, form, scope, parameterization, and performance of MMFF94, *J Comput Chem 17*, 490–519.
111. Halgren, T. A. (1996) Merck molecular force field. II. MMFF94 van der Waals and electrostatic parameters for intermolecular interactions, *J Comput Chem 17*, 520–552.
112. Halgren, T. A. (1996) Merck molecular force field. III. Molecular geometries and vibrational frequencies for MMFF94, *J Comput Chem 17*, 553–586.
113. Halgren, T. A., and Nachbar, R. B. (1996) Merck molecular force field. IV. Conformational energies and geometries for MMFF94, *J Comput Chem 17*, 587–615.
114. Halgren, T. A. (1996) Merck molecular force field. V. Extension of MMFF94 using experimental data, additional computational data, and empirical rules, *J Comput Chem 17*, 616–641.
115. Allinger, N. L., Chen, K. H., Lii, J. H., and Durkin, K. A. (2003) Alcohols, ethers, carbohydrates, and related compounds. I. The MM4 force field for simple compounds, *J Comput Chem 24*, 1447–1472.
116. Lii, J. H., Chen, K. H., Durkin, K. A., and Allinger, N. L. (2003) Alcohols, ethers, carbohydrates, and related compounds. II. The anomeric effect, *J Comput Chem 24*, 1473–1489.
117. Lii, J. H., Chen, K. H., Grindley, T. B., and Allinger, N. L. (2003) Alcohols, ethers, carbohydrates, and related compounds. III. The 1,2-dimethoxyethane system, *J Comput Chem 24*, 1490–1503.
118. Lii, J. H., Chen, K. H., and Allinger, N. L. (2003) Alcohols, ethers, carbohydrates, and related compounds. IV. Carbohydrates, *J Comput Chem 24*, 1504–1513.

Chapter 5

Automated Protein Structure Modeling with SWISS-MODEL Workspace and the Protein Model Portal

Lorenza Bordoli and Torsten Schwede

Abstract

Comparative protein structure modeling is a computational approach to build three-dimensional structural models for proteins using experimental structures of related protein family members as templates. Regular blind assessments of modeling accuracy have demonstrated that comparative protein structure modeling is currently the most reliable technique to model protein structures. Homology models are often sufficiently accurate to substitute for experimental structures in a wide variety of applications. Since the usefulness of a model for specific application is determined by its accuracy, model quality estimation is an essential component of protein structure prediction. Comparative protein modeling has become a routine approach in many areas of life science research since fully automated modeling systems allow also nonexperts to build reliable models. In this chapter, we describe practical approaches for automated protein structure modeling with SWISS-MODEL Workspace and the Protein Model Portal.

Key words: Protein structure prediction, Molecular models, Automation, Homology modeling, Comparative modeling, Quality estimation, SWISS-MODEL, Protein Model Portal, QMEAN

1. Introduction

Knowing a protein's three-dimensional structure is crucial for understanding its biological function at the molecular level. However, despite remarkable advances in protein structure determination by NMR and X-Ray crystallography, currently no experimental structural information is available for the vast majority of protein sequences resulting from large-scale genome sequencing and metagenomics projects. To overcome this knowledge gap, over the past decades, a wide variety of computational methods for predicting the structure of proteins have been developed. These methods differ significantly in their computational complexity, the range of proteins for which they can be applied, and the accuracy and reliability of the resulting models (1, 2). Here, we will focus on homology modeling

Andrew J.W. Orry and Ruben Abagyan (eds.), *Homology Modeling: Methods and Protocols*, Methods in Molecular Biology, vol. 857,
DOI 10.1007/978-1-61779-588-6_5,

(aka comparative or template-based modeling), where a model for a protein of interest is constructed using structural information from homologous proteins (1–6). Regular blind assessment of prediction techniques has shown that comparative protein structure modeling is currently the only technique which is able to reliably provide models of high quality over a wide range of size, while de novo prediction methods are limited to small proteins and peptides (7). On the other side, comparative modeling techniques are limited to cases for which suitable template structures can be identified. For example, this poses a major limitation when modeling membrane proteins, which are underrepresented in today's structure databases but embody the majority of pharmaceutically interesting drug targets (8). The usefulness of protein structure models has been demonstrated in a variety of biological applications (9–11), such as rational design of mutagenesis experiments (12), providing receptor models for virtual screening (13, 14), to develop strategies for protein engineering, or to support experimental structure solution by crystallography (15, 16) or electron microscopy (17–19).

Computational modeling has become a valuable tool to complement experimental elucidation of protein structures. To make three-dimensional information accessible to a broad community of biomedical researchers on a whole-genome scale, automated modeling pipelines had to be developed which were stable, reliable, accurate, and easy to use. Almost two decades ago, the first automated modeling server—SWISS-MODEL—was made available on the Internet (20). Since then, many more services have been developed to model the structures of proteins in an automated manner (21, 22), e.g., ModWeb (23), Robetta (24), HHpred (25), I-TASSER (26), Pcons (27), PHYRE (28), or M4T (29). Recent method developments aim to include additional experimental constraints into the modeling procedures (17–19, 30) and to establish methods specialized in certain protein families such as GPCRs (31, 32) or Antibodies (33, 34).

One main objective for automating the principal steps of comparative protein structure modeling—template selection, target–template alignment, model building, and model quality evaluation (Fig. 1)—is the need of making these technologies accessible to an audience of nonexperts in bioinformatics. This includes facilitating the usage of computational tools which otherwise required highly specialized technical skills, maintaining up-to-date modeling software, and managing large amounts of sequence and structural data stored in biological databases, which are needed to complete the modeling tasks. Secondly, due to the huge number of protein sequences whose structure has not yet been experimentally characterized, automated procedures are essential to cope with this flood of data, e.g., to increase the coverage of structural information for proteomes of whole organisms or families of proteins (20, 35–37).

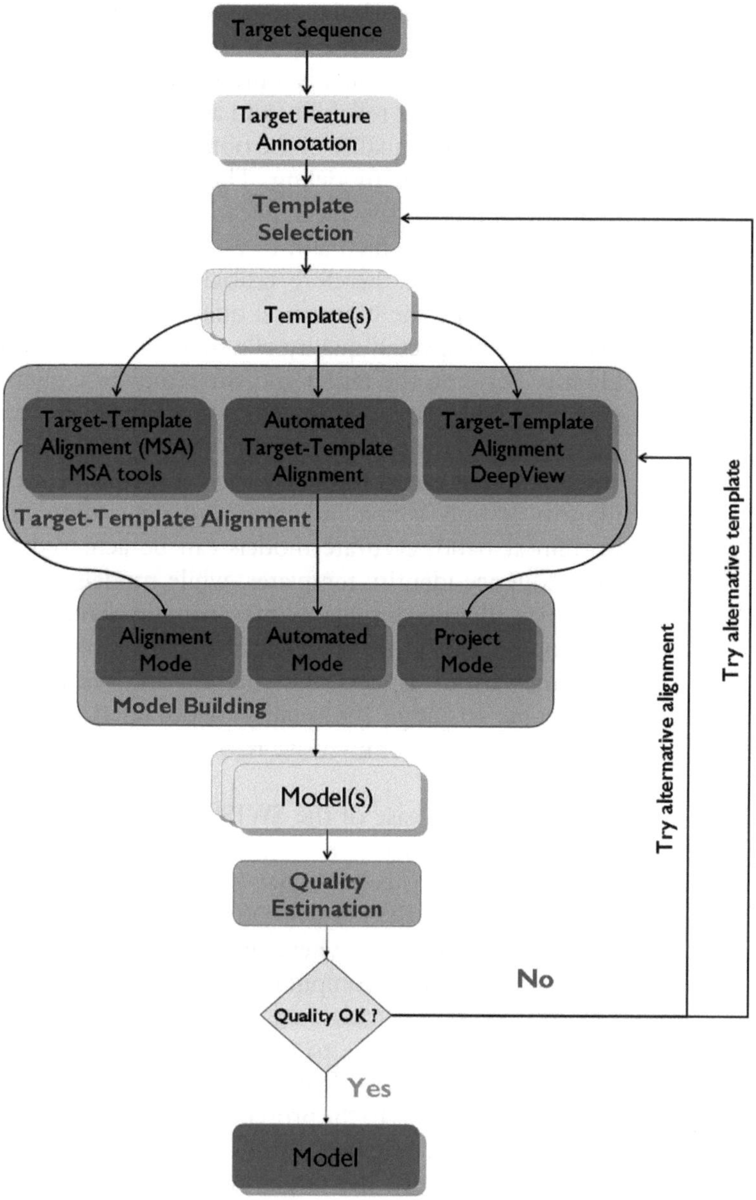

Fig. 1. SWISS-MODEL workflow. The flowchart illustrates the classical steps to construct a homology model of a target sequence as they are implemented in SWISS-MODEL Workspace. Starting from the sequence of the protein of interest (target) one or more related structures (templates) are identified (template selection). Annotation of the target sequence (feature annotation) can guide the choice of appropriate template(s). Based on the evolutionary distance between target and template(s) sequences, three different regimes of the target-template alignment step are available in the SWISS-MODEL Workspace: Automated, Alignment, or Project Mode. Target and template(s) sequences are aligned (target–template alignment) either in a fully automated fashion, by using external alignment tools, and (optionally) adjusted visually with the help of the DeepView program. The model is then constructed based on these alignments. Finally, the quality of the obtained model(s) can be estimated and verified and if necessary the procedure is repeated until a satisfactory result is obtained.

Finally, from a theoretical perspective, automatic procedures ensure the reproducibility of the modeling methods by excluding individual human bias, which is a prerequisite for the assessment and comparison of their reliability and accuracy (22, 38).

Validating the quality of the obtained models is a central aspect of protein structure modeling. The quality of models determines their usefulness for specific applications in life science research (9). Scoring functions which aim to estimate the expected accuracy of a protein model are, therefore, crucial to judge if it would be suitable to address a specific biomedical question. A well known first estimate for the expected quality of a structural model is the sequence identity between the target and the template sequences, where in general higher sequence similarity leads to more accurate models since the evolutionary structural divergence will be smaller (39) and alignment errors less likely to occur (40). However, sequence identity is only a first indicator and depending on the specific protein at hand, accurate models can be achieved based on very low sequence identity templates, while models based on medium sequence identity templates may contain significant errors. The development of more sophisticated scoring methods, taking into account various aspects of structural and sequence information to be able to judge the quality of obtained models (41–45), is currently a matter of intensive research.

1.1. The SWISS-MODEL Server

Since the first release of the SWISS-MODEL server, the resource has evolved to reflect advances of modeling algorithms as well as Internet and web-technologies (46). The most recent version of the server is the SWISS-MODEL Workspace (47), a web-based working environment, where users can easily compute and store the results of various computational tasks required to build homology models. In particular, the Workspace gives access to software and databases necessary to complete the four main steps of comparative modeling: (1) detection of experimental structures (templates) homologous to the protein of interest (target), (2) alignment of the target and template(s) protein sequences, (3) building of one or more models for the target protein, and (4) evaluation of the quality of the obtained model(s) (Fig. 1). In the fully "Automated" mode of the SWISS-MODEL Workspace, the amino acid sequence (or the database accession code) of the protein of interest is sufficient as input to compute a structural model in a completely automated fashion. For nontrivial modeling cases, however, where the evolutionary distance between target and template is large, it is advisable to use the "Alignment" mode of the server, where a curated multiple sequence alignment of target, template, and other family members of the protein can be submitted to compute the structural model. Similarly, the "Project" mode of the SWISS-MODEL Workspace allows the user to examine and manipulate the target–template alignment in its structural context within the DeepView (Swiss-Pdb

Viewer) visualization and structural analysis tool (20). The server will then build the coordinates of the model according to the target–template alignment specified by the user.

Programs like SWISS-MODEL generate the structural coordinates of the model based on the mapping between the target residues and the corresponding amino acids of the structural template(s). Regions of the protein, for which no template information is available, typically insertions and deletions in loop regions, are built by using libraries of backbone fragments (48) or by constraint space de novo reconstruction of these backbone segments (49). Local suboptimal geometry of the obtained model, e.g., distorted bonds, angles, and close atomic contacts due to imperfect combination of fragments from structural templates, is regularized by limited energy minimization using the Gromos96 force field (50). Finally, the quality of the overall model is validated using specialized model quality estimation tools (MQE) such as ANOLEA (44) or QMEAN (51). Often when building a structural model for a specific protein, it is useful to produce several models based on alternative target–template alignments, especially if the sequences are only distantly related. The expected quality of the produced models can then be predicted to identify which has(have) the highest probability of being the most accurate. Moreover, based on hypotheses about the functional mechanisms of a protein, the visualization of key residues in their structural context may facilitate deciding which models are the most useful for the biochemical application of interest. The SWISS-MODEL Workspace offers additional tools to support the building of protein 3D-model(s) such as programs for functional and domain annotation, template identification, and structure assessment (see Subheadings 2 and 3 for details).

1.2. Protein Model Portal

The goal of Protein Model Portal (PMP) (52) of the Nature PSI Structural Biology Knowledgebase (53) is to promote the efficient use of molecular models in biomedical research. PMP provides a comprehensive view of structural information for proteins by combining information on experimental structures and theoretical models from various modeling resources. When searching the PMP, data about experimental structures are derived from the latest version of the PDB databank (54), whereas comparative models are obtained from repositories of precompiled models (36, 37). It is not feasible to regularly precompute models for all protein sequences known today, and a more suitable template may have become available for a given protein of interest since it was initially modeled. Therefore, PMP provides an interface to simultaneously submit a modeling request to several state-of-the-art modeling resources (25, 29, 55, 56) to receive a set of up-to-date models by different homology modeling programs. Using different independent methods for modeling may indicate which parts of the protein structure model are expected to be more and which to be less reliable.

In other words, regions of the protein which are consistently predicted to be similar by different independent methods are considered more likely to be correct (57). Finally to estimate the quality of the obtained models, PMP provides an interface to submit models in parallel to several model quality estimation tools, e.g., ModEval (43), ModFold (58), and QMEAN (41, 51).

In this chapter, we illustrate the use of SWISS-MODEL and PMP for automated comparative protein structure modeling for a selection of examples.

2. Material

2.1. SWISS-MODEL Workspace

2.1.1. Access to the Service

1. A computer with a web browser and connection to the Internet to access the web address of the server: http://swissmodel.expasy.org/workspace/.
2. The Java runtime environment (JRE) installed on the computer to run Astex (59) a molecular graphics program accessible on the server web site. Java is typically installed on most computers. You can get the latest version at http://java.com.

2.1.2. Software

1. The DeepView (Swiss-PdbViewer) software (v4.0) (20) downloaded and installed from http://spdbv.vital-it.ch/. Microsoft Windows and Mac versions of the program are available.
2. To learn the basic handling of the program DeepView, we recommend following Gale Rhodes' tutorial at: http://spdbv.vital-it.ch/TheMolecularLevel/SPVTut/index.html.

2.1.3. Programs Accessible Through the Server

Several tools necessary to complete the modeling task are accessible through the server, i.e., they do not require local installation on the computer.

1. Protein sequence structure and function annotation programs: InterProScan (60) for protein domain motifs and families recognition, PsiPred (61) for secondary structure prediction, DisoPred (62) for disorder prediction, and MEMSAT (63) to predict transmembrane segments.
2. Database search programs for template selection: Blast (64), Iterative Profile Blast (64), and HHsearch (65).
3. Programs for protein structure and model quality evaluation: QMEAN (41), Gromos (50), and Anolea (44) to estimate the local (per residue) accuracy of the models; DFire (45) to estimate the global quality of the models; Whatchek (66) and Procheck (67) to verify the stereochemistry of protein structures and molecular models; and DSSP (68) and Promotif (69) to evaluate structural features, such as secondary and super-secondary structures elements.

2.2. PMP

2.2.1. Access to the Service

1. A computer with a web browser installed and a connection to the internet to access the web address of the server: http://proteinmodelportal.org/.
2. The JRE installed on the computer to run Jmol (70), a viewer for chemical structures embedded in the web site. Java is typically installed on most computers. You can get the latest version at http://java.com.

2.2.2. Participating Resources

Following resources are currently participating to the PMP:

1. The PDB (54) protein structure database.
2. Comparative models providers: Center for Structures of Membrane Proteins (CSMP) (71), Joint Center for Structural Genomics (JCSG) (72), Information System for G protein-coupled receptors (GPCRDB) (73), Northeast Center for Structural Genomics (NESG) (74), New York Structural Genomics Research Consortium (NYSGRC) (75), Joint Center for Molecular Modeling (JCMM) (76), ModBase (37), and SWISS-MODEL Repository (36) databases of comparative protein structure models.
3. Interactive services for model building: ModWeb (37), M4T (29), SWISS-MODEL (47), I-Tasser (56), and HHpred (25).
4. Model quality estimation tools: ModFOLD (58), QMEAN (51), and ModEval (43).

3. Methods

Please note that the examples used in this section to describe the usage and the results obtainable from the SWISS-MODEL Workspace and PMP represent the status of the these resources at the time of writing. Different results, in general better, may be obtained at a later point since more closely related experimental template structures might become available.

3.1. SWISS-MODEL Workspace

We use the *Caulobacter crescentus* protein PopA (UniProt accession code Q9A784 (77)) to demonstrate how to use the SWISS-MODEL Workspace to generate and analyze comparative models. PopA is a paralog in *C. crescentus* of PleD, a response regulator protein which is a component of the signal transduction pathway controlling transitions between motile and sessile lifestyles in eubacteria (78). PleD catalyzes the condensation of two GTP molecules to the cyclic dinucleotide di-GMP (c-di-GMP), an ubiquitous second messenger in bacteria (79). The diguanylate cyclase activity is harbored by the GGDEF (or DGC) domain of the protein. PleD also contains two response regulatory domains, CheY-like response regulator receiver (Rec, also called D1) domains.

3.1.1. User Account

1. The SWISS-MODEL Workspace is freely accessible at http://swissmodel.expasy.org. For each user, the results of their computations are organized in a personal account, a workspace. Each calculation is stored as a "work unit" of the Workspace, displaying title and status of the computation. Work units are automatically deleted after a week, unless the storage of the results is prolonged by the user.
2. Alternatively, occasional users have the possibility to use SWISS-MODEL without the need to create a personal account by bookmarking the results pages for future reference.

3.1.2. Target Sequence Feature Annotation

Tools to analyze the sequence of a protein and predict its functional and structural characteristics can be very useful in identifying the most probable structural template(s) (see Subheading 3.1.3). These programs are accessible in the "Domain Annotation" Tools section on the Workspace (Fig. 2). It is sufficient to provide the sequence or the UniProt accession code (80) of the protein of interest and select among a list of available tools:

1. InterProScan (60) queries protein sequences against the InterPro database (81) (see Note 1). In our example, InterProScan predicts the presence of a GGDEF domain in the C-terminal region of the PopA protein and two receiver domains in the N-terminal, respectively. Details about the location in the protein of different domains and signatures are graphically displayed and links to the InterPro database provide additional information about the protein classification and documentation about the signature annotations.
2. DISOPRED (62) detects intrinsically unstructured regions in protein, i.e., segments of protein with no defined three-dimensional structure in solution (see Note 2). Disordered residues are represented by asterisks (*), whereas ordered are shown with dots (.). PopA is predicted to contain no intrinsically disordered regions.
3. MEMSAT (63) predicts regions of proteins spanning cellular membranes, indicated with "X" in the output of the program. PopA appears to not contain any transmembrane segments.
4. PsiPred (61) predicts the occurrence of secondary structure elements, such as α-helixes, extended β-strands, or coil regions, which are graphically indicated by a letter H, E, and C respectively.
5. Comparing the functional annotations of the target protein with the protein features of possible templates can help deciding if a given structure can be used as scaffold to build a comparative model. A protein with a known 3D-structure sharing the same type of domains, or having a similar secondary structure elements arrangement can indicate an evolutionary

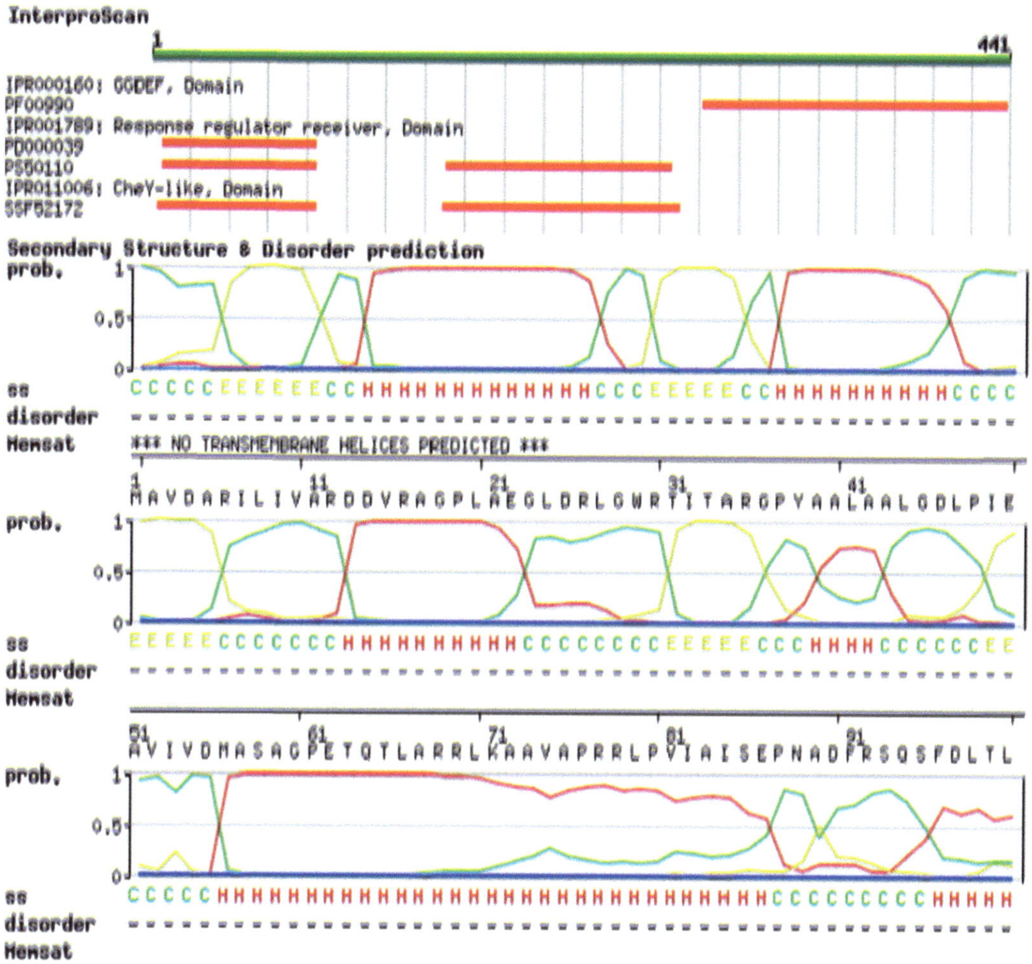

Fig. 2. SWISS-MODEL Workspace target sequence feature annotation. To predict functional and structural features of the target proteins, several annotation tools are available on the SWISS-MODEL Workspace. In this example, the *C. crescentus* PopA protein (represented as a *green bar on the top*) is predicted to contain a C-terminal GGDEF domain and two N-terminal receiver domains. The likelihood (between 0 and 1, where 1 means highest probability) of the occurrence of secondary structure elements are depicted as *curves* (*red* for α-helices, *yellow* for β-strands, and *green* for coiled regions). Prediction of disordered regions and transmembrane domains is also available. In particular, for PopA neither intrinsically unstructured regions nor portions of the protein spanning the membrane are detected.

relationship to the target protein. Indications about the presence of transmembrane domains or disordered regions are also valuable hints regarding the function and the domain architecture of the target protein and can be taken into account when evaluating if templates are available and for which region(s) of the protein of interest.

3.1.3. Template Detection

A prerequisite for building a homology model is the availability of one or more evolutionary-related proteins whose structure has been elucidated experimentally (see Note 3). For this purpose, the target protein sequence can be queried against a sequence library (SWISS-MODEL Template Library (SMTL)) extracted from known structures using increasingly sensitive search methods. The sequence (in FASTA or raw sequence format) or the corresponding UniProt AC can be submitted to the following search tools available in the Workspace "Template identification" tools section:

1. Blast (64), to detect evolutionarily closely related protein structures. Basic Blast standard parameters can be adjusted to regulate the sensitivity and the selectivity of the program (see Note 4).
2. Iterative Profile Blast (64) is used to identify more distantly related proteins (see Note 5).
3. HHSearch (65), an HMM-based profile–profile comparison tool, is a very sensitive search method to detect remotely related sequences (see Note 6).
4. A graphical synopsis of the search results is presented showing the region(s) of the related template protein(s) aligned to the query sequence. The matches are colored according to their statistical significance (Expectation- and/or Probability values, for details see Note 7), green color indicating more reliable hits. Domain boundaries according to InterPro annotations are also shown to guide the choice of suitable template with respect to functional domains. Details about the detected templates are accessible below the graphical representation, alongside with the alignment of the template sequence to the protein of interest.
5. In this example, Blast and Profile Blast template recognition tools detect three structures (PDB ID 1w25, 2wb4, and 2v0n) as possible templates for PopA. They represent structures of the paralog PleD protein in *C. crescentus* in complex with c-di-GMP, the activated form in complex with c-di-GMP and the activated form in complex with c-di-GMP and GTP-alpha-S, respectively (82, 83). HHsearch additionally detect the *Pseudomonas aeruginosa* diguanylate cyclase WspR (84) as potential template. All four structures span the full length of the target protein (see Note 8); three of them are paralogs whereas the WspR protein is an ortholog protein. Since all structures represent statistically significant hits (very low *E* values), users should decide based on templates annotations which is(are) the most suitable template(s) for building the comparative model for PopA. Typically, one would select a template with high sequence similarity (PDB IDs 1w25, 2wb4 or 2v0n (82, 83)), unless specific features are considered important for the planned application, i.e., using templates in active or inactive forms, bound to specific ligands, etc. (see Note 9).

6. If clustered versions of the templates library are searched using the template detection tools, all the structures of the same cluster can be retrieved by clicking the corresponding "show template cluster" link of the results list.

3.1.4. Target–Template Alignment

1. The target–template alignment generated by the template search tools can be used as starting point to create the correspondence between the residues of the target protein and the structure of the template, to ultimately produce the homology model. This is a critical step since standard homology modeling techniques will not recover from an incorrect input alignment, therefore special care should be addressed to this step.
2. The alignments in the output of the template identification tools can be retrieved as DeepView format file for further inspection. The file contains the target sequence aligned to the structure of the template. This allows the users to inspect the occurrence of amino acid insertions/deletions in the alignment in their structural context. For instance, it is more likely that during evolution an insertion/deletion has occurred in a flexible surface loop rather than in a well-structured secondary structure element such as an α-helix or a β-strand in the core of the structure. The alignment between target and template sequences can be modified using the DeepView program's "alignment window" and the changes visualized in the 3D environment of the structure. The "alignment window" also allows verifying if important residues of both target and template sequences (i.e., amino acids belonging to active sites) are correctly aligned. For this purpose, the DeepView function "scan for Prosite Patterns" (85) of the "Edit" menu can be applied.
3. Alternatively, pair wise or multiple sequence alignment between the target, the template and preferably related sequences, can be generated with other state-of-the-art alignments tools (see Note 10) and submitted to the server for computation of models (see Subheading 3.1.5).

3.1.5. Model Building

Three variations of the model generation step are available in Workspace: "Automated," "Alignment," and "Project" Modes. These are accessible in the "Modeling" section of the server.

1. The Automated Mode is recommended when the sequence similarity between target and template proteins is high, i.e., larger than 60%. It is sufficient to submit the target sequence (either in raw or Fasta format) and the SWISS-MODEL pipeline will select the template(s) based on a hierarchical procedure to search and select the most suitable structures (36). If several templates are available or a custom-made structure is required, the user can additionally specify to use a particular template by either indicating its PDB ID code or by uploading a file in PDB format of the structure (see Note 11).

2. The Alignment method is appropriate for more distantly related target and template sequences. Multiple sequence alignment algorithms and PSSM- or HHM-based profile–profile methods (86) will generate the reasonable alignments. However, often these alignments can be verified manually and improved using for instance, sequence alignment editors such as JalView (87). The alignment in one of the supported formats (FASTA, MSF, ClustalW, PFAM, and SELEX) can be subsequently submitted to the Workspace server. The alignment is checked for format compatibility and the user is required to indentify the sequences of the target and of the template protein and the PDB protein chain ID of the template structure (see Note 12) when submitting the alignment for the computation of models.
3. If the protein target–template sequence identity is close to the twilight zone (i.e., sequence identity below 20%) (88), particular care should be taken in manually curating the alignment between the target protein and the template structure prior computation of the comparative model. This is facilitated by the DeepView program (see Subheading 3.1.4, step 2). The target–template alignment is saved as DeepView "project file" and submitted for computation to the "Project Mode" of the server. The DeepView program also enables calculation of models using structures which are not part of the SMTL library (see Note 12).
4. Modeling of oligomeric proteins, i.e., a group of two or more associated polypeptide chains, is possible using DeepView and the "Project Mode" of the server. The prerequisite is to determine the correct quaternary structure of the template protein—which is typically not identical with the coordinates representing the asymmetric unit of a PDB entry. Prediction of the most likely biological assembly for a particular protein can be retrieved from the PISA database (89). A DeepView project file with the sequences of the homo-multimeric or hetero-multimeric protein target sequences and template structure is then created (for details see Note 13) and submitted to the server to obtain a model for the oligomeric complex.
5. After the computation of the structure for the macromolecule of interest is completed, the results are stored in a summary page of the workspace (Fig. 3) and users are notified by email.

Fig. 3. (continued) shown in this section. (**b**) Details of the target–template alignment are provided together with the secondary structure elements assignments. (**c**) Anolea (44) and Gromos energy (50) plots provide residue-based quality estimates of the model. Regions with positive energy values (*red bars*) indicate unfavorable interactions and regions of likely modeling errors. (**d**) Details about the modeling procedure are available at the end of the results. In the Automated Mode, an additional section regarding the template selection step will be shown.

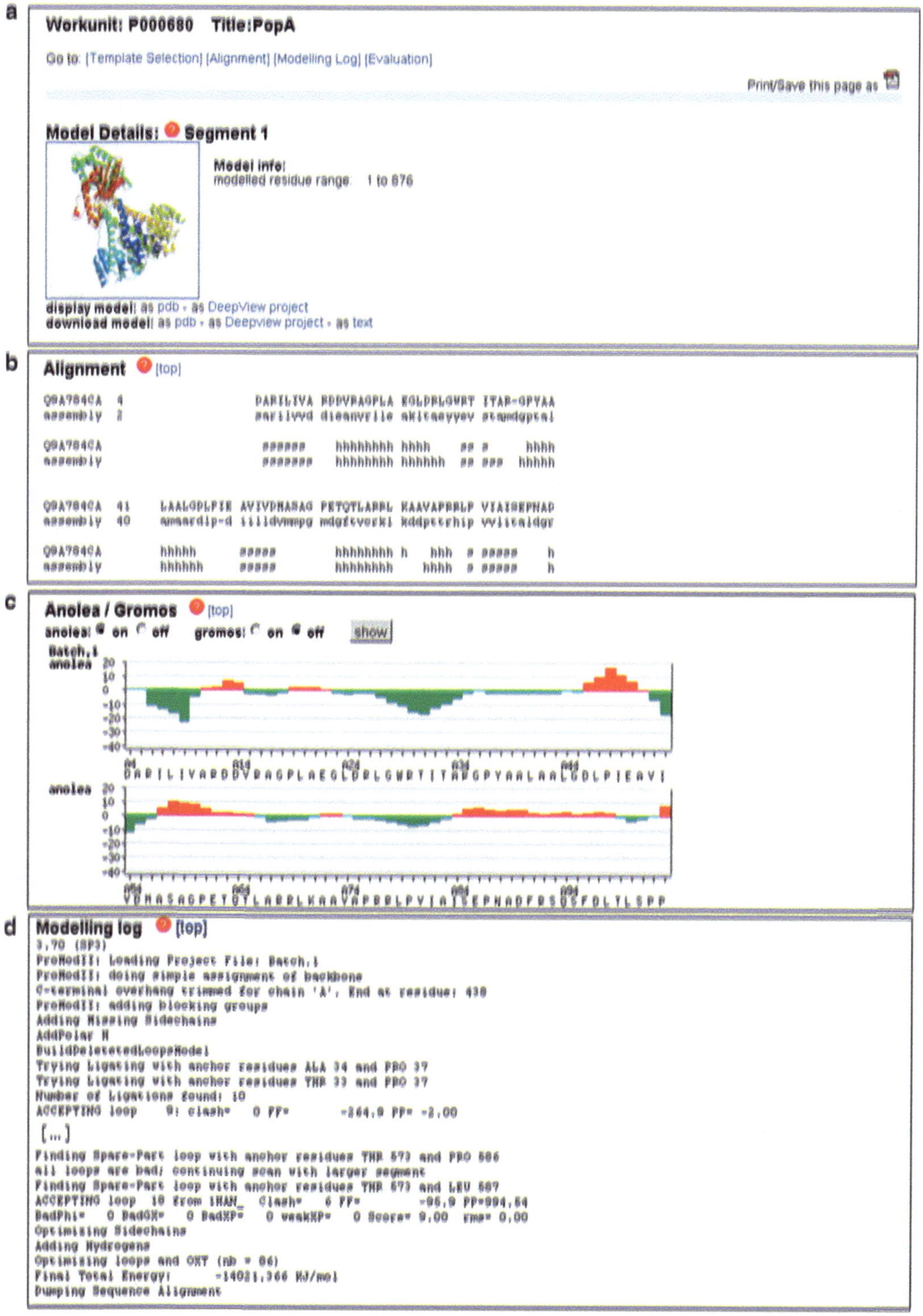

Fig. 3. Typical representation SWISS-MODEL Workspace modeling results. In this example, the *C. crescentus* PopA protein was modeled based on the structure of the paralog protein PleD (PDB ID 2wb4) using the Project Mode of the server. (**a**) The comparative model for PopA can be downloaded as PDB or DeepView project file. The model can be visualized directly on the web-page by clinking on the ribbon plot which will launch a java-based visualization tool. In the Automated Mode, additional information about the template and the statistical significance of the target–template alignment would be

6. Here we model the structure of PopA based on the structure of the activated diguanylate cyclase PleD in complex with c-di-GMP (PDB ID 2wb4). Activation of the PleD protein occurs upon phosphorylation-induced dimerization (90). For this reason, we model the structure of PopA based on the homodimer activated form of PleD. The most likely biological assembly of the template is downloaded from the PISA database (89). A DeepView project file of the target sequence aligned to the homodimeric template is created and the alignment carefully inspected. Particular attention is devoted in correctly aligning residues which constitute important functional sites, i.e., the catalytic A-site and the inhibitory I-site of the diguanylate cyclase (DGC or GGDEF) domain and the phosphor acceptor P-site in the receiver domain of both proteins (82, 91). Insertions and deletions in the target–template alignment are visually assessed in the context of the template PleD structure and also guided by the secondary structure element predictions of the target PopA sequence (see Subheading 3.1.2). Finally, the "Project file" containing the target–template alignment and the structure of the template is submitted to the server to calculate the comparative model for PopA.
7. The SWISS-MODEL Workspace's modeling results page is composed of different sections (Fig. 3). (1) In the "Model details" section, the structure of the computed macromolecule is available for download as PDB file or DeepView "Project file" for further analysis. The model can also be displayed directly from the web site by clicking on the model image which will launch the molecular graphics program Astex Viewer (59). In the fully Automated Mode, additional details are provided, i.e., the template on which the model was based (with a link to the corresponding PDB entry), the sequence identity and statistical significance of the target–template alignment (see Note 7). (2) The "Alignment" section contains the details of target–template alignment including secondary structure element assignments. (3) Estimation of model quality based on Anolea (44) and Gromos (50) is available as residue based graphical plot, to indicate parts of the model with unfavorable interactions. (4) Technical modeling details are accessible in the "Modeling Log" section. (5) If the "Automated" mode is applied, an additional "Template Selection Log" is present in the results section, providing information about the template selection step performed to search the SMTL for suitable templates.

3.1.6. Model Quality Estimation

Finally the quality of the obtained model(s) can be assessed and estimated using the programs available in the "Structure assessment" tools section of the Workspace. A list of quality estimation algorithms and programs to verify the structural quality of proteins can be applied to the obtained models. We distinguish between programs to predict the local (per residue) and the global expected accuracy of the computed models (see Subheading 2.1.3) and tools to verify the structure of the calculated models, e.g., structure geometries, packing quality, most probable side chain conformations, etc.

1. We analyze the quality of the homology model for PopA using QMEAN (41, 51) and Anolea (44) tools. The QMEAN scoring function estimates the local structural error at a given position in the protein. Regions in the model with low associated values are expected to be more reliably predicted. Anolea calculates pseudo energies based on potentials of mean force. Negative energy values indicate regions of the protein with favorable interatomic interactions. The sequence identity (~22%) between PopA and the template structure of PleD is close to the twilight zone of sequence alignments. For this reason is not surprising that the expected quality of some regions of the model is not high. However, we verified that functional important sites of the protein, e.g., the P- A-, and I-sites were better modeled than other loop regions of the protein (Fig. 4b).
2. The QMEAN *Z*-score is a quality estimate which relates structural features observed in a model to their expected distributions based on statistics for experimental protein structures of comparable size (54, 92). QMEAN *Z*-scores are normalized such that more positive values represent better model quality. Based on this measure, the quality of the obtained model for PopA of −1.59 lies within the expected range and is comparable to a medium resolution experimental structure (Fig. 4a).
3. We validate the predicted structure of PopA using the program Procheck (67). The analysis reveals a satisfactory quality of the model structure, e.g., in the Ramachandran plot (93) 91.1% of the PopA residues occupy the most favored regions, with only seven residues in disallowed areas of the plot.
4. Finally regions of the comparative models containing errors or of low quality can be further inspected and the corresponding segments in the target–template alignment adjusted to create a new model. The process (see Fig. 1) can be iterated until satisfactory results are obtained. This is facilitated by the use of the DeepView project files downloadable from the modeling results web site.

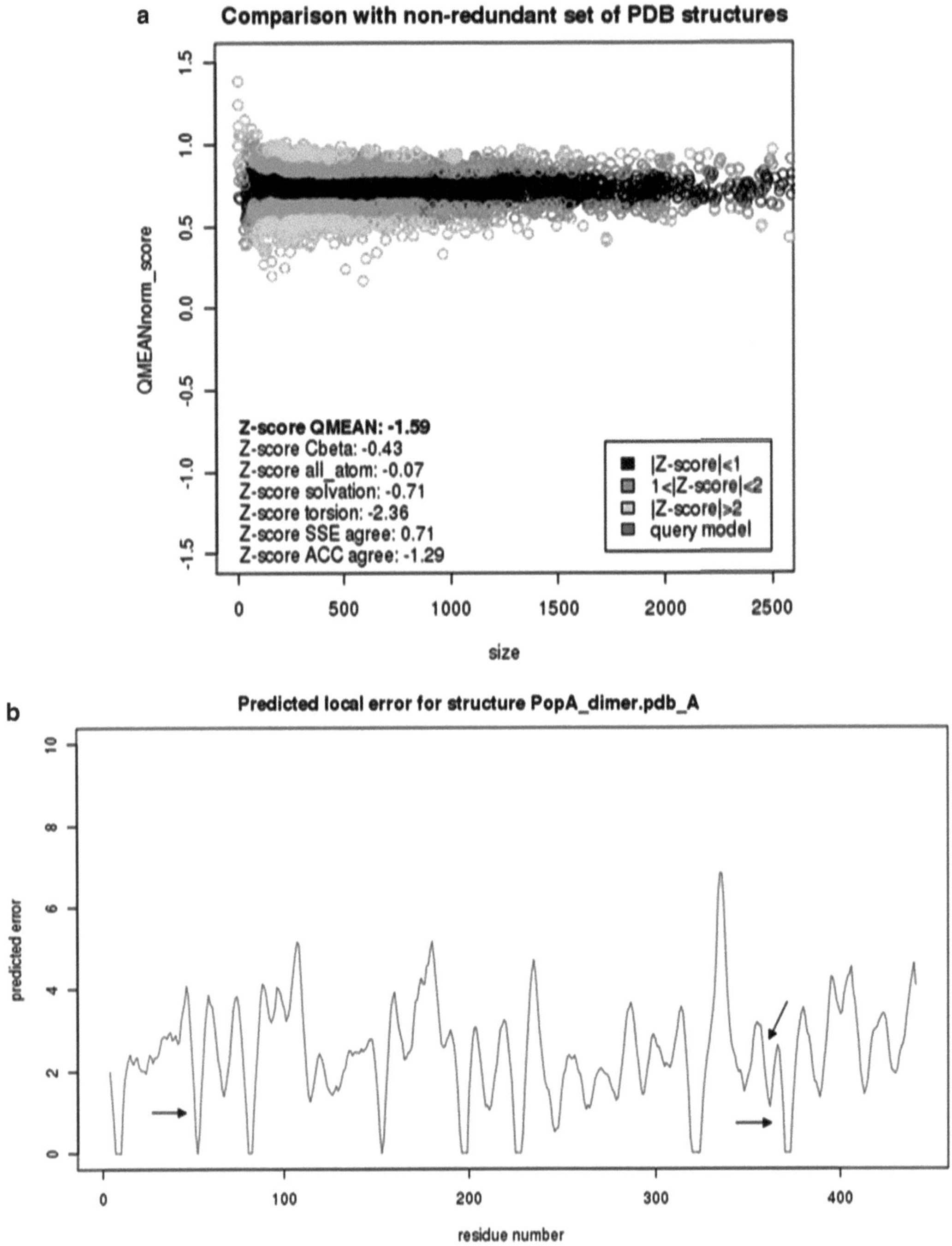

Fig. 4. Examples of SWISS-MODEL Workspace model quality estimation plots calculated using QMEAN. (**a**) The global estimated energy of the PopA model (*grey cross* in this figure and displayed as *red cross* in the online results of the server) is compared to the QMEAN energy estimates (51, 92) for a nonredundant set of high-quality experimental protein crystal structures of similar length, and their deviation from the expected distributions is represented as *Z*-scores. The QMEAN quality estimate for PopA lies within the expected range for models of this type and is comparable to a medium resolution experimental structure. (**b**) Local (per residue) plot of the QMEAN predicted errors for PopA. QMEAN scores for important functional sites (phosphorilation-, activation-, and inhibitory sites, respectively) are depicted as *arrows*, indicating that the local environment of these regions is not located in problematic segments of the predicted structure.

3.2. PMP

To illustrate how to access functional and structural information for a given protein using the PMP, we will use the example of the human Myeloid cell nuclear differentiation antigen protein (MNDA, UniProt accession code P41218). The MNDA protein is suggested to play a role in the granulocyte/monocyte cell-specific response to interferon (94–96).

3.2.1. Search Options

1. PMP can be queried by submitting the entire amino acid sequence of a protein or a fragment of it. UniProt (80) proteins with identical or very similar sequences will be identified and listed.
2. The portal can also be searched by database identifiers (e.g., UniProt, RefSeq (97), IPI (98), gi (99), Entrez (100)), or by keyword suggestions (e.g., “kinase”).
3. Models built based on a specific template structure can also be retrieved by entering either PDB accession codes (54) or structural genomics targets identifiers (101).

3.2.2. Results of the PMP Query

1. The results of the query are presented in a summary page (Fig. 5) with a graphical representation of the regions of the protein where structural information is available. Additionally functional annotation derived from UniProt and InterPro (81) (see Note 1) is provided. For the MNDA protein, an experimental protein structure exists for the N-terminal Pyrin domain (PDB ID 2DBG (102)), a putative protein–protein interaction domain (103). Whereas for the C-terminal domain of unknown function, three protein structure models have been precomputed by model resources accessible via PMP.
2. The graphical illustration of the matches is followed by a detailed list of the obtainable structural models for the protein of interest. Experimental protein structures in the PDB with more than 90% sequence identity to the target protein, are reported, if available.
3. Three models have been built for the MNDA protein by three resources accessible through the portal: ModBase (55), SWISS-MODEL Repository (36), and NESG (104). Each single model is tagged with a color coded (“traffic lights”) as first indication about its reliability. In this example, the models are based on a target–template alignment of about 60% sequence identity. Typically, models based on a target–template sequence alignment of this degree of similarity are largely correct (7, 105, 106). Search results can be sorted based on different attributes, e.g., models provider, template identifier, target–template percentage of sequence identity and region of the target covered.

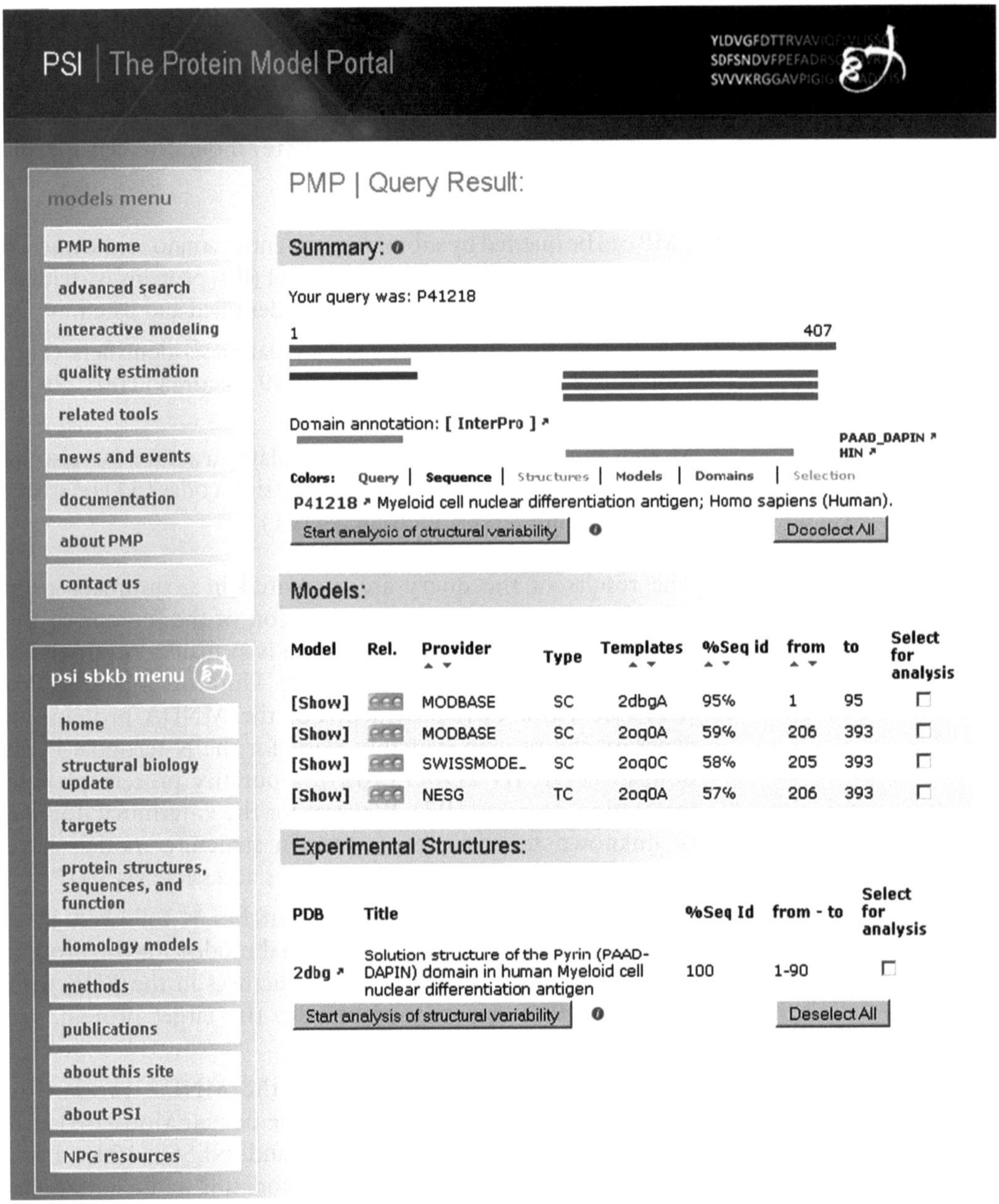

Fig. 5. Protein Model Portal (PMP) query results for the human myeloid cell nuclear differentiation antigen protein (UniProt P41218 (94, 95), upper bar numbered from 1 to 407). For the first 90 residues of this protein, an experimentally solved structure (*light grey bar* in this figure and displayed as a green bar in the online results of the server) is deposited in the PDB database (PDB ID 2dbg (102)). The protein structure corresponds to the PPAD_DAPIN N-terminal domain of the protein. For the C-terminal HIN domain, three homology models are obtainable from the PMP model providers ModBase, SWISS-MODEL, and NESG. Below the graphical representation a list of models and information about the structure is available. Additional information is accessible by clicking the corresponding model or PDB ID links. A subset of models or structures can be selected for further structural comparison.

4. For each model, the "Model Details" page provides further information (Fig. 6) about (1) the range of the modeled region, (2) the template used, (3) the target–template alignment the model was based on, (4) when the model was first created and verified, (5) the expected quality of the model, (6) a link to submit the model to quality estimation services, and (7) the URL to the model database to download the model coordinates file. The protein structure models can also be visualized using the web browser applet Jmol (70).
5. In case the model has not been updated for a while a sign warns that new structures may have become available which would allow building a more reliable model. The target protein can be submitted directly to the interactive modeling services to compute models based on the most recent templates library (Fig. 6). In our example, some models have not been updated for a while and some regions exist for which structural information is not available, it is worthwhile triggering a new round of calculations. As of 11 November 2010, the results of interactive modeling show that there are no new templates that could be used instead of 2OQ0 (107) to reliably model the C-terminal domain.

3.2.3. Protein Model and Structure Comparison

Models submitted by the different participating sites have been generated using various algorithmic approaches with different strengths and weaknesses. Also the quality of individual models highly depends on the evolutionary proximity to the selected structural templates. Finally, experimental structures may show structural variation due to domain motions, mobile loops, induced fit, etc. For these reasons, in the results page models and experimental structures spanning a common range can be selected to analyze their structural variability (Fig. 7a).

1. Differences within the ensemble of models and experimental structures can be identified using a matrix that shows the deviations of Cα distances of the collection of models (Fig. 7b).
2. In particular for each model or structure, regions of the protein that deviate more from the ensemble are shown in a plot (Fig. 7c).
3. The details of the superposed structures can also be visualized in page using Jmol (70) (Fig. 7d).

Whereas for the N-terminal domain of MNDA an experimental structure has been solved, for the C-terminal domain three structural models are available. As mentioned before the accuracy for these models are expected to be high and since all resources used the same template, the structural variations among them is

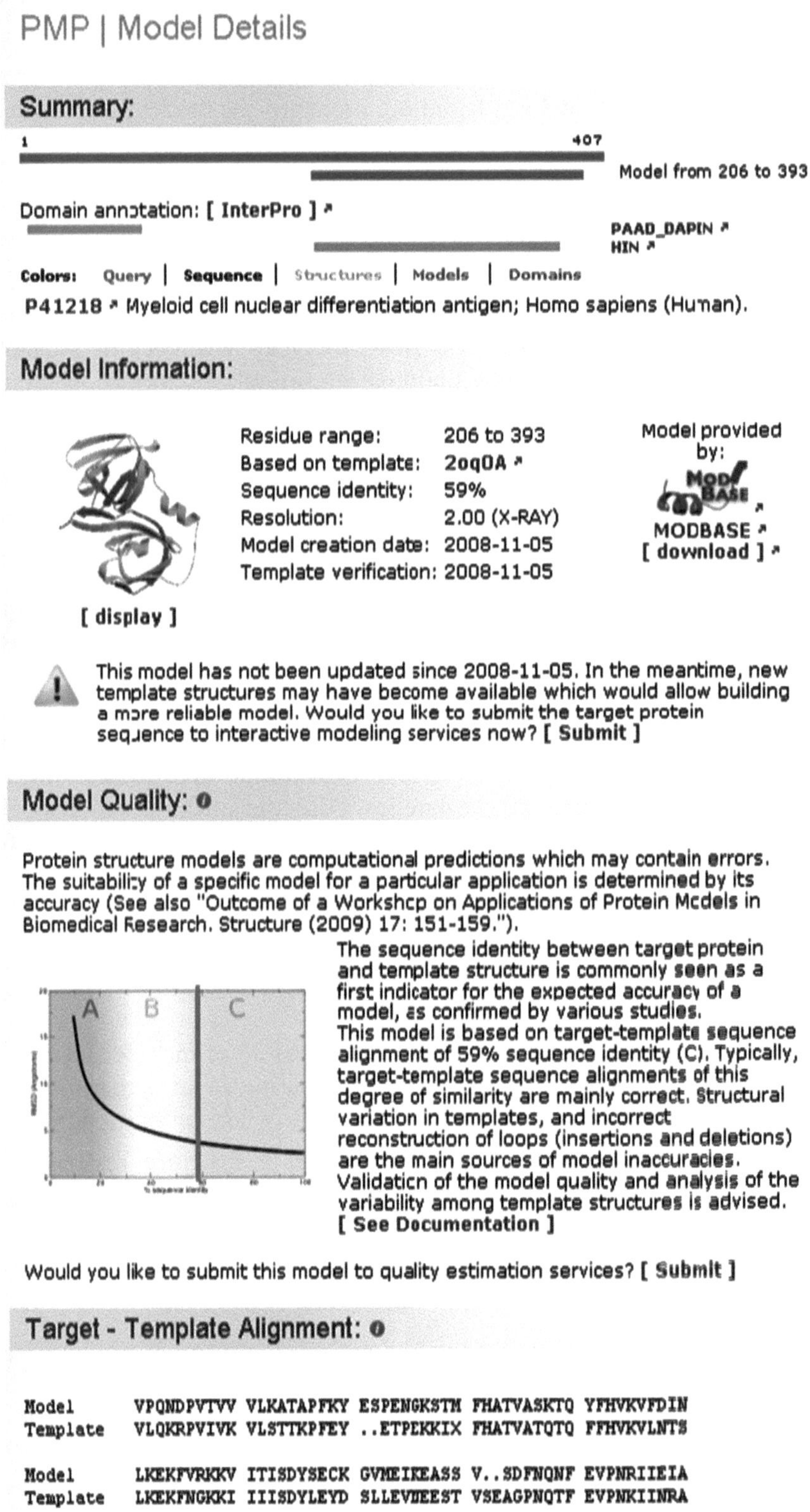
PMP | Model Details
Summary:
1
407
Model from 206 to 393
Domain annotation: [InterPro]
PAAD_DAPIN
HIN
Colors: Query | Sequence | Structures | Models | Domains
P41218 Myeloid cell nuclear differentiation antigen; Homo sapiens (Human).
Model Information:
Residue range: 206 to 393
Based on template: 2oq0A
Sequence identity: 59%
Resolution: 2.00 (X-RAY)
Model creation date: 2008-11-05
Template verification: 2008-11-05
Model provided by:
MODBASE
[download]
[display]
This model has not been updated since 2008-11-05. In the meantime, new template structures may have become available which would allow building a more reliable model. Would you like to submit the target protein sequence to interactive modeling services now? [Submit]
Model Quality:
Protein structure models are computational predictions which may contain errors. The suitability of a specific model for a particular application is determined by its accuracy (See also "Outcome of a Workshop on Applications of Protein Models in Biomedical Research. Structure (2009) 17: 151-159.").
A
B
C
The sequence identity between target protein and template structure is commonly seen as a first indicator for the expected accuracy of a model, as confirmed by various studies.
This model is based on target-template sequence alignment of 59% sequence identity (C). Typically, target-template sequence alignments of this degree of similarity are mainly correct. Structural variation in templates, and incorrect reconstruction of loops (insertions and deletions) are the main sources of model inaccuracies. Validation of the model quality and analysis of the variability among template structures is advised.
[See Documentation]
Would you like to submit this model to quality estimation services? [Submit]
Target - Template Alignment:
Model VPQNDPVTVV VLKATAPFKY ESPENGKSTM FHATVASKTQ YFHVKVFDIN
Template VLQKRPVIVK VLSTTKPFEY ..ETPEKKIX FHATVATQTQ FFHVKVLNTS
Model LKEKFVRKKV ITISDYSECK GVMEIKEASS V..SDFNQNF EVPNRIIEIA
Template LKEKFNGKKI IIISDYLEYD SLLEVHEEST VSEAGPNQTF EVPNKIINRA

expected to be low (Fig. 7). Some minor deviations are in fact observed around residues 230, 260, and 380 corresponding to loops region of the protein (Fig. 7d) which have been modeled differently by the various modeling servers.

3.2.4. Interactive Modeling

Model accuracy crucially depends on the availability of suitable template structures. Model repositories contain precompiled models based on the best available templates at the time of modeling. However, in the meantime better templates might have been released, which would allow for producing a higher quality model. Therefore, PMP provides a service interface (called "Interactive Modeling") where to submit target protein sequences to several established modeling services (29, 47, 55, 56, 108) and initiate a new template selection and modeling process for the protein of interest. Depending on the type of resource, protein structure models coordinate files are either sent as attachment to an e-mail or can be retrieved via the corresponding service website.

For the region of MNDA spanning residues ~90–200, at the time of writing there was no precomputed structural information available through PMP, however when submitting the target sequence to the interactive modeling services, ModWeb server calculates a new model structure based on template 3na7 (109) spanning residues 62–157. The sequence identity of the alignment used to build the model is relatively low (27%) and the results should be taken with caution and further analyzed by quality estimation tools.

3.2.5. Quality Estimation Resources

Various model quality estimation tools have been developed by the community to analyze different structural features of protein models to judge the correctness of structural predictions.

1. The accuracy of a precomputed model can be estimated using state-of-the-art model quality estimation tools (43, 51, 58), directly from the "Model Details" page.
2. Alternatively, any coordinate file (PDB format; see Note 11) can be submitted to the "Quality estimation" interface of the portal.

The three models generated for the C-terminal domains of the MNDA protein are estimated to be mainly correct with a medium

Fig. 6. PMP model details. For each model, target–template sequence identity, experimental annotation regarding the template, and cross-references to the model provider is available. A link allows users to automatically submit the protein sequence to interactive modeling servers for generating an updated prediction. The sequence alignment between the target and the template sequences is indicated, and a plot of the evolutionary distance between target and template gives an estimate about the expected accuracy of the model. Specialized model quality estimation tools can be automatically invoked for the model at hand to provide a more in depth assessment.

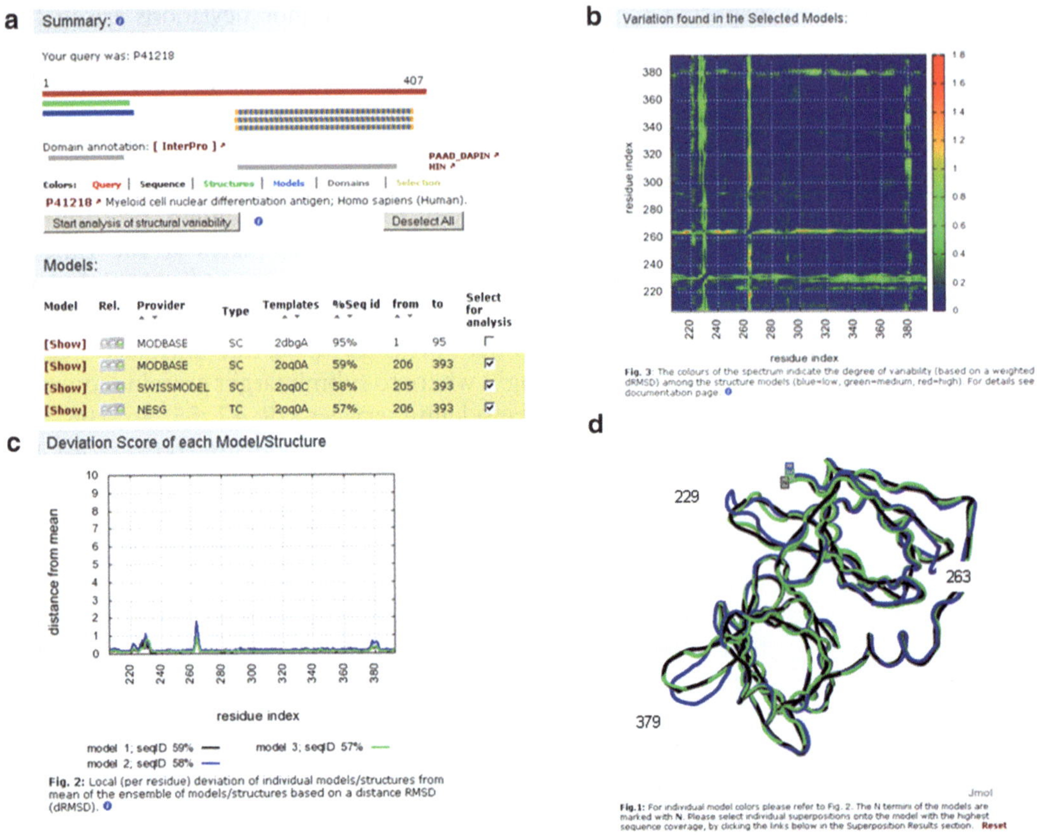

Fig. 7. PMP structure comparison results. Structural differences can be analyzed in case several structures or models are available for the same region of a protein. (**a**) The comparative models available for the C-terminal domain of the myeloid cell nuclear differentiation antigen protein were compared. A subset of models or structures can be selected either by clicking the corresponding bars in the graphical synopsis or by checking the boxes of the lists. (**b**) A two-dimensional matrix indicates which regions of the analyzed structures deviate most among each others (*blue*= low, *green*= medium, and *red*= high variability). For the comparative models of the antigen protein, these regions are located around residues 230, 260, and 380. (**c**) The *plot* shows the magnitude of the deviation (residue based) of individual models (or structures) from the mean of the ensemble of the analyzed macromolecules. (**d**) The variability among models or structures can be visualized as structural superposition. In *plots* (**c**) and (**d**) each comparative model is represented by a different color (*black*= ModBase, *blue*= SWISS-MODEL, and *green*= NESG models). As expected, regions of the models showing small differences around residues 230, 260, and 380 of the antigen protein are located in loop regions on the surface of the protein, which were reconstructed differently by the various modeling methods.

to high-quality scores especially for the β barrels core parts of the structure (Fig. 8). On the contrary, the model for the region spanning residues ~90–200 belongs to the low to bad quality range as expected for target–template sequence alignments below 30% sequence identity.

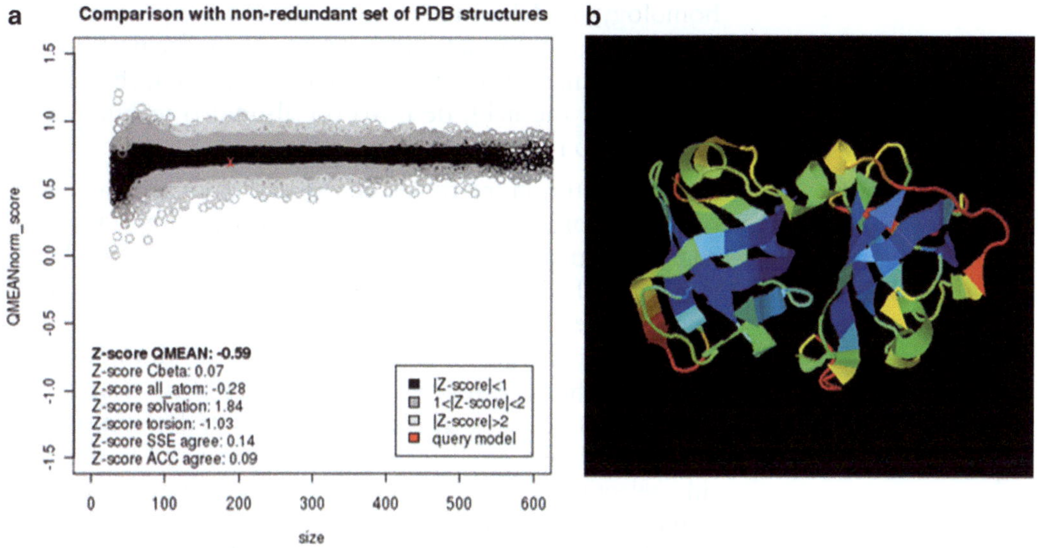

Fig. 8. Model quality estimation. The quality of the model of the C-terminal domain of the myeloid cell nuclear differentiation antigen protein was analyzed using one of the tools accessible from the PMP portal, the QMEAN scoring function. (**a**) The global estimated energy of the antigen protein (*red cross*) is compared to the QMEAN energy estimates (51, 92) for a nonredundant set of high-quality experimental protein crystal structures of similar length, and their deviation from the expected distributions is represented as *Z*-scores. The QMEAN quality estimate for a C-terminal model (Fig. 6) lies within 0–1 standard deviations from the mean values, suggesting overall a very good expected quality for this model, comparable to experimental structures. (**b**) The QMEAN method also allows predicting expected errors on a per residue basis. The model is colored according to the QMEAN score where *blue* regions represent regions predicted as reliable and *red* as potentially unreliable, respectively.

4. Notes

1. InterPro is a collection of protein "signatures" used for the classification and automatic annotation of proteins. InterPro classifies sequences at superfamily, family, and subfamily levels and predicts the occurrence of functional domains, repeats, and functional sites.

2. Intrinsically disordered regions in proteins have been associated with important biological functions involved for instance in cellular signaling and transcription regulation (110). Disordered regions often interfere with crystallization and are, therefore, typically missing in experimental structures (unless in complex with other partners). Attempts to model intrinsically disordered regions using comparative techniques are therefore in most cases not such a good idea.

3. In case no evolutionary-related template(s) for a given target protein can be found, it is not possible to reliably build a 3D structure model of this protein based on comparative/

homology modeling techniques. De novo approaches (i.e., without using information from homologous templates) may be applied instead. However, it should be noted that despite advances in the field, de novo (or ab initio) techniques are restricted to relatively small proteins.

4. The "substitution matrix" is one of the important parameters of Blast/Profile Blast algorithms. The matrix allows evaluating and calculating the score of two aligned protein (or DNA) sequences. Different substitution matrixes have been specifically designed to change the scope and tune sequence database search. In particular, the choice of the substitution matrix influences the sensitivity vs. the selectivity of the search. The sensitivity of a query is defined as the ability of detecting remote homologs, but possibly including false matches. On the other side, selectivity ensures a more stringent search minimizing the number of false positives, at the cost of missing some true homologs. In particular, for the BLOSUM type of substitution matrices, a higher index (e.g., BLOSUM 80) indicates a more selective type of search, whereas a lower index (e.g., BLOSUM 45) will results in a more sensitive query. For more information, see the BLAST documentation on the NCBI server (111).
5. Profile Blast consist of two main steps, in the first one a profile is constructed from closely related sequences detected by a standard Blast search against a nonredundant protein sequence database. The profile is a representation of the group of aligned homologous sequences. This step can be iterated to extend the profile with new, more distantly related sequences. In the second step, the profile is used to perform a Blast search of the SMTL sequence library to look for related proteins with known structure. The parameters of both steps can be adjusted to shift the balance between selectivity and sensitivity of the search (see Note 4).
6. In HMM–HMM-based alignment tools, both the query sequence and the sequences in the library are represented as HMM-based profiles. Therefore, the search is usually done against a culled version of the PDB database library, i.e., structures with similar sequences (e.g., 70% sequence identity) are clustered together.
7. In sequence database searches, the *E*- (or expected) value associated with the results indicates the statistical significance of a given match (or hit). Each match is associated with a score (*S*), with higher scores indicating better results. The *E* value estimates the probability of obtaining by chance a number of matches with this score (*S*) in a database of a particular size. In other words, the closer the *E* value is towards 0, the more

significant the alignment (between the query and the sequence found in the database) is. Similarly, the *P* (or probability) value describes the probability that an alignment with this score (*S*) occurs by chance in a database of this size. The closer the *P* value is towards 0, the better the alignment is.

8. In the best case scenario, one would detect a statistical significant template covering the entire length of the protein of interest. Very often, however, templates spanning only part of the query protein are detected. In this case, it is advisable to try to increase the sensitivity of the template detection methods, by additionally searching only those regions of the protein for which no templates were detected. Often, several noncontinuous structural templates are detected which allow to model the target protein in separate fragments. Prediction of the relative orientation of isolated domains with comparative modeling methods is only feasible if (a) one of the templates contains significant overlap with both domains and (b) their relative orientation is structurally well conserved.
9. The selection of the most suitable template should take into account not only the sequence similarity to the target protein, but also consider the quality of the experimental structure (e.g., resolution of the experimental technique), ligand molecules which may influence the local conformation of biding sites, or alternative conformations indicating structural variability observed within the protein family.
10. The development of sequence alignment algorithms is an active field of research in bioinformatics. For a (non-exhaustive) list of alignment tools employed in the field of protein structure prediction, see ref. 86.
11. A simple PDB-like file containing the coordinates of the template structure. For more information about PDB file format, refer to the corresponding documentation on the wwPDB website (112).
12. Please make sure when submitting a multiple sequence alignment that the names of the proteins specified in the alignment contain only alphanumerical characters. Use short names for the proteins (e.g., "Q9A784," "PopA_CAUCR," 2wb4) and verify that the alignment contains the sequence of the structure template. The selected template should be part of the SMTL library (see "Template library" Tools section of the server.)
13. A step by step tutorial how to use DeepView for oligomeric protein modeling is provided on the SWISS-MODEL server web site (http://swissmodel.expasy.org/) and (113).

Acknowledgments

The authors thank Konstantin Arnold for his dedicated support of the SWISS-MODEL service, Jürgen Haas for his commitment to new developments in PMP, and all members of the group for fruitful discussions.

Funding: The development and operation of SWISS-MODEL was supported by the SIB Swiss Institute of Bioinformatics; The PMP of the Nature PSI Structural Biology Knowledgebase was supported by the National Institutes of Health NIH as a subgrant with Rutgers University, under Prime Agreement Award Numbers: 3U54GM074958-04S2 and 1U01 GM093324-01.

References

1. Schwede, T., A. Sali, N. Eswar, and M.C. Peitsch, *Protein Structure Modeling.*, in *Computational Structural Biology*, T. Schwede and M.C. Peitsch, Editors. 2008, World Scientific Singapore. p. 3–35.
2. Baker, D. and A. Sali. (2001) Protein structure prediction and structural genomics. *Science.* **294**, 93–96.
3. Sali, A. and T.L. Blundell. (1993) Comparative protein modeling by satisfaction of spatial restraints. *J Mol Biol.* **234**, 779–815.
4. Sutcliffe, M.J., I. Haneef, D. Carney, and T.L. Blundell. (1987) Knowledge based modeling of homologous proteins, Part I: Three-dimensional frameworks derived from the simultaneous superposition of multiple structures. *Protein Eng.* **1**, 377–384.
5. Peitsch, M.C. (1996) ProMod and Swiss-Model: Internet-based tools for automated comparative protein modeling. *Biochem Soc Trans.* **24**, 274–279.
6. Fiser, A. Template-based protein structure modeling. *Methods Mol Biol.* **673**, 73–94.
7. Moult, J. (2005) A decade of CASP: progress, bottlenecks and prognosis in protein structure prediction. *Curr Opin Struct Biol.* **15**, 285–289.
8. Arinaminpathy, Y., E. Khurana, D.M. Engelman, and M.B. Gerstein. (2009) Computational analysis of membrane proteins: the largest class of drug targets. *Drug Discov Today.* **14**, 1130–1135.
9. Schwede, T., A. Sali, B. Honig, M. Levitt, et al. (2009) Outcome of a workshop on applications of protein models in biomedical research. *Structure.* **17**, 151–159.
10. Peitsch, M.C. (2002) About the use of protein models. *Bioinformatics.* **18**, 934–938.
11. Tramontano, A., *The biological applications of protein models.*, in *Computational Structural Biology*, T. Schwede and M.C. Peitsch, Editors. 2008, World Scientific Publishing. p. 111–127.
12. Junne, T., T. Schwede, V. Goder, and M. Spiess. (2006) The plug domain of yeast Sec61p is important for efficient protein translocation, but is not essential for cell viability. *Mol Biol Cell.* **17**, 4063–4068.
13. Grant, M.A. (2009) Protein structure prediction in structure-based ligand design and virtual screening. *Comb Chem High Throughput Screen.* **12**, 940–960.
14. Takeda-Shitaka, M., D. Takaya, C. Chiba, H. Tanaka, et al. (2004) Protein structure prediction in structure based drug design. *Curr Med Chem.* **11**, 551–558.
15. Das, R. and D. Baker. (2009) Prospects for de novo phasing with de novo protein models. *Acta Crystallogr D Biol Crystallogr.* **65**, 169–175.
16. Giorgetti, A., D. Raimondo, A.E. Miele, and A. Tramontano. (2005) Evaluating the usefulness of protein structure models for molecular replacement. *Bioinformatics.* **21 Suppl 2**, ii72–76.
17. Topf, M., M.L. Baker, M.A. Marti-Renom, W. Chiu, et al. (2006) Refinement of protein structures by iterative comparative modeling and CryoEM density fitting. *J Mol Biol.* **357**, 1655–1668.
18. Topf, M. and A. Sali. (2005) Combining electron microscopy and comparative protein structure modeling. *Curr Opin Struct Biol.* **15**, 578–585.
19. Zhu, J., L. Cheng, Q. Fang, Z.H. Zhou, et al. Building and refining protein models within

cryo-electron microscopy density maps based on homology modeling and multiscale structure refinement. *J Mol Biol.* **397**, 835–851.

20. Guex, N., M.C. Peitsch, and T. Schwede. (2009) Automated comparative protein structure modeling with SWISS-MODEL and Swiss-PdbViewer: a historical perspective. *Electrophoresis.* **30 Suppl 1**, S162–173.
21. Brazas, M.D., J.T. Yamada, and B.F. Ouellette. (2010) Providing web servers and training in Bioinformatics: 2010 update on the Bioinformatics Links Directory. *Nucleic Acids Res.* **38 Suppl**, W3–6.
22. Battey, J.N., J. Kopp, L. Bordoli, R.J. Read, et al. (2007) Automated server predictions in CASP7. *Proteins.* **69**, 68–82.
23. Pieper, U., B.M. Webb, D.T. Barkan, D. Schneidman-Duhovny, et al. (2011) ModBase, a database of annotated comparative protein structure models, and associated resources. *Nucleic Acids Res.* **39**, D465–474.
24. Chivian, D. and D. Baker. (2006) Homology modeling using parametric alignment ensemble generation with consensus and energy-based model selection. *Nucleic Acids Res.* **34**, e112.
25. Hildebrand, A., M. Remmert, A. Biegert, and J. Soding. (2009) Fast and accurate automatic structure prediction with HHpred. *Proteins.* **77 Suppl 9**, 128–132.
26. Zhang, Y. (2008) I-TASSER server for protein 3D structure prediction. *BMC Bioinformatics.* **9**, 40.
27. Larsson, P., M.J. Skwark, B. Wallner, and A. Elofsson. Improved predictions by Pcons.net using multiple templates. *Bioinformatics.* **27**, 426–427.
28. Kelley, L.A. and M.J. Sternberg. (2009) Protein structure prediction on the Web: a case study using the Phyre server. *Nat Protoc.* **4**, 363–371.
29. Fernandez-Fuentes, N., C.J. Madrid-Aliste, B.K. Rai, J.E. Fajardo, et al. (2007) M4T: a comparative protein structure modeling server. *Nucleic Acids Res.* **35**, W363–368.
30. Schneidman-Duhovny, D., M. Hammel, and A. Sali. (2011) Macromolecular docking restrained by a small angle X-ray scattering profile.*J Struct Biol* **173**, 461–471.
31. Vroling, B., M. Sanders, C. Baakman, A. Borrmann, et al. GPCRDB: information system for G protein-coupled receptors. *Nucleic Acids Res.* **39**, D309–319.
32. Zhang, Y., M.E. Devries, and J. Skolnick. (2006) Structure modeling of all identified G protein-coupled receptors in the human genome. *PLoS Comput Biol.* **2**, e13.
33. Marcatili, P., A. Rosi, and A. Tramontano. (2008) PIGS: automatic prediction of antibody structures. *Bioinformatics.* **24**, 1953–1954.
34. Sivasubramanian, A., A. Sircar, S. Chaudhury, and J.J. Gray. (2009) Toward high-resolution homology modeling of antibody Fv regions and application to antibody-antigen docking. *Proteins.* **74**, 497–514.
35. Schwede, T., A. Diemand, N. Guex, and M.C. Peitsch. (2000) Protein structure computing in the genomic era. *Res Microbiol.* **151**, 107–112.
36. Kiefer, F., K. Arnold, M. Kunzli, L. Bordoli, et al. (2009) The SWISS-MODEL Repository and associated resources. *Nucleic Acids Res.* **37**, D387–392.
37. Pieper, U., B.M. Webb, D.T. Barkan, D. Schneidman-Duhovny, et al. (2011) ModBase, a database of annotated comparative protein structure models, and associated resources. *Nucleic Acids Res* **39**, D465–D474.
38. Koh, I.Y., V.A. Eyrich, M.A. Marti-Renom, D. Przybylski, et al. (2003) EVA: Evaluation of protein structure prediction servers. *Nucleic Acids Res.* **31**, 3311–3315.
39. Chothia, C. and A.M. Lesk. (1986) The relation between the divergence of sequence and structure in proteins. *Embo J.* **5**, 823–826.
40. Peng, J. and J. Xu. (2010) Low-homology protein threading. *Bioinformatics.* **26**, i294–300.
41. Benkert, P., S.C. Tosatto, and T. Schwede. (2009) Global and local model quality estimation at CASP8 using the scoring functions QMEAN and QMEANclust. *Proteins.* **77 Suppl 9**, 173–180.
42. McGuffin, L.J. and D.B. Roche. (2010) Rapid model quality assessment for protein structure predictions using the comparison of multiple models without structural alignments. *Bioinformatics.* **26**, 182–188.
43. Eramian, D., N. Eswar, M.Y. Shen, and A. Sali. (2008) How well can the accuracy of comparative protein structure models be predicted? *Protein Sci.* **17**, 1881–1893.
44. Melo, F. and E. Feytmans, *Scoring Functions for Protein Structure Prediction.* Computational Structural Biology, ed. T. Schwede and M.C. Peitsch. 2008: World Scientific Publishing.
45. Zhou, H. and Y. Zhou. (2002) Distance-scaled, finite ideal-gas reference state improves structure-derived potentials of mean force for structure selection and stability prediction. *Protein Sci.* **11**, 2714–2726.
46. Guex, N. and M.C. Peitsch. (1997) SWISS-MODEL and the Swiss-PdbViewer: an

environment for comparative protein modeling. *Electrophoresis.* **18**, 2714–2723.

47. Arnold, K., L. Bordoli, J. Kopp, and T. Schwede. (2006) The SWISS-MODEL workspace: a web-based environment for protein structure homology modeling. *Bioinformatics.* **22**, 195–201.
48. Zhang, Y. and J. Skolnick. (2005) The protein structure prediction problem could be solved using the current PDB library. *Proc Natl Acad Sci U S A.* **102**, 1029–1034.
49. Peitsch, M.C. (1995) Protein modeling by E-Mail. *BioTechnology.* **13**, 658–660.
50. van Gunsteren, W.F., S.R. Billeter, A.A. Eising, P.H. Hünenberger, et al., *Biomolecular Simulations: The GROMOS96 Manual and User Guide.* 1996, Zürich: VdF Hochschulverlag ETHZ.
51. Benkert, P., M. Kunzli, and T. Schwede. (2009) QMEAN server for protein model quality estimation. *Nucleic Acids Res.* **37**, W510–514.
52. Arnold, K., F. Kiefer, J. Kopp, J.N. Battey, et al. (2009) The Protein Model Portal. *J Struct Funct Genomics.* **10**, 1–8.
53. Berman, H.M., J.D. Westbrook, M.J. Gabanyi, W. Tao, et al. (2009) The protein structure initiative structural genomics knowledgebase. *Nucleic Acids Res.* **37**, D365–368.
54. Berman, H., K. Henrick, H. Nakamura, and J.L. Markley. (2007) The worldwide Protein Data Bank (wwPDB): ensuring a single, uniform archive of PDB data. *Nucleic Acids Res.* **35**, D301–303.
55. Pieper, U., B.M. Webb, D.T. Barkan, D. Schneidman-Duhovny, et al. (2011) ModBase, a database of annotated comparative protein structure models, and associated resources. *Nucleic Acids Res.* D465–474.
56. Roy, A., A. Kucukural, and Y. Zhang. (2010) I-TASSER: a unified platform for automated protein structure and function prediction. *Nat Protoc.* **5**, 725–738.
57. Ginalski, K., A. Elofsson, D. Fischer, and L. Rychlewski. (2003) 3D-Jury: a simple approach to improve protein structure predictions. *Bioinformatics.* **19**, 1015–1018.
58. McGuffin, L.J. (2008) The ModFOLD server for the quality assessment of protein structural models. *Bioinformatics.* **24**, 586–587.
59. Hartshorn, M.J. (2002) AstexViewer: a visualisation aid for structure-based drug design. *J Comput Aided Mol Des.* **16**, 871–881.
60. Mulder, N. and R. Apweiler. (2007) InterPro and InterProScan: tools for protein sequence classification and comparison. *Methods Mol Biol.* **396**, 59–70.
61. Jones, D.T. (1999) Protein secondary structure prediction based on position-specific scoring matrices. *J Mol Biol.* **292**, 195–202.
62. Jones, D.T. and J.J. Ward. (2003) Prediction of disordered regions in proteins from position specific score matrices. *Proteins.* **53 Suppl 6**, 573–578.
63. Jones, D.T. (2007) Improving the accuracy of transmembrane protein topology prediction using evolutionary information. *Bioinformatics.* **23**, 538–544.
64. Altschul, S.F., T.L. Madden, A.A. Schaffer, J. Zhang, et al. (1997) Gapped BLAST and PSI-BLAST: a new generation of protein database search programs. *Nucleic Acids Res.* **25**, 3389–3402.
65. Soding, J. (2005) Protein homology detection by HMM-HMM comparison. *Bioinformatics.* **21**, 951–960.
66. Hooft, R.W., G. Vriend, C. Sander, and E.E. Abola. (1996) Errors in protein structures. *Nature.* **381**, 272.
67. Laskowski, R.A., M.W. MacArthur, D.S. Moss, and J.M. Thornton. (1993) PROCHECK: a program to check the stereochemical quality of protein structures. *J Appl Cryst.* **26**, 283–291.
68. Kabsch, W. and C. Sander. (1983) Dictionary of protein secondary structure: pattern recognition of hydrogen-bonded and geometrical features. *Biopolymers.* **22**, 2577–2637.
69. Hutchinson, E.G. and J.M. Thornton. (1996) PROMOTIF - a program to identify and analyze structural motifs in proteins. *Protein Sci.* **5**, 212–220.
70. Jmol: an open-source Java viewer for chemical structures in 3D. http://www.jmol.org/
71. Stroud, R.M., S. Choe, J. Holton, H.R. Kaback, et al. (2009) 2007 annual progress report synopsis of the Center for Structures of Membrane Proteins. *J Struct Funct Genomics.* **10**, 193–208.
72. Elsliger, M.A., A.M. Deacon, A. Godzik, S.A. Lesley, et al. (2010) The JCSG high-throughput structural biology pipeline. *Acta Crystallogr Sect F Struct Biol Cryst Commun.* **66**, 1137–1142.
73. Vroling, B., M. Sanders, C. Baakman, A. Borrmann, et al. (2011) GPCRDB: information system for G protein-coupled receptors. *Nucleic Acids Res.* **39**, D309–319.
74. Xiao, R., S. Anderson, J. Aramini, R. Belote, et al. (2010) The high-throughput protein sample production platform of the Northeast Structural Genomics Consortium. *J Struct Biol.* **172**, 21–33.

75. Bonanno, J.B., S.C. Almo, A. Bresnick, M.R. Chance, et al. (2005) New York-Structural GenomiX Research Consortium (NYSGXRC): a large scale center for the protein structure initiative. *J Struct Funct Genomics.* **6**, 225–232.
76. http://jcmm.burnham.org/.
77. Nierman, W.C., T.V. Feldblyum, M.T. Laub, I.T. Paulsen, et al. (2001) Complete genome sequence of Caulobacter crescentus. *Proc Natl Acad Sci U S A.* **98**, 4136–4141.
78. Aldridge, P., R. Paul, P. Goymer, P. Rainey, et al. (2003) Role of the GGDEF regulator PleD in polar development of Caulobacter crescentus. *Mol Microbiol.* **47**, 1695–1708.
79. Jenal, U. and J. Malone. (2006) Mechanisms of cyclic-di-GMP signaling in bacteria. *Annu Rev Genet.* **40**, 385–407.
80. Wu, C.H., R. Apweiler, A. Bairoch, D.A. Natale, et al. (2006) The Universal Protein Resource (UniProt): an expanding universe of protein information. *Nucleic Acids Res.* **34**, D187–191.
81. Hunter, S., R. Apweiler, T.K. Attwood, A. Bairoch, et al. (2009) InterPro: the integrative protein signature database. *Nucleic Acids Res.* **37**, D211–215.
82. Chan, C., R. Paul, D. Samoray, N.C. Amiot, et al. (2004) Structural basis of activity and allosteric control of diguanylate cyclase. *Proc Natl Acad Sci U S A.* **101**, 17084–17089.
83. Wassmann, P., C. Chan, R. Paul, A. Beck, et al. (2007) Structure of BeF3- -modified response regulator PleD: implications for diguanylate cyclase activation, catalysis, and feedback inhibition. *Structure.* **15**, 915–927.
84. De, N., M. Pirruccello, P.V. Krasteva, N. Bae, et al. (2008) Phosphorylation-independent regulation of the diguanylate cyclase WspR. *PLoS Biol.* **6**, e67.
85. Sigrist, C.J., L. Cerutti, E. de Castro, P.S. Langendijk-Genevaux, et al. (2010) PROSITE, a protein domain database for functional characterization and annotation. *Nucleic Acids Res.* **38**, D161–166.
86. Dunbrack, R.L., Jr. (2006) Sequence comparison and protein structure prediction. *Curr Opin Struct Biol.* **16**, 374–384.
87. Waterhouse, A.M., J.B. Procter, D.M. Martin, M. Clamp, et al. (2009) Jalview Version 2 – a multiple sequence alignment editor and analysis workbench. *Bioinformatics.* **25**, 1189–1191.
88. Rost, B. (1999) Twilight zone of protein sequence alignments. *Protein Eng.* **12**, 85–94.
89. Krissinel, E. and K. Henrick. (2007) Inference of macromolecular assemblies from crystalline state. *J Mol Biol.* **372**, 774–797.
90. Paul, R., S. Abel, P. Wassmann, A. Beck, et al. (2007) Activation of the diguanylate cyclase PleD by phosphorylation-mediated dimerization. *J Biol Chem.* **282**, 29170–29177.
91. Paul, R., S. Abel, P. Wassmann, A. Beck, et al. (2007) Activation of the diguanylate cyclase PleD by phosphorylation-mediated dimerization. *J Biol Chem.* **282**, 29170–29177.
92. Benkert, P., M. Biasini, and T. Schwede. (2011) Toward the estimation of the absolute quality of individual protein structure models. *Bioinformatics.* **27**, 343–350.
93. Ramachandran, G.N., C. Ramakrishnan, and V. Sasisekharan. (1963) Stereochemistry of polypeptide chain configurations. *J Mol Biol.* **7**, 95–99.
94. Briggs, R., L. Dworkin, J. Briggs, E. Dessypris, et al. (1994) Interferon alpha selectively affects expression of the human myeloid cell nuclear differentiation antigen in late stage cells in the monocytic but not the granulocytic lineage. *J Cell Biochem.* **54**, 198–206.
95. Briggs, R.C., J.A. Briggs, J. Ozer, L. Sealy, et al. (1994) The human myeloid cell nuclear differentiation antigen gene is one of at least two related interferon-inducible genes located on chromosome 1q that are expressed specifically in hematopoietic cells. *Blood.* **83**, 2153–2162.
96. Dawson, M.J., J.A. Trapani, R.C. Briggs, J.K. Nicholl, et al. (1995) The closely linked genes encoding the myeloid nuclear differentiation antigen (MNDA) and IFI16 exhibit contrasting haemopoietic expression. *Immunogenetics.* **41**, 40–43.
97. Pruitt, K.D., T. Tatusova, W. Klimke, and D.R. Maglott. (2009) NCBI Reference Sequences: current status, policy and new initiatives. *Nucleic Acids Res.* **37**, D32–36.
98. Kersey, P.J., J. Duarte, A. Williams, Y. Karavidopoulou, et al. (2004) The International Protein Index: an integrated database for proteomics experiments. *Proteomics.* **4**, 1985–1988.
99. Benson, D.A., I. Karsch-Mizrachi, D.J. Lipman, J. Ostell, et al. (2011) GenBank. *Nucleic Acids Res.* **39**, D32–37.
100. Baxevanis, A.D. (2008) Searching NCBI databases using Entrez. *Curr Protoc Bioinformatics.* **Chapter 1**, Unit 1 3.
101. Chen, L., R. Oughtred, H.M. Berman, and J. Westbrook. (2004) TargetDB: a target registration database for structural genomics projects. *Bioinformatics.* **20**, 2860–2862.

102. Saito, K., M. Inoue, S. Koshiba, T. Kigawa, et al. (2006) DOI:10.2210/pdb2dbg/pdb.
103. Fairbrother, W.J., N.C. Gordon, E.W. Humke, K.M. O'Rourke, et al. (2001) The PYRIN domain: a member of the death domain-fold superfamily. *Protein Sci.* **10**, 1911–1918.
104. http://www.nesg.org/.
105. Koh, I.Y., V.A. Eyrich, M.A. Marti-Renom, D. Przybylski, et al. (2003) EVA: Evaluation of protein structure prediction servers. *Nucleic Acids Res.* **31**, 3311–3315.
106. Kopp, J., L. Bordoli, J.N.D. Battey, F. Kiefer, et al. (2007) Assessment of CASP7 Predictions for Template-Based Modeling Targets. *Proteins: Structure, Function, and Bioinformatics.* **69**, 38–56.
107. Liao, J.C.C., R. Lam, M. Ravichandran, J. Ma, et al. (2007) DOI:10.2210/pdb2oq0/pdb.
108. Schwede, T., J. Kopp, N. Guex, and M.C. Peitsch. (2003) SWISS-MODEL: An automated protein homology-modeling server. *Nucleic Acids Res.* **31**, 3381–3385.
109. Caly, D.L., P.W. O'Toole, and S.A. Moore. (2010) The 2.2-Å structure of the HP0958 protein from Helicobacter pylori reveals a kinked anti-parallel coiled-coil hairpin domain and a highly conserved ZN-ribbon domain. *J Mol Biol.* **403**, 405–419.
110. Radivojac, P., L.M. Iakoucheva, C.J. Oldfield, Z. Obradovic, et al. (2007) Intrinsic disorder and functional proteomics. *Biophys J.* **92**, 1439–1456.
111. http://blast.ncbi.nlm.nih.gov/
112. http://www.wwpdb.org/docs.html.
113. Bordoli, L., F. Kiefer, K. Arnold, P. Benkert, et al. (2009) Protein structure homology modeling using SWISS-MODEL workspace. *Nat Protoc.* **4**, 1–13.

Chapter 6

A Practical Introduction to Molecular Dynamics Simulations: Applications to Homology Modeling

Alessandra Nurisso, Antoine Daina, and Ross C. Walker

Abstract

In this chapter, practical concepts and guidelines are provided for the use of molecular dynamics (MD) simulation for the refinement of homology models. First, an overview of the history and a theoretical background of MD are given. Literature examples of successful MD refinement of homology models are reviewed before selecting the Cytochrome P450 2J2 structure as a case study. We describe the setup of a system for classical MD simulation in a detailed stepwise fashion and how to perform the refinement described in the publication of Li et al. (Proteins 71:938–949, 2008). This tutorial is based on version 11 of the AMBER Molecular Dynamics software package (http://ambermd.org/). However, the approach discussed is equally applicable to any condensed phase MD simulation environment.

Key words: Molecular dynamics, Homology modeling, AMBER, Force fields, FF99SB

1. Introduction

Molecular recognition, signaling processes, atomic diffusion, catalysis phenomena, ion gating, and protein folding are just some of the biologically interesting events in which the motions of molecules play a crucial role. Simulations that provide a detailed atomistic understanding of such phenomena must, therefore, include a description of such motions. The most common method employed for in silico study of molecular flexibilities at the atomic level is the molecular dynamics (MD) method (1, 2). As described in more detail below, such methods numerically integrate Newton's second equation of motion to simulate how biological systems evolve as a function of time. Such simulations can be used to provide both statistical mechanics and thermodynamics properties.

Andrew J.W. Orry and Ruben Abagyan (eds.), *Homology Modeling: Methods and Protocols*, Methods in Molecular Biology, vol. 857,
DOI 10.1007/978-1-61779-588-6_6, © Springer Science+Business Media, LLC 2012

Since the first all-atom molecular dynamics (MD) simulation of an enzyme was described by McCammon et al. (3), in 1977, MD simulations have evolved to become an important tool in understanding the behavior of biomolecules. Since that first 10 ps long simulation of merely 500 atoms the field has grown to where small enzymes can be routinely simulated on the microsecond timescale (4–6). Simulations containing millions of atoms are now also considered routine (7, 8). While, somewhat heroic attempts have been made to fold entire, albeit small, proteins through the use of molecular dynamics simulation (9–11), the main use remains in the calculation of properties of folded peptides, which requires an initial folded protein structure. Typically this would be a crystal structure, from X-ray/neutron scattering, or a solution phase NMR structure such as those provided through the protein databank (http://www.pdb.org/).

When such initial structures are not available, one typically makes use of a homology model as an initial starting structure. One nonobvious use of MD simulations is actually the final stage refinement of homology models. It is this use of MD that we cover in this chapter.

It is known that an inefficient refinement method is one of the three major causes of errors affecting protein homology models, together with unsuitable template choice and inaccurate alignment (12). Describing the physical correctness of protein three dimensional (3-D) structures looks like the ideal task for physics-based methods and especially for MD simulations (13). In practice, MD techniques are generally ineffective at finding the native structure of all but the smallest proteins from scratch because of (1) the infeasibility of exploring, in its entirety, the vast conformational space and (2) the difficulty in distinguishing native geometries from other realistic yet nonnative conformations within the limitations of accuracy inherent in the description of the energy by the force field (14). In principle, the refinement of reasonably good quality 3-D protein models built by homology techniques is possible. This implies an efficient sampling method able to generate enough realistic native-like decoys from an initial template-based model and an evaluation function able to identify these decoys (14, 15).

The coupling of homology modeling with MD is useful in that it tackles the sampling deficiency of dynamics simulations by providing good quality initial guesses for the native structure. Indeed, comparative modeling relaxes the severe requirement of force fields to explore the huge conformational space of protein structures. The approach consists of replacing the exhaustive sampling of the hypersurface of energy with classical physics laws by important structural constraints from both 1-D alignment and 3-D superposition. It is worth noting that the sampling issues are, to some extent, linked to computer power and more complete conformational search is foreseen with the calculation capability explosion by

GPUs (16) and remotely accessible parallel computing via GRID or Cloud computing (17). However, the (short) history of computational chemistry teaches us that the optimistic and impatient molecular modeler community tends to use the always increasing computer power to design more complex systems and not to uphold the validity domain of models. In protein modeling, this behavior led to the impressive improvements in the description of protein environments at the atomic level: MD in explicit solvent boxes and detailed biphospholipidic membranes are now affordable to anyone having access to modern computational resources.

For homology modeling, refinement consists of solving the problem of making an already reasonably good quality 3-D structure prediction closer to the native form of the protein (hopefully from 3–4 Å to less than 1 Å C_α RMSD). In this context, suitably termed "the last mile of protein folding" (18), classical MD methods in explicit water have proven their performance in the CASP initiative (19) as well as in many examples found in the literature referring to the milestone article published in 2004 by Fan and Mark (20). In their work, the refinement of 60 small to medium-size protein structures (50–100 residues each) was evaluated by increasing the complexity of the description of the environment around proteins and the timescale of simulations. Of the methods tested involving constrained force-field minimization (here GROMACS (21, 22)) in explicit water (here the SPC model (23)) followed by unrestrained MD at 300 K for 10–100 ns was proven useful for homology-based protein structure refinement. However, the authors also rigorously gave detailed technical advice and depicted clear limitations of the methods that are not always accounted for in the numerous subsequent studies based on the given strategy. For example, they emphasized timescales of 10 ns, considered minimal for efficient sampling and noted that refinement is only possible if the native structure represents the global minimum for the force field, simulated in the particular environment. Indeed, the MD performance was satisfactory if the general fold of the small proteins was correct. For geometries less related to native, the protocol failed because of incomplete sampling and/or force-field deficiency in evaluation. So, as there is no guaranteed way to recognize the "best structure," it is often advised to take a geometric average over time as the final model.

Another aspect discussed was the use of explicit solvent, the increased degrees of freedom of which necessitate longer sampling. At the time, it was considered the best way to appropriately take electrostatic and solvation effects into account. This significant computational expense has since been questioned by advances made in implicit solvation such as the Generalized Born models (GB) and related evaluation functions (24). Chopra et al. have shown, for instance, that GB-based protocols performed better than simulations in periodic boxes of solvent on a large set of protein native and decoy geometries (25).

A modified CHARMM force field was developed by Chen et al. (26) accounting for implicit solvation parameters, emphasizing the benefit of incorporating reliable structural information into the MD refinement strategy by weakly imposing restraints to enforce secondary structures yet allowing enough flexibility for rearrangement.

Restrained MD simulations, in which parts of the systems are kept fixed according to known structural features, were also successfully applied. A specific case is the refinement of ion channel structures involving high degrees of symmetry (27). It was observed that free MD on a potassium channel tends to deviate from ideal symmetry because of thermal effect biases. In fact, the structure is somewhat perturbed in the first ps. A multistep protocol in NAMD (28) with the CHARMM force field was proposed in explicit water and membrane. The main contribution was the gradual application of symmetrical constraints to the oligomeric structure. Good improvement and better stability of the model were obtained for 8 ns simulations. It is worth stressing that the system was still stable after 16 ns but no further structural refinement was seen.

By carefully investigating the limitation of classical unrestrained MD, it was stated that failure should be related to the deviation during the free simulations rather than poor quality of the initial model to refine. In fact, a major weakness of MD may be that the native conformation is not necessarily the lowest free energy state in the simulation of the system as mentioned in a comprehensive AMBER benchmarking study (29).

Indeed, the second defect of molecular mechanics techniques, i.e., the inability to discriminate decoys from native geometries based on force-field energy, is maybe more critical and to some extent less directly related to computational power. Despite the continuous enhancement of force-field parameters, it remains challenging to obtain sensitive enough energy functions to discriminate decoys from near-native conformations. A way to overcome this intrinsic molecular mechanics deficiency is to implement knowledge-based parameters in a force field, as for example in YASARA (http://www.yasara.org/) (18, 30) which is derived from AMBER but with additional torsional terms optimized for the reproduction of a large set of high-resolution crystallographic structures.

Although at substantive computational cost, one of the distinct strong points of classical MD methodologies is that they rely on well-defined physical evaluation of structure and energy. This makes them potentially informative and easily interpretable for scientists (31). Moreover, and in spite of refinement protocols designed for their true aim (i.e., focusing on sampling and evaluation in the vicinity of the initial structure), carrying out MD can give important additional information on many biochemical and pharmacological processes involving protein flexibility or environmental features that may not be observed in experimental structures (solvents, ionic equilibriums, or biological membranes). These aspects require long timescale simulations of complex systems so again are directly related to the computational power (32). Furthermore, the perturbation observed in the first ps of unrestrained dynamics may be suitable to escape local energy minima and enable access to the active state of the protein even if the template is in an inactive state. Addition of knowledge-based features related to the protein itself or to a ligand with known effects permitted successful modeling of the GPCR active state (33, 34), for example.

Additionally, many methods exist to extend the conformational exploration, mainly involving altering the temperature of simulation. Straightforward increase in kinetic energy given to the system is generally hazardous, since it was reported to impact only slightly the refinement of close-to-native structures yet often resulting in major loss of the fold in cases in which the initial model was far from the desired result and not in a local potential energy well (20). More complicated protocols consist either of iterative cycles of heating–cooling processes (simulated annealing (35)), often used prior to classical simulations (36, 37), or in exploration of a range of temperatures by independent simultaneous simulations able to swap with each other at regular intervals (replica-exchange simulations (26, 38, 39)). The use of such methods improves the sampling by passing over high energy barriers, but the realistic physical description of the dynamic behavior of proteins, as in classical MD, is lost.

Instead of acting on temperature, an interesting method of pressure-guided dynamics was proposed to expand and optimize binding pockets by applying the so-called "balloon potential." The size expansion of small radii Lennard–Jones particles in a network to mimic increased pressure, whereas the backbone is constrained was employed in cavities of chemokine receptor-2 and yielded the discovery of two lead compounds (21). In doing so, the final binding site shape is unbiased towards any ligand, allowing more objective docking studies or virtual screening campaigns. This is a clear advantage in the drug-design context over the common methodology aiming at making room inside binding sites of proteins by the presence of known ligands (e.g., cocrystallized small molecules in the template structure) kept during some steps of the homology modeling process. A successful example of such approach is given where potential drug candidates were designed by structure-based methods within a ribosomal S6 kinase 2 (40).

In Subheading 3, later in this chapter, we give what is an inevitably incomplete list of examples of successful MD-based homology model refinement but one that attempts to provide sufficient detail for someone unfamiliar with the field to attempt such refinements. We then attempt to provide the reader with a detailed practical overview on how to use MD simulation techniques to refine a homology model. We focus on the use of the AMBER Molecular Dynamics Software (41); however, such techniques are transferable to any major MD package designed for the simulation of condensed phase biological systems, common examples being NAMD (28), GROMACS (21), CHARMM (42), and LAMMPS (43).

We begin by providing a short theoretical overview of MD, focusing on the key aspects of the technique.

2. Theoretical Background

Molecular dynamics methods are used in computational chemistry and molecular biology to simulate how biological systems evolve as a function of time. These methods, in their simplest form, evaluate the time evolution of a system by numerically integrating Newton's equations of motion. Specifically Newton's second law (Eq. 6.1):

$$a_i(t) = \frac{\mathrm{d}^2 x_i}{\mathrm{d}t^2} = \frac{F(x_i)}{m_i}, \tag{1}$$

where a_i is the acceleration of particle i at time t determined by the force $F(x_i)$ acting on particle i of mass m_i at position x_i.

The force $F(x_i)$ can be calculated in a number of ways using either quantum mechanical (QM) or molecular mechanical (MM) approaches. In the context of this chapter, we consider only MM (also termed "classical") approaches to computing the force. In this approach, $F(x_i)$ is calculated from the derivative of the expression for the potential energy as a function of position $V(x_i)$ which is described by a molecular mechanics force field, for example, the FF94 (44) or FF99SB (45) force fields. In these classical force fields, a molecule is considered to be a collection of balls corresponding to atoms with a fixed electronic distribution connected together by springs representing the bonds (46).

In the case of the AMBER force field, used in this section, the potential energy is a function of terms describing the bonds, angles, dihedrals, and nonbonded interactions in the system (Eq. 2):

$$V = \sum_{i=1}^{\mathrm{Natom}} V_{\mathrm{bond}}(i) + V_{\mathrm{angle}}(i) + V_{\mathrm{dihedral}}(i) + V_{\mathrm{non-bonded}}(i). \tag{2}$$

In its simplest form this equation can be expressed as follows (Eq. 6.3):

$$V(r^n) = \sum_{\mathrm{bonds}} K_r(r - r_{\mathrm{eq}})^2 + \sum_{\mathrm{angles}} K_\theta(\theta - \theta_{\mathrm{eq}})^2 + \sum_{\mathrm{dihedrals}} \frac{V_n}{2}[1 + \cos(n\phi - \gamma)] + \sum_{i<j}\left[\frac{A_{ij}}{R_{ij}^{12}} - \frac{B_{ij}}{R_{ij}^{6}} + \frac{q_i q_j}{\varepsilon_r R_{ij}}\right], \tag{3}$$

where the potential energy V is written as a function of the positions r of n atoms. $K_r, r_{eq}, K_\theta, \theta_{eq}, V_n, n, \gamma, A_{ij}, B_{ij}, \varepsilon_r, q_i$ and q_j are all empirically defined parameters. The first three terms of Eq. 6.3 correspond to the bond, angle, and dihedral terms, respectively, while the last term describes the nonbonded van der Waals and electrostatic interactions.

The velocity of individual atoms in a molecule at time t can be evaluated by integrating the classical equations of motion for every atom of the system at every time step $\mathrm{d}t$ prior to the current time. By the use of simple integrators (47, 48), the position of every atom in the system can be evaluated as a function of time. The computational cost and complexity in the practical implementation of MD simulations lies in the fact that the magnitude of the integration time step $\mathrm{d}t$ is limited by the Nyquist limit (49) which is determined by the fastest motions in the molecule. In the case of proteins, this corresponds to the stretching vibrations of bonds connecting hydrogen atoms to heavy atoms X–H ($\tau \approx 1\times10^{-14}\,\mathrm{s} \approx 10\,\mathrm{fs}$). To avoid errors in the integration over time the time step should be such that (Eq. 4).

$$\frac{\tau}{\mathrm{d}t} > \approx 20. \tag{4}$$

For proteins, this gives a maximum time step of $\approx 0.5\,\mathrm{fs}$. This makes long (nanosecond) MD simulations computationally expensive (2). One method for increasing the size of the time step, and so lowering the computational cost, is to constrain the bonds to hydrogen using an algorithm such as SHAKE (50). This keeps the X–H bond lengths constant at their equilibrium values and allows time steps of up to $2\,\mathrm{fs}$ to be used.

Practically MD simulations are typically carried out in four steps under isothermal-isobaric conditions (Fig. 1).

In the first stage, the system to be simulated in an explicit solvent environment with an initial structure derived from NMR, X-ray, or homology modeling is placed in a periodic lattice and then prepared for simulation by adding missing atoms, assigning charges, and atom types, which are ultimately translated into the parameters in Eq. 3, and then eventually adding solvent molecules. The system is then typically subjected to one or more rounds of structural minimization to relieve any high energy strains in the initial model. The system is then slowly heated, typically within the NVT ensemble, over a period of approximately 20–100 ps. Next the system is equilibrated, often in the NPT ensemble, to allow the system density to converge and for the structure to relax away from any initial high energy state implied by the initial structure and any added atoms or solvent molecules. At this stage, time-dependent system properties such as energy, density, temperature, pressure, and RMSD to the initial structure are checked for convergence.

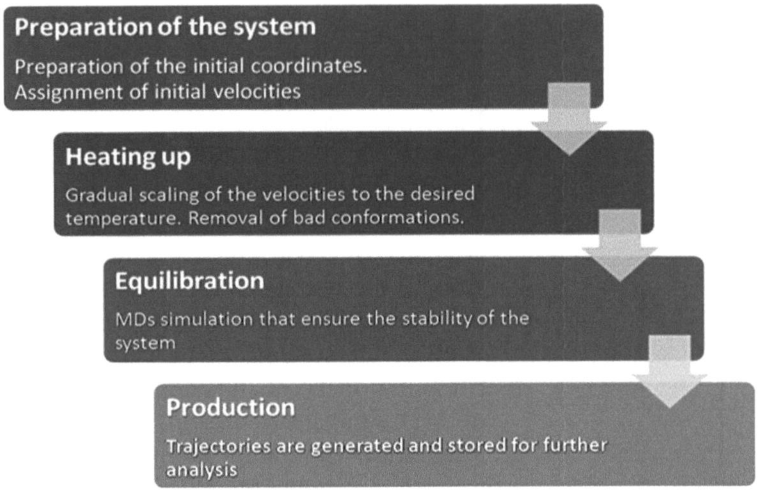

Fig. 1. A general protocol for running MD simulations.

Once equilibrium is reached, a production phase, in any one of the three microcanonical ensembles, is conducted in which structural and energetic data is collected at specific time intervals. This data collection typically includes atomic positions, velocities, and other physical properties of the simulated system as a function of time.

The goal of the production phase is generally to generate enough representative conformations in a trajectory to satisfy the *ergodic hypothesis*, which states that the average values over time of physical quantities characterizing a system are equal to the statistical average values of these quantities. If enough representative conformations are sampled, relevant biophysical properties, both average and time dependent, can then be calculated.

3. Applications of MD to Homology Modeling Refinement in Drug-Design Strategies

High-quality 3-D protein structures are of critical importance for rational drug design and many structure-based methodologies were developed to help identifying novel pharmacological targets, assessing the druggability of cavities and finally discovering new bioactive molecules (51). In cases where sufficient biostructural information is known but the 3-D structure is not solved, homology modeling approaches have been successfully employed. Specific examples of homology methodologies involving MD-based refinement protocols that have shown significant successes in the various steps of structure-based drug-design strategies are highlighted here.

Despite the apparently infinite variations in the refinement techniques described in the scientific literature, the majority of drug-design oriented homology model refinement strategies involve classical MD coupled with molecular docking.

Drug-design based on homology models was and still is massively used for G-protein-coupled receptors (GPCRs), mainly because this family of membrane proteins is the biotarget of many classes of drugs and part of numerous and various physiological processes. GPCRs are structurally diverse especially at the ligand binding sites. New GPCR structures have recently been solved and publicly available (52–54).

An example is the construction by homology of the Mu opioid receptor in the InsightII (http://www.accelrys.com/) environment. Model refinement included decreasing restrained optimization ending with short (200 ps) MD simulations in a complete explicit membrane–aqueous matrix at 310 and 330 K. The final receptor model was then used to manually dock Naltrexone, a potent antagonist drug. A second round of very short (11 ps) partly constrained MD was run for the reformed drug–protein complex. This let the structure shift from an inactive GPCR to an active conformation providing additional dynamical information on the activation process (34).

Another GPCR homology model was the human gonadotropin-releasing hormone receptor. Meticulous, detailed, and long MD (160 ns) was carried out using GROMACS at 310 K in explicit water (SPC model (23)) and membrane environment by relaxing different parts of the structure one after the other. The final structure was then subjected to six more independent simulations at 310 and 350 K aimed at assessing its geometry. Stability of the entire system after 35 ns of unrestrained simulations was considered sufficient for validation (55).

Numerous other examples of GPCR models involving MD stages have been published with many of them reviewed elsewhere (52, 54–56).

Other proteins of crucial importance for pharmaceutical research are the cytochromes P450 (CYP450). Among this large superfamily of heme-containing proteins (60 different isoenzymes in human), considered as the major metabolizers of drugs and other xenobiotics as well as endogenous molecules (57), some may be drug targets.

Li et al. produced a model of CYP2J2, a CYP450 involved in physiological metabolism and potentially a novel biotarget for cancer and cardiovascular disease therapy. The 3-D structure, initially built and minimized in InsightII/Modeler (58), is the case study detailed in Subheading 4.

A similar strategy was followed in another CYP450 drug design-focused homology modeling work. Mouse CYP2C38 and CYP2C39 were constructed focusing on the structure of their binding cavities to understand the diverse substrate selectivity profiles of both enzymes, despite their high level of homology

(92% sequence identity). Models were constructed and minimized in the InsightII modeling environment. The Discover module, also by Accelrys, was then used to subject both structures to unrestrained MD refinements with the CVFF force field (59) and TIP3P explicit water (60) at 298 K for 500 ps. The average geometries over the last 300 ps were selected as structural targets for parallel docking of selective and nonselective ligands. The binding modes and predicted energies helped identify key residues for ligand binding and selectivity (61).

The orphan CYP4A22 is also a potential CYP450 drug target involved in regulating blood pressure. Identification of cavities and assessment of their druggability was made possible on a homology model built and minimized with Accelrys's Discovery Studio and refined with 3 ns unrestrained MD in GROMACS with explicit water (SPC model (23)). The final model was considered not as an average but as the geometry with the lowest potential energy. Docking with ligandFit (62) of two possible substrates, arachidonic acid and erythromycin, followed by simulated annealing cycles allowed the selection of amino acid positions for targeted mutations (63).

Recently, the biochemical synthesis and fate of prostaglandins have emerged as an important research area for new classes of future drugs aimed at curing inflammation among other pathologies (64).

Hamza et al. have established a homology-based protocol to generate 3-D models of two distinct microsomal proteins involved in the prostaglandin biochemistry, i.e. prostaglandin E synthase-1 (mPGES) and phosphodiesterase-2 (PDE2). The former has not been crystallized yet and the construction of a homology-based trimeric structure allows the docking of known ligands with predicted affinities that are reasonably correlated with binding experiments. One X-ray structure of the latter protein is available (65), but its binding pockets turned out to be unsuitable for explaining the binding of known ligands.

Both models were constructed with InsightII/Modeler (58) and the first refinement involved simulated annealing with the CHARMM force field. The ligand charges used for manual docking and subsequent MD were calculated by quantum mechanics techniques (HF/6.31G*). Explicit solvent (TIP3P water (60)) and membrane simulations (POPC model (66)) were achieved in AMBER for 1.6 ns at 300 K with constraints on the C_{α}. The MD trajectory was further analyzed to propose the final structure of reformed complexes as the average of the last 500 ps and to estimate binding free energies with GBSA models (67, 68).

The design of antimicrobial agents has also gained from homology models, e.g., for tackling parasitic multidrug resistance faced in tuberculosis therapy.

The assessment of *Mycobacterium tuberculosis* 1-deoxy-D-xylulose-5-phosphate reductoisomerase (MtDXR) as a potential drug target implied the generation of a homology structure with InsightII/Modeler, a first minimization in the CVFF force field (59) and reformation of the complexes by manual docking of known binders. These ligand-constrained structures were considered as input for 1.2 ns MD simulations in explicit water with the same force field. The model was validated by the agreement with experimental point mutations and the excellent agreement with the later published crystal structure. Moreover, the additional information provided by MD on the induced-fit behavior upon ligand binding provided a good example of the complementarity between dynamics simulations and the static information extracted from X-ray structures (69).

Recently, MurC ligase, another protein involved in the peptidoglycan biosynthesis in *M. tuberculosis*, was assessed as a putative novel drug target. Similar to the previous example, a dual protocol involving docking and unrestrained MD of 5 ns in explicit water in GROMACS allowed the identification of some structural features important for molecular recognition, starting points for the rational design of novel antibiotics (69). Daga et al. recently published a homology model of the Hepatitis B virus DNA polymerase constructed in the Swiss-Pdb Viewer 3.7/SwissModel environment (70, 71) and the docking studies augmented with flexibility information from MD simulations. After a stepwise minimization gradually relaxing the structural constraints on the initial model, known ligands were docked with the GOLD engine (72) into the main cavity of the viral protein. The reformed complexes were then submitted to 5 ns unrestrained AMBER simulations in explicit water and redocked with the same ligands. The conformational changes observed in pre- and post-MD reformed complexes helped explain the better affinity of inhibitors compared to substrates. This analysis also allowed the generation of hypotheses on the importance of the binding site plasticity in the resistance pattern of experimental mutants (73).

Academic life science has a specific interest for neglected or tropical diseases, for instance malaria. Molecular modeling makes its contribution, of course. A fragment of merozoite surface protein-1 of *Plasmodium vivax* (PvMSP-1) was constructed with homology techniques (InsightII) and refined with classical MD of very short timescale (5 ps) in explicit solvent. The final model was not considered by averaging the structures but by taking the last generated conformation of the simulation and minimizing it with the CVFF force field (59). The usefulness of this model lies in the description of a cavity on the surface with properties suitable for both proteins and small molecule recognition. This provides perspective for new modes of action, antimalaric agent design, as well as better understanding of the biochemical principle of antibody interactions with this parasitic protein (74).

4. Methods

The refinement of models derived from comparative studies is necessary because loop and side chain conformations of a protein model represent only one of all the possible conformations and the low energy structure found by minimization algorithms corresponds only to one nearby local minimum. To detect the energetically most favored 3-D structure of a system, a modified strategy is needed for searching the conformational space more thoroughly (46). MD simulations offer an effective way to solve this problem, especially for molecules characterized by many torsion angles, moreover additionally taking account of solvent effects.

AMBER is a user-friendly program composed of a set of molecular mechanics force fields for the simulation of biomolecules and a package of molecular simulation programs useful, together with AmberTools, for setting up, running and analyzing MD simulations (41). The following tutorial assumes the use of AMBER v11 (see Note 1). Use of other versions may have subtle differences to the approach and format described here. The various input and output files used in this book chapter are available via the URL described in Note 1.

To provide useful guidelines and a practical example of refining homology models using the AMBER software, the unrefined homology model of the Cytochrome P450 2J2 will be used as starting structure (75). The 3-D structure was obtained by using the homology modeling package Modeler (58) beginning with the primary sequence of the human Cytochrome P450 2C9 in complex with warfarin, showing a sequence identity of 42%. The system is composed of 457 amino acid residues and a heme cofactor, for a total of 3,767 atoms. No hydrogen atoms are included with the model.

To perform the MD refinement, in explicit water, the essential steps listed herein, and adapted from (75) are described in detail:

- Generation of the molecular topology/parameter and initial coordinate files necessary for performing minimizations and MD simulations of the homology model.
- Creation of the input files necessary for running minimizations and MD simulations of the homology model.
- Running minimization steps as necessary.
- Running MD simulations to equilibrate the system (heating and equilibration phases).
- Running MD simulations, collecting trajectories (production phase).
- Calculating the average structure from the collected trajectories for subsequent analyses.

- Performing basic analysis of the trajectories, such as calculating root-mean-squared deviations (RMSD) and plotting various energy terms as a function of time.
- Evaluation of the final and optimized structure with respect to its geometry and energy.

Throughout this section, all filenames, command lines, input files, and program names will be written in *italic*. The various input files discussed below are provided in the supplemental material. Before running any of the programs provided with AMBER, the UNIX shell environment variable that specifies where AMBER is installed should be set properly.

export AMBERHOME=/usr/local/amber11

4.1. Setting Up the System: Cytochrome P450 2J2

The first step of refinement using an MD approach is to create the necessary input files for performing minimization and simulation. This requires:

- A file containing a description of the molecular topology and the force-field parameters (default file extension: *prmtop*).
- A file containing a description of the atom coordinates and the current periodic box dimensions (default file extension: *inpcrd*).
- The input files consisting of a series of name lists, a FORTRAN language extension for allowing unformatted reading of a series of variables, defining control variables that determine the options and type of simulation to be run (default file extension: *mdin*).

A number of different force field variants are supplied with AMBER. In previous versions of the AMBER molecular dynamics package, the default was the Cornell et al. or FF94 (44) force field. With AMBER v11, the force field recommended for the simulation of proteins and nucleic acids in explicit solvent is the version FF99SB (see Note 2). In this example, the FF99SB all-atom force field will be used, in which standard amino acid residues are parameterized and consequently recognized by the *XLEaP* module of the AmberTools package. *XLEaP* is required not only for producing the files by reading the force-field parameters from the defined libraries but also for visualizing the input structures. A *PDB* file of the homology model is needed for generating the necessary input files for running the MD simulation refinement. Such structures, compared to the ones obtained through experimental methods, typically require more elaborate minimization and equilibration steps prior to the production of dynamics simulation trajectories.

The unrefined homology model considered in this example contains a cofactor, the heme group: the modeled protein belongs to the superfamily of heme-containing cytochrome P450 monooxygenase.

The heme porphyrin is considered as a nonstandard residue by AMBER: it is not recognized by *XLEaP* since it is not parameterized in the FF99SB force field. It requires structural information and additional force-field parameters that have to be provided before creating the topology and coordinate files of the whole system (see Note 3). However, parameters for the most common cofactors, carbohydrates, lipids, nucleic acids, organic molecules, and ions are archived and freely available from the web site (http://www.pharmacy.manchester.ac.uk/bryce/amber/). For the heme group, two files are already provided: the *prep* file, containing all the information about connectivity and charges of each atom of the cofactor, and the *frcmod* file, a parameter file that can be loaded into *XLEaP* to add missing force-field parameters. Thanks to both files, the cofactor is considered as a single parameterized residue named HEM.

Let us take a look at the Cytochrome P450 2J2 model (*homology_model.pdb*) provided with the supplemental information by editing the PDB file and by eventually modifying it (see Note 4). The first step is to start up *XLEaP* (see Note 5):

$*AMBERHOME/exe/xleap –s –f* $*AMBERHOME/dat/leap/cmd/leaprc.ff99SB*

Through this command line, the *XLEaP* window is opened as well as the series of libraries and parameter files that define the FF99SB force-field parameters to be used. The "*–s*" switch tells *XLEaP* to ignore any user defined defaults, while the second part of the command tells *XLEaP* to execute the start-up script for the FF99SB force field. In this case, the files characterizing the cofactor need to also be loaded to supplement the current force field. To load them, the commands:

loadamberparams heme_all.frcmod

loadamberprep heme_all.prep

should be typed in the *XLEaP* window. The heme cofactor is now part of the FF99SB force field description currently loaded into *XLEaP*.

Using the *loadpdb* command, the PDB file of the homology model can now be loaded into *XLEaP* that will add missing hydrogen atoms to the system, indicating the number of atoms added as well as the global charge and will create a new unit called *2j2:*

2j2=loadpdb homology_model.pdb

The final input files to be created are the parameter/topology and the coordinate files for the biological system that should be solvated, containing explicit neutralizing counterions. The *addions* command implemented in *XLEaP* builds a Coulombic potential on a 1.0 Å grid and then places counterions one at a time at the points of lowest/highest electrostatic potential.

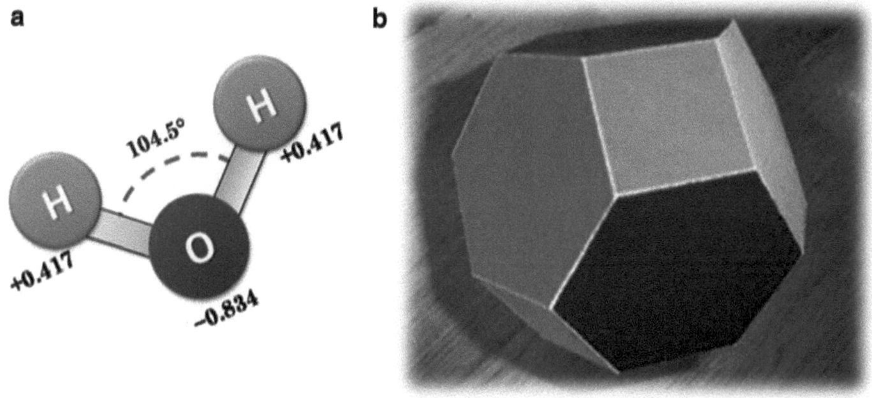

Fig. 2. TIP3P water model (a) and the truncated octahedral box full of water molecules, commonly used in MD simulations for solvating the solute atoms.

addions 2j2 Na+ 0

This command, in which "0" means "neutralize," should add a total of 2 sodium ions to counteract the –2 charge of the homology model (see Note 6).

A realistic biological system is always expected to be located in a hydrated environment. Thus, the system is next embedded in a box of explicit water molecules. Several water models have been developed, but one of the simplest and most widely used is the TIP3P model (60). It is a rigid model, characterized by three interaction sites corresponding to the three atoms of a water molecule. A point charge is assigned to each atom along with Lennard–Jones parameters from the FF99SB libraries (Fig. 2a). To reduce the problem of solute rotation normally found in classical rectangular boxes, an efficient box shape, the truncated octahedron, is used (Fig. 2b). The command *solvateoct* will add a 10 Å buffer of TIP3P water molecules around the system in each direction, forming a truncated octahedral shaped ice cube.

solvateoct 2j2 TIP3PBOX 10

XLEaP will then add sufficient solvent molecules around the starting structure such that there is at least 10 Å distance between an atom in the starting structure and the edges of the water box. The *prmtop* and *inpcrd* files can be now saved:

saveamberparm 2j2 homology_model.prmtop homology_model.inpcrd

and used for running minimizations and MD in AMBER. The system, with added water and ions, now comprises 44,470 atoms, 7,496 belonging to the solute, 12,324 water molecules, and 2 sodium atoms. All of the previous steps are summarized in Fig. 3.

Useful considerations before starting the MD refinement are reported in the Notes 7–9.

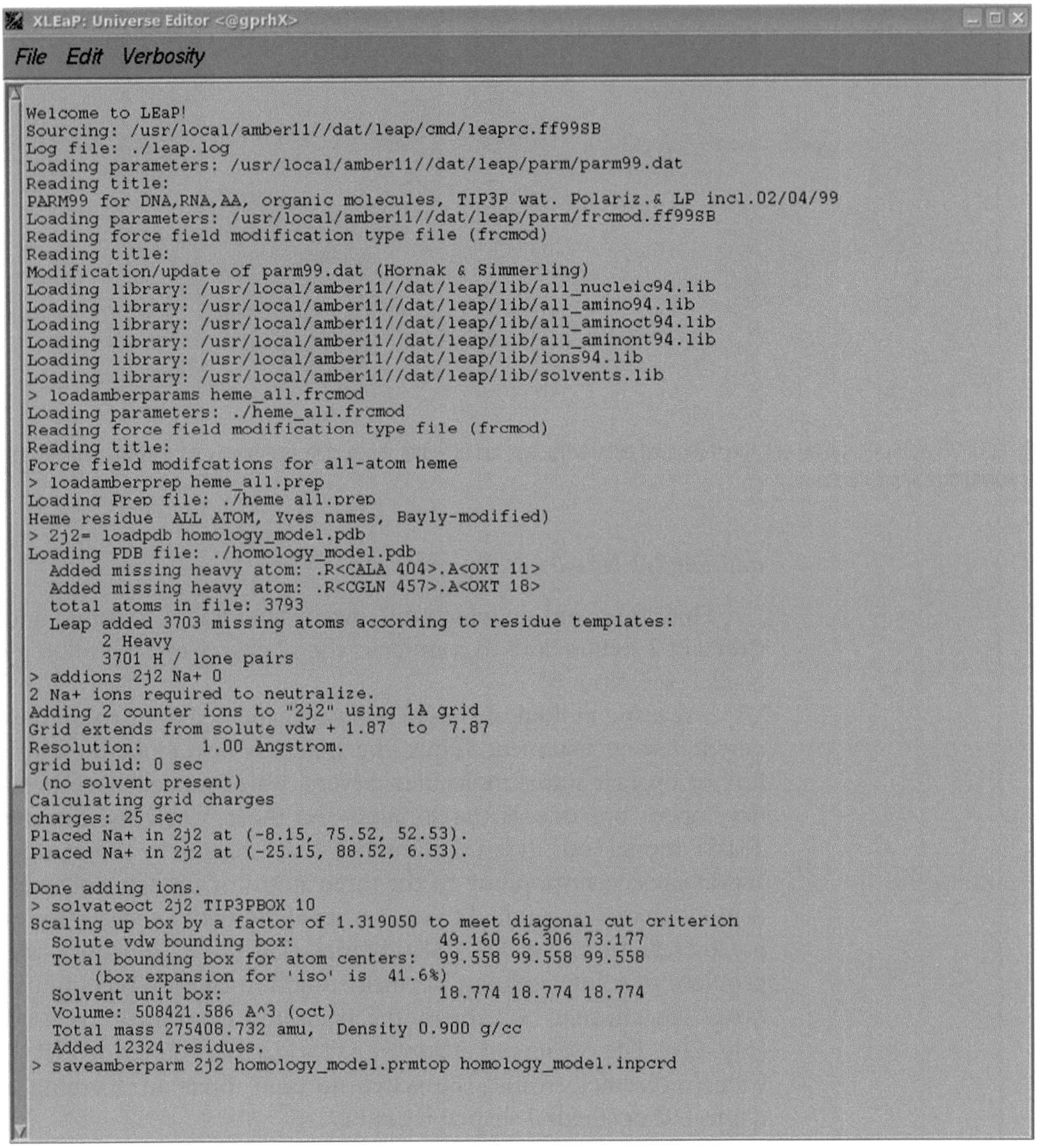

```
XLEaP: Universe Editor <@gprhX>
File Edit Verbosity

Welcome to LEaP!
Sourcing: /usr/local/amber11//dat/leap/cmd/leaprc.ff99SB
Log file: ./leap.log
Loading parameters: /usr/local/amber11//dat/leap/parm/parm99.dat
Reading title:
PARM99 for DNA,RNA,AA, organic molecules, TIP3P wat. Polariz.& LP incl.02/04/99
Loading parameters: /usr/local/amber11//dat/leap/parm/frcmod.ff99SB
Reading force field modification type file (frcmod)
Reading title:
Modification/update of parm99.dat (Hornak & Simmerling)
Loading library: /usr/local/amber11//dat/leap/lib/all_nucleic94.lib
Loading library: /usr/local/amber11//dat/leap/lib/all_amino94.lib
Loading library: /usr/local/amber11//dat/leap/lib/all_aminoct94.lib
Loading library: /usr/local/amber11//dat/leap/lib/all_aminont94.lib
Loading library: /usr/local/amber11//dat/leap/lib/ions94.lib
Loading library: /usr/local/amber11//dat/leap/lib/solvents.lib
> loadamberparams heme_all.frcmod
Loading parameters: ./heme_all.frcmod
Reading force field modification type file (frcmod)
Reading title:
Force field modifcations for all-atom heme
> loadamberprep heme_all.prep
Loading Prep file: ./heme_all.prep
Heme residue  ALL ATOM, Yves names, Bayly-modified)
> 2j2= loadpdb homology_model.pdb
Loading PDB file: ./homology_model.pdb
  Added missing heavy atom: .R<CALA 404>.A<OXT 11>
  Added missing heavy atom: .R<CGLN 457>.A<OXT 18>
  total atoms in file: 3793
  Leap added 3703 missing atoms according to residue templates:
       2 Heavy
       3701 H / lone pairs
> addions 2j2 Na+ 0
2 Na+ ions required to neutralize.
Adding 2 counter ions to "2j2" using 1A grid
Grid extends from solute vdw + 1.87  to  7.87
Resolution:      1.00 Angstrom.
grid build: 0 sec
 (no solvent present)
Calculating grid charges
charges: 25 sec
Placed Na+ in 2j2 at (-8.15, 75.52, 52.53).
Placed Na+ in 2j2 at (-25.15, 88.52, 6.53).

Done adding ions.
> solvateoct 2j2 TIP3PBOX 10
Scaling up box by a factor of 1.319050 to meet diagonal cut criterion
  Solute vdw bounding box:              49.160 66.306 73.177
  Total bounding box for atom centers:  99.558 99.558 99.558
      (box expansion for 'iso' is  41.6%)
  Solvent unit box:                     18.774 18.774 18.774
  Volume: 508421.586 A^3 (oct)
  Total mass 275408.732 amu,  Density 0.900 g/cc
  Added 12324 residues.
> saveamberparm 2j2 homology_model.prmtop homology_model.inpcrd
```

Fig. 3. How to prepare files for MD simulations using the XLEaP module of AmberTools 1.4: the Cytochrome P450 2J2 example.

4.2. Relaxing the System Prior to MD: Minimization of the Solvent

The minimization procedure for the solvated homology model consists of a two stage approach. In the first stage, the protein is kept rigid and only the positions of water molecules and ions are be optimized. In the second stage, the whole system is minimized. AMBER supports different minimization algorithms: the most commonly used are steepest descent and conjugate gradient. In general, the steepest descent algorithm is good for quickly removing the largest strains in the system but converges slowly when close to a minimum.

Harmonic positional restraints are used in the initial minimization to keep the protein fixed by specifying the initial structure as a reference structure. This can be seen as a spring attached to each of the solute atoms connected to their initial positions. Moving each restrained atom from the starting position produces a force that tends to restore it to the initial position. By varying the magnitude of the force constant, this effect can be increased or decreased (see Note 10). The *Sander* input file for the initial minimization of solvent and ions (***min1.in***) should be prepared as follows:

```
P450_2j2:      initial     minimization
solvent + ions
 &cntrl
  imin    = 1,
  maxcyc  = 1000,
  ncyc    = 500,
  ntb     = 1,
  ntr     = 1,
  cut     = 8.0,
 /
Hold the solute fixed
50.0
RES 1 458
END
END
```

where

- IMIN = 1: minimization is turned on.
- MAXCYC = 1,000: conduct a total of 1,000 steps of minimization.
- NCYC = 500: initially do 500 steps of steepest descent minimization followed by 500 steps (MAXCYC–NCYC) steps of conjugate gradient minimization.
- NTB = 1: use constant volume periodic boundaries.
- CUT = 8.0: use a cutoff of 8 Å.
- NTR = 1: use position restraints based on the atoms expressed in the last 5 lines of the input file. In this example, a force constant of 50 kcal/mol Å^2 and restrain residues 1 through 458 (the solute). This means that the water and counterions are free to move.

The PME method is performed by default (see Note 9). The minimization can be run by using the *homology_model.prmtop* and *homology_model.inpcrd* files created before and by typing (on a single line):

$AMBERHOME/exe/sander –O –i min1.in –o min1.out –p homology_model.prmtop –c homology_model.inpcrd –r homology_model_min1.rst –ref homology_model.inpcrd

This should take no more than 5–10 min to run and will produce *min1.out* and *homology_model_min1*.rst as output. Note that, on the command line, the option "*–ref*" specifies the reference structure (*homology_model.inpcrd*) to consider for the atomic position restraints. Runtime could be reduced by running the simulation in parallel; however, this is beyond the scope of this tutorial.

Inspecting the *min1.out* file reveals that there are initially rather high van der Waals and electrostatics energies (VDWAALS, 1–4 VDW and EEL terms) which reveal bad contacts in both the water and the solute. These rapidly decrease as the solvent positions are minimized.

4.3. Relaxing the System Prior to MD: Minimization of the Solute

The next stage of minimization consists of minimizing the entire system using a combination of steepest descent and conjugate gradient methods. In this case, 3,000 steps of unrestrained minimization will be performed. Since minimization is generally very quick, it is often recommended to run more minimization steps than strictly necessary. Here, 3,000 cycles should be enough as described in the paper used as reference (75). The input file (*min2.in*) for the minimization and the command used to run it are as follows:

```
P450_2j2:  initial  minimization  of  the
whole system
 &cntrl
  imin   = 1,
  maxcyc = 3000,
  ncyc   = 1500,
  ntb    = 1,
  ntr    = 0,
  cut    = 8.0,
 /
$AMBERHOME/exe/sander -O -i min2.in -o min2.out -p
homology_model.prmtop -c homology_model_min1.rst -r
      homology_model_min2.rst
```

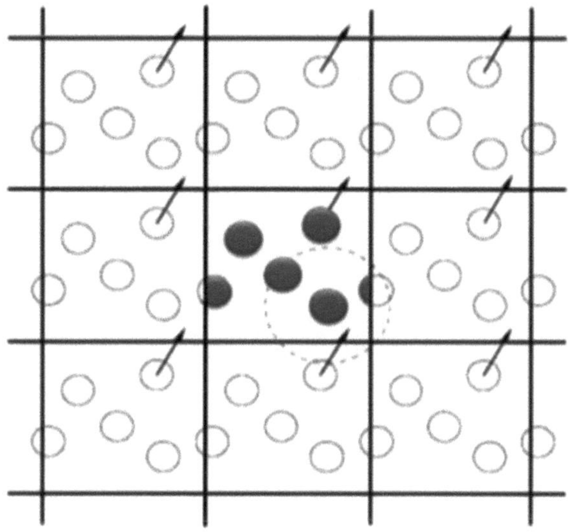

Fig. 4. Two-dimensional representation of periodic boundary conditions. The cut-off for treating the nonbonded interaction for a particle is represented with a *dashed line*.

This should complete within 20–30 min. The *homology_model_min1.rst* file from the previous run, which contains the last structure from the first stage of minimization, was used as the input structure (-c) for this minimization stage. If desired it is now possible to create a PDB file of the minimized structure:

$AMBERHOME/exe/ambpdb -p homology_model.prmtop < homology_model_min2.rst > homology_model_min2.pd

VMD (76), Chimera (77) or other molecular modeling software can be used to visualize this PDB (Fig. 4a). This can also be compared to the initial structure (Fig. 4b).

4.4. Molecular Dynamics (Heating) with Restraints on the Solute

The next stage of the refinement protocol is heating the minimized system to 300 K. A thermostat is used for maintaining and equalizing the system temperature, in this case the Langevin thermostat (78). Langevin dynamics simulate both the effect of molecular collisions and the resulting dissipation of energy that occurs in real solvent by adding a frictional force to model dissipative losses and a random force to model the effect of collisions. Since the input structure is a homology model, it is advisable to use weak positional restraints on the solute during heating. Remember that the final aim of our MD simulation is running production phases at constant temperature and pressure, mimicking laboratory conditions: it would seem prudent to run the heating in an NPT ensemble. At the low temperatures, during the first few picoseconds of the heating phase, the calculation of pressure is inaccurate and the response of the barostat can distort the system. Thus, the first 60 ps of heating is run at constant volume. Once the system has reached

300 K, the restraints can be removed and the ensemble switched to constant pressure before running a further 100 ps of equilibration at 300 K (see Note 11).

Here is the input file for the heating phase (*md1.in*), 60 ps of dynamics simulation with weak positional restraints on the solute. We use SHAKE constraints to fix hydrogen atom bond lengths allowing us to run with a 2 fs time step (50):

```
P450_2j2: heating phase
 &cntrl
  imin    = 0,
  irest   = 0,
  ntx     = 1,
  ntb     = 1,
  cut     = 8.0,
  ntr     = 1,
  ntc     = 2,
  ntf     = 2,
  tempi   = 10.0,
  temp0   = 300.0,
  ntt     = 3,
  gamma_ln = 1.0,
  nstlim = 30000, dt = 0.002,
  ntpr  =  100,  ntwx  =  100,  ntwr  =
1000, ig=-1,
 /
Keep  the  solute  fixed  with  weak
  restraints
  10.0
  RES 1 458
  END
  END
```

and the command to launch it. This time, the command *pmemd* is used since it provides higher performance (see Note 7):

$AMBERHOME/exe/pmemd –O –i md1.in –o md1.out –p homology_model.prmtop –c homology_model_min2.rst –r homology_model_md1.rst –x homology_model_md1.mdcrd –ref homology_model_min2.rst

The file *homology_model_min2.rst* containing the coordinates of the final minimized structure is used not only as the starting point for the heating phase but also as the reference to restrain the solute. This run will take several hours to complete so you may want to leave it running overnight. Alternatively, if you have a multicore machine and the parallel version of AMBER installed, you can run the calculation on multiple cores to speed up the calculation, e.g., *mpirun –np 8 $AMBERHOME/exe/pmemd.MPI –O –i ....*)

The meaning of each of the terms of the *md1.in* input file are as follows:

- IMIN = 0: minimization is turned off, molecular dynamics is run.
- IREST = 0, NTX = 1: only the coordinates of the system are read from the *homology_model_min2.rst* file. Previous velocities are not used to restart the simulation.
- NTB = 1: use constant volume periodic boundaries.
- CUT = 8.0: use a cutoff of 8 Å for the van der Waals interactions.
- NTR = 1: use position restraints based on the information given in the input file. In this case, we will restrain the solute with a force constant of 10.0 kcal/mol $Å^2$.
- NTC = 2, NTF = 2: the SHAKE algorithm is turned on and used to constrain bonds involving hydrogen.
- TEMPI = 10.0, TEMP0 = 300.0: the simulation will start with a temperature of 10 K, allowing it to heat up to 300 K.
- NTT = 3, GAMMA_LN = 1.0: Langevin dynamics is used to control the temperature using a collision frequency of 1.0 ps^{-1}.
- NSTLIM = 30,000, DT = 0.002: a total of 30,000 molecular dynamics steps with a time step of 2 fs per step are run, to give a total simulation time of 60 ps.
- NTPR = 100, NTWX = 100, NTWR = 1,000: write to the output file (NTPR) every 100 steps (200 fs), to the trajectory file (NTWX) every 100 steps and write a restart file (NTWR), in case the job crashes, every 1,000 steps.
- IG = –1: This tells *pmemd* to seed the random number generator using the wall clock time in microseconds. It is recommended this always be set when running Langevin dynamics.

4.5. Molecular Dynamics (Equilibration) Without Restraints on the Solute

After the system has been successfully heated up at constant volume with weak restraints on the solute, the next stage is to run with constant pressure conditions allowing the density of the system to equilibrate. This phase will be run for 100 ps, giving the density time to reach equilibrium. This is the *md2.in* input file:

```
P450_2j2: equilibration phase
 &cntrl
  imin = 0, irest = 1, ntx = 5,
  ntb = 2, pres0 = 1.0, ntp = 1,
  taup = 2.0,
  cut = 8.0, ntr = 0,
  ntc = 2, ntf = 2,
  temp0 = 300.0,
  ntt = 3, gamma_ln = 1.0,
  nstlim = 50000, dt = 0.002,
  ntpr = 100, ntwx = 100, ntwr =
1000, ig=-1,
 /
```

The meaning of each of the terms that have changed is as follows:

- IREST = 1, NTX = 5: this time the simulation will be restarted after the 60 ps of constant volume simulation. IREST tells *sander/pmemd* to restart a simulation, so the time is not reset to zero but will start at 60 ps. Previously, NTX was set at the default of 1 which meant only the coordinates were read from the *rst* file. This time, NTX is 5 meaning that the coordinates, velocities, and box information will be read from the *rst* file.
- NTB = 2, PRES0 = 1.0, NTP = 1, TAUP = 2.0: use constant pressure periodic boundary conditions with an average pressure of 1 atm (PRES0). Isotropic position scaling is used to maintain the pressure (NTP = 1) and a relaxation time of 2 ps is used (TAUP = 2.0).
- NTR = 0: no positional restraints are applied.
- NSTLIM = 50,000, DT = 0.002: a total of 50,000 molecular dynamics steps are run, with a time step of 2 fs per step, to give a total simulation time of 100 ps.

Using the following command, the equilibration is run. The *rst* file from the heating stage is used to start this step since this contains the final coordinates, velocities, and box information from the previous heating run.

$AMBERHOME/exe/pmemd –O –i md2.in –o md2.out –p homology_model.prmtop –c homology_model_md1.rst –r homology_model_md2.rst –x homology_model_md2.mdcrd

4.6. Analysis of Trajectories: Has an Initial Equilibrium Been Reached?

Before starting the production phase of the MD refinement, it is essential to check that the system has reached an initial equilibrium. There are a number of system properties that should be monitored to assess the quality of the 160 ps of heating and equilibration.

These include the potential, kinetic and total energies, the temperature, the pressure, the density, and the RMSD. The various properties from both output files *md1.out*, *md2.out* should be extracted. For this, a *perl* script *process_mdout.perl* is provided in *$AMBERHOME/AmberTools/src/etc/*. This can be run as follows:

perl $AMBERHOME/AmberTools/src/etc/process_mdout.perl md1.out md2.out

This process outputs a series of summary files that can be plotted to evaluate if the various properties have reached an initial equilibrium. The files *summary.EPTOT*, *summary.EKTOT*, and *summary.ETOT* give information about the energies. These are plotted in Fig. 5a. Here, the black line (positive) is the kinetic energy, the red line is the potential energy (negative), and the blue line is the total energy. It can be seen that all of the energies increased during the very first ps, corresponding to the heating from 10 to 300 K. The kinetic energy then remained constant implying that the thermostat, which acts on the kinetic energy, was working correctly. The potential energy, and consequently the total energy, initially increased and then plateaued during the constant volume stage (0–60 ps) before decreasing as the system relaxed when the restraints were switched off and the box volume allowed to vary during the constant pressure run (60–80 ps). The potential energy then leveled off and remained constant for the remainder of the simulation (80–160 ps), indicating that the initial relaxation away from the starting structure was successful.

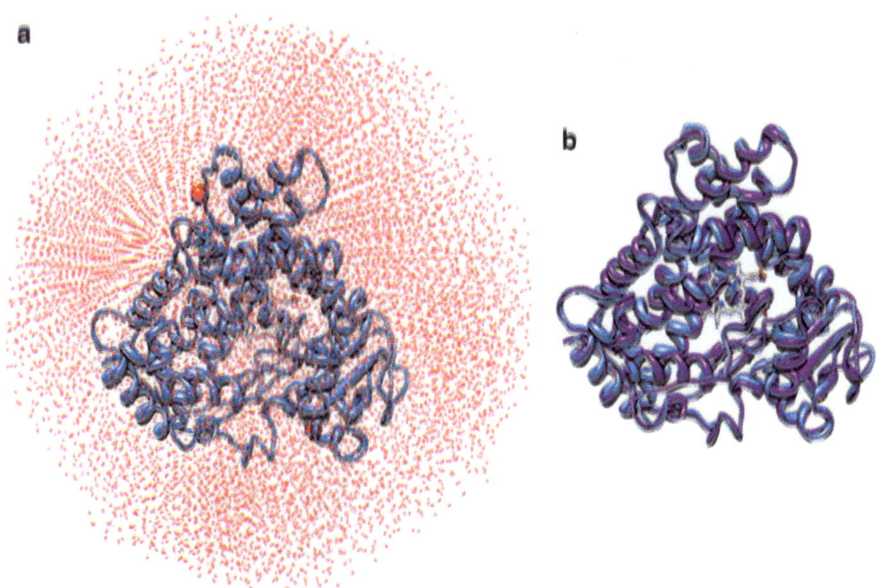

Fig. 5. Visualization of the solvated initial minimized Cytochrome P450 2J2 homology model (**a**) and superposition of the initial structure and the structure after the minimization (**b**).

Figure 5b shows the system temperature as a function of simulation time. This started at 10 K and then increased to 300 K over a period of about 5 ps. The temperature then remained more or less constant for the remainder of the simulation indicating the use of Langevin dynamics for temperature regulation was successful.

The pressure plot (Fig. 6c) is slightly different than the previous plots. For the first 60 ps the pressure is zero. This is to be expected since a constant volume simulation was run in which the pressure was not evaluated. At 60 ps, the constant pressure simulation allowed the volume of the box to change, at which point the pressure dropped sharply becoming negative. The negative pressures correspond to a *force* acting to decrease the size of the box, while the positive pressures correspond to a *force* acting to increase it. The important point here is that while the pressure graph seems to show that the pressure fluctuated wildly during the simulation the mean pressure stabilized around 1 atm after about 50 ps of simulation.

Finally, the density (Fig. 6d) is expected to mirror the volume. The density is not written to the output file during constant volume simulations and so is only reported from 60 ps onwards. It can be seen from Fig. 6d that the system has equilibrated at a density of approximately 1.04 g/cm^3. This is reasonable since the density of pure liquid water at 300 K is approximately 1.00 g/cm^3.

A final question is: have the structural features remained reasonable? One useful measure to consider is the root mean square deviation (RMSD) from the starting structure. The program *ptraj*, part of AmberTools, can be used to calculate the RMSD as a function of time. Here the RMSD of the alpha-carbons will be calculated from the final structure of the minimization (*homology_model_min2.pdb*). Using the following input file (*rmsd.in*) and the following command line, *ptraj* will calculate the RMSD as a function of the simulation time:

```
trajin homology_model_md1.mdcrd
trajin homology_model_md2.mdcrd
reference homology_model_min2.pdb
rms reference out backbone.rmsd
@CA,C,N time 0.2
/
```

The time is set to 0.2 ps corresponding to the frame rate in the trajectory (mdcrd) file (100 steps × 2 fs per step).

$AMBERHOME/exe/ptraj_homology_model.prmtop < rmsd.in > rmsd.out

The output file, *backbone.rmsd*, can be plotted (Fig. 6). From Fig. 6, it can be seen that the RMSD of the backbone atoms

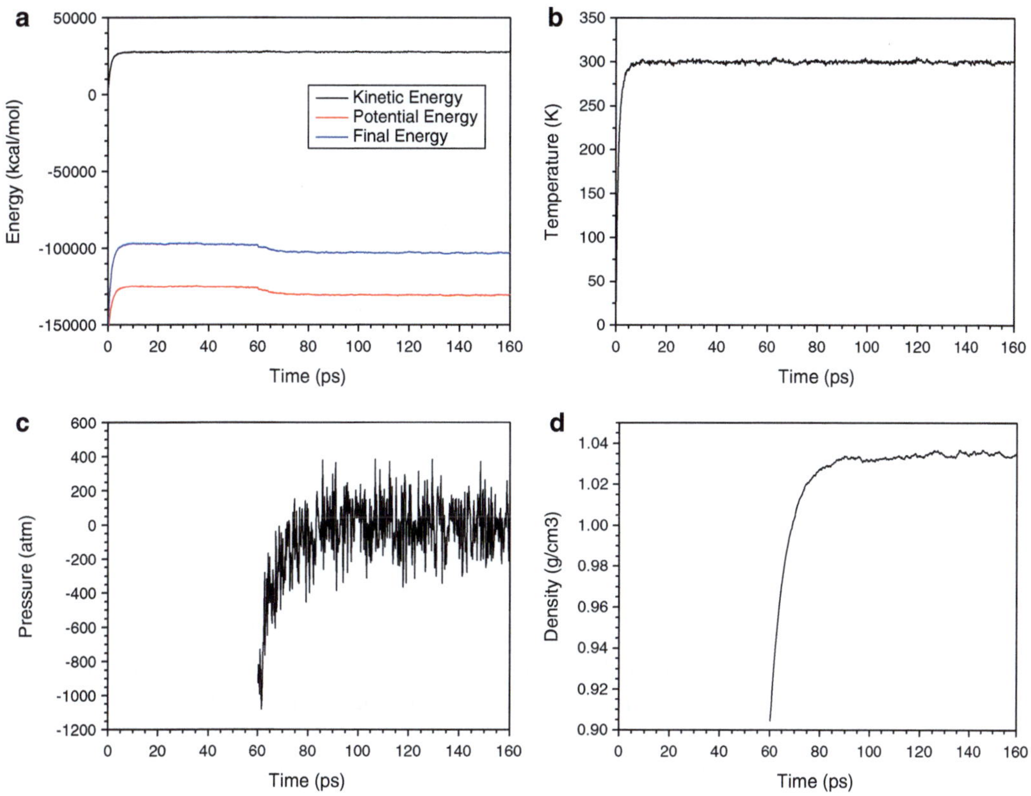

Fig. 6. Plots against time for the heating and equilibration phases of the energies (**a**), temperature (**b**), pressure (**c**), and density (**d**).

remained low for the first 60 ps, due to the restraints applied on the solute. Upon removing the restraints, the RMSD increased as the molecule relaxed within the solvent. The RMSD initially plateaued but then continued to rise towards the end of the equilibration phase. This continued small rise in RMSD suggests that the simulation has not yet reached an initial equilibrium. However, the absence of any sudden jumps in the RMSD indicates that the simulation is stable and, as will be explained below the first 800 ps of production can be considered as additional equilibration and so it is okay to proceed with the production phase of the MD refinement (see Note 12).

4.7. Molecular Dynamics Refinement Production Phase

Once an initial equilibrium has been reached, with the temperature and density stable, the final stage of the simulation can be run. This consists of running a production simulation at 300 K. Since we are following the protocol in the Li et al. (75) paper, 1 ns of simulation at 300 K will be run. For this the following input file can be used (*md3.in*):

```
P450_2j2: production phase
 &cntrl
  imin = 0, irest = 1, ntx = 5,
  ntb = 2, pres0 = 1.0, ntp = 1,
  taup = 1.0,
  cut = 8.0, ntr = 0,
  ntc = 2, ntf = 2,
  tempi = 300.0, temp0 = 300.0,
  ntt = 3, gamma_ln = 0.5,
  nstlim = 500000, dt = 0.002,
  ntpr = 100, ntwx = 100, ntwr =
1000, ig=-1,
/
```

This stage consists of 500,000 steps (NSTLIM) with a 2 fs time step (DT) yielding 1 ns of MD production. Given the system now appears to be stable and the temperature equilibrated the degree of thermostat coupling can now be reduced (GAMMA_LN=0.5). The command for launching the production phase is:

$AMBERHOME/exe/pmemd –O –i md3.in –o md3.out –p homology_model.prmtop –c homology_model_md2.rst –r homology_model_md3.rst –x homology_model_md3.mdcrd

This will take several days to run on a single CPU core so in practice should be run in parallel using the MPI version of *pmemd* (*pmemd.MPI*).

4.8. How to Obtain the Refined Homology Model from the Simulation

The final stage of the homology model refinement is to process the production trajectory to obtain a representative structure that can then be minimized to provide a refined homology model. For the purposes of this tutorial, the Cartesian averaging, followed by minimization, approach utilized in the Li et al. paper will be used (see Note 13).

First a mass-weighted backbone RMSD fit of every frame of the trajectory collected during the production phase to the first frame is performed: this removes rotation and translation aspects of the solute during the simulation. Second, the last 200 ps of the production trajectory where the average structure may be more meaningful, since the system has had more time to explore phase space, are considered for the calculation of the average Cartesian structure. At the same time, the water and ions can be removed. This can be accomplished with *ptraj* using the input file, *average.in*:

```
trajin homology_model_md3.mdcrd 4001
5000
strip :WAT
strip :Na+
rms first @C,CA,N
average average.pdb PDB
/
```

and the command for running it:

$AMBERHOME/exe/ptraj homology_model.prmtop <average.in >average.out

This creates the file *average.pdb* containing the averaged Cartesian coordinates of the last 200 ps (frame 4,001–5,000) of solute from the production MD simulation. Figure 7 shows the result.

As can be seen from Fig. 7, some parts of the structure appear very small, notably some of the hydrogen bonds lengths are tiny. As explained in Note 13, this is a limitation of averaging in Cartesian space and this is why the use of a snapshot from MD production or clustering, although more complex, may be more appropriate in some cases. The distorted parts of the average structure suggest that these residues are very dynamic and able to freely rotate during this section of the trajectory. What can be seen from Fig. 8 though is that the backbone is well formed, indicating that the

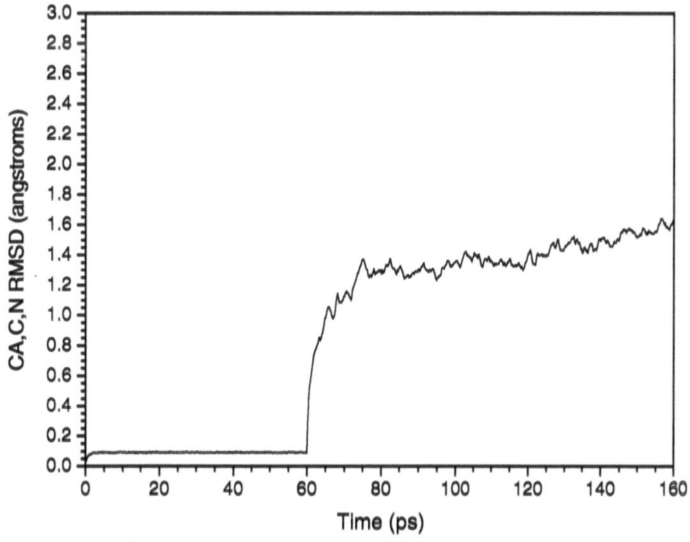

Fig. 7. Backbone (CA, C, N) RMSD vs. time for the heating and equilibration phase of the MD refinement.

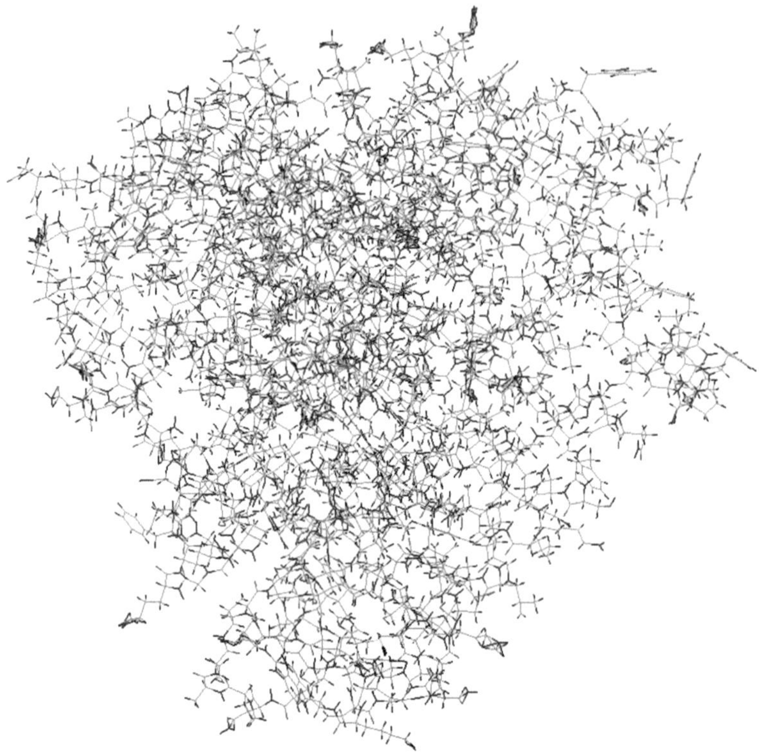

Fig. 8. Average structure from the last 1,000 steps (800–1,000 ps) of the production MD simulation.

folded part of the structure stays well defined between 800 and 1 ns. This corresponds with the RMSD plot of the production phase calculated with *ptraj* (*prod_rmsd.in*):

```
  trajin homology_model_md3.mdcrd
  reference homology_model_min2.pdb
  rms reference out prod_backbone.rmsd
  @CA,C,N time 0.2
  /
$AMBERHOME/exe/ptraj homology_model.prmtop
      < prod_rmsd.in >prod_rmsd.out
```

To complete the refinement, the final step is to minimize the averaged structure. In following the approach used in ref. 75, a total of 5,000 cycles of conjugate gradient minimization will be run. In ref. 75, it is not clear how solvation was dealt with during this final minimization stage, however, for the purposes of this tutorial a Generalized Born implicit solvation model will be used (79).

This avoids the complexities of trying to minimize either the averaged solvent, which does not provide a meaningful structure, or new solvent which would be added by *XLEaP*.

The first stage is to build a topology and coordinate file for the averaged structure. This can be done using *XLEaP* as described above. This time skipping the addition of counter ions and solvent:

$AMBERHOME/exe/xleap –s –f$AMBERHOME/dat/leap/cmd/leaprc.ff99SBloadamberparams heme_all.frcmodloadamberprep heme_all.prep2j2=loadpdb average.pdbsaveamberparm 2j2 average.prmtop average.inpcrd

The following input file (*average_min.in*) can then be used to minimize the averaged structure:

```
P450_2j2: Final averaged structure minimization
 &cntrl
  imin    = 1,
  maxcyc  = 5000,
  ncyc    = 0,
  ntb     = 0,
  ntr     = 0,
  igb     = 1,
  cut     = 9999.0,
 /
```

where:

- NTB = 0: the simulation is not a periodic one.
- IGB = 1: The Generalized Born implicit solvent model will be used.
- CUT = 9,999.0: No cutoff will be used since this is an implicit solvation model. Setting CUT to larger than the system size ensures this.

Running the minimization with:

$AMBERHOME/exe/pmemd –O –i average_min.in –o average_min.out –p average.prmtop –c average.inpcrd –r average_min.rst

yields the final refined homology model as *average_min.rst*. This can then be converted to a pdb file:

$AMBERHOME/exe/ambpdb –p average.prmtop < average_min.rst > 2j2_refined_model.pdb

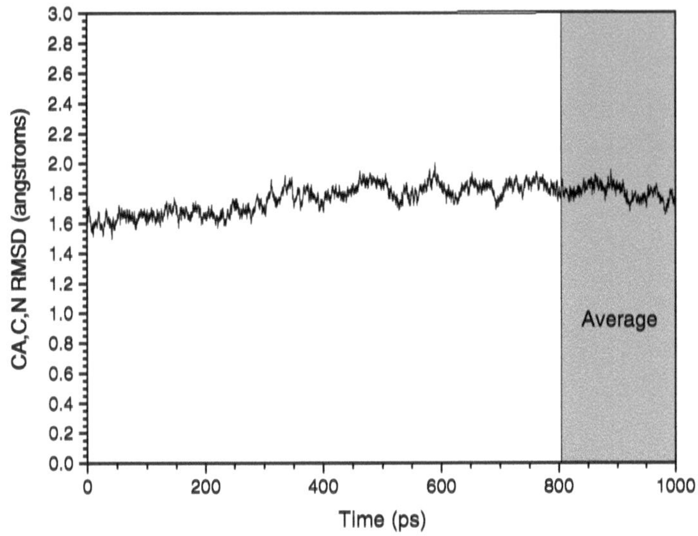

Fig. 9. Backbone (CA, C, N) RMSD vs. time for the production phase of the MD refinement.

This structure can then be used as the starting structure for a range of studies such as additional MD simulations, docking or other drug design studies. As before, various molecular modeling programs can be used to visualize the final structure. Figure 9 shows cross eyes stereo images of the final refined structure of Cytochrome P450 2J2 (A) and the final refined structure overlaid with the initial homology model (B).

5. Notes

1. AMBER 11 and AmberTools are available from the following web site: (http://ambermd.org/). Installation instructions can be found in the documentation available at: (http://ambermd.org/doc11/). The various input and output files used in this book chapter are available at: (http://ambermd.org/tutorials/homology_modelling_humana_2011/).
2. FF99SB contains several improvements compared to the older versions (45). The most notable changes are updated torsion terms for Phi–Psi angles which fix the overestimation of alpha helices that occurs when using the older force fields. For homology model refinement such improvements are clearly critical for obtaining accurate results.
3. To build and parameterize nonstandard molecules, a tutorial is available at the AMBER web site (http://ambermd.org/tutorials/basic/tutorial4b/).

4. The names used for all the residues in the PDB files must match those defined in the *XLEaP* force field library files or in user defined library files. *XLEaP* expects that all atoms of each residue in the PDB file are listed in the same order as in the corresponding libraries. The TER separator should be added for ending a protein chain and beginning a new one as well as for separating proteins from ligands or other elements of the system. Information about the structural features, origin of the protein, and connectivity, normally described at the top and at the end of a PDB file, should be removed. It is important to remember these details before creating the input files for the simulation.
5. Dysfunctional *XLEaP* menus may be linked to NumLock toggled on.
6. It is also helpful to view the new structure to ensure that the charges have been placed as intended by using the *edit* command. The new unit *2j2* can be viewed using the edit command of *XLEaP* (*edit 2j2*).
7. AMBER v11 contains two dynamics engines. The first is called *Sander*, this supports all standard and advanced MD methods implemented in AMBER, however, because of this it is not highly optimized for speed. The second, called *pmemd*, supports a subset of the functionality of *Sander*, but is significantly faster both in serial and in parallel. In this example, we use *Sander* for the minimizations. However, for a faster computation of the MD trajectories, *pmemd* will be used.
8. The first problems typically encountered when performing MD refinement of homology models are the close contacts between protein atoms, after *XLEaP* added hydrogens and solvent. As the homology model does not include solvent, the solvation process can give very large initial van der Waals and electrostatic forces. Additionally, while a truncated octahedral box of pre-equilibrated TIP3P water molecules was created to solvate the system, the initial water positions were not influenced by the electrostatic field of the solute. Moreover, there may be gaps between solvent and solute as well as between solvent and box edges. Unfortunately, such void space can lead to the formation of *vacuum* bubbles and subsequent instability in the MD simulation. Thus, a meticulous minimization is typically needed before slowly heating the system to 300 K. It is also advisable to allow the water box to relax during an equilibration stage prior to running the production: by keeping the pressure constant (in an NPT ensemble), the volume of the box will change. This approach lets the water molecules around the solute and the system's density to equilibrate.
9. During the simulation in which everything is free to move, the biological system, placed in a box of water molecules, includes some atoms belonging to solvent and/or solute at the edge, in contact with the surrounding *vacuum*.

To avoid this artificial situation and to ensure a complete immersion of the solute in the solvent during the simulation, periodic boundary conditions are employed. In this way, the system will be surrounded with replicas of itself in all directions to yield a periodic lattice of identical cells. When a particle moves in the central cell, its periodic image will move in the same manner in the other cells. When it is found at the edge, it will leave the central cell, entering from the opposite side of the same cell (Fig. 10). The computational costs of this method can be reduced by introducing appropriate approximations for treating the van der Waals and electrostatic interactions. In periodic boundary conditions, all charged particles of a system interact with each other in the central box and in all image boxes following Coulomb's law modified by the appropriate translation vectors. By employing the Particle Mesh Ewald (PME) method, it is possible to obtain the infinite electrostatics by dividing the calculation up between a real space component and a reciprocal space component (80). PME is applied by default in *Sander* and *pmemd* and should always be used for explicit solvent simulations. Since van der Waals interactions fall off quickly with distance, they can be truncated at a specific *cut-off* distance. For most calculations, the ideal range is

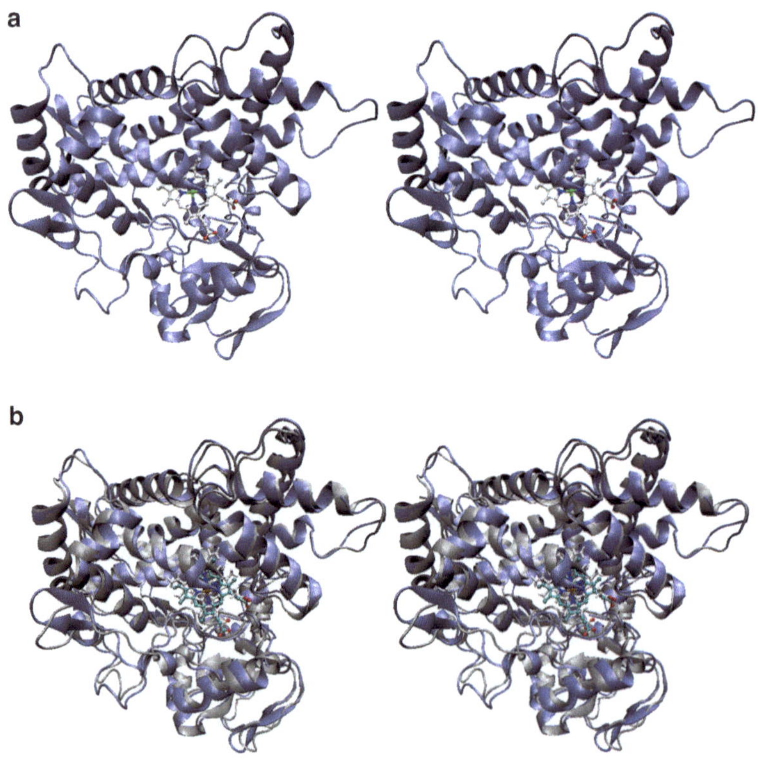

Fig. 10. Cross-eyed stereo images of the final refined structure of Cytochrome P450 2J2 (**a**) and the final structure overlaid with the initial homology model (**b**).

between 8 and 10 Å. One should never reduce this below 8 Å for periodic boundary PME calculations.

10. Harmonic positional restraints during the minimization steps can be especially useful in refinement of homology models which may be far from the equilibrium. Minimization and MD can be run stepwise with restraint forces gradually reduced.
11. We start the simulation at 10 K, instead of 0 K to provide the system with a very small set of initial velocities, generated as a Boltzmann distribution. This is not critical but it can help in creating uncorrelated trajectories when running multiple simulations, with different initial random seeds.
12. One can also start collecting data, for averaging, from the very beginning of the production phase. In this case, it would likely be necessary to first extend the equilibration step.
13. There are a number of approaches by which this can be done. One of the simplest, together with the extraction of the last snapshot from the MD production, is to calculate the average structure, in Cartesian space, over a portion of the production trajectory. This is the method used by Li et al. (75). It works well in the majority of cases but it may cause problems if parts of the protein are disordered since a simple average of the Cartesian space sampled will yield nonphysical structures for these parts of the protein. Similar issues can occur with groups that are free to rotate, for example methyl groups. A more robust approach, yet beyond the scope of this tutorial, would be to perform clustering analysis on the production trajectory. This would generate a number of centroids representing specific clusters of structures sampled during the 1 ns production run. The trajectory snapshot with RMSD closest to each of the centroids could then be subjected to minimization providing a series of refined homology models, similar to the collection of structures typically obtained from NMR refinement.

Acknowledgments

This work was supported in part by grant 09-LR-06-117792-WALR from the University of California Lab Fees program (RCW) and grant NSF1047875 from the US National Science Foundation (RCW). We additionally thank the NSF TeraGrid (award TG-MCB090110) for providing supercomputer time in support of this work. We would also like to thank Weihua Li and Yun Tang of the School of Pharmacy, East China University of Science and Technology for their fast response and willingness to share with us their P450 2J2 homology structure. We thank Pr. Pierre-Alain Carrupt (School of Pharmaceutical Sciences, University of Geneva, University of Lausanne) for technical support.

References

1. Becker, O. M. (2001) *Computational biochemistry and biophysics* CRC, New York.
2. Cramer, C. J. (2004) *Essentials of computational chemistry: theories and models* John Wiley & Sons Inc, New York.
3. McCammon, J. A., Gelin, B. R., and Karplus, M. (1977) Dynamics of folded proteins, *Nature 267*, 585–590.
4. Duan, Y. and Kollman, P. (1998) Pathways to a protein folding intermediate observed in a 1-microsecond simulation in aqueous solution, *Science 282*, 740–744.
5. Yeh, I. C. and Hummer, G. (2002) Peptide loop-closure kinetics from microsecond molecular dynamics simulations in explicit solvent, *J. Am. Chem. Soc 124*, 6563–6568.
6. Klepeis, J. L., Lindorff-Larsen, K., Dror, R. O., and Shaw, D. E. (2009) Long-timescale molecular dynamics simulations of protein structure and function, *Current opinion in structural biology 19*, 120–127.
7. Sanbonmatsu, K. Y., Joseph, S., and Tung, C. S. (2005) Simulating movement of tRNA into the ribosome during decoding, *Proceedings of the National Academy of Sciences of the United States of America 102*, 15854–15859.
8. Freddolino, P. L., Arkhipov, A. S., Larson, S. B., McPherson, A., and Schulten, K. (2006) Molecular dynamics simulations of the complete satellite tobacco mosaic virus, *Structure 14*, 437–449.
9. Simmerling, C., Strockbine, B., and Roitberg, A. E. (2002) All-atom structure prediction and folding simulations of a stable protein, *J. Am. Chem. Soc 124*, 11258–11259.
10. Lei, H., Wu, C., Liu, H., and Duan, Y. (2007) Folding free-energy landscape of villin headpiece subdomain from molecular dynamics simulations, *Proceedings of the National Academy of Sciences 104*, 4925–4930.
11. He, Y., Chen, C., and Xiao, Y. (2009) United-Residue (UNRES) Langevin Dynamics Simulations of trpzip2 Folding, *Journal of Computational Biology 16*, 1719–1730.
12. Larsson, P., Wallner, B., Lindahl, E., and Elofsson, A. (2008) Using multiple templates to improve quality of homology models in automated homology modeling, *Protein Science 17*, 990–1002.
13. Krieger, E., Joo, K., Lee, J., Lee, J., Raman, S., Thompson, J., Tyka, M., Baker, D., and Karplus, K. (2009) Improving physical realism, stereochemistry, and side-chain accuracy in homology modeling: Four approaches that performed well in CASP8, *Proteins: Structure, Function, and Bioinformatics 77*, 114–122.
14. Xiang, Z. (2006) Advances in homology protein structure modeling, *Current protein & peptide science 7*, 217–227.
15. Stumpff-Kane, A. W., Maksimiak, K., Lee, M. S., and Feig, M. (2008) Sampling of near-native protein conformations during protein structure refinement using a coarse-grained model, normal modes, and molecular dynamics simulations, *Proteins: Structure, Function, and Bioinformatics 70*, 1345–1356.
16. Xu. D, Williamson. M J, Walker. R C. (2010) Advancements in Molecular Dynamics Simulations of Biomolecules on Graphical Processing Units, in *Ann.Rep.Comp.Chem 6*, pp 2–19.
17. Koehler, M., Ruckenbauer, M., Janciak, I., Benkner, S., Lischka, H., and Gansterer, W. (2010) Supporting Molecular Modeling Workflows within a Grid Services Cloud, *Computational Science and Its Applications, ICCSA 2010* 13–28.
18. Krieger, E., Joo, K., Lee, J., Lee, J., Raman, S., Thompson, J., Tyka, M., Baker, D., and Karplus, K. (2009) Improving physical realism, stereochemistry, and side-chain accuracy in homology modeling: Four approaches that performed well in CASP8, *Proteins: Structure, Function, and Bioinformatics 77*, 114–122.
19. Kryshtafovych, A., Fidelis, K., and Moult, J. (2009) CASP PROGRESS REPORTS, *Proteins 77*, 217–228.
20. Fan, H. and Mark, A. E. (2004) Refinement of homology based protein structures by molecular dynamics simulation techniques, *Protein Science 13*, 211–220.
21. Berendsen, H. J. C., van der Spoel, D., and Van Drunen, R. (1995) GROMACS: a message-passing parallel molecular dynamics implementation, *Computer Physics Communications 91*, 43–56.
22. Lindahl, E., Hess, B., and van der Spoel, D. (2001) GROMACS 3.0: a package for molecular simulation and trajectory analysis, *Journal of Molecular Modeling 7*, 306–317.
23. Berendsen, H. J. C., Postma, J. P. M., van Gunsteren, W. F., and Hermans, J. (1981) Interaction models for water in relation to protein hydration, *Intermolecular forces* 331–342.
24. Im, W., Lee, M. S., and Brooks III, C. L. (2003) Generalized born model with a simple smoothing function, *Journal of Computational Chemistry 24*, 1691–1702.
25. Chopra, G., Summa, C. M., and Levitt, M. (2008) Solvent dramatically affects protein structure refinement, *Proceedings of the National Academy of Sciences 105*, 20239–20244.

26. Chen, J. and Brooks III, C. L. (2007) Can molecular dynamics simulations provide high resolution refinement of protein structure?, *Proteins: Structure, Function, and Bioinformatics 67*, 922–930.
27. Anishkin, A., Milac, A. L., and Guy, H. R. (2010) Symmetry-restrained molecular dynamics simulations improve homology models of potassium channels, *Proteins: Structure, Function, and Bioinformatics 78*, 932–949.
28. Phillips, J. C., Braun, R., Wang, W., Gumbart, J., Tajkhorshid, E., Villa, E., Chipot, C., Skeel, R. D., Kale, L., and Schulten, K. (2005) Scalable molecular dynamics with NAMD, *Journal of Computational Chemistry 26*, 1781–1802.
29. Wroblewska, L. and Skolnick, J. (2007) Can a physics based, all atom potential find a protein's native structure among misfolded structures? I. Large scale AMBER benchmarking, *Journal of Computational Chemistry 28*, 2059–2066.
30. Krieger, E., Koraimann, G., and Vriend, G. (2002) Increasing the precision of comparative models with YASARA NOVA - a self parameterizing force field, *Proteins: Structure, Function, and Bioinformatics 47*, 393–402.
31. Cavasotto, C. N. and Phatak, S. S. (2009) Homology modeling in drug discovery: current trends and applications, *Drug discovery today 14*, 676–683.
32. Klepeis, J. L., Lindorff-Larsen, K., Dror, R. O., and Shaw, D. E. (2009) Long-timescale molecular dynamics simulations of protein structure and function, *Current opinion in structural biology 19*, 120–127.
33. Floquet, N., M'Kadmi, C., Perahia, D., Gagne, D., Berge, G., Marie, J., Baneres, J. L., Galleyrand, J. C., Fehrentz, J. A., and Martinez, J. (2010) Activation of the ghrelin receptor is described by a privileged collective motion: a model for constitutive and agonist-induced activation of a sub-class A G-protein coupled receptor (GPCR), *Journal of molecular biology 395*, 769–784.
34. Zhang, Y., Sham, Y. Y., Rajamani, R., Gao, J., and Portoghese, P. S. (2005) Homology modeling and molecular dynamics simulations of the mu opioid receptor in a membraneûaqueous system, *Chembiochem 6*, 853–859.
35. Aarts, E. H. L. and Van Laarhoven, P. J. M. (1985) Statistical cooling: A general approach to combinatorial optimization problems, *Philips J. Res. 40*, 193–226.
36. Meng, X. Y., Zheng, Q. C., and Zhang, H. X. (2009) A comparative analysis of binding sites between mouse CYP2C38 and CYP2C39 based on homology modeling, molecular dynamics simulation and docking studies, *Biochimica et Biophysica Acta (BBA)-Proteins & Proteomics 1794*, 1066–1072.
37. Speranskiy, K., Cascio, M., and Kurnikova, M. (2007) Homology modeling and molecular dynamics simulations of the glycine receptor ligand binding domain, *Proteins: Structure, Function, and Bioinformatics 67*, 950–960.
38. Sugita, Y. and Okamoto, Y. (1999) Replica-exchange molecular dynamics method for protein folding, *Chemical Physics Letters 314*, 141–151.
39. Zhu, J., Fan, H., Periole, X., Honig, B., and Mark, A. E. (2008) Refining homology models by combining replica exchange molecular dynamics and statistical potentials, *Proteins: Structure, Function, and Bioinformatics 72*, 1171–1188.
40. Nguyen, T. L., Gussio, R., Smith, J. A., Lannigan, D. A., Hecht, S. M., Scudiero, D. A., Shoemaker, R. H., and Zaharevitz, D. W. (2006) Homology model of RSK2 N-terminal kinase domain, structure-based identification of novel RSK2 inhibitors, and preliminary common pharmacophore, *Bioorganic & medicinal chemistry 14*, 6097–6105.
41. Case, D. A., Darden, T., Cheatham III, T. E., Simmerling, C., Wang, J., Duke, R. E., Luo, R., Walker, R. C., Zhang, W., Merz, K. M., B.Roberts, B.Wang, S.Hayik, A.Roitberg, G.Seabra, I.Kolossváry, K.F.Wong, F.Paesani, , J. V., J.Liu, X.Wu, , S. R. B., T.Steinbrecher, H.Gohlke, Q.Cai, X.Ye, J.Wang, M.-J.Hsieh, G.Cui, D.R.Roe, D.H.Mathews, , M. G. S., C.Sagui, V.Babin, T.Luchko, S.Gusarov, and , A. K. (2010) Amber 11, University of California (San Francisco).
42. Brooks, B. R., Bruccoleri, R. E., and Olafson, B. D. (1983) CHARMM: A program for macromolecular energy, minimization, and dynamics calculations, *Journal of Computational Chemistry 4*, 187–217.
43. Plimpton, S. (1995) Fast parallel algorithms for short-range molecular dynamics, *Journal of Computational Physics 117*, 1–19.
44. Cornell, W. D., Cieplak, P., Bayly, C. I., Gould, I. R., Merz, K. M., Ferguson, D. M., Spellmeyer, D. C., Fox, T., Caldwell, J. W., and Kollman, P. A. (1995) A second generation force field for the simulation of proteins, nucleic acids, and organic molecules, *Journal of the American Chemical Society 117*, 5179–5197.
45. Wickstrom, L., Okur, A., and Simmerling, C. (2009) Evaluating the performance of the ff99SB force field based on NMR scalar coupling data, *Biophysical journal 97*, 853–856.
46. Holtje, H. D., Sippl, W., Rognan, D., and Folkers G. (2008) *Molecular modeling: basic principles and applications* WILEY-VCH, Weinheim.

47. Verlet, L. (1968) Computer experiments on classical fluids. ii. equilibrium correlation functions, *Phys. Rev 165*, 201–214.
48. Honeycutt, R. W. (1970) The potential calculation and some applications, *Methods in Computational Physics 9*, 136–211.
49. Grenander, U. (1959) *Probability and statistics: the Harald Cramer volume* Almqvist & Wiksell.
50. Ryckaert, J. P., Ciccotti, G., and Berendsen, H. J. C. (1977) Numerical integration of the Cartesian equations of motion of a system with constraints: molecular dynamics of n-alkanes, *J. comput. Phys 23*, 327–341.
51. Wyss, P. C., Gerber, P., Hartman, P. G., Hubschwerlen, C., Locher, H., Marty, H. P., and Stahl, M. (2003) Novel dihydrofolate reductase inhibitors. Structure-based versus diversity-based library design and high-throughput synthesis and screening, *J. Med. Chem 46*, 2304–2312.
52. Bortolato, A., Mobarec, J. C., Provasi, D., and Filizola, M. (2009) Progress in elucidating the structural and dynamic character of G Protein-Coupled Receptor oligomers for use in drug discovery, *Current pharmaceutical design 15*, 4017–4025.
53. Costanzi, S., Siegel, J., Tikhonova, I. G., and Jacobson, K. A. (2009) Rhodopsin and the others: a historical perspective on structural studies of G protein-coupled receptors, *Current pharmaceutical design 15*, 3994–4002.
54. Mobarec, J. C. and Filizola, M. (2008) Advances in the development and application of computational methodologies for structural modeling of G-protein-coupled receptors, *Expert Opin. Drug Discov. 3*, 343–355.
55. Valadez, E., Ulloa-Aguirre, A., and Pin eiro, A. (2008) Modeling and molecular dynamics simulation of the human gonadotropin-releasing hormone receptor in a lipid bilayer, *The Journal of Physical Chemistry B 112*, 10704–10713.
56. Yarnitzky, T., Levit, A., and Niv, M. Y. (2010) Homology modeling of G-protein-coupled receptors with X-ray structures on the rise, *Current opinion in drug discovery & development 13*, 317–325.
57. Nebert, D. W. and Russell, D. W. (2002) Clinical importance of the cytochromes P450, *The Lancet 360*, 1155–1162.
58. Sali, A., Potterton, L., Yuan, F., van Vlijmen, H., and Karplus, M. (1995) Evaluation of comparative protein modeling by MODELLER, *Proteins: Structure, Function, and Bioinformatics 23*, 318–326.
59. Dauber-Osguthrop, P., Roberts, V. A., Osguthorpe, D. J., Wolff, J., Genest, M., and Hagler, A. T. (1988) Structure and energetics of ligand binding to proteins: Escherichia coli dihydrofolate reductase trimethoprim, a drug receptor system, *Proteins: Structure, Function, and Bioinformatics 4*, 31–47.
60. Jorgensen, W. L., Chandrasekhar, J., Madura, J. D., Impey, R. W., and Klein, M. L. (1983) Comparison of simple potential functions for simulating liquid water, *The Journal of chemical physics 79*, 926–935.
61. Meng, X. Y., Zheng, Q. C., and Zhang, H. X. (2009) A comparative analysis of binding sites between mouse CYP2C38 and CYP2C39 based on homology modeling, molecular dynamics simulation and docking studies, *Biochimica et Biophysica Acta (BBA)-Proteins & Proteomics 1794*, 1066–1072.
62. Venkatachalam, C. M., Jiang, X., Oldfield, T., and Waldman, M. (2003) LigandFit: a novel method for the shape-directed rapid docking of ligands to protein active sites, *Journal of Molecular Graphics and Modelling 21*, 289–307.
63. Gajendrarao, P., Krishnamoorthy, N., Sakkiah, S., Lazar, P., and Lee, K. W. (2010) Molecular modeling study on orphan human protein CYP4A22 for identification of potential ligand binding site, *Journal of Molecular Graphics and Modelling 28*, 524–532.
64. Houslay, M. D., Schafer, P., and Zhang, K. Y. J. (2005) Keynote review: phosphodiesterase-4 as a therapeutic target, *Drug discovery today 10*, 1503–1519.
65. Pandit, J., Forman, M. D., Fennell, K. F., Dillman, K. S., and Menniti, F. S. (2009) Mechanism for the allosteric regulation of phosphodiesterase 2A deduced from the X-ray structure of a near full-length construct, *Proceedings of the National Academy of Sciences 106*, 18225–18230.
66. Heller, H., Schaefer, M., and Schulten, K. (1993) Molecular dynamics simulation of a bilayer of 200 lipids in the gel and in the liquid crystal phase, *The Journal of Physical Chemistry 97*, 8343–8360.
67. Hamza, A., AbdulHameed, M. D. M., and Zhan, C. G. (2008) Understanding microscopic binding of human microsomal prostaglandin E synthase-1 with substrates and inhibitors by molecular modeling and dynamics simulation, *The Journal of Physical Chemistry B 112*, 7320–7329.
68. Hamza, A. and Zhan, C. G. (2009) Determination of the Structure of Human Phosphodiesterase-2 in a Bound State and Its Binding with Inhibitors by Molecular Modeling, Docking, and Dynamics Simulation, *The Journal of Physical Chemistry B 113*, 2896–2908.

69. Singh, N., Avery, M. A., and McCurdy, C. R. (2007) Toward Mycobacterium tuberculosis DXR inhibitor design: homology modeling and molecular dynamics simulations, *Journal of Computer-Aided Molecular Design 21*, 511–522.
70. Guex, N. and Peitsch, M. C. (1997) SWISS MODEL and the Swiss Pdb Viewer: an environment for comparative protein modeling, *Electrophoresis 18*, 2714–2723.
71. Kiefer, F., Arnold, K., Kunzli, M., Bordoli, L., and Schwede, T. (2009) The SWISS-MODEL Repository and associated resources, *Nucleic acids research 37*, D387–D392.
72. Verdonk, M. L., Cole, J. C., Hartshorn, M. J., Murray, C. W., and Taylor, R. D. (2003) Improved proteinûligand docking using GOLD, *Proteins: Structure, Function, and Bioinformatics 52*, 609–623.
73. Daga, P. R., Duan, J., and Doerksen, R. J. (2010) Computational model of hepatitis B virus DNA polymerase: Molecular dynamics and docking to understand resistant mutations, *Protein Science 19*, 796–807.
74. Serrano, M. L., Perez, H. A., and Medina, J. D. (2006) Structure of C-terminal fragment of merozoite surface protein-1 from Plasmodium vivax determined by homology modeling and molecular dynamics refinement, *Bioorganic & medicinal chemistry 14*, 8359–8365.
75. Li, W., Tang, Y., Liu, H., Cheng, J., Zhu, W., and Jiang, H. (2008) Probing ligand binding modes of human cytochrome P450 2J2 by homology modeling, molecular dynamics simulation, and flexible molecular docking, *Proteins: Structure, Function, and Bioinformatics 71*, 938–949.
76. Humphrey, W., Dalke, A., and Schulten, K. (1996) VMD: visual molecular dynamics, *Journal of molecular graphics 14*, 33–38.
77. Pettersen, E. F., Goddard, T. D., Huang, C. C., Couch, G. S., Greenblatt, D. M., Meng, E. C., and Ferrin, T. E. (2004) UCSF Chimera-a visualization system for exploratory research and analysis, *Journal of Computational Chemistry 25*, 1605–1612.
78. Izaguirre, J. A., Catarello, D. P., Wozniak, J. M., and Skeel, R. D. (2001) Langevin stabilization of molecular dynamics, *The Journal of chemical physics 114*, 2090–2099.
79. Still, W. C., Tempczyk, A., Hawley, R. C., and Hendrickson, T. (1990) Semianalytical treatment of solvation for molecular mechanics and dynamics, *Journal of the American Chemical Society 112*, 6127–6129.
80. Darden, T., York, D., and Pedersen, L. (1993) Particle mesh Ewald: An N log (N) method for Ewald sums in large systems, *The Journal of chemical physics 98*, 10089–10092.

Chapter 7

Methods for Accurate Homology Modeling by Global Optimization

Keehyoung Joo, Jinwoo Lee, and Jooyoung Lee

Abstract

High accuracy protein modeling from its sequence information is an important step toward revealing the sequence–structure–function relationship of proteins and nowadays it becomes increasingly more useful for practical purposes such as in drug discovery and in protein design. We have developed a protocol for protein structure prediction that can generate highly accurate protein models in terms of backbone structure, side-chain orientation, hydrogen bonding, and binding sites of ligands. To obtain accurate protein models, we have combined a powerful global optimization method with traditional homology modeling procedures such as multiple sequence alignment, chain building, and side-chain remodeling. We have built a series of specific score functions for these steps, and optimized them by utilizing conformational space annealing, which is one of the most successful combinatorial optimization algorithms currently available.

Key words: Homology modeling, Protein structure prediction, Global optimization, Energy function, Multiple sequence alignment, Side-chain modeling, Conformational space annealing

1. Introduction

Recently, protein structure prediction by homology modeling has become a basic tool that is routinely used in structural biology and bioinformatics (1, 2). Although many computational methods have been developed in this field, high accuracy protein modeling still remains as a challenging problem. For example, it is rather difficult to generate protein models which are more accurate than what one can get by simply copying the best available homologus protein (out of the templates used for homology modeling).

In the recent CASP experiments (CASP7 and CASP8) for protein structure prediction, the high-accuracy template-based

Andrew J.W. Orry and Ruben Abagyan (eds.), *Homology Modeling: Methods and Protocols*, Methods in Molecular Biology, vol. 857, DOI 10.1007/978-1-61779-588-6_7, © Springer Science+Business Media, LLC 2012

modeling (HA-TBM) category is considered separately along with template-based modeling (TBM) and free modeling (FM) categories, and there were many examples where protein models were more accurate than the best available templates in terms of accuracies of backbone structure, side-chain orientation, hydrogen bonding, and usefulness for molecular replacement in X-ray crystallography (3, 4).

Three major steps of the standard homology modeling protocol are multiple sequence alignment (MSA), 3D (three-dimensional) model building, and side-chain remodeling, and recently, we have incorporated the global optimization method called conformational space annealing (CSA) to these three procedures to generate highly accurate protein models. In detail, the protocol of homology modeling using CSA consists of the following five steps: (1) fold recognition (finding homologus templates from known protein structures), (2) multiple sequence/structure alignment by global optimization, (3) 3D structure modeling, (4) assessment of protein models and alignments, (5) side-chain remodeling by global optimization.

Fold recognition is to find homologus templates to the target protein from known protein structures in the PDB, and this step of identifying similar structures in the PDB is the most crucial one for successful homology modeling. Many sequence-based fold recognition methods incorporate properties of sequence similarity, profile similarity, and secondary structure similarity between proteins. Often, multiple templates are obtained by fold recognition, and the next step is to extract as much useful structural information from them, typically by performing multiple alignment between the target protein and templates.

In the second step, to generate more useful MSAs, we developed a method, called MSACSA, which explores the diverse alignment space to search rigorously low-energy alignments of given templates based on a consistency-based score function (5). In the following steps, we generate many candidate alignments, and construct initial 3D models using MODELLER, and assess the quality of the alignments by assessing those of the 3D models by using a support vector regression (SVR) machine. Here, preferred combinations of templates as well as choices for multiple alignment out of many alternative solutions are determined. For 3D model building from a few selected alignments, we optimize the MODELLER energy function as rigorously as possible to generate protein structures satisfying as much spatial restraints derived from its alignment as well as proper stereochemistry of proteins (6). For side-chain remodeling, again we adopt the global optimization method of CSA to determine the orientations of side chains both in the surface and inside the core area of protein structures (4). Here the backbone-dependent rotamer library of SCWRL 3.0 is used. Below, we describe each step of the protocol to generate highly accurate protein models by global optimization.

2. Materials

For protein structure modeling, various bioinformatics and 3D modeling-related tools should be first installed in your computer system. They include PSI-BLAST, PSIPRED, MODELLER, the backbone-dependent rotamer library of SCWRL 3.0, DFIRE, DSSP, TM-align, and SPICKER. PSI-BLAST program is a basic tool to generate sequence profile by searching protein sequence databases (e.g., *nr* database from NCBI) (7). Secondary structure of a protein sequence is predicted by PSIPRED (8). MODELLER is a 3D structure building program by using templates and an alignment as inputs (2). The backbone-dependent rotamer library of SCWRL 3.0 program (9) can be downloaded from Dr. Dunbrack's webpage (10). DFIRE, an energy function to assess the quality of a given protein structure can be obtained by email request to the authors (11). DSSP program calculates secondary structures, solvent accessibility, and other structural properties for a given protein 3D structure (12). TM-align calculates structural similarity for two given protein structures, and SPICKER is a clustering program to select a few representative structures from many (~100) predicted models.

For optimization of energy functions for MSA and 3D model building, parallel computing resources are recommended to reduce computation time, and parallel algorithms of CSA method have to be implemented on a parallel computing system (e.g., a cluster system). A few implementations of CSA can be found from the literature (13, 14) and a recent CHARMM package containing the CSA routine, which will be available soon (15). Here we explain briefly how CSA steps are composed of.

2.1. A Brief Description of Conformational Space Annealing

Recently, CSA method is implemented in CHARMM, and the source code of CSA is available (15). The CSA method searches the whole conformational space in its early stages and narrows the search to smaller regions with low energy as the distance cutoff, D_{cut}, which defines a (varying) threshold for the similarity between two solutions, is reduced. As in genetic algorithms, it starts with a preassigned number (50 in this work) of randomly generated and subsequently energy-minimized solutions. This pool of solutions/conformations is called the *bank*. At the beginning, the bank is a sparse representation of the entire conformational space. In the following, the meaning of conformation depends on the context where CSA is used. For MSA optimization, a conformation means an alignment. For 3D structure modeling, it presents a protein 3D structure model, and for side-chain remodeling, it refers to a set of side-chain conformations for a given fixed back-bone structure. For implementation of CSA, we need a series of new concepts. They are (1) an energy function to minimize, (2) a distance measure between two conformations, (3) a local minimizer of a given conformation, (4) ways to combine two parent conformations to generate a daughter one. For details, see each section of the methods. Equipped with these four concepts, CSA proceeds as follows:

1. Generate 50 conformations which are randomly generated and subsequently energy minimized by a local minimizer.
2. Calculate D_{ave} as the average distance between all pairs of the 50 conformations, and set D_{cut} as $D_{ave}/2$.
3. Select 30 distinct conformations called seeds which have not yet been used.
4. For each seed, perturb the conformation and subsequently energy minimize the perturbed conformation to generate a daughter conformation. If we generate 20 daughter conformations per seed, a total of $30 \times 20 = 600$ daughter conformations are prepared.
5. Update the existing 50 conformations using the 600 daughters by a special update scheme as described below.
6. Reduce D_{cut} by a fixed ratio $r = 0.997$ (see Note 1).
7. Go to the seed selection step until all seeds are used.
8. When all seeds are used, one iteration is completed. Set all conformations as unused, and repeat another iteration of the search.
9. If the second iteration completes, and the number of the pool is not 100, add additional 50 random and subsequently energy-minimized conformations to the pool. Set $D_{cut} = D_{ave}/2$, and go to the seed selection step once again. If the second iteration completes, and the number of pool is 100, it completes the CSA.

Energy minimization: For continuous function with gradient available, conjugate gradient minimization is used. For a discrete function to optimize as in the case of multiple alignment and side-chain remodeling, we used a quench procedure as follows. Perturb a conformation and compare its energy with original one, and take the lower energy one. Repeat this process by a fixed number of trials.

Update scheme: For each daughter conformation, α, the closest conformation A in terms of the corresponding distance measure (see each section of the methods) is determined. Let us denote the distance as $D(\alpha,A)$. If $D(\alpha,A) \leq D_{cut}$, α is considered similar to A; in this case α replaces A in the pool of conformations provided that it is lower in energy. If α is not similar to A, but its energy is lower than that of the highest-energy conformation in the bank, B, α replaces B. In neither of the above conditions holds, α is rejected.

2.2. Model Validation

To assess the quality of a given 3D model (see Subheading 3.3), you should build in advance an SVR machine using the following four steps.

1. Prepare a set of decoy structures with known structural quality in terms of TM-score.
2. For each model, calculate the following five feature components. In the following, N_{res} is the number of residues of the given model.
   (a) $SS_{score} = -\sum_{i=1}^{N_{res}} P(\text{SSTYPE}(i))$, where $P(.)$ is the probability value from PSIPRED and SSTYPE(i) is the secondary structure type of the ith residue.
   (b) $SA_{score} = \sum_{k=1}^{25} \sum_{i=1}^{N_{res}} D_k(i)\left(\text{RSA}_{\text{model}}(i) - \text{RSA}_k(i)\right)^2$, where $D_k(i)$ is the weighted Euclidean distance between profiles from the query and the kth nearest neighbor in the database, $\text{RSA}_{\text{model}}(i)$ is the relative solvent accessible surface area (SASA) of the ith residue of the model, and $\text{RSA}_k(i)$ is the relative SASA of the ith residue of the kth neighbor.
   (c) $HP_{score} = \sum_{i=1}^{N_{res}} \text{DsspACC}(i) \times \text{HP}(i)$, where DsspACC($i$) is the SASA of residue i from DSSP and HP(i) is the HP-table value for the ith residue (see Note 2).
   (d) DFIRE energy of the model.
   (e) MODELLER energy of the model.
3. We are now prepared with a table which contains TM-scores and five feature components for all decoy structures.
4. Build an SVR machine using the table by LIBSVM (16, 17).
   Now you can predict TM-score of a given model by SVR machine using following procedure.
5. For a given model, calculate the five feature components described above.
6. Predict TM-score of the given model using the prebuilt SVR machine.
7. For each template combination, we assign the quality of the list/alignment by the average of the predicted TM-scores of the 3D models.

3. Methods

3.1. Fold Recognition

Fold recognition is the starting point of homology modeling. We have used an in-house profile–profile comparing method, called FOLDFINDER to rank templates of known structures from PDB (4). We have built a profile database of protein chains by using PSI-BLAST with standard parameters (E-value cutoff is set to 0.0001 and the procedure is iterated three times). For example, for CASP7 experiment, we built a profile database of 11,914 chains obtained from PISCES culling server (18) at 95% sequence identity level with sequence length in the range of 50–1,000 residues. 11,914 chains include X-ray and NMR structures but not EM structures. We also built secondary structure profiles for chains in the database by using DSSP program (coil, helix and extended states are represented by vectors (1,0,0), (0,1,0), and (0,0,1), respectively).

1. For each chain in the database, its pair-wise sequence alignment with the target sequence is obtained by dynamic programming using the following match score: $S_{ij} = S_{ij}^{\mathrm{p}} + 0.4 \times S_{ij}^{\mathrm{h}} + 0.01$, where S_{ij}^{p} is the Pearson's correlation coefficient between the *i*th row vector of the target sequence profile and the *j*th row vector of the template profile. S_{ij}^{h} is the Pearson's correlation coefficient between the *i*th row vector of the predicted secondary structure probability by PSIPRED and the *j*th row vector of the secondary structure profile of the template. Dynamic programming is performed using the affine gap penalty function of $w(k) = -(1.5 + 0.07 \times k)$, where k is the gap length. End-gaps are not penalized (global-local alignment) (see Note 3).
2. All template chains of the database are sorted according to their alignment scores, and the statistical significance of an alignment score is measured by its *z*-score and *p*-value. An example of the FOLDFINDER output is shown in Table. 1.
3. Considering top-scoring templates with *z*-score typically greater than 4.0 (see Note 4), structurally redundant templates (TM-score > 0.98) are removed. With these templates, we further perform structural clustering by using TM-align considering all pairs of templates. We consider a subset of templates where TM score < 0.5 between all members. We prepare typically 5–10 sets of template combinations. Each combination is called a list and it is used as an input to the subsequent step of multiple alignment. In the CASP experiments, the number of templates ranges 1–15 for one list (see Note 5).

3.2. Multiple Sequence/Structure Alignment

We perform multiple sequence/structure alignment by using MSACSA method (5). For each list of template combination, we execute the following steps to obtain low-energy multiple alignments by CSA optimization. Optimization by CSA is repeatedly applied in this chapter. The general procedures are described in Subheading 2.1, and in the following, we describe the step-specific elements of CSA.

1. Preparation of pair-wise restraint library: For each template in the list, we carry out profile–profile alignment with the target sequence using FOLDFINDER as described in the fold recognition step. Matched residue pairs are stored into the pair-wise

Table 1
An example of the FOLDFINDER output for the target T0506 of CASP8 experiment is shown. Templates with z-score > 4.0 are considered to be significant hits for a target sequence

#	Chain	Nc	Nt	Aln	Score	SeqID	Gap	Z	p-val	nd	Annotation
1	2arzA	247	254	269	281.44	21.98	2.9	7.2	.11E-06	1	b.45.1.1 All beta protei
2	1xhnA	184	254	270	173.10	21.43	9.2	3.4	.29E-03	1	b.45.1.1 All beta protei
3	1vl7A	157	254	259	163.49	15.13	7.9	3.0	.58E-03	1	b.45.1.1 All beta protei
4	2dkhA	639	254	648	147.80	12.25	35.2	2.5	.17E-02	?	? OXIDOREDUCTASE Crystal
5	2fr1A	486	254	498	144.82	9.92	37.8	2.4	.22E-02	2	c.2.1.2 c.2.1.2 Alpha an
6	2r0cA	549	254	554	142.84	12.05	38.8	2.3	.25E-02	?	? OXIDOREDUCTASE Structu
7	2pajA	492	254	492	142.69	14.96	36.3	2.3	.25E-02	?	? HYDROLASE Crystal stru
8	1pn0A	665	254	667	142.63	12.70	49.0	2.3	.25E-02	3	c.3.1.2 c.47.1.10 d.16.1
9	2p9bA	458	254	460	142.59	12.70	32.4	2.3	.25E-02	?	? HYDROLASE Crystal stru
10	2qa2A	499	254	500	141.47	11.46	42.9	2.3	.27E-02	?	? OXIDOREDUCTASE Crystal

Chain, protein chain; Nc, template length; Nt, target length; Aln, alignment length; Score, alignment score; SeqID, sequence identity; Gap, gap percent in the alignment; *z*, *z*-score; nd, number of domain according to SCOP classification; Annotation, annotation of the template according to SCOP and PDB descriptions

restraint library. In addition, for all pairs of templates in the list, pair-wise structure alignment is carried out using TM-align, and the matched residue pairs are also added into the pair-wise restraint library. For each residue pair in the restraint library, the sequence identity between two sequences to which the two residues belong is assigned as the weight w to be used in the score function below.

2. We define an energy function for a given multiple alignment A, as the measure of consistency of A with the restraint library. With N sequences and M aligned columns, it becomes:

$$E(A) = -100 \times \frac{\sum_{i,j=1,i<j}^{N} w_{ij} \sum_{k=1}^{M} \delta_{ij}^{k}(A)}{\sum_{i,j=1,i<j}^{N} w_{ij} L_{ij}}, \quad (1)$$

where $\delta_{ij}^{k}(A) = 1$ if the aligned residues between the ith and the jth sequences at the kth column are in the library, otherwise $\delta_{ij}^{k}(A) = 0$. L_{ij} and w_{ij} are the pair-wise alignment length and the sequence identity between the ith and the jth sequences, respectively.

3. Define the distance measure between two given multiple alignments as the number of residue mismatches considering all pair-wise sequence alignments between the two given multiple alignments.
4. Local optimization to minimize the energy value of a given multiple alignment is carried out by a series of perturbation of the alignment for up to t times. Typically, we set $t = 10NL_{max}$, where L_{max} is the length of the largest sequence in the list. Perturbations are performed by local moves of gaps in the alignment (see Note 6).
5. Combination of two multiple alignments: we generate a daughter alignment by replacing a part of a seed alignment by the corresponding part of another alignment. We limit the replacing part within 40% of the seed alignment.
6. With the preparation steps of steps 3–5, it is straightforward to carry out CSA to optimize $E(A)$ defined in Eq. 1 to generate a total of 100 multiple alignments (see Subheading 2.1).

An example of the lowest-energy alignment and the energy landscape of the multiple alignment are shown in Fig. 1. This step is the key process for modeling highly accurate protein 3D structures. A total of 100 MSAs obtained from this step for each list of templates are used as the input for the next step.

3.3. Assessment of Alignment/3D Structure Modeling

In this step, we select 5–10 alignments by applying an assessment method. The assessment is carried by a machine trained by SVR for feature vectors which are extracted from 3D protein models generated by MODELLER. Details of the prebuilt assessment method is described in Subheading 2.2. Selected alignments are used to generate higher-quality 3D protein models by applying CSA method to optimize the MODELLER energy function (6).

1. For the assessment of an alignment, we first generate 25 protein 3D models using MODELLER and the alignment under evaluation.
2. The quality of each 3D model is evaluated using the assessment method, and the quality of each alignment is estimated by the average 3D model quality from 25 initial models.
3. Five to ten top alignments are selected to proceed with the subsequent procedures.
4. For each alignment selected, we generate 100 protein 3D models by further optimization of MODELLER energy function using the CSA method, which we call as MODELLERCSA (6).
5. To execute MODELLERCSA, one needs to provide a few preliminary procedures: distance measure between two protein 3D models is defined as the C_α RMSD value between them. For local energy minimization, we used what is already imple-

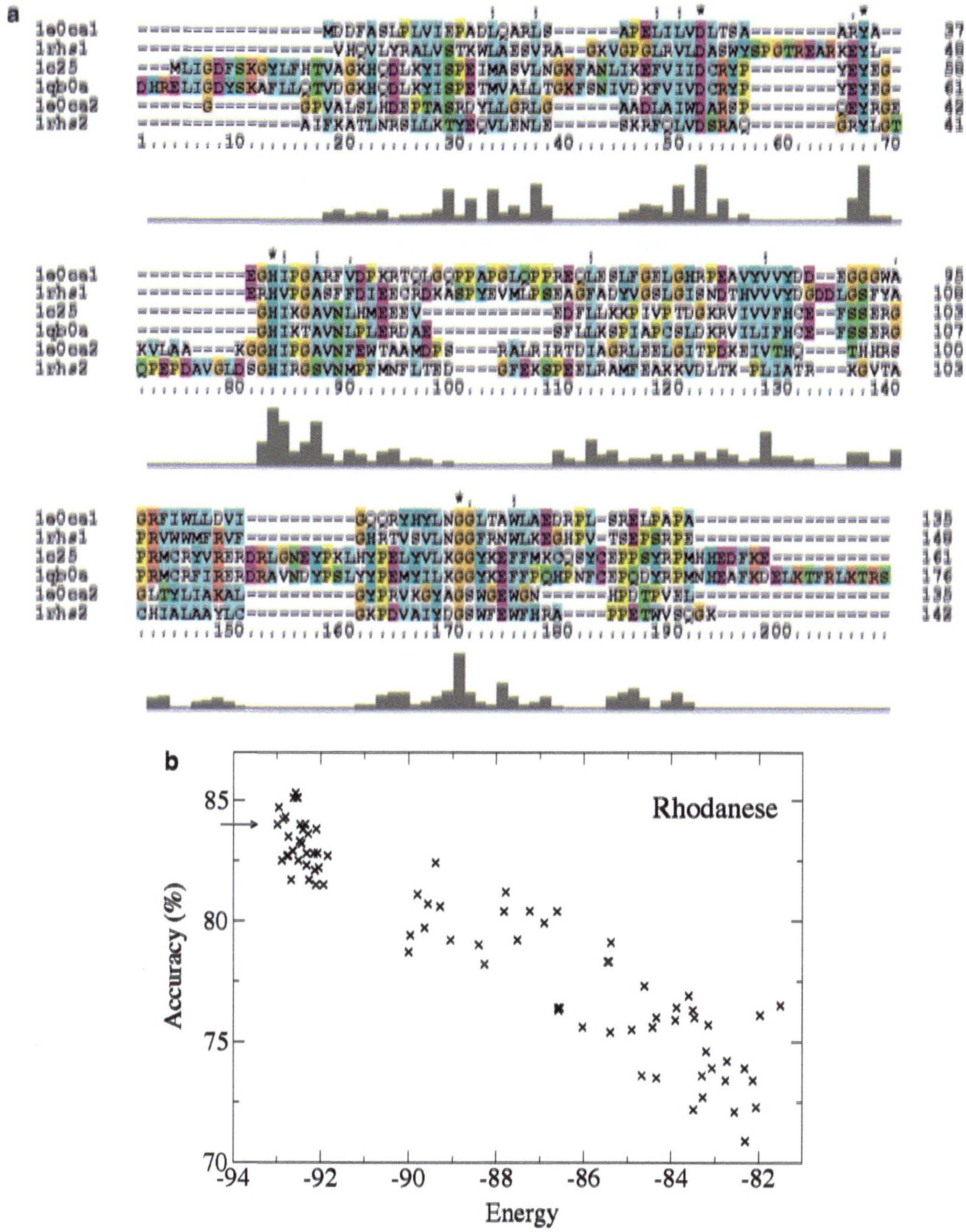

Fig. 1. An example of the lowest-energy multiple sequence alignment (**a**) and the energy landscape (**b**) of the alignment for *Rhodanese* family from the HOMSTRAD database is shown. The *Rhodanese* family consists of six structurally homologous proteins, and the level of sequence similarities is shown as a histogram in (**a**). Alternative alignments as well as the lowest-energy alignment are obtained by optimizing $E(A)$ of Eq. 1 by MSACSA. Each symbol in the energy landscape represents an alternative alignment generated by MSACSA. The x-axis represents the value of $E(A)$, and the y-axis represents the alignment accuracy relative to the reference alignment constructed by human inspection of six protein structures. In (**b**), the lowest-energy alignment is indicated by an *arrow*, and it should be noted that it does not correspond to the most accurate alignment relative to the reference. Therefore, one should consider several low-energy alternative alignments to generate accurate protein models. Figure (**a**) is generated by clustalX program.

mented in the MODELLER package (conjugate-gradient minimization method). To generate a daughter model by crossover, we replace a part of the seed model by the corresponding part of another model. The replacement is limited up to 40% of the seed model as before (see Note 7 and Subheading 2.1).

It is shown (6) that the quality of a protein 3D model improves as its MODELLER energy is optimized. The comparison of 3D model qualities between structures generated by MODELLER and MODELLERCSA is shown in Fig. 2. Backbone accuracies as well as side-chain accuracies are

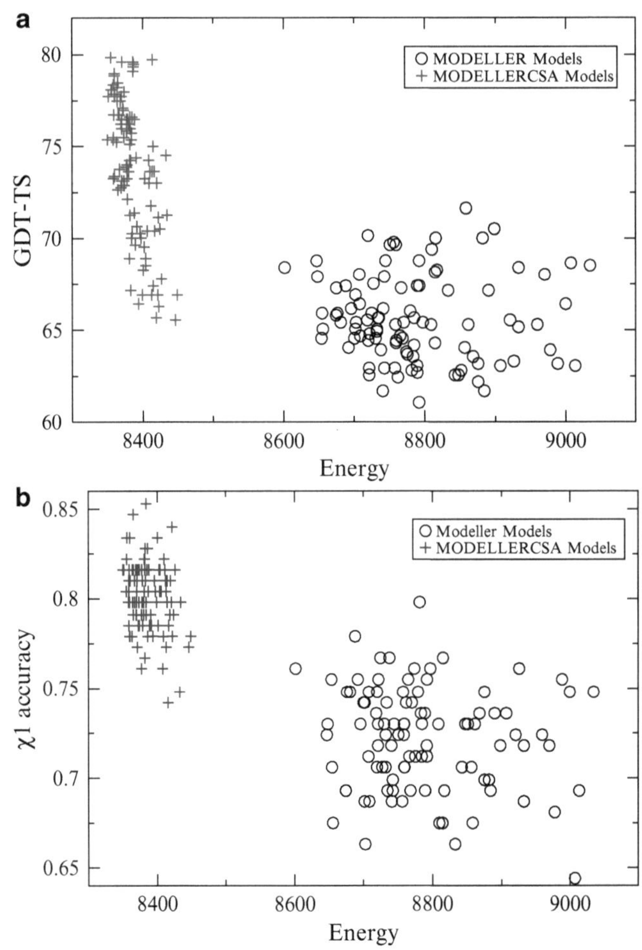

Fig. 2. Backbone accuracies (**a**) and side-chain accuracies (**b**) are plotted in terms of MODELLER energy for MODELLER generated models and MODELLERCSA generated models of *sodfe* family from HOMSTRAD database. The backbone accuracy is measured by GDT-TS, which is used in CASP assessment as a standard measure. The side-chain accuracy is measured by χ_1, which is the percentage of correct rotamer within 30° from the native structure.

plotted in terms of the MODELLER energy. Five representative models among 100 optimized models are selected by reassessment of the models and clustering them into five groups. These five models are used for side-chain remodeling in the next procedure.

6. By using the same assessment method used above, we select top alignments and five models generated by MODELLERCSA.
7. By using SPICKER clustering method, we select representative models from cluster centers. Typically, we select a total of five models (see Note 8).

3.4. Side-Chain Modeling

We have used the backbone-dependent rotamer library of SCWRL 3.0 (9) to remodel side chains of a given protein 3D model. For each 3D model selected from the previous step, we have built a target-specific rotamer library based on the consistency of the side chain conformations:

1. For each residue i, we calculate the average (m_i) and the standard deviation (σ_i) of χ_i^1 angles of 100 models.
2. If $\sigma_i \leq 15°$, we add ten sets of all χ_i^1 angles closest to m_i into the rotamer library.
3. If $\sigma_i > 15°$, we use the backbone-dependent rotamer library SCWRL 3.0 for the residue.

   Rotamers are optimized by CSA, which is called ROTAMERCSA, to remodel side chains of a selected model using the rotamer library and the energy function below.
4. An energy function E is defined for side-chain optimization: $E = E_{\mathrm{SCWRL}} + E_{\mathrm{DFIRE}}$, where E_{SCWRL} is the score function used in SCWRL 3.0 and E_{DFIRE} is the DFIRE energy (11).
5. Distance measure between two sets of side-chain conformations are defined as the sum of Euclidean distance for corresponding rotamer angles.
6. Local minimization is carried out by stochastic quenching as in the case of MSACSA.
7. A daughter conformation is generated by replacing a part of seed model's rotamers by the corresponding part of another model's rotamers.
8. Now, run CSA (see Subheading 2.1).

Figure 3 shows side-chain accuracies of 27 HA-TBM targets from CASP7 obtained by ROTAMERCSA. Results by MODELLER as well as MODELLERCSA are also shown for comparison. It illustrates step-by-step improvement of the side-chain modeling (see Note 9). An example of the final 3D model after side-chain remodeling is shown in Fig. 4.

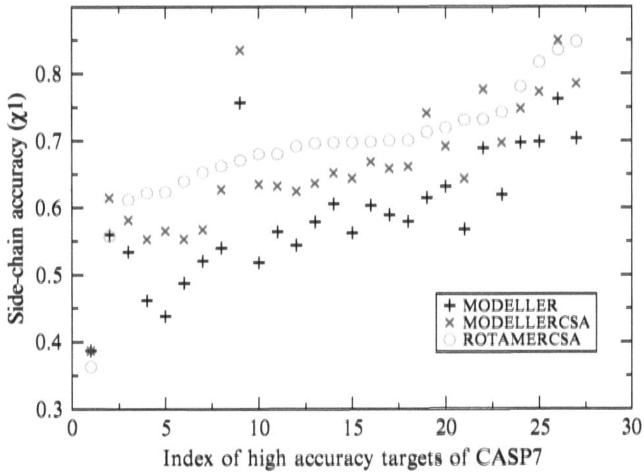

Fig. 3. Side-chain accuracies for 27 high-accuracy TBM targets of CASP7 are shown. *Plus symbols* correspond to the models generated simply by executing MODELLER program. *Times symbols* (×) correspond to the models obtained by MODELLERCSA. *Open circles* correspond to the models where backbones are kept identical to the MODELLERCSA results, and side chains are remodeled by ROTAMERCSA. Overall side-chain accuracy improves gradually by applying more sophisticated methods than simple MODELLER chain building. Executing additional ROTAMERCSA after MODELLERCSA improves χ_1 accuracy, although there are cases where best χ_1 accuracy is achieved by MODELLERCSA (5 of 27).

4. Notes

1. The value of D_{cut} is kept constant after it reaches $D_{ave}/5$.
2. We have used the hydrophobicity values of 0.74, 0.91, 0.62, 0.62, 0.88, 0.72, 0.78, 0.88, 0.52, 0.85, 0.85, 0.63, 0.64, 0.62, 0.64, 0.66, 0.70, 0.86, 0.85, 0.76 for residue types A, C, D, E, F, G, H, I, K, L, M, N, P, Q, R, S, T, V, W, Y (19).
3. Parameters were obtained by optimizing the average accuracy of sequence alignments for 388 references with sequence identity ≤40% from HOMSTRAD database.
4. In the fold recognition step, when the top scoring template by FOLDFINDER is not so prominent in terms of *z*-score (*z*-score < 3.0), additional template candidates by other methods are also considered. Other fold recognition web servers include 3D-jury (http://bioinfo.pl/~3djury) (20) and HHsearch (21) provided from web server.
5. Selecting templates should be carefully considered in aspects of alignment length, sequence identity, and consistency of secondary structure between target and templates. Also, if there are gap regions especially in the target sequence of multiple alignment, it is good to consider templates which can cover gap regions in the alignment.

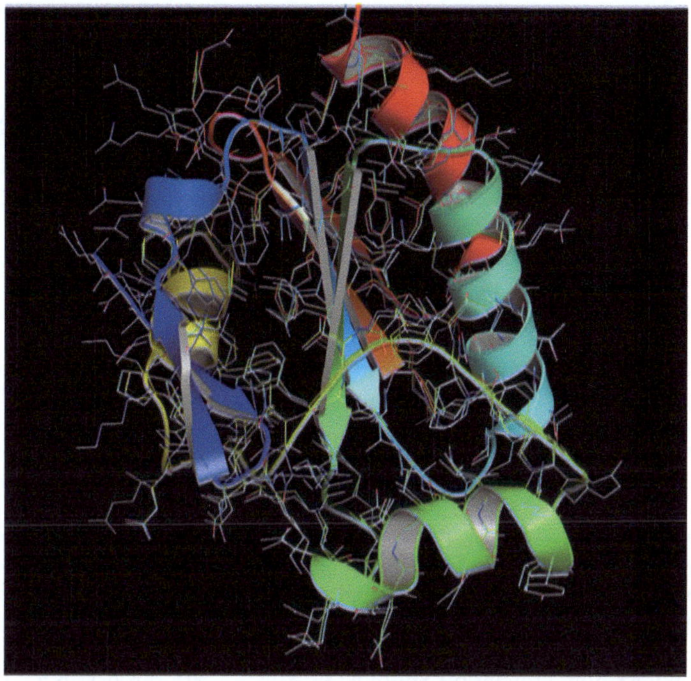

Fig. 4. The superposition between the native structure of T0345 (PDB ID: 2he3) and the lowest energy model generated by the full CASP7 procedure is shown. The model was constructed and submitted as the LEE model (model 1) prior to the release of the native structure. Backbone heavy atom RMSD between the model and the native structure is about 1.6 Å for the entire chain of 173 residues. The GDT-TS score is 96.0. The cartoon figures represent the native backbone structure and the model backbone structure, indistinguishable from each other. The χ_1 angle accuracies are improved through the steps discussed in this chapter from the value of 70.4 (MODELLER), to 78.6 (MODELLERCSA) and finally to 84.8 (ROTAMERCSA). Aromatic residues in the core region are well predicted. Some exposed side chains, especially lysine side chains, do not agree between the two structures. The figure is generated by pymol.

6. These moves consist of random insertion, deletion, and relocation of gap(s) (22, 23).
7. In the MODELLERCSA, a daughter model is combined by using internal variables of two parent 3D models (such as bond angles, bond length, and dihedral angles). A consecutive part of one parent's internal coordinates are replaced by the corresponding internal coordinates of the other parent, and resulting structure is subject to subsequent energy minimization. As a result, daughter structures partially inherit bond angles, bond lengths, and backbone, and side-chain dihedral angles of their parents.
8. SPICKER uses distance cut value of 3.5 Å for clustering. We have used a variable distance cut value in the range 1.0–3.5 Å.
9. Accuracies of side chain for target solved in NMR experiment are relatively lower than solved in X-ray crystallography.

Acknowledgments

This work was supported by Creative Research Initiatives (Center for in silico Protein Science, 2009-0063610) of MEST/KOSEF. We thank KIAS Center for Advanced Computation for providing computing resources.

References

1. Baker, D., Sali, A. (2001) Protein structure prediction and structural genomics. Science **294** (5540), 93–96
2. Sali, A., Blundell, T.L. (1993) Comparative protein modelling by satisfaction of spatial restraints. J. Mol. Biol. **234**(3), 779–815
3. Read, R.J., Chavali, G. (2007) Assessment of casp7 predictions in the high accuracy template-based modeling category. Proteins **69 Suppl 8**, 27–37
4. Joo, K., Lee, J., Lee, S., et al. (2007) High accuracy template based modeling by global optimization. Proteins **69 Suppl 8**, 83–89
5. Joo, K., Lee, J., Kim, I., et al. (2008) Multiple sequence alignment by conformational space annealing. Biophys. J. **95** (10), 4813–4819
6. Joo, K., Lee, J., Seo, J., et al. (2009) All-atom chain-building by optimizing modeller energy function using conformational space annealing. Proteins **75**, 1010–1023
7. Altschul, S.F., Madden, T.L., Schaffer, A.A., et al. (1997) Gapped blast and psi-blast: a new generation of protein database search programs. Nucleic Acids Res. **25**(17), 3389–402
8. Jones, D.T. (1999) Protein secondary structure prediction based on position-specific scoring matrices. J. Mol. Biol. **292** (2), 195–202
9. Canutescu, A.A., Shelenkov, A.A., Dunbrack, R.L. (2003) A graph-theory algorithm for rapid protein side-chain prediction. Protein Sci. **12** (9), 2001–2014
10. Dunbrack, R.L., Karplus, M. (1993) Backbone-dependent Rotamer Library for Proteins: Application to Side-chain prediction. *J. Mol. Biol.* **230**, 543–574 (http://dunbrack.fccc.edu/bbdep/index.php)
11. Zhou, H., Zhou, Y. (2002) Distance-scaled, finite ideal-gas reference state improves structure-derived potentials of mean force for structure selection and stability prediction. Protein Sci. **11**(11), 2714–2726
12. Kabsch, W., Sander, C. (1983) Dictionary of protein secondary structure: pattern recognition of hydrogen-bonded and geometrical features. Biopolymers **22** (12), 2577–2637
13. Lee, J., Scheraga, H.A., Rackovsky, S. (1997) New optimization method for conformational energy calculations on polypeptides: Conformational space annealing. J. Comput. Chem. **18**(9), 1222–1232
14. Lee, J., Lee, I.H., Lee, J. (2003) Unbiased global optimization of lennard-jones clusters for $n \leq 201$ using the conformational space annealing method. Phys. Rev. Lett. **91**, 080201
15. Brooks, B.R., Bruccoleri, R.E., Olafson, B.D., et al. (1983) Charmm: A program for macromolecular energy, minimization, and dynamics calculations. J. Comput. Chem. **4** (2), 187–217
16. Chang, C.C., Lin, C.J. (2001) LIBSVM: a library for support vector machines. Software available at *http://www.csie.ntu.edu.tw/~cjlin/libsvm*
17. Fan, R.E., Chen, P.H., Lin, C.J. (2005) Working set selection using second order information for training support vector machines. J. Mach. Learn. Res. **6**, 1889–1918
18. Wang, G., Dunbrack, R.L. (2005) Pisces: recent improvements to a pdb sequence culling server. Nucleic Acids Res. **33**(Web Server issue)
19. Rose, G.D., Geselowitz, A.R., Lesser, G.J., et al. (1985) Hydrophobicity of amino acid residues in globular proteins. Science **229**(4716), 834–838
20. Ginalski, K., Elofsson, A., Fischer, D., et al. (2003) A simple approach to improve protein structure predictions. Bioinformatics **19** (8), 1015–1018
21. Söding, J. (2005) Protein homology detection by hmm-hmm comparison. Bioinformatics **21**(7), 951–960
22. Ishikawa, M., Toya, T., Hoshida, M., et al. (1993) Multiple sequence alignment by parallel simulated annealing. Comput. Appl. Biosci. **9** (3), 267–73
23. Kim, J., Pramanik, S., Chung, M.J. (1994) Multiple sequence alignment using simulated annealing. Comput. Appl. Biosci. **10** (4), 419–26

Chapter 8

Ligand-Guided Receptor Optimization

Vsevolod Katritch, Manuel Rueda, and Ruben Abagyan

Abstract

Receptor models generated by homology or even obtained by crystallography often have their binding pockets suboptimal for ligand docking and virtual screening applications due to insufficient accuracy or induced fit bias. Knowledge of previously discovered receptor ligands provides key information that can be used for improving docking and screening performance of the receptor. Here, we present a comprehensive ligand-guided receptor optimization (LiBERO) algorithm that exploits ligand information for selecting the best performing protein models from an ensemble. The energetically feasible protein conformers are generated through normal mode analysis and Monte Carlo conformational sampling. The algorithm allows iteration of the conformer generation and selection steps until convergence of a specially developed fitness function which quantifies the conformer's ability to select known ligands from decoys in a small-scale virtual screening test. Because of the requirement for a large number of computationally intensive docking calculations, the automated algorithm has been implemented to use Linux clusters allowing easy parallel scaling. Here, we will discuss the setup of LiBERO calculations, selection of parameters, and a range of possible uses of the algorithm which has already proven itself in several practical applications to binding pocket optimization and prospective virtual ligand screening.

Key words: Homology models, Internal coordinate mechanics, Ligand docking, Virtual screening, Binding pocket, Drug discovery

1. Introduction

Traditional homology modeling involves starting from a known homologue and relying on an energy function and restraints to predict the differences in the modeled protein. However, the energy function alone does not provide unambiguous discrimination between multiple low energy conformations. Knowing the ligands that are supposed to bind to a pocket of the model may help the modeling in two different ways: (1) generate a more relevant ensemble of models by including one or several "seed" ligands with restraints into the sampling (1) and (2) use a panel of active and decoy ligands to rank models by their ability

Andrew J.W. Orry and Ruben Abagyan (eds.), *Homology Modeling: Methods and Protocols*, Methods in Molecular Biology, vol. 857, DOI 10.1007/978-1-61779-588-6_8, to discriminate between actives and decoys after docking and scoring of the panel to each trial pocket (2). Prediction of the ligand–receptor interactions requires high accuracy of the protein models and, therefore, may lead to a more accurate model if the sampling procedure can find it. Even small ≈1–2 Å variations of the atomic positions in the binding pocket can prevent the formation of the critical hydrogen bonds or create steric clashes precluding correct ligand docking in a rigid protein model (3, 4). As recent large-scale cross-docking experiments suggest (5, 6) such deviations are rather common even in high resolution structures of protein–ligand complexes, allowing correct docking for only about 50% of ligand–receptor pairs on average. The problem is even more pronounced for models built by homology, especially those with moderate (<50%) to low (<35%) levels of sequence identity to the target, where not only significant deviations of side chain atoms but also shifts in protein backbone position are expected. Energy-based refinement of the protein model itself is often insufficient, and special treatment of the binding pocket is required for improved predictions of ligand binding. In lieu of such optimization, docking applications resort to using softer and less specific potentials and impose knowledge-derived restrains to position the ligand (e.g., ref. 7).

In practice, some preexisting knowledge of specific small molecule ligands is available for many clinically relevant targets and can provide additional guidance for optimization of the binding pocket model. In a simplest form, the ligand-guided optimization involves direct co-refinement of flexible side chains of the pocket in a presence of one or several known "seed" ligands (1). This approach, however, has serious limitations since the ligand pose cannot be unambiguously predicted unless some key interactions of the ligand are known a priori. More sophisticated ligand-guided algorithms exploit extensive sampling of conformational states of the binding pocket, with or without ligands, to create a comprehensive collection of plausible conformers. Selection of the best conformers is then performed by testing for enrichment with actives after docking and scoring of the active/decoy panel. The first application of this method was reported in refs. 8 and 9. However, these studies did not account for the possibility that ligand binding may require some conformational changes in the protein backbone.

We have recently introduced a more automated ligand-guided backbone ensemble receptor optimization (LiBERO) framework which allows multiple generations of models and uses normal mode analysis (NMA) to generate the backbone conformation ensembles. The algorithm is based on two key steps: (1) generation of multiple receptor conformers—with or without seed ligands and (2) selection of the conformers according to docking/VLS performance. These two steps are repeated iteratively until the models reaches optimal VLS performance. LiBERO has proved to be useful in several applications including optimization of homology models for $A_{2A}AR$ (10) and other adenosine receptor subtypes (11). It was also tested for prediction of conformational changes in binding pocket induced by specific classes, including full and partial agonists of the β_2-adrenergic receptor (12, 13). Moreover, the receptor models optimized by the ligand-guided technology have been validated in prospective screening studies, making possible discovery of novel ligand chemotypes for human androgen receptor (8), melanin-concentrating hormone receptor MHC-R1 (9), and adenosine A2a receptor (14).

2. Theory

Figure 1 illustrates a general outline of the LiBERO algorithm (10, 15). The algorithm takes as input one or several initial protein structures, which can be homology models from multiple templates or distinct conformations found in multiple crystal structures. The other source of input comes from the ligand dataset consisting of target-specific ligands which can be divided into small "seed" subsets, possibly accompanied by experimental distance restraints, and a large "training" test.

2.1. Generation of Protein Conformations

The goal of this step is to produce a large number of nonredundant energetically feasible receptor conformations starting from one or several initial models. Several alternative techniques are used to generate receptor conformations, depending on the extent and nature of expected deviations from the starting model(s).

2.1.1. Multiple Homology Models

When multiple initial homology models are available based on different structural templates or alternative plausible alignments to a single template, it is advisable to test them as initial "candidates" for the ligand-based optimization. Inclusion of multiple templates is most practical for classes of receptors and enzymes, which undergo well-described large-scale conformational changes in the binding pocket as a part of their functional mechanism (e.g., protein kinases (16)).

2.1.2. Side Chain Sampling with Known Ligands

In its simplest form conformational sampling involves only side chains of a receptor binding pocket, while the protein backbone is kept fixed. This can be preferable when modeling is based on close homology within a protein family (>50% identical residues) and minimal backbone deviations from the template are expected (11). The binding pocket residues are roughly defined by the vicinity of a ligand in the original homology template or can also be defined by ICM PocketFinder algorithm (17). To prevent collapse of the binding pocket, the conformational sampling can be performed

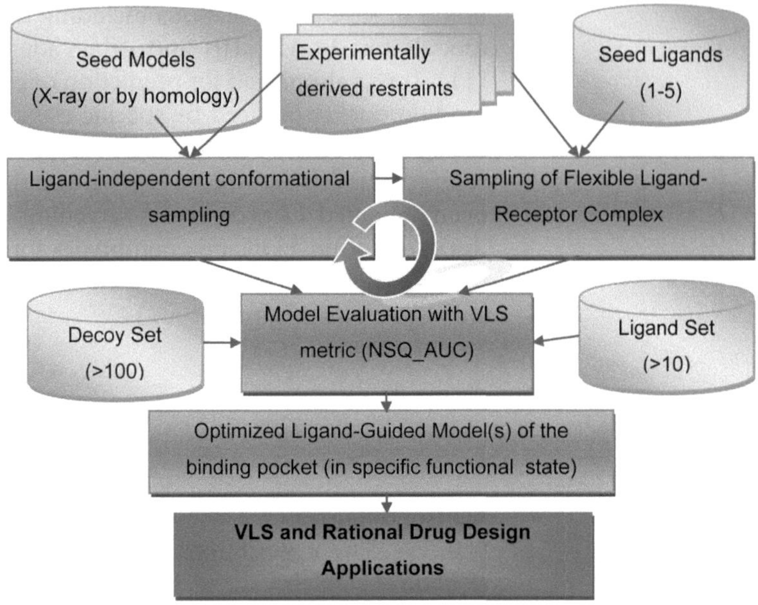

Fig. 1. General outline of the LiBERO algorithm for rational drug discovery applications. The algorithm starts with (1) one or several initial "seed" models built by homology or adopted from a crystal structure in a specific functional state, (2) one or few representative seed ligands, and (3) if available, additional experimentally derived restraints. Two procedures for sampling possible conformational states of the model are used. The first one with emphasis on large-scale movement of the backbone (e.g., NMA), the second using energy-based sampling of a seed ligand in the all atom flexible model of the binding pocket. The two sampling methods can be used consecutively or in parallel; the first method can be skipped in cases when large backbone movements are not expected (e.g., for close subfamily homologs). The generated models are then evaluated in a docking/VLS benchmark according to their ability to separate representative ligands of the receptor from decoy nonbinding compounds using a balanced NSQ_AUC metric. The procedure is iterated through a sampling-evaluation step until convergence of VLS performance is achieved. The optimized model of the binding pocket representing specific functional and conformational states can be effectively used for VLS and Drug Design applications. Multiple models can be generated by using different subsets of ligands if these subsets require a different induced fit in the model.

with a seed ligand placed in the pocket. The trial ligand placement is performed by docking into the flexible receptor starting from multiple ligand orientations, as described previously (1). Alternatively, a blob of repulsive potential can be used to maintain volume of the pocket (6).

We use biased probability Monte Carlo (BPMC) minimization (18) in ICM internal coordinates (19) for sampling of side chain torsion variables, while leaving polypeptide covalent geometry and protein backbone fixed. These algorithms allow extensive conformational sampling of a small molecule ligand with a limited number of flexible side chains in the binding pocket. To improve sampling efficiency, soft distance restraints can be introduced in some cases in the models to account for residue–residue contacts and/or residue–ligand contacts validated by site-directed mutagenesis. While some experimental restraints have been well characterized for certain ligand and receptor classes (e.g., a salt bridge between the charged amino group of ligand and Asp3.32 in all aminergic G-protein-coupled receptors), in general, mutagenesis-derived restraints should be used with caution as indirect effects of mutations can often be mistaken for a direct contact (15). If experimental data do not support any specific interatomic restraints, simple nonspecific volume restraints can enforce ligand docking within a known binding pocket.

2.1.3. Conformers with Backbone Variations

Side chain optimization alone may be insufficient for accurate ligand recognition in many cases, especially for protein models built with low level of homology to the structural template (<30%) or conformational states that require large backbone deviations. In those cases, the procedure will benefit from allowing variations in the protein backbone. Adequate backbone sampling remains a challenging goal for molecular mechanics and molecular dynamics (MD) applications due to the sheer size of the systems, the complexity of the energy landscape and the inaccuracies of the energy function. For some protein families, the problem can be simplified by focusing on possible backbone variations in specific regions of exceptional structural plasticity/flexibility, deduced experimentally and/or from analysis of family structure and function. One prominent example of conformational flexibility in the binding pockets involves DFG-in and DFG-out states of the activation loop in protein kinases (16), while variations in extracellular loops and the tips of the transmembrane helices exemplify structural plasticity within the GPCR superfamily (15). Backbone variations in these regions can be modeled by extensive conformational sampling (20), rigid body movements of the secondary structure elements (12, 13), or local NMA (21) techniques.

Elastic network NMA (EN-NMA) (22) is a fast and versatile sampling approach that allows generating large variations in protein backbone, often not observed in the range of timescales accessible by other sampling techniques such as MD. As described elsewhere (23), in our approach, the interaction energy between two heavy atoms is described by a harmonic potential where the initial distances are taken to be at the energy minimum, and the spring constant is assigned according to inverse exponent of the interatomic distances (24). Diagonalization of the Hessian yields the eigenvectors (i.e., the collective direction of atomic motion), and the eigenvalues, which give the energy cost of deforming the system along the eigenvectors. The Cartesian ensemble is built by generating "random" displacements along the normal mode "important subspace" so that it represents the overall equilibrium dynamics of the protein, or alternatively, along a few normal modes representing an expected transition. Conformations obtained by EN-NMA slightly distort the covalent geometry of the model, so it should be refined using physical energy-based minimizations.

Some of models generated both by side chain sampling or NMA can be very similar to each other, and this redundancy of the conformer set can be reduced by its clustering according to the ligand and contact residue conformations. The clustering criteria, however, must be sensitive to any small local deviations in the pocket since even single atom variation can impact the model performance in VLS.

2.2. Selection of the Ligand and Decoy Sets

Information on specific ligands for a vast majority of clinically relevant human proteins is available in literature and general (e.g., ChEMBL and KiDB) or protein family-specialized ligand databases (GLIDA and kinase), or come directly from in-house HTS programs. Adequate ligand selection for the seed and training sets is important for quality of the resulting models and their suitability for particular drug discovery applications.

2.2.1. Ligand Training Set

Higher affinity ligands are generally preferable for the ligand set, as their binding is more likely to optimally represent most common key interactions with receptors. Also, preference should be usually given to larger ligands filling a major part of the pocket, as smaller ligands may guide optimization towards a smaller pocket, which is usually detrimental for VLS performance (**25**).

Selection of a ligand training set also depends on the particular application of the resulting model. Thus, it is preferable to have rather diverse optimization set for a model intended for initial VLS, where a consensus "one-size-fits-all" model that binds a large number of diverse ligands is most desirable. On the other hand, if the model is intended for rational optimization of a specific lead series, more accurate scaffold-specific model can be achieved by using only ligands based on this particular scaffold or isosteric scaffolds. Also, one should avoid excessive redundancy in the ligand set, as inclusion of many highly similar ligands will not only consume more computational resources, but more importantly, may bias the optimization towards this particular ligand subset.

For many receptors, ligands can be classified in certain groups according to known functional and conformational selectivity (e.g., agonists vs. antagonists in nuclear receptors and GPCRs or type I and type II inhibitors in kinases). In this case, receptor optimization can be performed separately for each of these function-specific ligand sets. This will lead to different conformations of the pocket, potentially reflecting changes characteristic for binding of these ligand classes. The method overall is rather tolerant to the presence in the training set of lower affinity ligands or ligands that require a special induced fit, but its performance may start to deteriorate if too many inappropriate ligands are present.

2.2.2. Seed Ligands

In some cases, reduction of the sampling space and faster convergence of the optimization procedure can be achieved by all-atom ligand–receptor co-refinement using few selected ligands as seed compounds. Usually, seed ligands are those with the highest binding affinity and availability of reliable mutagenesis information that can be used to set soft binding restrains. Seed ligands should be excluded from the training set to avoid over-fitting.

2.2.3. Decoy Set

The decoy set for assessment of VLS performance should be selected to represent chemical diversity and approximately match distribution of physicochemical properties of the ligand set of "actives." Techniques for the selection of relevant decoy sets have been described recently and may help to improve accuracy of the resulting models. In most cases, a set of 10–30 ligands and 100–1,000 decoys is adequate.

2.3. Ligand Docking and Scoring

To evaluate each nonredundant conformer, the ligand and decoy sets of compounds should be routinely docked into the binding pocket of each receptor conformer, which requires a fast docking procedure. The fast ICM ligand docking uses a BPMC optimization of the ligand internal coordinates in the set of grid potential maps of the receptor (1, 19, 26). Flexible ligands are automatically placed into the binding pocket in several random orientations used as starting points for Monte Carlo optimization. The optimized energy function includes the ligand internal strain and a weighted sum of the grid map values in ligand atom centers. To improve convergence of docking predictions, three independent runs of the docking procedure are usually performed, and the best scoring pose per compound is stored. The ligand binding poses are evaluated with all-atom ICM ligand binding score that has been derived from a multi-receptor screening benchmark as a compromise between approximated Gibbs-free energy of binding and numerical errors (27, 28). The score is calculated as:

$$S_{\text{bind}} = E_{\text{int}} + T\Delta S_{\text{Tor}} + E_{\text{vw}} + \alpha_1 E_{\text{el}} + \alpha_2 E_{\text{hb}} + \alpha_3 E_{\text{hp}} + \alpha_4 E_{\text{sf}}, \quad (1)$$

where E_{vw}, E_{el}, E_{hb}, E_{hp}, and E_{sf}, respectively, are van der Waals, electrostatic, hydrogen bonding, nonpolar, and polar atom solvation energy differences between bound and unbound states, E_{int} is the ligand internal strain, ΔS_{Tor} is its conformational entropy loss upon binding, $T = 300$ K, and α_i are ligand- and receptor-independent constants.

As the receptor optimization approach heavily relies on docking as a model assessment tool, reasonable reproducibility of the binding mode is vital for successful application of the method. ICM fast grid docking as one of the most robust and reproducible docking algorithms (28) is an ideal choice for such evaluative screening.

For suboptimal pocket conformations in the intermediate stages of optimization, however, several (usually 3) independent docking runs are needed to reliably reproduce ligand conformations. Low reproducibility of ligand poses in multiple runs even after several iterative steps is also a strong indicator that the system is not moving towards convergence. This could happen, for example, when compounds in the ligand set have a complex undefined stereochemistry, which can be dealt with by either defining active isomers, or allowing sampling of isomeric states in docking.

2.4. Selection of the Best Protein Conformers with NSQ_AUC Metric

Performance in docking/VLS (i.e., the ability of the receptor conformer model to separate true ligands from nonbinding decoys (8, 9, 13, 14)) is defined by the distribution of the binding scores for the ligand and decoy set. Some of the commonly used metrics of VLS performance include the median rank of the ligand scores, the hit rate, enrichment factor, or the "area-under-the-curve" (AUC). The curve, known as receiver operator curve (ROC), is a plot of the "true-positive rate" versus the "false-positive rate" for varying value of the docking score threshold. While ROC curve by itself is very indicative of the VLS performance, the above cumulative measures has its shortcomings which are discussed in literature (see, e.g., ref. 29). Recently, we introduced a normalized square root AUC (NSQ_AUC) metric, which puts a soft emphasis on "early" hit enrichment in screening results while retaining contribution for overall selectivity and sensitivity of the model (14). Similar to standard AUC, value of NSQ_AUC is based on calculation of the area under the ROC curve. The difference is that the effective area (AUC*) is defined for the ROC curve plotted with X coordinate calculated as square root of "false-positive rate," $X = \text{Sqrt(FP)}$. The NSQ_AUC is then calculated as:

$$\text{NSQ\_AUC} = 100\left(\frac{\text{AUC}^* - \text{AUC}^*_{\text{random}}}{\text{AUC}^*_{\text{perfect}} \,/\, \text{AUC}^*_{\text{random}}}\right).$$

Thus, the value of NSQ_AUC is more sensitive to initial enrichment than the commonly used linear AUC. The NSQ_AUC measure returns the value of 100% for any perfect separation of signal from noise and values close to zero for a random subset of noise.

2.5. Iterative Ligand-Guided Refinement

Early applications of ligand-guided receptor optimization methodology used only one run of the sampling-selection procedure. While a large set of generated conformers, for example 800 in ref. 9, increased the chance of finding a model with improved VLS performance, we observed that multiple iterations of the procedure introduced by LiBERO provided significant advantages.

Thus, detailed analysis of intermediate results in refs. 10 and 11 showed that on each iteration of the LiBERO procedure, the probability of finding an improved model significantly increased. This effect is a result of inheritance of some advantageous conformational features in the pocket from the previous generation model, combined with newly found features. Another important advantage is that multiple iterations also allow monitoring of the progress of the VLS performance, and thus establishing criteria for convergence for receptor optimization.

2.6. Criteria for Optimization Convergence

Quality of the modeling systems can be monitored by both (1) average ICM ligand-binding scores for the ligand "active" set and (2) NSQ_AUC calculated for ligand/decoy sets. When the values of these parameters max out and do not change significantly over several iterations, this likely indicates convergence of the system (see Fig. 2). Additional criteria for filtering may include consistency of the binding poses for the same ligands (i.e., as measured by conserved ligand–protein contacts) and/or ligands based on similar scaffolds. The pose convergence in ICM can be evaluated by an automatic procedure that checks for the presence of "anchor interactions" or certain binding motifs of the docked ligands.

Separation of ligands and decoys in the final optimized models does not need to reach 100% NSQ_AUC, as some of the compounds

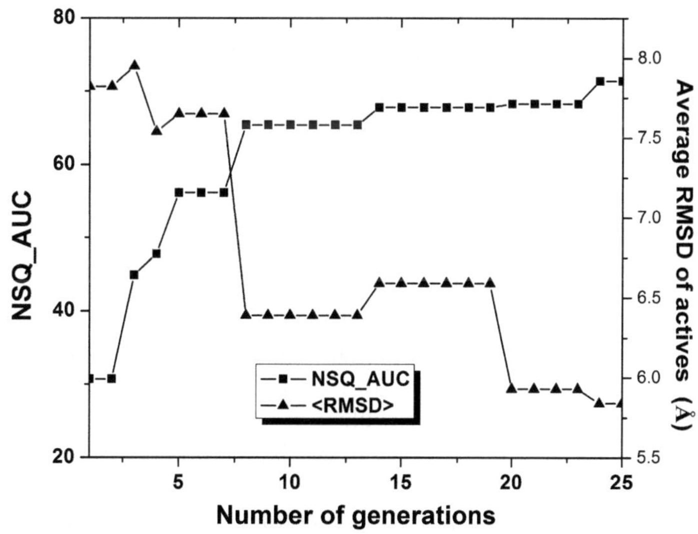

Fig. 2. Improvement in VLS performance (as measured by NSQ_AUC) obtained with ALiBERO for an A2A receptor homology model. Note that the average ligand RMSD values with respect to the crystal (ligand ZMA in PDB: 3eml; RMSD performed on common scaffold for the 23 actives used in this run) decrease as the NSQ_AUC values improve (see RMSD scale at right *y*-axis).

in the diverse ligand set may still not be docked and/or scored correctly. The acceptable values of converged average ICM score are usually better than −30 kJ/mol and NSQ_AUC exceeding 70%, though this may vary for different receptors and ligand/decoys sets. While some of the "outlier" ligands may be just less amenable for the docking procedure (e.g., compounds with complex nonaromatic ring systems), others may require a different conformer for adequate docking and scoring. For the latter cases repeating the LiBERO procedure for only a specific subset of similar "outlier" ligands may result in identification of an alternative receptor conformation optimal for binding of a distinct class of ligands.

2.7. Requirements and Limitations of the Method

While LiBERO method has proved useful in a number of virtual ligand screening and drug discovery applications, it is important to understand some requirements for the modeling system. The *first* and most critical requirement is availability of information about high-affinity ligands. For many human targets in GPCRs, kinases, proteases, and other protein families, dozens of selective high-affinity ligands are known, sufficient for an adequate ligand set. However, other targets in early stages of validation may have very limited number of ligands/substrates known, or lack this information at all (e.g., orphan receptors). For these cases, and also cases of putative allosteric pockets, one can attempt other pocket optimization methods (e.g., SCARE (30) or "fumigation" (6) approaches that do not require a known ligand set).

The second requirement is the availability of a relatively close 3D structural template homolog(s) to ensure adequate quality of the initial homology model. While well-behaved binding pocket models for VLS can be obtained even in some cases when the target backbone deviates as much as 3–4 Å from the template (10, 31), such cases require availability of an exceptionally good quality—in terms of both affinity and diversity—ligand sets.

Modeling systems that do not satisfy these requirements may run a risk of over-fitting. Thus, small ligand sets lacking diversity may result in a binding pocket tightly closed around this particular ligand type, but not accepting other ligands (though in case of lead optimization this may be acceptable). If large-scale movements of the backbone are allowed, the pocket model becomes too adaptable and the complexity of the problem becomes comparable to the problem of protein folding.

We must also emphasize that while the backbone movements in LiBERO help to improve ligand–receptor contacts, the method does not guarantee significantly improved backbone placement in the receptor, as measured by RMSD. Though an optimized structure may remain "skewed" as compared to the "true" experimental receptor structure, the key improvement is the number of correctly predicted ligand–receptor contacts (32). As we have shown recently, the latter model quality metric is correlated with VLS performance and is thus more relevant to docking applications (10). Also, effective prediction of ligand–receptor interactions is important for practical applications and allows further validation of the model through point mutation experiments.

3. Methods

The LiBERO method presented in the previous sections has been recently implemented in a fully automated fashion (ALiBERO), on which the sampling-selection steps are performed without user intervention. ALiBERO version of the method has been able to reproduce and improve some our previously published results with optimized models and is currently being used with other GPCRs and other protein families. The next section we describe the major steps needed for setting up and running a calculation, while additional details of the method development are presented elsewhere (Rueda et al, submitted).

3.1. Computational Setup

ALiBERO is implemented as an iterative algorithm, on which a large population of conformers is generated (i.e., via EN-NMA), and the conformer displaying the best screening performance is selected for the next generation. The default *fitness function* is calculated as the normalized square root of the area under the ROC curve (NSQ_AUC). Alternatively, the fitness function can be the average ICM score or the area under the ROC curve (AUC).This iterative process is repeated until a termination condition has been reached, such as reaching a threshold NSQ_AUC, or when successive iterations no longer produce better results.

ALiBERO script was implemented in Perl (v5.8.8), and runs on a "master" node using internal parallel threads involving ICM software (26) for ligand–receptor docking and ligand–receptor refinement calculations. In its current implementation, ALiBERO uses 1 CPU per each VLS run. The programs allow submission of the VLS threads either locally (i.e., a standard Linux multi-core CPU Desktop) or to Linux-based clusters running the PBS/Torque queue system (see Note 1).

3.2. Input Parameters

ALiBERO needs an *input file*, which specifies the location of the initial homology model file and the ligand/decoy dataset, as well as parameters for the iterative procedure as shown in the example below.

```
----ALiBERO input file---------------------------------

pbs         triton

inputob     ~/AA2AR/INPUT/RECEPTORS/aa2ar.ob

projdir     ~/AA2AR/AA2AR_nsa_300K_529L_elion_ref_x3

temperature  300

function   nsa

elitism     on

nligands   529

inx         ~/AA2AR//INPUT/LIGANDS/AA2AR_Ligands_wDecoys.inx

sdf         ~/AA2AR//INPUT/LIGANDS/AA2AR_Ligands_wDecoys.sdf

macrodir   ~/AA2AR//INPUT/MACROS

refinement on

ntop       10

repeat     3

laziness5

------------------------------------------------------------

--
```

In this example, used for the Adenosine A2A receptor homology model optimization, the calculation was submitted to a PBS queue system on "Triton" at the San Diego Supercomputer Center. The location of the initial homology model file in ICM object format is specified by "inputob" parameter. The "sdf" and "inx" parameters define location of the ligand/decoy set in SDF format; note that the SDF file must have a column named "Active," which specifies active with value "1" and decoys with value "0." In this case, a training set consisting of 29 actives + 500 decoys was used. The "projdir" value specifies location of the output files and "macrodir" is a directory containing the ICM macro files to be used. The VLS performance was measured by the NSQ_AUC fitness function (function "nsa") (see Note 2). As commented in

Subheading 2 above, some receptors may benefit from the use of soft distance restraints (*drestraint* in ICM scripting language). Such restraints can be specified in the provided ICM macro dedicated to the all atom Monte Carlo refinement step.

The temperature was set to 300 K for the EN-NMA procedure, which corresponds to about 1 Å RMSD average backbone variations. The docking calculations were repeated three times independently to ensure reproducible docking and an additional all atom energy-based refinement was done for the top 10 scoring ligand–receptor complexes obtained in the docking step.

3.3. ALiBERO Runs

As a rule of thumb, we recommend performing a small-scale calculation (i.e., using small number of CPUs and a small ligand set) before performing full production runs. The objective of such tests is to monitor the changes in the fitness function values and to visually check reproducibility of the ligand binding modes within pockets.

For a quick comparison of model performance, one can simply use as fitness function the average ICM binding score for the ligand set (or rather portion of the ligand set to allow for possible outliers). This alternative objective function does not require docking and evaluation of decoys, and thus may be employed to avoid extensive docking computations in the initial steps of the optimization procedure when performance gains are large and obvious. However, more robust absolute measures such as NSQ_AUC are required in later stages for adequate evaluation of the models.

According to our experience, the performance is greatly improved when testing a large number of conformers on each generation. A large number of conformers improve the likelihood of finding a good performing model, while keeping the number of generations small. Overall, we have found that more reliable optimization results are achieved when using between 50 and 100 conformers on each generation. However, in many cases, optimizations measured by NSQ_AUQ were achieved with as few as ten conformers and without replicating VLS runs. It is also a good idea to set the parameter "elitism" to "on"; this only accepts the best conformation in the current iteration if it improves the fitness function. One reliable way of validating the predictions in "real case scenarios" is by repeating ALiBERO full runs, and by checking for consistency of fitness function values among runs, as well as for consistency in binding modes and ligand–protein conserved contacts. If enough ligand data is available, it is possible to remove some ligands from the training set and try to recover them as actives in VLS after the optimization steps.

An full ALIBERO run consisting of ten generations (100 conformers, 500 ligands VLS, 3× repetitions) takes about 2–3 days using ~300 Intel Nehalem 2.4 GHz cores on the "Triton" cluster at the San Diego Super Computer Center. The calculations that were interrupted or failed to reach desired values of the fitness function can be easily restarted from the last iteration step (see also Note 3). It is worth mentioning that the most time consuming part of the method is the docking/VLS, whereas the rest of the steps (EN-NMA, calculation of grid maps, calculation of NSQ_AUC, selection of models, etc.) only represent a minor percentage of the total CPU time (see Note 4).

3.4. Output Presentation and Analysis

The performance of ALiBERO depends on the quality of the initial homology models, the ligand dataset, as well as the parameters used. Thus, although the automatic protocol will do its best to optimize *any* model, a bad combination of protein/ligand/parameters may lead to suboptimal models. For this reason, it is highly recommend to visually inspecting the results. On every generation, ALiBERO generates an ICM binary file consisting of the 3D ligand poses for best performing protein conformers, as well as tables, ROC curves, and all the information needed for browsing the solutions (see Fig. 3).

If the complexity of the optimization is high, like that of working with GPCRs, several stages of ALiBERO may be required. For

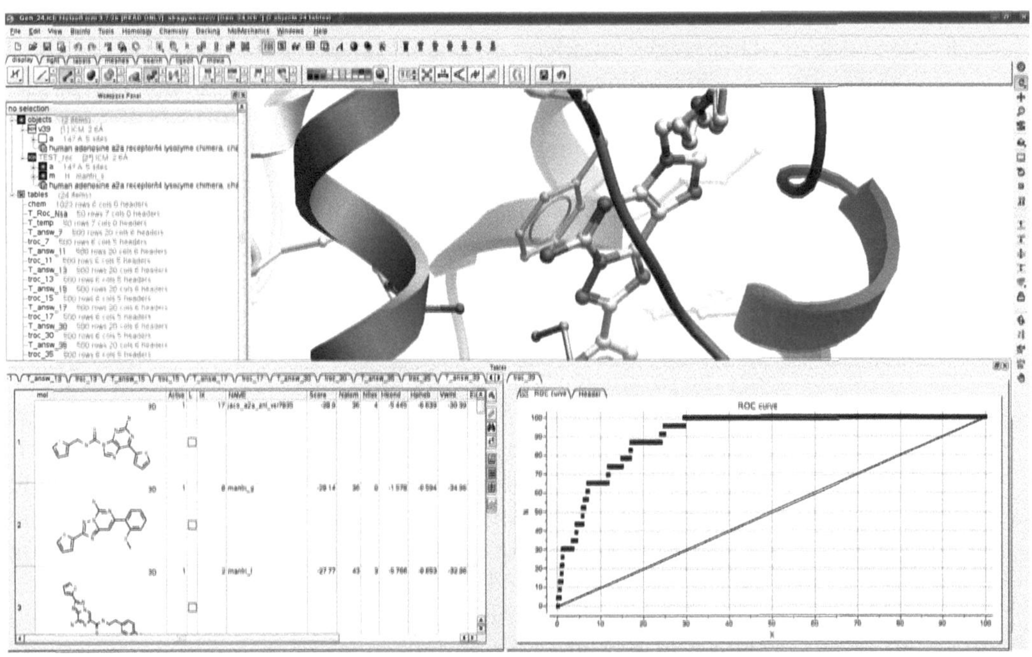

Fig. 3. Example of ALiBERO output as viewed with ICM software. On every generation, ALiBERO generates an ICM binary file containing all the information needed for browsing the docking solutions.

instance, larger backbone displacements may be needed only at the beginning of the optimization, while smaller ones may be needed at the later stages. Also, additional "anchor interactions" (if available) in conjunction with NSQ_AUC may be quite helpful in the later stages. The final optimized models resulting from ALiBERO are then ready for use in large-scale VLS, on which thousands or even millions of compounds may be screened.

4. Conclusions

Performance of 3D receptor models in virtual ligand screening and other drug discovery applications can be dramatically improved by ligand-guided receptor optimization, where a set of known ligands is used to optimize the shape of the binding pocket. Presented here LiBERO methodology expands applications of the ligand-guided approach to models that require backbone adjustment in the binding pocket. LiBERO also introduces an iterative process, where in each step of iteration, the protein conformations are generated by NMA and/or energy-based sampling followed by the selection of the best conformers using a specially developed VLS performance metric (NSA-AUC) as a cumulative fitness function. This approach has proved successful in a growing number of applications, which include prediction of agonist-induced conformational changes in the receptor pocket, ligand interactions within a homology models and prospective structure-based ligand screening for drug discovery. This algorithm, based on the ICM docking/VLS screening platform, is implemented as ALiBERO, a script that allows automatic highly parallel distributed execution on a Linux computer cluster managed by the PBS queuing system. The ALiBERO script is available from the authors upon request as an add-on to ICM (Molsoft LLC) molecular modeling package for the Linux platform.

5. Notes

1. ALiBERO can be executed either in a single workstation mode (`PBS no`) or in on a cluster mode (`PBS Name_of_the_Cluster`). Execution on the cluster requires a site ICM-VLS license for the cluster and an automated user login to the cluster master node.
2. To speed up calculation in the first iterations, one can use a simplified objective function (`function score`) which is based on docking score of ligands only and does not require docking of decoys. The full ligand/decoy selectivity benchmark

(`function nsa`) is still strongly recommended in the final steps of refinement and evaluation of the model. In the latter case, it is important to keep relatively high number (~200) and diversity of decoys to prevent model selection against specific decoys.

3. "Laziness" is a technical parameter in ALiBERO input file that controls parallel execution of multiple docking jobs on a cluster. Since some of the docking jobs may be lost in the cluster environment or executed much slower than the others, setting "laziness," for example at 5%, allows the master program to start execution of the next iteration of the optimization procedure without waiting for the last 5% of the docking results.
4. In its current implementation, the program is optimized for execution in a cluster queue with homogeneous core performance, performance in a more heterogeneous computational environment (e.g., CPU cloud computing can be suboptimal).

Acknowledgment

The authors thank Chris Edwards for help with manuscript preparation.

References

1. Totrov, M. and R. Abagyan, *Flexible protein-ligand docking by global energy optimization in internal coordinates.* Proteins, 1997. **Suppl 1**: p. 215–20.
2. Totrov, M. and A. R., *Derivation of sensitive discrimination potential for virtual ligand screening.* (RECOMB 99) Lyon France, ACM Press. , 1999: p. 312–7.
3. Erickson, J.A., et al., *Lessons in molecular recognition: the effects of ligand and protein flexibility on molecular docking accuracy.* J Med Chem, 2004. **47**(1): p. 45–55.
4. Brylinski, M. and J. Skolnick, *What is the relationship between the global structures of apo and holo proteins?* Proteins, 2008. **70**(2): p. 363–77.
5. Bottegoni, G., et al., *Four-dimensional docking: a fast and accurate account of discrete receptor flexibility in ligand docking.* J Med Chem, 2009. **52**(2): p. 397–406.
6. Abagyan, R. and I. Kufareva, *The flexible pocketome engine for structural chemogenomics.* Methods Mol Biol, 2009. **575**: p. 249–79.
7. Marcou, G. and D. Rognan, *Optimizing fragment and scaffold docking by use of molecular interaction fingerprints.* J Chem Inf Model, 2007. **47**(1): p. 195–207.
8. Bisson, W.H., et al., *Discovery of antiandrogen activity of nonsteroidal scaffolds of marketed drugs.* Proc Natl Acad Sci, 2007. **104**(29): p. 11927–32.
9. Cavasotto, C.N., et al., *Discovery of novel chemotypes to a G-protein-coupled receptor through ligand-steered homology modeling and structure-based virtual screening.* J Med Chem, 2008. **51**(3): p. 581–8.
10. Katritch, V., et al., *GPCR 3D homology models for ligand screening: lessons learned from blind predictions of adenosine A2a receptor complex.* Proteins, 2010. **78**(1): p. 197–211.
11. Katritch, V., I. Kufareva, and R. Abagyan, *Structure based prediction of subtype-selectivity for adenosine receptor antagonists.* Neuropharmacology, 2011. **60**(1): p. 108–15.
12. Katritch, V., et al., *Analysis of full and partial agonists binding to beta2-adrenergic receptor suggests a role of transmembrane helix V in agonist-specific conformational changes.* J Mol Recognit, 2009. **22**(4): p. 307–18.

13. Reynolds, K.A., V. Katritch, and R. Abagyan, *Identifying conformational changes of the beta(2) adrenoceptor that enable accurate prediction of ligand/receptor interactions and screening for GPCR modulators.* J Comput Aided Mol Des, 2009. **23**(5): p. 273–88.
14. Katritch, V., et al., *Structure-based discovery of novel chemotypes for adenosine A(2A) receptor antagonists.* J Med Chem, 2010. **53**(4): p. 1799–809.
15. Reynolds, K., R. Abagyan, and V. Katritch, Structure and Modeling of GPCRs: Implications for Drug Discovery, in *GPCR Molecular Pharmacology and Drug Targeting: Shifting Paradigms and New Directions, A. ed. Gilchrist, Editor.* 2010, *Wiley & Sons, Inc: Hoboken, NJ.* p. 385–433.
16. Kufareva, I. and R. Abagyan, *Type-II kinase inhibitor docking, screening, and profiling using modified structures of active kinase states.* J Med Chem, 2008. **51**(24): p. 7921–32.
17. An, J., M. Totrov, and R. Abagyan, *Pocketome via comprehensive identification and classification of ligand binding envelopes.* Mol Cell Proteomics, 2005. **4**(6): p. 752–61.
18. Abagyan, R. and M. Totrov, *Biased probability Monte Carlo conformational searches and electrostatic calculations for peptides and proteins.* J Mol Biol, 1994. **235**(3): p. 983–1002.
19. Abagyan, R.A., M.M. Totrov, and D.A. Kuznetsov, *Icm: A New Method For Protein Modeling and Design: Applications To Docking and Structure Prediction From The Distorted Native Conformation.* J. Comp. Chem. , 1994. **15**: p. 488–506.
20. Arnautova, Y.A., R.A. Abagyan, and M. Totrov, Development of a new physics-based internal coordinate mechanics force field and its application to protein loop modeling. *Proteins.* 79: 477–98, 2011. PMCID: 3057902
21. Cavasotto, C.N., J.A. Kovacs, and R.A. Abagyan, *Representing receptor flexibility in ligand docking through relevant normal modes.* J Am Chem Soc, 2005. **127**(26): p. 9632–40.
22. Tirion, M.M., *Large Amplitude Elastic Motions in Proteins from a Single-Parameter, Atomic Analysis.* Phys Rev Lett, 1996. 77(9): p. 1905–8.
23. Rueda, M., G. Bottegoni, and R. Abagyan, *Consistent improvement of cross-docking results using binding site ensembles generated with elastic network normal modes. J Chem Inf Model.* 49: 716–25, 2009. PMCID: 2891173
24. Kovacs, J.A., M. Yeager, and R. Abagyan, *Damped-dynamics flexible fitting.* Biophys J, 2008. **95**(7): p. 3192–207.
25. Rueda, M., G. Bottegoni, and R. Abagyan, *Recipes for the Selection of Experimental Protein Conformations for Virtual Screening.* J Chem Inf Model, 2009.
26. Abagyan, R.A., et al., *ICM Manual.* 2009, MolSoft LLC: La Jolla, CA.
27. Schapira, M., M. Totrov, and R. Abagyan, *Prediction of the binding energy for small molecules, peptides and proteins.* J Mol Recognit, 1999. **12**(3): p. 177–90.
28. Bursulaya, B.D., et al., *Comparative study of several algorithms for flexible ligand docking.* J Comput Aided Mol Des, 2003. **17**(11): p. 755–63.
29. Truchon, J.F. and C.I. Bayly, *Evaluating virtual screening methods: good and bad metrics for the "early recognition" problem.* J Chem Inf Model, 2007. **47**(2): p. 488–508.
30. Bottegoni, G., et al., *A new method for ligand docking to flexible receptors by dual alanine scanning and refinement (SCARE).* J Comput Aided Mol Des, 2008.
31. Michino, M., et al., *Community-wide assessment of GPCR structure modelling and ligand docking: GPCR Dock 2008.* Nat Rev Drug Discov, 2009. **8**(6): p. 455–63.
32. Rueda, M., et al., *SimiCon: a web tool for protein-ligand model comparison through calculation of equivalent atomic contacts.* Bioinformatics, 2010. **26**(21): p. 2784–5.

Chapter 9

Loop Simulations

Maxim Totrov

Abstract

Loop modeling is crucial for high-quality homology model construction outside conserved secondary structure elements. Dozens of loop modeling protocols involving a range of database and ab initio search algorithms and a variety of scoring functions have been proposed. Knowledge-based loop modeling methods are very fast and some can successfully and reliably predict loops up to about eight residues long. Several recent ab initio loop simulation methods can be used to construct accurate models of loops up to 12–13 residues long, albeit at a substantial computational cost. Major current challenges are the simulations of loops longer than 12–13 residues, the modeling of multiple interacting flexible loops, and the sensitivity of the loop predictions to the accuracy of the loop environment.

Key words: Protein loops, Loop simulation, Loop modeling, Conformational sampling

1. Introduction

Enormous bulk of sequence data produced by high-throughput genomics efforts and the complexity of experimental protein structure determination continue to maintain a large gap between the number of identified genes and proteins with solved 3D structures (2–3 orders of magnitude, i.e., UniRef100 database has >11 million entries, Protein Data Bank (PDB) has ~39,000 entries with nonidentical sequences). Despite certain progress in ab initio protein structure prediction, the examples of successful protein folding starting from sequence alone remain isolated and the practical utility of current methods is unclear. By contrast, comparative modeling based on homology to a protein with solved 3D structure is widely used and the approach is largely successful in predicting the overall tertiary structure, providing practically useful information on the localization of specific amino acid residues on the protein surface, in the functionally important sites, or the protein core (1). For a close homolog the quality of the models

Andrew J.W. Orry and Ruben Abagyan (eds.), *Homology Modeling: Methods and Protocols*, Methods in Molecular Biology, vol. 857, DOI 10.1007/978-1-61779-588-6_9, © Springer Science+Business Media, LLC 2012

can approach atomic resolution. However, the accuracy of modeling varies significantly between the secondary structure elements (α-helixes and β-strands), where rigid backbone approximation is usually acceptable, and the loops which tend to be more mobile. This is especially true when insertions or deletions appear in the template/target alignment. Many homology modeling programs currently in use can generate the loops with acceptable covalent geometry, typically by database search, but finding a near-native conformation has proven difficult, and the loops are consistently the most inaccurate parts of the homology models (2).

On the other hand, loops often form parts of the functionally important binding or enzymatic sites. As an extreme but highly practically important example, antibodies bind antigens via their complementarity-determining regions (CDRs) which are essentially sets of six variable loops (CDR1–CDR3 on both light and heavy chains) on a well-conserved scaffold of the immunoglobulin (Ig) domain core. Loops also can be functionally mobile, with the conformational switch regulating activity, as illustrated by the so-called DFG loop in the tyrosine kinases, which has the "in" (active) and "out" (inactive) conformations (3, 4).

Loops also present an interesting model system for theoretical studies of protein energetics and conformational analysis. The same energy contributions that stabilize particular conformations of loops ultimately should also guide folding of entire proteins. While full exploration of the conformational space and energy hypersurface of a protein remains prohibitively expensive for all but a few smallest folded protein domains, near-exhaustive conformational sampling and thorough comparison of different energy approximations can now be performed on large sets of loops.

2. Methods

Loop prediction problem can be formulated as generation and identification of a near-native loop conformation, given the structure (exact experimental coordinates or, more practically important, an inexact model) of the rest of the protein. Significant efforts over last several decades have been dedicated to the development of accurate loop prediction methods, and dozens of algorithms have been proposed. Two main groups of prediction methods can be distinguished, knowledge based and ab initio, with some methods utilizing elements of both approaches (Fig. 1). Knowledge-based methods use databases of experimentally observed polypeptide chain conformations, typically extracted from the PDB (5). Loop segments that geometrically match the terminal residue positions are identified and further scored according to their fit with the rest

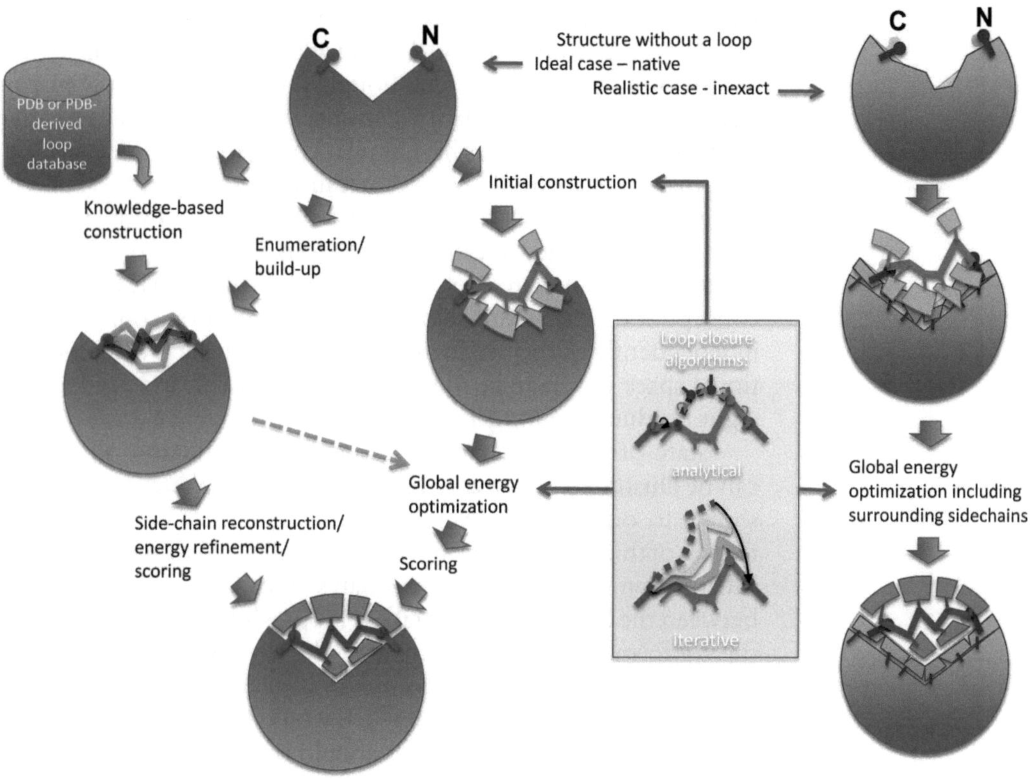

Fig. 1. Key algorithms, protocols, and concepts in loop simulations.

of the structure and/or sequence similarity to the target loop. On the other hand, ab initio methods are based on various forms of conformational sampling. Although knowledge-based loop modeling methods are typically much faster, they are limited by the available amount of experimental data, whereas ab initio approaches in principle can predict novel structures never observed previously. Theoretically, the conformational space of a loop expands exponentially with the loop length and therefore its coverage by any fixed loop database becomes increasingly sparse for longer loops. Estimates (now 10–15 years old) suggested that experimental data provide sufficient sampling for loops up to 5–6 residues long (6, 7). To some extent, more relaxed termini superposition cutoffs can improve coverage, while an energy minimization stage can be used to resolve associated distortions of terminal junctions (8). Still, most of the knowledge-based methods reported (8–11) perform well only for shorter loops.

Either combinatorial construction from the shorter loop fragments or additional ab initio-like conformational search maybe necessary for knowledge-based reconstruction of near-native conformations for long loops. The situation might be changing with the rapid expansion of the PDB, and more recent analysis suggested that the loop conformational space may be saturated up to the length of 12 residues (12), although this conclusion was in part based on sequence similarity considerations, i.e., assuming that loops of similar sequences have similar conformations. The assumption may be statistically correct because local sequence similarity correlates with overall homology and therefore fold similarity, but may not hold when locally homologous loop occurs within the context of an unrelated fold. Very recent analysis that applied the concept of the "structural alphabet" to classify loop conformations independently of their sequences indicates that the loop conformational space coverage in PDB structures is still sparse for loops of eight residues and longer (13).

State-of-the-art database search loop prediction algorithms can be illustrated by the new version of FREAD, which was recently shown to outperform several ab initio methods (14). Distinctive feature of the method is the use of the so-called environment-specific substitution score, which evaluates local sequence similarity between the query and the database loops while taking into account the conformational "environment." The method has an impressive speed advantage over ab initio methods, taking only minutes even for long loops, predictions for which would likely take days or even weeks of ab initio simulations. It should be noted that FREAD has a rather high failure rate (situations where no prediction at all is produced; ~50% for longer loops) and thus simple RMSD comparisons may not be entirely fair. Also, in general the assessment of the predictive ability of methods that use database search is complicated by the necessity to "jackknife" the training data to remove the benchmark targets and entries closely related to them, the definition of "closely related" being highly subjective.

To utilize empirical data without sacrificing coverage, shorter fragments found in the database may be assembled into longer loops, potentially creating novel conformations, previously unobserved experimentally but sharing segments with experimental structures and thus likely energetically favorable. Fragment assembly loop construction method based on ROSETTA (15) uses nine-residue segment libraries to sample longer loops (16). However, recently developed ROSETTA-based ab initio loop construction was shown to outperform this older knowledge-based approach (17).

2.1. Ab Initio Loop Modeling Methods

Native conformation of the loop should represent the global minimum of its free energy. Thus, ab initio methods identify the near-native structures via some form of global energy optimization. Success of an ab initio loop prediction method depends on two main factors: the ability of the conformational search algorithm to locate lowest energy minima of the energy (scoring) function and the accuracy of the scoring function, i.e., its ability to rank near-native solutions over the various decoys. The search and the scoring may be separated into distinct stages of the modeling protocol, or combined within an iterative optimization algorithm. Separate search and scoring approach is conceptually attractive due to the simplicity, modularity, and apparent possibility to assess and choose independently the best options for the two stages. However, it should be noted that in reality the performance of the scoring function depends on the quality of the ensemble. If the "native-like" solutions in the ensemble have some distortions, they may preclude recognition of these solutions by the scoring function. For example, even sub-angstrom deviations in the structure may result in significant steric clashes which would severely affect scoring using force-field energy. The conformation generation algorithm that is "aware" of the scoring could perform an energy minimization, resolving clashes and likely producing better results on the scoring stage. On the other hand, a more "tolerant" scoring function may give good scores to near-native solutions that have significant distortions (unfortunately, likely at the cost of other artifacts).

A subclass of ab initio methods that clearly separate sampling and scoring can be designated as enumeration methods. One of the first enumeration methods was described by Moult and James (2). A more recent exhaustive enumeration algorithm, PETRA (18), utilizes a virtual database (APD, or ab initio polypeptide database) of all possible polypeptide fragments with 10 ϕ/ψ pairs that are allowed to adopt eight discrete combinations, for a total of 10^8 entries. Good coverage was demonstrated for short (five residue) loops. Clearly, combinatorial explosion constrains this approach both in terms of loop length and the number of ϕ/ψ states, which ultimately limits accuracy. Tosatto et al. proposed a "divide-and-conquer" algorithm utilizing a pre-generated database of artificial loop segments containing only median and terminal residue positions (19). A query for a given pair of terminal positions and loop length yields possible middle residue positions, which are used as new C- or N-termini for queries of half-length loops, etc., until full loop is reconstructed. Sufficiently dense coverage of the loop space by the pre-generated database is clearly critical, and even 1,000,000 entries appeared to be insufficient for loops longer than six residues. Since the database is computer generated, in principle it can be expanded if ample memory and disk space is available.

Another enumerative method, LOOPER (20) applies two-state amino acid residue model, alpha-helix like and extended/strand like (four states for glycine residues) for exhaustive discrete sampling of conformational space of the two half-loops, which are then reconnected combinatorially and energy minimized to obtain an ensemble of closed low-energy conformations for the complete loop.

A significant difficulty in separating sampling and scoring is that sufficient sampling without any guidance from some form of scoring function is only feasible for relatively short loops where terminal restraints largely define loop conformations. At a minimum, steric avoidance has to be considered during conformation generation for longer loops to eliminate vast numbers of geometrically possible but unphysical structures.

The procedure proposed by Galaktionov et al. (21) utilizes more detailed 5-state model (8 states for glycine) of the polypeptide backbone. All possible combinations of these states were modeled and conformations that span the gap (within certain tolerance) between residues flanking the loop at the N- and C-terminal were energy minimized with harmonic restraints. To avoid exponential explosion in the number of conformation to be evaluated for longer loops, build-up procedure that adds residues one by one from the N terminus was developed. At each step the procedure eliminated backbone "trajectories" that clash with themselves or the body of the protein, or wander too far from the C terminus to reconnect, given the number of remaining residues to be built.

Further focusing on physically relevant conformations is necessary to perform efficient enumeration for longer loops. This can be achieved by the introduction of a scoring function during loop generation or sampling, but detailed atomistic representation of the loop and calculation of energy terms can be computationally costly. A common theme in many modern ab initio loop prediction methods is the use of multiple stages, where initially some form of simplified representation of the polypeptide chain is used to rapidly sample the broad conformational space of the loop, and then refine the most promising solutions in more detail on the later stage(s).

For example, Rapp and Friesner generated initial set of loop conformations on a simplified model with C_β atoms only, using random starting loop geometries closed via optimization of end-point geometry (22). These initial conformations were refined in atom–atom representation via a combination of energy minimizations and molecular dynamics runs. Olson et al. proposed a "multiscale" approach where initial sampling is performed using cubic lattice-based low-resolution model with one center per amino acid residue located at the center of mass of the side chain (MONSSTER (23)); on the second stage the models are refined using replica-exchange molecular dynamics and scored using CHARMM and GB solvation model (24). Significant improvement in RMSD (by more than 1 Å on average) of the native-like solutions was observed upon all-atom refinement. Several other protocols discussed in the subsequent sections also take advantage of multistage approach.

2.2. Loop Closure

A key aspect of loop conformational sampling is the requirement of loop closure: since both N- and C-termini are assumed to be statically attached to the rigid parts of the protein fold, conformational search should be constrained to the subspace of main-chain conformations which have correct covalent geometry at the terminal junctions.

In the knowledge-based sampling methods, loop closure represents the principal filter: typically the chain segments in the database that match (within a certain tolerance) the desired positions of the termini are selected. In the ab initio methods on the other hand, new loop conformations are generated in the course of the simulation, and therefore it is more efficient to steer or constrain conformation generation process to closed loops rather than filter out non-closed conformations later. In principle, if a complete force-field energy including bonded terms (i.e., bond stretching and bond bending) is used, energy minimization will enforce correct loop closure. However, this "brute-force" approach can be highly inefficient because a lot of the energy calculation cycles will be spent on restoring reasonable covalent geometry, instead of optimization of weaker non-covalent interactions. Therefore, a large variety of methods have been developed to generate new polypeptide chain conformations that match the fixed terminal positions. Three classes of loop closure methods can be distinguished: analytical, iterative optimization, and build-up. In the analytical methods, the search algorithm can alter a subset of polypeptide chain's degrees of freedom (DoFs, such as certain ϕ/ψ torsions), while the remaining DoFs are automatically recalculated so that the loop remains closed. In the iterative optimization methods, closure constraints are expressed as a function which is optimized to achieve closure, often in combination with other terms. In build-up methods, the loop is constructed by sequentially adding residues starting from one or both termini.

2.2.1. Analytical Methods

Analytical loop closure was first investigated in the classical work by Go and Scheraga (25), where it was formulated as a system of six equations in the six dihedral angles. Extensive analysis by Wedemeyer and Scheraga showed how these equations can be reduced to a polynomial solved analytically and how the longer loops for which the problem becomes under-determined can be treated (26). Analytical methods solve what is sometimes called reverse kinematic problem (27), which concerns finding six angles that would make a chain of vectors reach from a given starting point to a given end point in a specified orientation. Similar algorithms have been developed in robotics to evaluate rotations in the joints of a mechanical arm consisting of multiple rigid limbs so that its tip can reach desired points in space.

Rapid generation of the perturbed backbone loop conformations without disruption of covalent geometry is most useful within the context of stochastic sampling methods such as Monte Carlo simulation. Thus, large rearrangements of the backbone are performed by triaxial loop closure (TLC) method (28) in the Hierarchical Monte Carlo sampling (29) protocol, applied to assess mobility of flexible loops in protein structures rather than for the more common native conformation prediction. In the Local Move

Monte Carlo (LMMC) method, after a single backbone torsion is randomly modified, six other torsions are recalculated to maintain loop continuity (30). Mandell et al. incorporated kinematic closure (KIC) steps in their ROSETTA-based Monte Carlo loop modeling protocol (17). Enhanced sampling as compared to the previous, knowledge-based protocol was demonstrated, and the algorithm overall achieved impressive accuracy.

Apparent advantages of the analytical methods are their accuracy and speed. However, analytical closure solutions may not exist for many (perhaps large majority of) combinations of independent variables. Therefore, multiple closure attempts with different sets of values for independent variables may have to be performed before a new solution is found, essentially making the algorithm iterative. Furthermore, because analytical solution is unaware of physical steric constraints on the polypeptide chain, some of the ϕ/ψ angle pairs from an analytic solution are likely to fall into unfavorable regions of the Ramachandran plot (31), again requiring multiple attempts to find a physically acceptable solution.

An analytical/iterative method, cyclic coordinate descent (32) consists of steps that analytically set a single torsion to the value that best satisfies closure constraints. The method appears to be more robust than fully analytical closure and can be biased toward low-energy ϕ/ψ angle combinations using probabilistic acceptance criterion of the analytical steps, based on Ramachandran plot.

The accuracy advantage of the analytical closure is less clear when one considers the fact that the underlying rigid covalent geometry model is in itself an approximation. Most analytical closure methods may represent the loop as excessively rigid because typically only ϕ/ψ torsions are considered as flexible, while keeping all bond lengths and bond angles fixed at standard values (ω torsions are also usually kept at 180°, i.e., *trans*-amide conformer overwhelmingly prevalent for most amino acids; note that *cis*-prolines are actually not uncommon, an exception that is often ignored). A recent analysis (33) of a nonredundant set of ultra-high-resolution protein structures confirmed the earlier observations (34, 35) that the backbone covalent geometry should not be considered as completely fixed and context independent because it varies systematically as a function of the ϕ and ψ backbone dihedral angles. The largest (from 107.5 to 114.0 for non-proline/glycine residues) variations within the most populated regions of the Ramachandran map occur for $\angle NC_{\alpha}C$ angle.

Analytical closure algorithms can be modified to allow bond angle variations (36). More recent analytical loop closure methods including TLC (28) also incorporate small degree of bond length flexibility. Full cyclic coordinate descent (FCCD) (37), a variation on the CCD method was developed to close loops in C_{α}-only representation, where much larger variations of the pseudo bond angles occur.

2.2.2. Build-Up Methods

Build-up methods attempt to sequentially (residue by residue) construct an approximately closed loop that can be refined using some form of iterative optimization method. Often build-up is performed as a part of enumerative sampling approaches discussed above. In another example, Protein Local Optimization Program (PLOP) (38, 39) generates closed loops by independent build-up of the polypeptide chain from both N- and C-termini followed by identification of matching half-loop pairs which meet each other at the central "closure" residue within certain tolerance and satisfy appropriate criteria for the planar and dihedral angles at the closure point. Subsequent energy optimizations refine the closure. Different conformations are generated by selecting representative ϕ/ψ "rotamer" states from detailed (5° step) Ramachandran maps for each residue during build-up.

2.2.3. Iterative Methods

Iterative loop closure methods typically start with a complete loop in a conformation that is far from closed and/or is otherwise highly distorted, and arrive at a closed conformation via a series of iterations, while also maintaining or restoring correct covalent geometry. Numeric/iterative methods are generally more flexible and can easily incorporate additional constraints as well as some of the physical energy terms or even the full force-field energy. Among the earliest implementations of the iterative approach is the Random Tweak (40), which starts with a random loop conformation and achieves closure via iterative small changes of ϕ/ψ angles optimizing the closure constraints. Enhanced version of the algorithm, the Direct Tweak (41) supplements closure constraints with a simple steric repulsion potential to produce clash-free closed loop conformations.

Scaling relaxation technique starts with the loop closure by scaling bond lengths in the loop, with simultaneous scaling of bond stretching parameters of the force field (42). Subsequently, energy minimization is performed, with the parameters gradually reverted back to their regular values, allowing the loop to recover correct covalent geometry.

Iterative loop closure can be performed in conjunction with discrete conformational state representations used in enumerative sampling approaches. For example, RAPPER (43) constructs the loop in backbone ϕ/ψ torsions-only representation using fine-grained residue-specific ϕ/ψ state sets derived from a nonredundant set of high-resolution protein structures. So-called Round Robin Scheduling algorithm is used to iteratively construct conformations that satisfy gap closure and steric exclusion constraints. The authors of the algorithm compared performance of their fine-grained ϕ/ψ state sets with a number of coarse-grained representations (2, 18, 44, 45) that use 4–11 states per residue. They found that inverse relationship exists between the number of states in a particular ϕ/ψ state set and the lowest RMSD as well as the rate of failures to close the loop. Thus, the most dense 5° fine-grained set with more than 2,000 ϕ/ψ states was recommended for use in RAPPER.

Loop modeling protocol in MODELLER (46) starts with a random distribution of all loop atoms in the region between the termini. Optimization of the energy function via a series of gradient minimizations and molecular dynamics runs restores local covalent geometry and eventually produces a low-energy closed loop structure. Multiple independent runs of the protocol produce an ensemble of solutions from which the best answer is selected. Somewhat similar method also starting with random arrangement of loop atoms was recently proposed by Liu et al. (47), but instead of relying on bonded force-field terms to restore covalent geometry, iterative distance adjustments and superpositions of rigid template fragments of amino acid residues are applied.

Local torsional deformation (LTD) (48) method iteratively perturbs several torsions along the polypeptide backbone. The deformations remain local because only the atom defining the torsion is rotated, with more remote parts of the molecular tree remaining static. Resulting distortions of covalent geometry are resolved during subsequent force-field energy (GROMOS) (49) minimization. Perturbation/minimization steps are repeated iteratively within a Monte Carlo with minimization (MCM) procedure.

When torsion-space optimization is used, the force-field terms normally do not include bond bending and bond stretching and thus do not enforce loop closure. Thus, explicit additional constraints are necessary, such as harmonic constraints between dummy atoms attached to the loop and their real counterparts in the body of the protein, as in the work of Zhang et al. (50). Monte Carlo with simulated annealing was used to simultaneously optimize the closure constraints and a simple softcore steric repulsion potential.

2.3. Scoring Functions

Irrespective of the sampling algorithm, candidate loop conformations need to be ranked so that a putative near-native conformation can be selected. In principle, an obvious choice for the scoring function is the physics-based force-field energy. However, force fields have certain drawbacks. Physical terms are "noisy," i.e., only slightly different conformations can have widely different energies because electrostatics and particularly van der Waals terms have very steep dependencies on atom positions at atomic contact distances. Furthermore, prohibitive cost of explicit solvent (water) simulations means that empirical implicit solvation terms have to be used, undermining somewhat the consistency of the physical energy function. Even with implicit solvent, calculations of pairwise terms and in particular, accurate solvation electrostatics for all-atom models remain computationally challenging. These difficulties with force-field-based energy functions led a number of groups to explore the alternative, knowledge-based or statistical potentials. It remains to be seen whether simplified energy functions can achieve sufficient accuracy to compete with force fields in loop modeling.

2.3.1. Scoring Functions: Knowledge-Based Potentials

Knowledge-based, or statistical potentials are based on the idea that the observed distributions of interatomic distances or frequencies of contacts between particular kinds of atoms in experimentally solved protein structures should reflect the energetics of interaction between these atoms. The attractive aspect of this approach is that potentially it can account for poorly understood or even yet unknown interaction terms that contribute to the conformational energy of the polypeptide in solution, as long as examples of such interactions are seen in the database. Statistical potentials also tend to be much smoother than physical force fields, a property that is desirable for efficient optimization. Nevertheless, a direct comparison of force-field-based scoring (Amber/GBSA (51, 52)) and an implementation of statistical potential (RAPDF (53)) in loop simulations showed that force-field potentials outperformed statistical potential across all loop lengths in the benchmark (54). There has been some progress in the development of statistical potentials, and Zhang et al. reported that their distance-scaled finite ideal-gas reference state (DFIRE (55)) statistical potential performed at least as well as several versions of force-field scoring in a loop prediction benchmark, at a fraction of computational cost (56). More recent application of DFIRE to select native-like conformations from an ensemble of conformations of two flexible interacting loops showed that in this more difficult setup the statistical potential was able to select native-like conformation only in 31% of cases (57). When true (X-ray) native loop conformations were included in selection, 78% of them were picked by DFIRE as top ranking, which may mean that the near-native solutions found via sampling may have been simply too crude to be recognized (solutions closer than 2 Å backbone RMSD were considered as near-native in this study).

An interesting variation on the knowledge-based approach to scoring is a statistical backbone torsion potential, based on the frequencies of ϕ/ψ angle pairs instead of pairwise distances. The distribution of all ϕ/ψ angle pairs forms the classical Ramachandran plot (31), broadly useful in the assessment of protein structure quality but insufficient by itself to segregate native structures from decoys. Rata et al. extended this concept to amino acid residue doublets, deriving ϕ/ψ and ψ/ϕ probability distributions for all specific consecutive residue pairs in the form of dihedral probability density functions (DPDFs) (58). The issue of the relative sparseness of data available for the 400 residue pairs was alleviated using iteratively constructed Gaussian representation of the density functions. When evaluated on the Coil Decoy Set, DPDF-based potential was able to select the native loop conformation at or near the top of the distribution, which is particularly remarkable because this type of potential only accounts for "local" interactions within residues and between adjacent ones.

Interestingly, MODELLER (46, 59) combines force-field terms (CHARMM (60)) for treatment of bonded interactions, with statistical mean force potential (MFP (61)) for nonbonded interactions and a function mimicking Ramachandran plot (31) preferences for backbone ϕ/ψ angles or rotamer states (62) for side-chain χ angles.

2.3.2. Force-Field-Derived Scoring Functions

The majority of recent loop modeling methods include force fields as a part of scoring function at least in the late stages of simulation protocol (16, 38, 46, 54, 63, 64). All-atom force fields that are used in loop modeling include OPLS (65), CHARMM (60), AMBER (51), and ECEPP (66, 67). Protein loops are typically highly exposed to solvent (water) and thus adequate treatment of solvent interactions is essential for accurate scoring. Core force-field parameterizations typically do not account for solvation effects unless solvent (water) is explicitly included in the simulations. Due to the high computational cost, extensive loop sampling with explicit solvent remains in general impractical. Instead, force fields have been combined with a variety of implicit solvation and continuum solvent electrostatic models. Generalized Born (GB) model, in particular, has been the method of choice in many recent studies, because its accuracy can approach that of the Poisson equation solvers at a fraction of computational cost. While GB model is based on a single key equation expressing charge–charge and charge–solvent interactions as a function of the generalized Born radii of atoms, specific implementations differ in the way the conformation-dependent GB radii are estimated. Several different GB implementations were compared in loop modeling simulations (68): PLOP (39)-based prediction protocol was combined with electrostatic terms using simple distance-dependent dielectric (69); surface-based GB with nonpolar interaction term (SGB/NP) (70); analytic GB with constant surface tension (AGB-γ); analytic GB with nonpolar interaction term (AGBNP) (71); and a modification of the latter that corrected for excessively favorable salt bridge interactions in GB model (AGBNP+). The last model performed best, while distance-dependent dielectric (a non-GB model) performed worst. It was also shown that the accuracy of loop predictions can be increased by optimizing solvation parameters specifically for protein loops (72). Parameterization is carried out using the assumption that the optimal parameter set should stabilize the native loop conformation against a set of loop decoys. Thus, Das and Meirovitch (72, 73) optimized parameters of the simple distance-dependent dielectric models ($\varepsilon = nr$) combined with SA model using a "training" group of nine loops. The approach was further refined by using more accurate Generalized Born electrostatic model instead of simplistic $\varepsilon = nr$, although the authors concluded that GB model did not improve the results significantly (74). By comparison, Zhu et al. (38) achieved high accuracy predictions with GB model supplemented with an additional empirical pair-wise hydrophobic contact term.

Taken alone, $\varepsilon = nr$ electrostatic model is inferior because it only accounts for solvent screening but not for the charge–solvent interactions. This shortcoming can be at least partially addressed if it is combined with atom-type-specific surface energy densities in the SA model such as proposed by Wesson and Eisenberg (75). Indeed, by tuning these surface energy densities, very good performance in loop simulations can be achieved (76).

An interesting modification of the force-field energy was proposed by Xiang et al., who developed the so-called colony energy concept (41). Colony energy term reflects the density of other conformations in the vicinity of a given conformation and thus rewards broader low-energy regions over singular minima, introducing entropy-like contribution in the scoring function. Small but consistent improvement in average RMSD was demonstrated across a range of loop lengths.

2.4. Use of Internal Coordinates

Efficient and extensive search of the conformational space in ab initio loop simulations can greatly benefit from the advantages of the internal coordinate representation of the polypeptide, which naturally separates the degrees of freedom that need to be thoroughly explored (torsions, primarily ϕ/ψ pairs) and those that can be either kept fixed or allowed minimal variation (bond lengths and bond angles). Internal coordinate representation not only reduces dimensionality of the optimization problem (up to ten-fold), but also accelerates energy calculations by eliminating unnecessary calculation of bonded terms and improves convergence radius of local gradient minimizations (77).

The internal coordinate representation for polypeptides was originally introduced in the ECEPP algorithm and corresponding force field (66, 67, 78, 79), used for conformational energy computations of peptides and proteins. Since then, many ab initio loop simulation methods employed torsional representation at least on some stages, in particular initial loop construction.

Internal coordinate-based modeling is at the core of the ICM program (77, 78), an integrated molecular modeling and bioinformatics system. ICM-based loop simulation protocol (76) actually combines energy minimizations and loop closure by imposing quadratic constraints on the pairs of terminal atoms: at each of the two junctions, the backbone chain is broken across C_α–C bond; the N-terminal part ends with a virtual C atom constrained to a real C atom in the C-terminal part and conversely, the C-terminal part begins with a virtual C_α that is constrained to the real C_α in the

N-terminal part. While in this setup the closure may require more computational time, the efficiency of the gradient minimizer greatly reduces the number of steps needed to achieve convergence, and simultaneous minimization of physical energy and closure constraints produces clash-free, low-energy closed loop conformations directly. The protocol employs two-step approach: on the first stage, conformational space of the loop backbone is broadly explored using simplified glycine–alanine–proline (GAP, all other residues reduced to alanine) model; on the second stage, full side chains of non-GAP residues are restored and best representative conformations from the GAP-generated ensemble are refined. Solvent accessible surface (SAS)-based solvation term optimized specifically for loop simulations is used.

Table 1 presents the loop modeling results reported in the literature by various groups and obtained with ab initio or with combination modeling methods. It should be emphasized that the results shown in Table 1 are intended to give a general idea about state-of-the-art in loop modeling. Direct comparison of the methods employed to obtain these results is difficult because different loop sets were used by the majority of authors and the effect of crystal packing was taken into account in some of the studies. Data from Table 1 show that conformations of short loops (<7–8 residues) can be predicted with high accuracy (39, 41). Longer (11–13 residue) loops may require consideration of the crystal contacts (38) (PLOP and PLOP II), although the sophisticated hierarchical loop prediction method (HLP (63)) demonstrated certain success for longer loops even without the help of crystal contact data. ICM also performed well across the range of loop lengths.

2.5. Loop Prediction in Inexact Environment

Realistic scenario of loop refinement in comparative models, where the conformation of the rest of the protein may still contain significant structural inaccuracies, would require prediction of, at least, side-chain conformations of the residues surrounding a given loop. The N- and C-terminal attachment points on the protein core would also deviate from their ideal native positions/orientations. However, large majority of loop prediction methods have been evaluated for their ability to reconstruct a loop in its native environment, in some cases even including crystal contacts. Thus, it is likely that the accuracy of loop modeling in the real-world applications will be often lower than the benchmark results reported. However, some of the recent studies investigated the performance of several methods in a realistic setup of inexact loop environment.

Evaluation of the MODELLER loop simulation protocol included a test where the environment of the loop was distorted via an MD simulation at high temperature (46). Dependence of the loop prediction accuracy on the amplitude of the distortion (up to 3 Å) was investigated. Approximately linear increase in

Table 1
Accuracy [average (median) RMSD, Å] of different loop prediction methods

Loop length	4	5	6	7	8	9	10	11	12	13
Modeller[a]	0.7	1.1	1.7	2.0	2.5	3.5	3.5	5.5	6.0	–
LOOPY[b]	–	0.85	0.92	1.23	1.45	2.68	2.21	3.52	3.42	
RAPPER[c]	0.47	0.90	0.95	1.37	2.28	2.41	3.48	4.94	4.99	
Rosetta[d]	0.69	–	–	–	1.45	–	–	–	3.62	
LoopBuilder[e]	–	–	–	–	1.31 (0.97)	1.88 (1.17)	1.93 (1.64)	2.50 (1.95)	2.65 (2.41)	
PLOP[f]	0.24 (0.20)	0.43 (0.21)	0.52 (0.26)	0.61 (0.28)	0.84 (0.43)	1.28 (0.42)	1.22 (0.53)	1.63 (1.24)	2.28 (2.06)	
PLOP II[g]	–	–	–	–	–	–	1.00 (0.62)	1.15 (0.60)	1.25 (0.76)	1.28 (0.72)
HLP[h]	–	–	0.70 (0.30)	–	1.20 (0.6)	–	0.60 (0.40)	–	1.20 (0.60)	–
Rosetta KIC[i]	–	–	–	–	–	–	–	–	1.90 (1.00)	–
ICMFF	0.25 (0.21)	0.51 (0.27)	0.55 (0.34)	0.66 (0.33)	0.84 (0.46)	0.98 (0.44)	0.88 (0.50)	1.45 (1.00)	1.16 (0.73)	1.67 (0.74)

[a] From Fig. 9 of Fiser et al. (46)
[b] From Table I of Xiang et al. (41)
[c] From Table III of de Bakker et al. (54)
[d] From Tables IV and VV of Rohl et al. (16)
[e] From Table V of Soto et al. (64)
[f] From Table IV of Jacobson et al. (39)
[g] From Table II of Zhu et al. (38)
[h] From Table I of Sellers et al. (63)
[i] From Supplementary Table II of Mandel et al. (17)

RMSD was observed, although no pronounced dependence was seen for the longest (12 residue) loops, perhaps because accuracy for these loops was poor from the start.

FREAD (14) was tested on a highly realistic benchmark of 212 loops extracted from the models submitted to the critical assessment of structure prediction methods (CASP (79)) experiment. The method showed significantly better results than several ab initio algorithms, probably owing to the lesser dependence of the knowledge-based approach on the loop environment.

Sellers et al. (63) examined how loop refinement accuracy is affected by the errors in conformations of the surrounding side chains. The HLP (38) method, based on the previously developed PLOP (39), was tested on a set of 6-, 8-, 10-, and 12-residue loops within the native structure and within the perturbed structure where side chains adjacent to the loop were repacked around a random nonnative loop conformation. RMSDs of the predicted loop conformations increased dramatically (on average fourfold) when modeled within perturbed environment, and less than 50% of the loops where predicted correctly (within 1.5 Å backbone RMSD from native structure), as compared to 80% of loops correctly predicted in the native context. Modification of the HLP protocol, HLP with surrounding side chains (HLP-SS), allowed concurrent optimization of the side chains located within a certain cutoff from the loop. HLP-SS achieved a significant overall improvement in accuracy, largely eliminating "sampling errors" where HLP was unable to generate near-native conformations because of the obstruction by the perturbed side chains. At the same time, there was a significant increase in the number of "energy errors" where nonnative conformations scored better than near-native. This observation illustrates a difficult trade-off involved in more realistic loop simulations including the environment: additional degrees of freedom associated with the conformational sampling beyond the loop itself expand the search space, potentially bringing into play many new artifacts of the energy function. Thus, not only more powerful sampling algorithms but also more accurate scoring functions are necessary to model reliably the loop and its environment.

Another oft-overlooked aspect of the realistic loop modeling exercises is that in practice the "loop" may not be necessarily devoid of any secondary structure: some of its residues can extend preceding or following β-strands or α-helixes. Such cases may present difficulties, in particular, for the knowledge-based methods that use databases focused on the coiled regions in experimental structures. In the case of ab initio methods, the scoring function needs to be able to account for an appropriate stabilization energy of the residues that become parts of secondary structure elements.

2.6. Modeling of the Multiple Interacting Loops

While the majority of prediction methods focus on individual loops, practical modeling scenarios may involve two or more adjacent loops with unknown conformations which can affect each other. Notable example is antibody CDRs.

Danielson and Lill (57) proposed a method for simultaneously predicting interacting loop regions. Individual loops are first sampled independently using LoopyMod algorithm(64). Resulting ensembles are combined and sterically incompatible combinations of loop conformations removed. Finally, side chains are repacked and the resulting conformations scored using DFIRE (55). The method was tested on seven pairs of interacting loops from a single protein structure (trypsin), selecting flexible segments of 6, 9, or 12 residues for each loop. Only for the pairs of two 6-residue loops or 6- and 9-residue loops the method was able to locate near-native conformations with RMSDs on average better than 2 Å among top ten solutions. Both the sampling power of the search algorithm and the selectivity of the score appeared to be insufficient when both loops were nine residues or longer.

Protocols for multiple loop simulations targeting relatively narrow protein classes, such as GPCRs (80) and antibodies (81), have been proposed, taking advantage of the system-specific knowledge. These studies had exploratory character, i.e., the GPCR study concentrated on probing the possible conformations of the extracellular loops rather than making specific predictions, and in the case of antibodies, predictions for CDR3 loops in the realistic inexact environment proved to be of low accuracy.

2.7. Loop Modeling in Ligand-Binding Sites

There are numerous cases where loop motions alter configurations of binding sites allowing ligand-binding modes associated with higher affinity and specificity. Thus, prediction of alternative conformations for flexible loops in the active sites or other ligand interaction sites on proteins can be highly valuable in ligand design. Simultaneous modeling of loop flexing and ligand association is challenging due to a greatly expanded conformational space of the combined system. However, it is likely that many of the flexible loops can only access a small number of low-energy conformations at normal conditions, and binding of a ligand shifts the equilibrium within this ensemble toward the conformation that has optimal interactions with the ligand (so-called conformational selection hypothesis (82)).

This hypothesis suggests that one can sample the loop in a free protein first and then dock the ligand into an ensemble of representative structures. Wong and Jacobson (83) investigated this approach to modeling of flexible loops for the active sites of six proteins. Loop conformations were initially sampled using replica-exchange molecular dynamics simulations using *apo* (ligand-free) structures, followed by clustering of the conformations extracted from the MD trajectories and refinement of representative structures using PLOP (39). For five of the six systems, the protocol produced conformations closer than 2 Å backbone RMSD to the holo (ligand-bound) structure. These modeled conformations also showed improved performance in VLS experiments.

Loops engaged in interactions with protein partners were simulated using the Rosetta KIC method in the Mandell et al. study (17). The results show that loop simulations in most cases could capture the induced-fit effects, predicting loop conformations closer to those experimentally observed in complex with the specific partner protein used in the simulation as compared to the complexes with alternative partners. It should be noted that this modeling protocol assumes that the configuration of the complex is known prior to the loop simulation. In a realistic scenario, it may or may not be possible to predict (presumably by docking) the overall complex structure without considering the loop.

2.8. Online Resources

Several loop prediction methods are currently available as online servers (Table 2). These are mostly the knowledge-based algorithms, while ab initio methods are underrepresented, clearly due to the high computational cost.

2.9. Future Directions

Loop simulation field continues to evolve rapidly. Progress in sampling algorithms and the availability of greater computing power now allows several ab initio methods to achieve reliably good

Table 2
On-line loop prediction servers

Server	Method description	URL	References
ArchPRED	Knowledge based: loop library search with a series of filters followed by gradient minimization	http://manaslu.aecom.yu.edu/loopred/	(84)
MODLOOP	Ab initio algorithm from MODELER	http://modbase.compbio.ucsf.edu/modloop/	(85)
SuperLooper	Knowledge based: search in LIP or LIMP databases, the latter specifically built for modeling membrane proteins	http://bioinf-applied.charite.de/superlooper/	(10)
Wloop	Knowledge based: search in a database of PDB fragments connecting secondary structure elements	http://psb00.snv.jussieu.fr/wloop/loop.html	(86)

accuracy for loops of up to 12–13 residues. Yet much longer loops can be found in protein structures. Also, commonly used in the field formal definition of the loop as a segment of polypeptide chain between two elements of secondary structure is perhaps too restrictive from the practical standpoint. In real-life problems, loops more often than not emerge as simply the regions of unknown structure that may include extensions of existing secondary structure elements, or contain additional ones like β-hairpins or short helixes. Co-simulation of several flexible regions also remains challenging. More efficient sampling and in particular, better accuracy of energy functions will be necessary to expand the applicability of existing ab initio methods.

3. Notes

There are two distinct classes of errors that typically occur in loop prediction: energy (or scoring function) errors and sampling errors. The first type occurs when the energy function used by the loop modeling method assigns a better score (lower energy) to a nonnative conformation. To improve confidence in ranking, reevaluation of energies with a different scoring function can be recommended. True near-native conformation will likely remain the best ranked across multiple scoring schemes. The second type of errors (i.e., sampling) occur when near-native conformations are not explored by the sampling algorithm. One way to ensure sufficient sampling is to establish convergence by running multiple independent simulations and comparing the results. Identical or similar top-ranked conformations from several simulations indicate (but do not guarantee) sufficient sampling. Note that this is only applicable to the methods with a stochastic component, since fully deterministic algorithms always produce the same result.

Some cases of loops may require special consideration. Disulfide bonds are often not taken into account by loop sampling algorithms, therefore additional filtering of the generated loop conformations to select those that allow disulfide formation may be necessary. Many methods assume that only *trans*-conformation of the peptide bond is allowed. While for most amino acids occurrence of *cis*-conformation is exceedingly rare, *cis*-prolines are fairly common; thus, if the loop under study contains proline, possibility of *cis*-conformer should be considered.

Generally, accuracy of models tends to be higher for the relatively less exposed loops, on which the bulk of the protein imposes significant steric constraints.

References

1. Jaroszewski, L. (2009) Protein structure prediction based on sequence similarity, *Methods Mol Biol 569*, 129–156.
2. Moult, J., and James, M. N. (1986) An algorithm for determining the conformation of polypeptide segments in proteins by systematic search, *Proteins 1*, 146–163.
3. Schindler, T., Bornmann, W., Pellicena, P., Miller, W. T., Clarkson, B., and Kuriyan, J. (2000) Structural mechanism for STI-571 inhibition of abelson tyrosine kinase, *Science 289*, 1938–1942.
4. Kufareva, I., and Abagyan, R. (2008) Type-II kinase inhibitor docking, screening, and profiling using modified structures of active kinase states, *J Med Chem 51*, 7921–7932.
5. Berman, H. M., Westbrook, J., Feng, Z., Gilliland, G., Bhat, T. N., Weissig, H., Shindyalov, I. N., and Bourne, P. E. (2000) The Protein Data Bank, *Nucleic acids research 28*, 235–242.
6. Fidelis, K., Stern, P. S., Bacon, D., and Moult, J. (1994) Comparison of systematic search and database methods for constructing segments of protein structure, *Protein Eng 7*, 953–960.
7. Deane, C. M., and Blundell, T. L. (2001) CODA: a combined algorithm for predicting the structurally variable regions of protein models, *Protein Sci 10*, 599–612.
8. van Vlijmen, H. W., and Karplus, M. (1997) PDB-based protein loop prediction: parameters for selection and methods for optimization, *J Mol Biol 267*, 975–1001.
9. Wojcik, J., Mornon, J. P., and Chomilier, J. (1999) New efficient statistical sequence-dependent structure prediction of short to medium-sized protein loops based on an exhaustive loop classification, *J Mol Biol 289*, 1469–1490.
10. Michalsky, E., Goede, A., and Preissner, R. (2003) Loops In Proteins (LIP) – a comprehensive loop database for homology modelling, *Protein Eng 16*, 979–985.
11. Burke, D. F., and Deane, C. M. (2001) Improved protein loop prediction from sequence alone, *Protein Eng 14*, 473–478.
12. Fernandez-Fuentes, N., and Fiser, A. (2006) Saturating representation of loop conformational fragments in structure databanks, *BMC Struct Biol 6*, 15.
13. Regad, L., Martin, J., Nuel, G., and Camproux, A. C. (2010) Mining protein loops using a structural alphabet and statistical exceptionality, *BMC bioinformatics 11*, 75.
14. Choi, Y., and Deane, C. M. (2010) FREAD revisited: Accurate loop structure prediction using a database search algorithm, *Proteins 78*, 1431–1440.
15. Simons, K. T., Kooperberg, C., Huang, E., and Baker, D. (1997) Assembly of protein tertiary structures from fragments with similar local sequences using simulated annealing and Bayesian scoring functions, *J Mol Biol 268*, 209–225.
16. Rohl, C. A., Strauss, C. E., Chivian, D., and Baker, D. (2004) Modeling structurally variable regions in homologous proteins with rosetta, *Proteins 55*, 656–677.
17. Mandell, D. J., Coutsias, E. A., and Kortemme, T. (2009) Sub-angstrom accuracy in protein loop reconstruction by robotics-inspired conformational sampling, *Nat Methods 6*, 551–552.
18. Deane, C. M., and Blundell, T. L. (2000) A novel exhaustive search algorithm for predicting the conformation of polypeptide segments in proteins, *Proteins 40*, 135–144.
19. Tosatto, S. C., Bindewald, E., Hesser, J., and Manner, R. (2002) A divide and conquer approach to fast loop modeling, *Protein Eng 15*, 279–286.
20. Spassov, V. Z., Flook, P. K., and Yan, L. (2008) LOOPER: a molecular mechanics-based algorithm for protein loop prediction, *Protein Eng Des Sel 21*, 91–100.
21. Galaktionov, S., Nikiforovich, G. V., and Marshall, G. R. (2001) Ab initio modeling of small, medium, and large loops in proteins, *Biopolymers 60*, 153–168.
22. Rapp, C. S., and Friesner, R. A. (1999) Prediction of loop geometries using a generalized born model of solvation effects, *Proteins 35*, 173–183.
23. Kolinski, A., and Skolnick, J. (1998) Assembly of protein structure from sparse experimental data: an efficient Monte Carlo model, *Proteins 32*, 475–494.
24. Olson, M. A., Feig, M., and Brooks, C. L., 3rd. (2008) Prediction of protein loop conformations using multiscale modeling methods with physical energy scoring functions, *J Comput Chem 29*, 820–831.
25. Go, N., and Scheraga, H. A. (1970) Ring Closure and Local Conformational Deformations of Chain Molecules, *Macromolecules 3*, 178–187.
26. Wedemeyer, W. J., and Scheraga, H. A. (1999) Exact analytical loop closure in proteins using polynomial equations, *Journal of Computational Chemistry 20*, 819–844.
27. Kolodny, R., Guibas, L., Levitt, M., and Koehl, P. (2005) Inverse Kinematics in Biology: The

Protein Loop Closure Problem., *Int J Robotics Research 24*, 151–163.
28. Coutsias, E. A., Seok, C., Jacobson, M. P., and Dill, K. A. (2004) A kinematic view of loop closure, *J Comput Chem 25*, 510–528.
29. Nilmeier, J., Hua, L., Coutsias, E. A., and Jacobson, M. P. (2011) Assessing Protein Loop Flexibility by Hierarchical Monte Carlo Sampling, *Journal of Chemical Theory and Computation 7*, 1564–1574.
30. Cui, M., Mezei, M., and Osman, R. (2008) Prediction of protein loop structures using a local move Monte Carlo approach and a grid-based force field, *Protein Eng Des Sel 21*, 729–735.
31. Ramachandran, G. N., Ramakrishnan, C., and Sasisekharan, V. (1963) Stereochemistry of polypeptide chain configurations, *J Mol Biol 7*, 95–99.
32. Canutescu, A. A., and Dunbrack, R. L., Jr. (2003) Cyclic coordinate descent: A robotics algorithm for protein loop closure, *Protein Sci 12*, 963–972.
33. Berkholz, D. S., Shapovalov, M. V., Dunbrack, R. L., Jr., and Karplus, P. A. (2009) Conformation dependence of backbone geometry in proteins, *Structure 17*, 1316–1325.
34. Schaefer, L., and Cao, M. (1995) Predictions of protein backbone bond distances and angles from first principles, *Journal of Molecular Structure: THEOCHEM 333*, 201–208.
35. Karplus, P. A. (1996) Experimentally observed conformation-dependent geometry and hidden strain in proteins, *Protein Sci 5*, 1406–1420.
36. Bruccoleri, R. E., and Karplus, M. (1985) Chain closure with bond angle variations, *Macromolecules 18*, 2767–2773.
37. Boomsma, W., and Hamelryck, T. (2005) Full cyclic coordinate descent: solving the protein loop closure problem in Calpha space, *BMC bioinformatics 6*, 159.
38. Zhu, K., Pincus, D. L., Zhao, S., and Friesner, R. A. (2006) Long loop prediction using the protein local optimization program, *Proteins 65*, 438–452.
39. Jacobson, M. P., Pincus, D. L., Rapp, C. S., Day, T. J., Honig, B., Shaw, D. E., and Friesner, R. A. (2004) A hierarchical approach to all-atom protein loop prediction, *Proteins 55*, 351–367.
40. Shenkin, P. S., Yarmush, D. L., Fine, R. M., Wang, H. J., and Levinthal, C. (1987) Predicting antibody hypervariable loop conformation. I. Ensembles of random conformations for ringlike structures, *Biopolymers 26*, 2053–2085.
41. Xiang, Z., Soto, C. S., and Honig, B. (2002) Evaluating conformational free energies: the colony energy and its application to the problem of loop prediction, *Proc Natl Acad Sci U S A 99*, 7432–7437.
42. Zheng, Q., Rosenfeld, R., Vajda, S., and DeLisi, C. (1993) Determining protein loop conformation using scaling-relaxation techniques, *Protein Sci 2*, 1242–1248.
43. DePristo, M. A., de Bakker, P. I., Lovell, S. C., and Blundell, T. L. (2003) Ab initio construction of polypeptide fragments: efficient generation of accurate, representative ensembles, *Proteins 51*, 41–55.
44. Park, B. H., and Levitt, M. (1995) The complexity and accuracy of discrete state models of protein structure, *J Mol Biol 249*, 493–507.
45. Rooman, M. J., Kocher, J. P., and Wodak, S. J. (1991) Prediction of protein backbone conformation based on seven structure assignments. Influence of local interactions, *J Mol Biol 221*, 961–979.
46. Fiser, A., Do, R. K., and Sali, A. (2000) Modeling of loops in protein structures, *Protein Sci 9*, 1753–1773.
47. Liu, P., Zhu, F., Rassokhin, D. N., and Agrafiotis, D. K. (2009) A self-organizing algorithm for modeling protein loops, *PLoS Comput Biol 5*, e1000478.
48. Baysal, C., and Meirovitch, H. (1999) Free energy based populations of interconverting microstates of a cyclic peptide lead to the experimental NMR data, *Biopolymers 50*, 329–344.
49. Scott, W. R. P., Hünenberger, P. H., Tironi, I. G., Mark, A. E., Billeter, S. R., Fennen, J., Torda, A. E., Huber, T., Krüger, P., and van Gunsteren, W. F. (1999) The GROMOS Biomolecular Simulation Program Package, *The Journal of Physical Chemistry A 103*, 3596–3607.
50. Zhang, H., Lai, L., Wang, L., Han, Y., and Tang, Y. (1997) A fast and efficient program for modeling protein loops, *Biopolymers 41*, 61–72.
51. Ponder, J. W., and Case, D. A. (2003) Force fields for protein simulations, *Adv Protein Chem 66*, 27–85.
52. Bashford, D., and Case, D. A. (2000) Generalized born models of macromolecular solvation effects, *Annu Rev Phys Chem 51*, 129–152.
53. Samudrala, R., and Moult, J. (1998) An all-atom distance-dependent conditional probability discriminatory function for protein structure prediction, *J Mol Biol 275*, 895–916.
54. de Bakker, P. I., DePristo, M. A., Burke, D. F., and Blundell, T. L. (2003) Ab initio construction of polypeptide fragments: Accuracy of

loop decoy discrimination by an all-atom statistical potential and the AMBER force field with the Generalized Born solvation model, *Proteins 51*, 21–40.

55. Zhou, H., and Zhou, Y. (2002) Distance-scaled, finite ideal-gas reference state improves structure-derived potentials of mean force for structure selection and stability prediction, *Protein Sci 11*, 2714–2726.
56. Zhang, C., Liu, S., and Zhou, Y. (2004) Accurate and efficient loop selections by the DFIRE-based all-atom statistical potential, *Protein Sci 13*, 391–399.
57. Danielson, M. L., and Lill, M. A. (2010) New computational method for prediction of interacting protein loop regions, *Proteins 78*, 1748–1759.
58. Rata, I. A., Li, Y., and Jakobsson, E. (2010) Backbone statistical potential from local sequence-structure interactions in protein loops, *J Phys Chem B 114*, 1859–1869.
59. Sali, A., and Blundell, T. L. (1993) Comparative protein modelling by satisfaction of spatial restraints, *J Mol Biol 234*, 779–815.
60. MacKerell, A. D., Bashford, D., Bellott, Dunbrack, R. L., Evanseck, J. D., Field, M. J., Fischer, S., Gao, J., Guo, H., Ha, S., Joseph-McCarthy, D., Kuchnir, L., Kuczera, K., Lau, F. T. K., Mattos, C., Michnick, S., Ngo, T., Nguyen, D. T., Prodhom, B., Reiher, W. E., Roux, B., Schlenkrich, M., Smith, J. C., Stote, R., Straub, J., Watanabe, M., Wiórkiewicz-Kuczera, J., Yin, D., and Karplus, M. (1998) All-Atom Empirical Potential for Molecular Modeling and Dynamics Studies of Proteins, *The Journal of Physical Chemistry B 102*, 3586–3616.
61. Melo, F., and Feytmans, E. (1997) Novel knowledge-based mean force potential at atomic level, *J Mol Biol 267*, 207–222.
62. Ponder, J. W., and Richards, F. M. (1987) Tertiary templates for proteins. Use of packing criteria in the enumeration of allowed sequences for different structural classes, *J Mol Biol 193*, 775–791.
63. Sellers, B. D., Zhu, K., Zhao, S., Friesner, R. A., and Jacobson, M. P. (2008) Toward better refinement of comparative models: predicting loops in inexact environments, *Proteins 72*, 959–971.
64. Soto, C. S., Fasnacht, M., Zhu, J., Forrest, L., and Honig, B. (2008) Loop modeling: Sampling, filtering, and scoring, *Proteins 70*, 834–843.
65. Kaminski, G. A., Friesner, R. A., Tirado-Rives, J., and Jorgensen, W. L. (2001) Evaluation and Reparametrization of the OPLS-AA Force Field for Proteins via Comparison with Accurate Quantum Chemical Calculations on Peptides, *The Journal of Physical Chemistry B 105*, 6474-6487.
66. Scheraga, H. A., and Gold, V. (1968) Calculations of Conformations of Polypeptides, in *Advances in Physical Organic Chemistry*, pp 103–184, Academic Press.
67. Némethy, G., Gibson, K. D., Palmer, K. A., Yoon, C. N., Paterlini, G., Zagari, A., Rumsey, S., and Scheraga, H. A. (1992) Energy parameters in polypeptides .10. Improved geometrical parameters and nonbonded interactions for use in the ECEPP/3 algorithm, with application to praline-containing peptides *Journal of physical chemistry 96*, 6472.
68. Felts, A. K., Gallicchio, E., Chekmarev, D., Paris, K. A., Friesner, R. A., and Levy, R. M. (2008) Prediction of Protein Loop Conformations using the AGBNP Implicit Solvent Model and Torsion Angle Sampling, *J Chem Theory Comput 4*, 855–868.
69. Pickersgill, R. W. (1988) A rapid method of calculating charge-charge interaction energies in proteins, *Protein Eng 2*, 247–248.
70. Levy, R. M., Zhang, L. Y., Gallicchio, E., and Felts, A. K. (2003) On the Nonpolar Hydration Free Energy of Proteins: Surface Area and Continuum Solvent Models for the Solute-Solvent Interaction Energy, *Journal of the American Chemical Society 125*, 9523–9530.
71. Gallicchio, E., and Levy, R. M. (2004) AGBNP: an analytic implicit solvent model suitable for molecular dynamics simulations and high-resolution modeling, *J Comput Chem 25*, 479–499.
72. Das, B., and Meirovitch, H. (2001) Optimization of solvation models for predicting the structure of surface loops in proteins, *Proteins 43*, 303–314.
73. Das, B., and Meirovitch, H. (2003) Solvation parameters for predicting the structure of surface loops in proteins: transferability and entropic effects, *Proteins 51*, 470–483.
74. Szarecka, A., and Meirovitch, H. (2006) Optimization of the GB/SA solvation model for predicting the structure of surface loops in proteins, *J Phys Chem B 110*, 2869–2880.
75. Wesson, L., and Eisenberg, D. (1992) Atomic solvation parameters applied to molecular dynamics of proteins in solution, *Protein Sci 1*, 227–235.
76. Arnautova, Y. A., Abagyan, R. A., and Totrov, M. (2011) Development of a new physics-based internal coordinate mechanics force field and its application to protein loop modeling, *Proteins 79*, 477–498

77. Abagyan, R., Totrov, M., and Kuznetsov, D. (1994) ICM-A new method for protein modeling and design: Applications to *J Comp Chem 15*, 488–506.
78. Abagyan, R., and Totrov, M. (1994) Biased probability Monte Carlo conformational searches and electrostatic calculations for peptides and proteins, *J Mol Biol 235*, 983–1002.
79. Kryshtafovych, A., Venclovas, C., Fidelis, K., and Moult, J. (2005) Progress over the first decade of CASP experiments, *Proteins 61 Suppl 7*, 225–236.
80. Nikiforovich, G. V., Taylor, C. M., Marshall, G. R., and Baranski, T. J. (2010) Modeling the possible conformations of the extracellular loops in G-protein-coupled receptors, *Proteins 78*, 271–285.
81. Sellers, B. D., Nilmeier, J. P., and Jacobson, M. P. (2010) Antibodies as a model system for comparative model refinement, *Proteins 78*, 2490–2505.
82. Tsai, C. J., Kumar, S., Ma, B., and Nussinov, R. (1999) Folding funnels, binding funnels, and protein function, *Protein Sci 8*, 1181–1190.
83. Wong, S., and Jacobson, M. P. (2008) Conformational selection in silico: loop latching motions and ligand binding in enzymes, *Proteins 71*, 153–164.
84. Fernandez-Fuentes, N., Zhai, J., and Fiser, A. (2006) ArchPRED: a template based loop structure prediction server, *Nucleic acids research 34*, W173–176.
85. Fiser, A., and Sali, A. (2003) ModLoop: automated modeling of loops in protein structures, *Bioinformatics (Oxford, England) 19*, 2500–2501.
86. Alland, C., Moreews, F., Boens, D., Carpentier, M., Chiusa, S., Lonquety, M., Renault, N., Wong, Y., Cantalloube, H., Chomilier, J., Hochez, J., Pothier, J., Villoutreix, B. O., Zagury, J. F., and Tuffery, P. (2005) RPBS: a web resource for structural bioinformatics, *Nucleic acids research 33*, W44–49.

Chapter 10

Methods of Protein Structure Comparison

Irina Kufareva and Ruben Abagyan

Abstract

Despite its apparent simplicity, the problem of quantifying the differences between two structures of the same protein or complex is nontrivial and continues evolving. In this chapter, we described several methods routinely used to compare computational models to experimental answers in several modeling assessments. The two major classes of measures, positional distance-based and contact-based, are presented, compared, and analyzed.

The most popular measure of the first class, the global RMSD, is shown to be the least representative of the degree of structural similarity because it is dominated by the largest error. Several distance-dependent algorithms designed to attenuate the drawbacks of RMSD are described. Measures of the second class, contact-based, are shown to be more robust and relevant. We also illustrate the importance of using combined measures, utility-based measures, and the role of the distributions derived from the pairs of experimental structures in interpreting the results.

Key words: Protein structure comparison, Modeling, Docking, Accuracy, Assessment, Root mean square deviation, Atomic contacts, Residue contacts, Naïve model, *Z*-score, Cumulative distribution function, VLS enrichment

1. Introduction

Applications of protein structures comparison methods. The majority of the proteome is made by amino acid sequences that, due to evolutionary selection, reliably and reproducibly form essentially the same three-dimensional structure. This observation formed a basis of the "one sequence–one structure" paradigm that dominated the protein science for a long time. However, the growing redundancy of protein structure databases, i.e., the increase in the number of structures per protein (1–3), made it clear that these fascinating molecules possess a lot more than a simple, unique rigid structure, and that varying degrees of the inherent flexibility of proteins

Andrew J.W. Orry and Ruben Abagyan (eds.), *Homology Modeling: Methods and Protocols*, Methods in Molecular Biology, vol. 857, DOI 10.1007/978-1-61779-588-6_10, © Springer Science+Business Media, LLC 2012

are critical for their functioning. Consequently, quantifying the structural differences in a sensible way becomes essential.

Structure comparison methods have been actively developed and used in the field of computational modeling assessments for quantitative evaluation of correctness of predicted models. Since 1994, a community-wide experiment called CASP (Critical Assessment of techniques for protein Structure Prediction, (4)) provides the modeling community with the possibility to evaluate their methods in blind prediction of structures of newly solved (but unpublished at the moment of the assessment) proteins. The submitted models are compared to an experimental structure using various criteria specifically developed for this task (5). In the recent years, other initiatives of this kind have emerged, including critical assessment of predicted interactions (CAPRI (6)), GPCR Dock (7), the assessment of modeling and docking methods for human G-protein-coupled receptor targets, and the assessment of the docking and scoring algorithms (8, 9).

Despite the fact that the methods presented in this chapter were originally developed for comparison of computational models to the experimental answers, their applicability is not limited to the modeling assessments. They now find their use in identification, evaluation, understanding, and prediction of protein conformational changes which constitute the fundamental basis of their biological functioning.

Properties of an ideal protein similarity measure. An ideal measure should allow both a single "summary" number within a fixed range (e.g., 0–100%) and an underlying detailed vector or matrix representation. The single number must distinguish well between related (correct) and nonrelated (incorrect) structure pairs, i.e., its distributions on the two sets must overlap to a minimal possible degree. It has to be relevant, i.e., capture the nature of protein folding or protein interaction determinants rather than satisfy simple geometric criteria. It has to have the minimal number of parameters, which in turn need to be well justified and understandable. It has to be stable and robust against minor or fractional (affecting a small fraction of the model) experimental and modeling errors; such changes in the structures should not lead to major leaps in the calculated similarity measure values. It has to capture the similarities or differences between the structures at any given level of accuracy/resolution. Ideally, it should have an intuitive visual interpretation. Although the complex nature of the problem prevents a universally acceptable single solution, some consensus measures are definitely emerging.

Characterization of protein structure comparison measures on protein structure pair datasets. In this chapter, we present an overview of several superimposition (distance)-based and contact-based measures and characterize them by calculating their distribution on three sets of protein structure pairs. The first set consists of 130,000 pairs of experimentally determined structures of identical proteins in PDB. It includes molecules related by non-crystallographic symmetry, structures determined in different crystal forms and in composition with different protein or small-molecule binding partners. The second and third sets are made of models of two G-protein-coupled receptors, dopamine D3 receptor and chemokine receptor CXCR4, generated by participants of the community-wide GPCR Dock assessment (10) in summer 2010 prior to release of the experimental coordinates of these receptors in complex with small molecule (D3 and CXCR4) and peptide (CXCR4) modulators (11, 12). The second and third sets are representative of the average modeling accuracy that can be achieved when the experimentally determined structures of closely related homologs are available (~40% of sequence identity as in the case of dopamine receptor D3 with previously solved β_1 and β_2 adrenergic receptors) or when the homology with existing structures is more distant (~25% of sequence identity as in the case of chemokine receptor CXCR4).

2. Methods

2.1. Main Types of Comparison Measures

Sequence-dependent vs. sequence-independent methods. Sequence-dependent methods of protein structure comparison assume strict one-to-one correspondence between target and model residues. In sequence-independent methods, structural superimposition is performed independently, followed by the evaluation of residue correspondence obtained from such superimposition. The usefulness of the sequence-independent approach is limited to cases where a model approximately captures the correct target fold but the amino acid sequence threading within this fold is incorrect, e.g., when one turn shift of an alpha-helix occurs. An example of an alignment-independent measure is the AL0 score routinely used in CASP model evaluation (13). AL0 score measures model accuracy by counting the number of correctly aligned residues in the sequence-independent superposition of the model and the reference target structure. A model residue is considered to be correctly aligned if the C_α atom falls within 3.8 Å of the corresponding atom in the experimental structure, and there is no other experimental structure C_α atom nearer. AL0 score values are clearly dependent on the superimposition; in its original implementation used for CASP model assessment, the score is calculated using the so-called LGA (local/global alignment (14)) superimposition of the two structures. A variety of sequence-independent structural alignment methods have been developed in the field: CE (15), DALI (16), DejaVu (17), MAMMOTH (18), Structal (19), FOLDMINER (20), KENOBI/K2 (21), LSQMAN (22), Matras (23, 24), PrISM (25), ProSup (26), SSM (27), and others.

The results of alignment-dependent and alignment-independent structure comparison are highly correlated with the exception of very distant homology cases. For the rest of this chapter, we, therefore, focus on alignment-dependent methods of protein structure comparison and methods of identification of subtle similarities and differences between the models and the reference structures in rather accurate modeling applications.

Evaluation of local vs. global similarity. Identification of global vs. local similarity represents two orthogonal directions in comparison of protein structures, i.e. structures that are most similar globally may not be the best in terms of local similarity. Flexible or disordered fragments such as long loops and/or termini are often poorly predicted and may significantly compromise the otherwise good similarity between structures. Relative domain movements observed in multi-domain proteins can also contribute to the poor global similarity scores. Focusing on local similarity helps to avoid these issues. Local similarity can be interpreted as a cumulative similarity score for all regions of the protein or, otherwise, can focus on a specific region such as, for example, ligand binding pocket, while ignoring the remaining parts of the protein.

Superimposition-based vs. superimposition-independent methods. Any method that relies on distance measurements between reference points in the model and their respective counterparts in the reference template requires prior superimposition of the model onto template, with the results of the comparison clearly dependent on the superimposition. Finding an optimal superimposition is an ambiguous task that has multiple solutions optimizing specific parameters, therefore, all superimposition-dependent methods suffer from this ambiguity. A superimposition that minimizes the global root mean square deviation (RMSD) of the model to the template may not necessarily be the best solution for the reasons described above: such superimposition is often compromised by a small number of significantly deviating fragments. Superimposition of a specific subset may not resolve this issue because the choice of the subset is subjective and ambiguous. A method that iteratively optimizes the superimposition of two protein structures by assigning lower weights to most deviating fragments and, in this way by finding the largest superimposable core of the two proteins, is described below. However, even in this approach, the choice of weight decay function is rather arbitrary and subjective which may lead to multiple solutions introducing ambiguity in any similarity score derived from these superimpositions. Superimposition-independent methods, such as contact-based measures, are devoid of this ambiguity.

2.2. Distance-Based Measures of Protein Structure Similarity

RMSD is the most commonly used quantitative measure of the similarity between two superimposed atomic coordinates. RMSD values are presented in Å and calculated by

$$\text{RMSD} = \sqrt{\frac{1}{n}\sum_{i=1}^{n} d_i^2},$$

where the averaging is performed over the n pairs of equivalent atoms and d_i is the distance between the two atoms in the ith pair. RMSD can be calculated for any type and subset of atoms; for example, C_α atoms of the entire protein, C_α atoms of all residues in a specific subset (e.g., the transmembrane helices, binding pocket, or a loop), all heavy atoms of a specific subset of residues, or all heavy atoms in a small-molecule ligands.

The main disadvantage of the RMSD lies in the fact that it is dominated by the amplitudes of errors. Two structures that are identical with the exception of a position of a single loop or a flexible terminus typically have a large global backbone RMSD and cannot be effectively superimposed by any algorithm that optimizes the global RMSD. An example of such a pair is given by the active and inactive conformations of an estrogen receptor α (ERα) which are only different by the movement of a single helix 12 (Fig. 1). By global backbone RMSD, this pair is virtually indistinguishable from the pair of albumin structures where multiple smaller scale rearrangements occur. The colored map in Fig. 1 shows the distribution of the protein backbone RMSD for a large number of experimentally determined structure pairs of identical proteins in the PDB. It demonstrates that for the majority of pairs, the RMSD ranges from 0 to 1.2 Å, due to inherent protein flexibility and experimental resolution limits. Figure 1 also presents the results of comparison of most accurate GPCR Dock 2010 models to their respective reference (answer) structures. It is clear that the backbone RMSD values are distributed around 2.3 Å for the easier homology modeling case, D3, and around 4.5 Å for the distant homology modeling case, CXCR4. It is, however, important to realize that these RMSD distributions do not reflect the true model accuracy because they are largely affected by flexible and poorly defined regions such as C-termini and extracellular loops in both GPCRs.

An important extension of the RMSD measure, the weighted RMSD (wRMSD), allows focusing on selected atomic subsets, for example, downplaying the regions known to be inherently unstructured:

$$\text{wRMSD} = \sqrt{\frac{\sum_{i=1}^{n} w_i d_i^2}{\sum_{i=1}^{n} w_i}}.$$

Internal symmetry, ambiguities, and RMSD. Any kind of RMSD-based measurement requires prior assignment of atom correspondences.

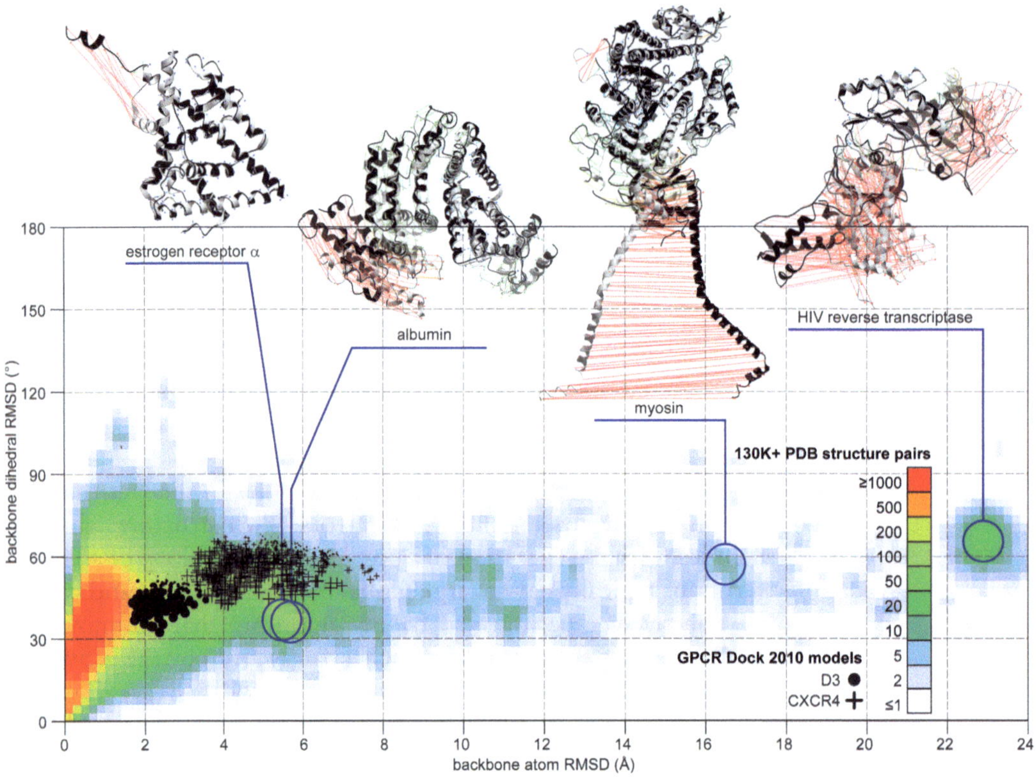

Fig. 1. Distribution of backbone atom RMSD/backbone dihedral RMSD values for a large number of experimentally determined pairs of protein structures in PDB. Representative structure pairs are shown. Computational models of dopamine D3 receptor (filled circle) and chemokine receptor CXCR4 (plus sign) are presented on the experimental structure pair background.

In the case of C_α, RMSD between two structures of the same protein, atom pair correspondence is established trivially via sequence alignment, however, measuring all heavy atom RMSD usually requires careful consideration of internal symmetry: the atom pair correspondence in such cases cannot be established unambiguously because some atoms within each structure are topologically equivalent to one another. For example, $C_{\delta 1}$ and $C_{\delta 2}$ atoms in a single phenylalanine (Phe) residue are topologically equivalent and therefore can be mapped into $C_{\delta 1}$ and $C_{\delta 2}$ atoms of the corresponding Phe residue of a different structure in two ways. The list of residues that cannot be mapped unambiguously includes: Arg, Asp, Glu, Leu, Phe, Tyr, and Val. Fortunately, the complexity of finding the optimal correspondence minimizing the overall side-chain RMSD is linear with respect to the number of residues. Figure 2a illustrates the distribution of heavy atom RMSD for a large set of small-ligand binding pocket pairs in PDB, calculated with and without side-chain rotamer enumeration. While on average, finding the optimal atom correspondence reduces pocket RMSD by less than 0.1 Å, this effect is largely unpredictable and for extreme cases, can reach 0.5 Å (Fig. 2b).

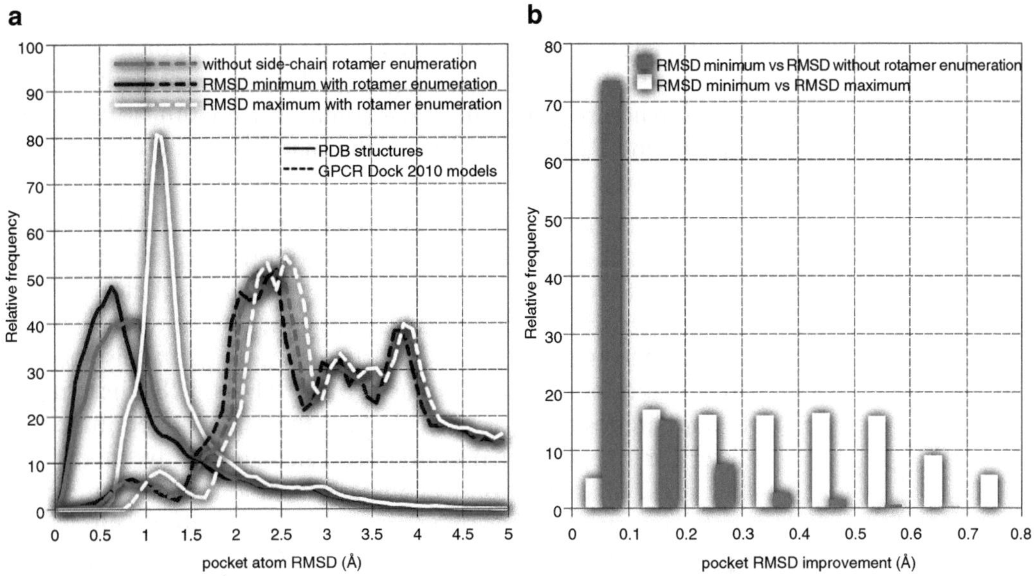

Fig. 2. Full atom RMSD between two identical sets of protein residues depends on the atom correspondence that, due to internal side chain symmetry, can be established in multiple ways (**a**). Equivalent rotamer enumeration lowers the calculated pocket RMSD by ~0.07 Å on average, and by as much as 0.5 Å in extreme cases (**b**). Statistics collected from a set of 65,000 PDB pocket pairs is presented as well as the results of analysis of GPCR Dock 2010 models.

RMS of dihedral angles. An approach complementary to Cartesian backbone RMSD is based on the representation of the protein structure in the internal coordinates that include bond lengths, planar bond angles, and dihedral torsion angles. For example, the geometry of a polypeptide chain backbone is described by the set of pairs of dihedral angle values, φ and ψ, which provides a way for a superimposition-independent structure comparison with the dihedral angle RMS used as the similarity scoring function. The dihedral angle RMS is complementary to the atom RMSD in the sense that it captures a different, less intuitive aspect of protein structure similarity. Modification of a small number of backbone dihedral angles can distort the global structure and packing beyond recognition while having only marginal effect on the dihedral angle RMS. At the same time, very similar structures are sometimes characterized by significant variations in their dihedral angles simply because these variations may partially cancel each other (e.g., peptide flips). These phenomena are well illustrated by the experimental distribution of backbone dihedral angle RMSD as compared to backbone RMSD (Fig. 1). For example, none of the three representative outlier clusters in terms of backbone RMSD, estrogen receptor, albumin, and myosin, is characterized by dihedral angle RMS deviating by more than two standard deviations from the experimental structure pair average. Similarly, the distribution of dihedral angle RMS of the GPCR Dock models to their respective reference structures lies in a region well populated by the experimental structure pairs, while their common sense similarity in terms of backbone RMSD remains on the margins of the experimental distribution.

Global distance test (GDT). As described above, RMSD heavily depends on the precise superimposition of the two structures and is strongly affected by the most deviated fragments. A clever way to overcome both shortcomings was implemented in the two methods routinely used for CASP model evaluation, global distance test (GDT), and longest continuous segment (LCS) (28): here multiple superimpositions, each including the largest superimposable subset for one of the residues, are calculated between the two structures. In application to comparison of a model to an experimental answer, it means that for each residue from the model, the largest continuous (LCS) or arbitrary (GDT) set of the model residues is found that contains the residue and superimposes with the corresponding set in the reference structure under a selected RMSD (LCS) or distance (GDT) cutoff. The maximal residue set for each cutoff is chosen, followed by averaging over several fixed cutoffs (e.g. 1, 2, 4, and 8 Å). The output of a GDT calculation represents a curve that plots the distance cutoff against the percent of residues that can be fitted under this distance cutoff. A larger area under the curve corresponds to more accurate prediction.

The distribution plot of GDT total score on the large set of experimental structure pairs is shown in Fig. 3a. Unlike the global backbone RMSD, the GDT measure recognizes structural similarity very well for the absolute majority of experimental pairs (GDT-TS > 50%). It is also more robust against small fragments movements. In particular, it effectively distinguishes the pair of active and inactive conformations of ERα which differ only in helix 12 conformation (GDT-TS = 93%), from the pair of albumin structures with multiple smaller scale domain distortions (GDT-TS = 60%).

TM score. Another problem that one runs into when using RMSD to compare protein structures is that the RMSD distribution also depends on the size of the protein. This becomes important when the models of several different size proteins are evaluated in comparison with one another. The dependence of RMSD on the protein size can be eliminated by calculating the so-called TM score (29):

$$\text{TM score} = \max\left[\frac{1}{L_{\text{target}}}\sum_{i=1}^{L_{\text{aligned}}}\frac{1}{1+(D_i / D_0(L_{\text{target}}))^2}\right].$$

Here L_{target} and L_{aligned} are the number of residues in the reference structure and the aligned region of the model, respectively, and $D_0(L_{\text{target}}) = 1.24\sqrt[3]{L_{\text{target}} - 15} - 1.8$ is a distance scale derived from

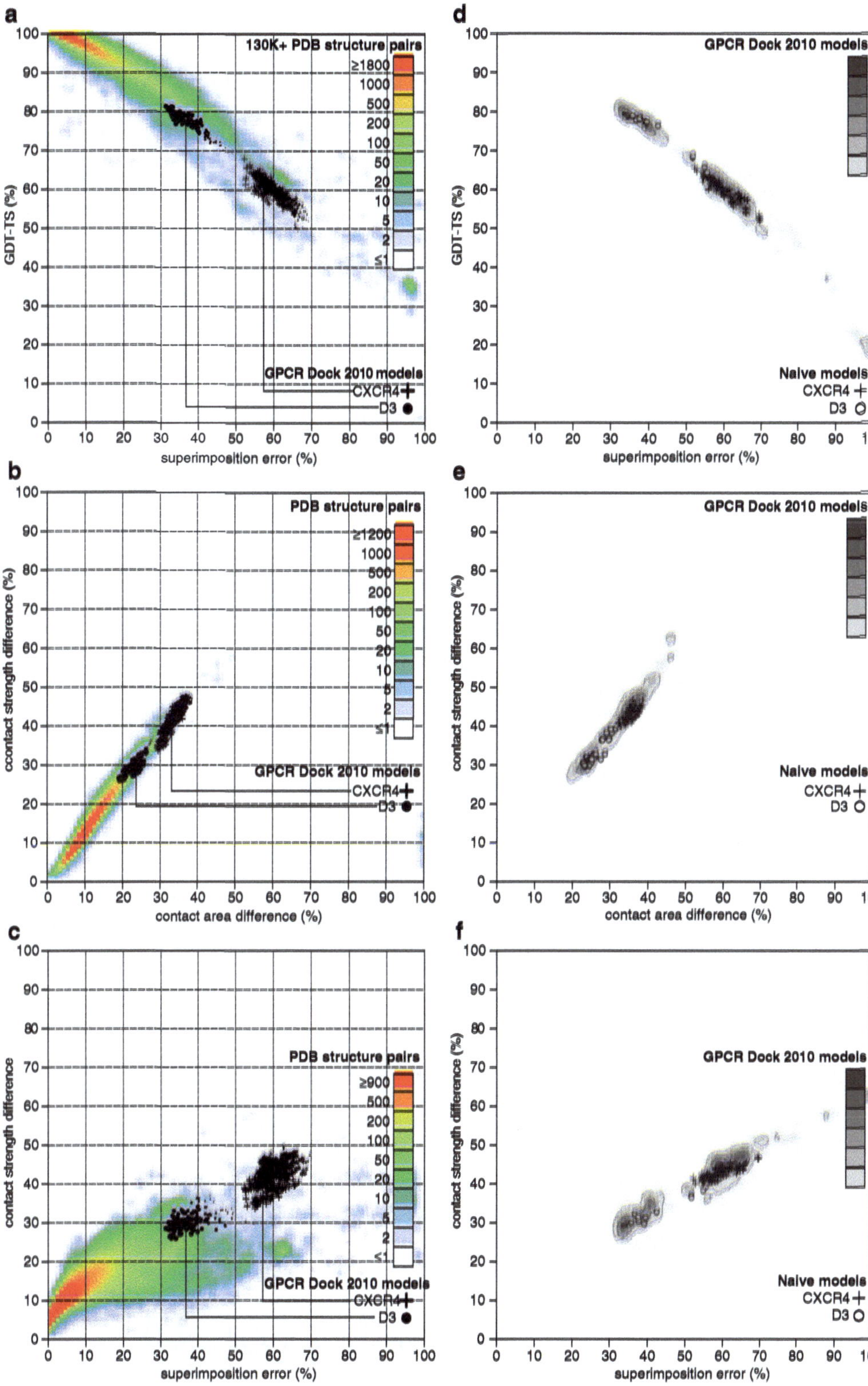

Fig. 3. Distribution of different measures of protein structure similarity for a set of 130,000 protein structure pairs in PDB (**a–c**, heat map), GPCR Dock 2010 models (**a–c**, filled circle for D3 and plus sign for CXCR4; **d–f**, heat map), and naïve models of GPCR Dock 2010 targets (**d–f**, open circle for D3 and plus sign for CXCR4). Only the top half of each GPCR Dock model set is shown: models less accurate than average are eliminated.

the analysis of large subsets of related and unrelated structures that is used to normalize the distances. Through its dependence on the L_{target}, the dependence of the obtained score on the target size is eliminated.

Iterative weighted superimposition and the associated superimposition error. The main CASP measure, GDT, is dependent on several arbitrarily chosen fixed distance cutoffs. This dependence is replaced by a continuous distance-dependent weight in the iterative weighted superimposition algorithm (30). By unbiased weight assignment to different atomic subsets, this algorithm gradually finds the better superimposable core between the two structures. It includes the following steps:

1. The atomic equivalences are established between the two structures and a vector of per-atom weights $\{W_1, W_2, \ldots, W_n\}$ is set to $\{1, 1, \ldots, 1\}$.
2. The weighted superimposition is performed (31) and weighted RMSD is calculated as described above.
3. The deviations $\{d_1, d_2, \ldots, d_n\}$ are calculated for all atom pairs, and their X-quantile, d_X is determined. The quantile X is an input parameter for the procedure that defines the minimal size of the superimposable core to be found; by default it is equal to 50%.
4. The new weights are calculated according to the formula.

$$W_i = \exp\left(-d_i^2 / d_X^2\right)$$

   The well superimposed atoms are assigned weights close to 1, while the weights associated with strongly deviating atom pairs get progressively smaller.

5. Steps 2–4 are iterated until the weighted RMSD value stops improving or the specified maximum number of iterations is reached.

Following this superimposition, the similarity of the two structures can be evaluated by the weighted RMSD or by taking the average of weights recalculated for the structure according to step 4 with d_X set to a fixed value, e.g., 2 Å. The complement of this number, denoted superimposition error (E_{super}), ranges from 0 to 100% with lower values corresponding to more similar structure pairs:

$$E_{super} = 100\% \times \left(1 - \frac{1}{n}\sum_{i=1}^{n} \exp\left(-\frac{d_i^2}{d_X^2}\right)\right).$$

The presence of a minority of strongly deviating atoms does not compromise the superimposition error, while large discrepancies are accurately captured and quantified (Fig. 4).

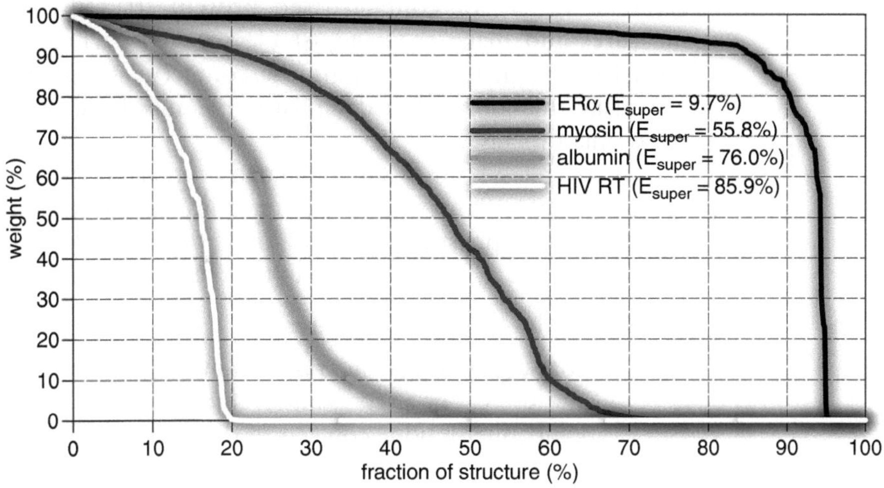

Fig. 4. Calculation of superimposition quality and superimposition error for representative structure pairs from Fig. 1. Superimposition quality is calculated as the area under the weight curve; superimposition error (E_{super}) is its complement to 100%. Essentially identical structure pairs like active/inactive conformation pair of ERα receive high weight for the majority of the structure and, consequently, low value of superimposition error.

The algorithm resembles the one published by Damm and Carlson (32) with a few modifications, including the adaptable standard deviation for the Gaussian distribution (step 4 of the algorithm) and the way the weighted RMSD is calculated (normalization by the sum of weights). The adaptable denominator in the distribution ensures a better quality superposition.

Figure 3a represents the distribution of GDT-TS vs. superimposition error for the experimental structure pair set and for the two sets of GPCR Dock 2010 models. The two measures are highly correlated. The adaptive nature of the GDT-TS measure that combined *multiple* superimpositions for different parts of poorly superimposable structures makes it more permissive; for the absolute majority of experimental structure pairs its value exceeds 50%. In contrast, superimposition error quantifies the structural deviations for a single weighted superimposition based on the largest common substructure; when two structures lack a significant common superimposable domain, superimposition error values may exceed 80%.

2.3. Contact-Based Measures of Protein Structure Similarity

Contact-based measures rely on comparison of pairwise distances and/or interactions within one of the protein structures with the corresponding distances/interactions in the other structure rather than on finding the distances between the corresponding points in the two structures. They, therefore, possess the advantage of being superimposition-independent. Pairwise contact matrices found many applications as a method of 2D representation of 3D protein structure (33–36). Multiple possible contact definitions create a variety of contact-based protein structure similarity measures and make them adjustable to each particular subject area. In particular, by changing the contact distance cutoff one can make the contact-based similarity measure local or global to the evaluator's taste.

In general, a *contact* can be defined as an arbitrary continuous function of two points in a protein structure, not necessarily representing the true physical interaction of these points. Point selection defines the "grain" or resolution in the contact definition:

- Residue level contact measures.
  - Coarse grain residue representation: single point per residue, e.g., C_α, C_β, or a representative side chain "center of mass" point. Inter-residue contacts in the form of C_α–C_α distances were used, for example, in DALI algorithm for alignment-independent protein structure comparison (16).
  - Full atom residue representation.
- Residue fragment level contact measures (same as above but each residue is divided into fragments, usually the backbone made of N, C_α, C, O atoms and the side chain).
- Atom level contact measures (contacts are calculated between the individual atom pairs).

Ways of determining the contact function include:

- Algebraic functions of interpoint distance (discontinuous, e.g., Heaviside step function or continuous/smooth).
- Functions based on physical principles (contact surface area, interaction energy, etc.).
- Tabulated physics-based contact strengths as a function of interpoint distance and geometry.

Contact area and contact strength difference. In their contact area difference (CAD) paper (37), Abagyan and Totrov came up with a contact definition that directly correlates with the strength of physical interactions, namely, they defined a residue contact as the difference in accessible surface area when calculated for a pair of residues separately or together. This *contact area* measure provides the most realistic assessment of fold similarity between the two structures, because it requires specific residue pairs to be in contact with about the same area. If the side chains are not packed correctly even with roughly similar fold, the distance will be large.

Contact functions based solely on C_α–C_α or C_β–C_β pairwise distances do not require correct (matching) residue–residue packing provided the backbones are similar. Given two residues whose C_α or C_β atoms are located at the distance of d Å, the residue *contact strength* can be calculated as

$$f(d)=\begin{cases}1 & \text{if} d<d_{\min}\\ \dfrac{d_{\max}-d}{d_{\max}-d_{\min}} & \text{if} d_{\min}<d<d_{\max}\\ 0 & \text{if} d>d_{\max}\end{cases},$$

where d_{min} and d_{max} are predefined distance margin boundaries. The values of d_{min} and d_{max} can be chosen in such a way that the corresponding contact strengths are correlated with the pairwise residue contact areas which in turn describe the real physical residue interactions. C_β–C_β contacts approximate contact areas more accurately than C_α–C_α, because on average, C_β atoms are closer to the centers of mass of the residues they belong to. In ref. 38, this approach was further improved by replacing C_β atoms by virtual points, C'_β, located in the direction of C_α–C_β bonds at the distance of $1.5 \times d(C_\alpha, C_\beta)$ from the C_α atom of each residue. This was shown to further improve the correlation between the calculated contact strengths and residue contact areas with the optimal margin boundaries found to be $d_{min} = 4$ Å and $d_{max} = 8$ Å.

When comparing two structures by their contacts, one builds two matrices of atomic contact strengths: $C^R_{n\times n}$ for the first structure and $C^M_{n\times n}$ for the second structure or model. The contact similarity matrix $C^{R\cap M}$ is constructed using $C^{R\cap M}[i,j] = \mathrm{Min}(C^R[i,j], C^M[i,j])$; its weight is found as $|C^{R\cap M}| = \Sigma_{i,j} C_{R\cap M}[i,j]$. This weight can be compared to one of three quantities: the weight of the reference contact matrix, $|C^R|$, model contact matrix, $|C^M|$, or the union of the two, $|C^{R\cup M}|$, defined by $C^{R\cup M}[i,j] = \mathrm{Max}(C^R[i,j], C^M[i,j])$ or $C^{R\cup M}[i,j] = (C^R[i,j] + C^M[i,j])/2$. The three approaches result in quantities ranging from 0 to 100% and reflecting recall, precision, and accuracy with which the model reproduces the reference structure contacts. Alternatively, one may choose to report the contact *differences* which simply complement the above similarity measures to 1 or 100% (contact distance or difference = 1 − contact similarity).

Figure 3b shows that for a large subset of PDB structure pairs, as well as for GPCR Dock 2010 models, contact strength differences calculated using the virtual C'_β points are highly correlated with CAD. For most pairs of experimentally determined structures of the same protein, protein flexibility and experimental errors lead to the contact strength differences of 5–20%. Small flexible fragments or even large domain movements have only minor effect on the contact strength matrices making the contact strength measures robust to elastic large-scale deformations. At the same time, these measures are sensitive to major changes in packing occurring as a result of modeling errors: the best GPCR Dock models appear to be about 30% different from the reference structure in the case of D3 and about 40% different in the case of CXCR4.

Further developments of contact strength definitions may include their parameterization according to the interacting residue types, complementation of the C_β–C_β distances with other parameters to better capture the dependence of the contact strength or likelihood on the relative residue orientation, and elimination of the trivial contacts occurring due to the covalent linkages between the neighboring residues. These research topics are, however, beyond the scope of the present chapter.

The importance of multiple criteria analysis. The location of computational model populations on the plots of distance-based and contact-based measures of protein similarity in Figs. 3a and 3b shows that in both cases, the models occupy the outskirts of the experimental distribution, with models built by closer homology (D3) being more accurate than distant homology models (CXCR4). The biggest insight, however, is gained when distance-based and contact-based measures are plotted against one another (Fig. 3c). In these coordinates, it becomes clear that for the experimental structure, pairs may often differ in conformation (as reflected by superimposition error) or in contacts (as reflected by contact strength difference), but rarely in both. In contrast, computational models differ from their respective answers by both parameters simultaneously, especially in the more difficult modeling case of CXCR4. This observation stressed the importance of applying complementary structure similarity measures that combine distance-based and contact-based approaches.

2.4. Comparing Protein–Protein and Protein–Ligand Complexes

Protein structure similarity measures presented above had the goal of comparing two structures of a single protein; however, the same general principles apply to evaluation of the predictions of molecular *interactions*. In 2002, the CAPRI (Critical Assessment of Predicted Interactions) experiment started with the focus on protein docking (39). Other initiatives followed including the GPCR Dock assessment started in 2008 and focused on small molecule docking to GPCR targets (7) as well as the recent assessment of ligand docking and virtual screening organized by Open-Eye (8, 9).

The task of *docking* is defined as prediction of the geometry and interactions in a *complex* of the given protein with either another protein (*protein docking*) or a small-molecule ligand (*small molecule docking*). In its pure form, the docking problem is based on the assumption that the structures of the unbound components are available. However, in real-life applications, it is rarely the case; even when such structures do exist, they may not be directly usable for complex geometry prediction because of the induced fit effect (40, 41) and uncertainties in amino acid tautomerization, protonation, and hydration (42). If the unbound structures do not exist they must be generated by homology for proteins and by 2D to 3D conversion for small molecules which introduces an additional level of difficulty in the docking protocol.

Methods that are used for the evaluation of docking predictions are largely based on the same principles as the methods of comparison of protein structures described above. However, because the focus is on the intermolecular interactions, one must ensure that the unrelated discrepancies in the structures of the individual interaction partners have minimal effect on the evaluation outcome.

Let us assume for simplicity that the complex of interest consists of only two molecules and that one of them (a protein) can be classified as a receptor, while the other one (another protein, a peptide, or a small molecule) is a ligand. In protein–protein complex prediction, the designation of one of the partners as a receptor is rather arbitrary and may be performed based on the size, rigidity, availability of structural information, or other criteria.

The most common way to evaluate the correctness of the docking geometry is to measure the RMSD of the ligand from its reference position in the answer complex after the optimal superimposition of the receptor molecules. The choice of this optimal superimposition is the first subjective decision that the evaluator has to make, especially in the case when the receptor had to be modeled and therefore the reference and the modeled receptor structures are significantly different. To reduce the effect of the irrelevant incorrectly modeled receptor parts, it is important that the receptor superimposition is performed by a smaller subset of atoms that includes the immediate binding interface (or binding pocket in case of a small molecule docking problem). Criteria for the selection of the binding interface residues should be carefully formulated and stated upfront; the usual procedure involves selection of residues located at a certain distance from the ligand in the reference structure followed by expansion of this selection through the sequence so that the short discontinuous stretches of residues are either merged or eliminated. The final selection must consist of continuous sequence stretches of at least 4–5 residues each to ensure that they can be properly aligned between the model and the reference structure. The interface selection must be derived from the reference structure and propagated to each complex model by the alignment-derived residue correspondence.

The interface atoms or pocket residues must now be superimposed for each model onto the reference structure. While the standard superimposition approach is the optimization of the selection heavy atom RMSD, flexible side chains, loops, and termini may compromise the superimposition quality and therefore one of the more robust superimposition methods described above is preferred. Once the superimposition is performed, the time comes to measure the RMSD between the ligand atoms in the model and the reference structures. The spectrum of caveats and challenges here is similar to that described in the previous paragraphs about RMSD, with the important distinction that whether the atoms in direct contact with the receptor constitute a minor or a major part of the ligand, they should remain the primary focus of the RMSD calculation. On the contrary, parts of the ligand distant from the interface or not in direct contact with the receptor must be down-weighted or disregarded in such an evaluation. For example, the contribution of the solvent-exposed parts of the ligand to the overall similarity score was eliminated in the GPCR Dock 2008 assessment (the solvent exposed phenoxy group of the adenosine A_{2A} receptor antagonist (7, 43, 44) (Fig. 5a). In protein docking, elimination of the effects of ligand parts not directly involved in the interaction with the receptor becomes critical (Fig. 5b).

Due to these caveats and ambiguities, positional distance-based measures need to be complemented with the contact measures of docking complexes. Contact definitions for protein–protein complexes are identical to the single protein case but are applied to intermolecular residue contacts only. Contacts are calculated between each pair of residues in the receptor and in the ligand and can involve C_α–C_α, C_β–C_β, virtual $C'_\beta - C'_\beta$ distances as well as the actual residue contact areas. In case of small molecule ligands, because the scope of the problem is smaller and because atomic-level interactions become primarily important, the definition of contact strengths should be extended to allow calculation of the interatomic instead of the inter-residue contacts.

The definition of an atomic contact used for scoring protein–ligand complexes in the GPCR Dock 2008 modeling and docking assessment (7) involved a step-wise function of interatomic distance equal to 1 below the specified distance cutoff (4 Å) and 0 otherwise (Fig. 6a, black curve). In other words, each of the models was

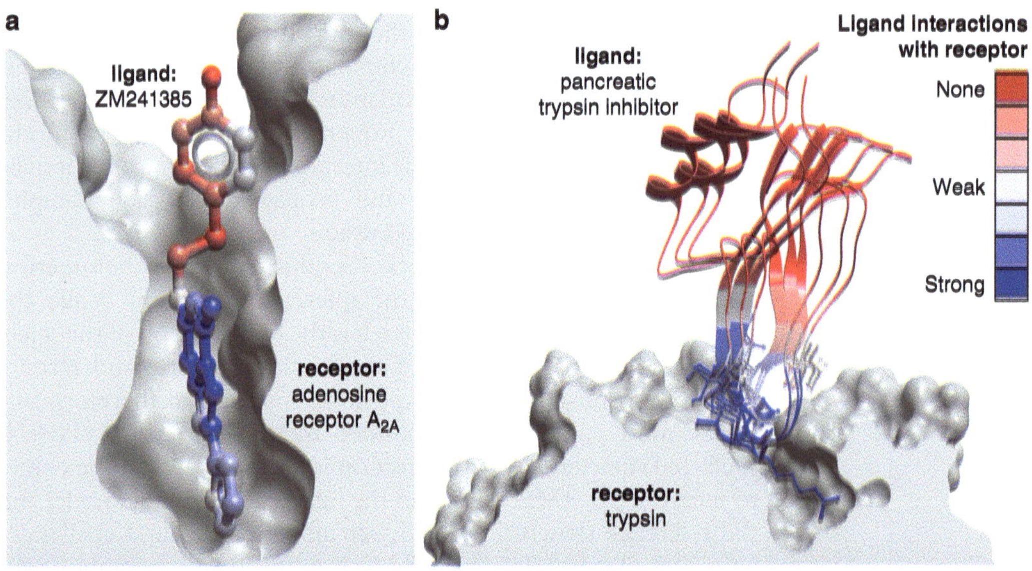

Fig. 5. Distance-based evaluation of protein–ligand (**a**) or protein–protein (**b**) complexes must be focused on ligand parts that are in direct contact with the receptor and not on the entire ligand molecule. Because position and conformation of solvent exposed parts is only approximately defined by the interaction within the complex, such parts must be either excluded or down-weighted in docking complex evaluation.

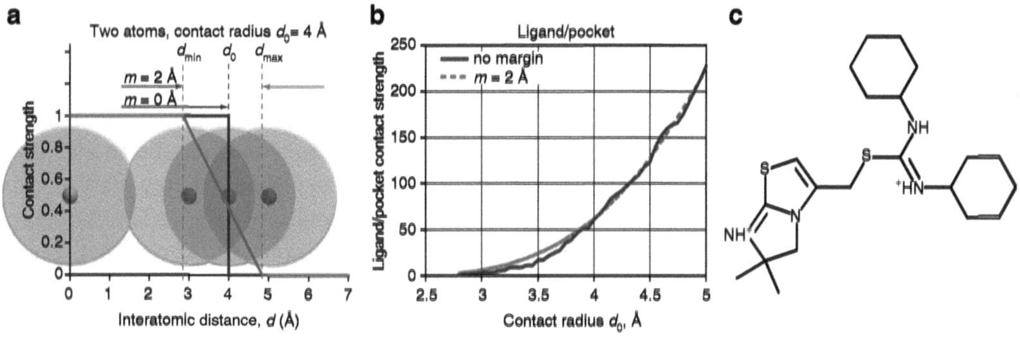

Fig. 6. Issues in evaluation of atomic contacts in protein complexes with small molecules: (a) definition of atomic contact strength with and without the continuous decrease margin; (b) hard distance cutoff (no margin) definition of the atomic contact leads to unstable behavior of the contact strength as a function of contact radius; (c) example of a small molecule with high degree of internal symmetry. Topologically equivalent atom permutations need to be enumerated when evaluating RMSD or comparing contacts of this molecule with its copy in a different structure.

characterized by the set of all ligand–receptor atom pairs located at the distance of ≤4 Å; this set was compared to the corresponding atom pair set in the reference structure (45). While simple conceptually and computationally, this "hard distance cutoff" approach leads to unstable and discontinuous behavior of the contact difference function, because minor changes in the ligand and side-chain conformation may result in large leaps in the number of matching contacts (Fig. 6b). To avoid this problem, the ligand–receptor atomic contact definition was refined in GPCR Dock 2010 with the *continuous decrease margin* approach in the spirit of (38). Instead of abruptly dropping to zero at the single cutoff of d_0, the contact strength gradually decreased between two distances, d_{min} and $d_{max} = d_{min} + m$, where m is the margin size. The margin boundaries, d_{min} and d_{max}, were adjusted so that the average number of contacts calculated with and without the margin is the same using the following equation:

$$d_{min} = d_0 - r \times m; \quad d_{max} = d_0 + (1 - r) \times m,$$

where r was calculated as $r = 0.49 + 0.17 \times m / d_0$. This equation was obtained by linear regression on the large number of complex structures.

The atomic contact definition can be further improved by making it atom-type dependent and/or orientation dependent; this will allow, for example, automatic assignment of higher weight to correctly predicted hydrogen bonds between the ligand and the protein.

Interatomic contact strength matrices can be calculated for the model and the reference structure. Taking the element-wise minima produces the matrix of correctly identified contact strengths which can be further compared to the reference matrix to give contact recall, model matrix for contact precision, or a combination of the two to give some form of contact accuracy. In cases where the physical atom–atom contacts are measured, contact precision can usually be disregarded: molecular geometry and van der Waals interactions impose natural constraints onto precision values because they limit the number of physical contacts that can be made.

The phenomenon of internal molecular symmetry may become a serious hurdle for the evaluation of similarity of a predicted docking complex to the experimentally derived answer by either distance-based or contact-based measures. If the ligand possesses any symmetrical groups, all topologically equivalent mappings of its atom set onto itself must be considered. For example, because the resonance-stabilized thiol form of the thiourea group is symmetric, as many as 16 atom permutations in the compound IT1t (Fig. 6c) result in exactly the same ligand covalent geometry and bond topology; all of these have to be tested when determining either RMSD or contact similarity of this compound to its copy in a different structure. In combination with the internal symmetry of neighboring side chains, this may easily lead to exponential growth of computational complexity.

2.5. Combining Measures for Ranking a Model Population

As described above, the concept of protein structure similarity involves multiple criteria leading to a very different ranking of models. Combining these criteria into a single numerical score seeks a fair balance between complimentary measures each representing an important part of the whole picture. However, the uncertainties of this combination (which terms to use and now to normalize them) often create even more confusion. An approach that is routinely used in CASP is based on the analysis of the distribution of scores calculated for each individual assessment criterion and each individual modeling target. Score mean and standard deviation (SD) are calculated for each criterion after which the score is converted into the *intrapopulation* Z-score by taking

$$Z_S = \frac{S - \mu_S}{\sigma_S},$$

where μ_S and σ_S are the average and standard deviation of the score S. Z-scores can be easily modified so that a larger value corresponds to a higher level of accuracy. In many cases, it is beneficial to remove the lowest accuracy outliers in the set so that they do not significantly affect the overall distribution. The intrapopulation Z-scores calculated in this way for the multiple assessment criteria (e.g., RMSD and contacts) are then averaged to obtain a single Z-score that is used to rank the models for the given target.

The intrapopulation Z-score approach allows bringing multiple differentially distributed criteria onto the same scale. In this way, it enables a fair comparison of the models for a given target protein without giving preferences to any of the assessment criteria and provides a way to determine the most accurate models in the population. The approach, however, is not devoid of drawbacks. Most importantly, it gives no information about *how* accurate the most accurate models are; therefore, *Z*-scores appear incomparable between different targets of varying difficulty. For a challenging target, even a model with the highest *Z*-scores is often extremely far from truth, while for targets with closer homology to the existing templates lower *Z*-score values may correspond to very accurate predictions. Furthermore, the choice of measures to be included in the *Z*-score is not only subjective, but often also is decided only at the evaluation stage. Combining correlated criteria implicitly gives them higher weight in the overall *Z*-score. Finally, because not all assessment criteria are normally distributed, conversion of these values into *Z*-scores creates somewhat distorted statistics, in this case probabilities (a.k.a., the *p* values) or their logarithms calculated for specific distributions make better contributions to the score (however, they cannot be mixed with the *Z*-scores).

The main problem of the intrapopulation *Z*-score approach is the absence of information about how close the models are to the correct answer. Even within a population of completely incorrect models, some model will be the "best." To overcome this problem, a better method is to compare the predictions with the distribution of the natural structural differences between "correct," i.e., experimentally determined structure pairs. With the wealth of protein structure information growing exponentially (1), it is easy to calculate, for example, the distribution of ligand RMSD values between multiple structures of the same complex. After that, one can normalize a model ligand RMSD value from the reference structure by determining what fraction of experimental structure pairs are characterized by the same or higher ligand RMSD (cumulative distribution function, CDF). In principle, it is possible to calculate the *Z*-score of each model in the reference experimental value distribution, however, caution is necessary for criteria with non-normal distributions. The flipside of the CDF approach is that in difficult cases the majority of the models may appear far too distant from the real target structure to receive a non-zero CDF score; therefore, the model population ranking may become impossible.

To illustrate the concept of CDF percentiles, we calculated their values for the sets of D3 and CXCR4 models in GPCR Dock 2010 (Table 1). For example, in comparison with the most favorable reference (answer) structure, an average model in the top half of the D3 set was better than 5.24% of experimental pairs by superimposition error, while an average CXCR4 model was only better than 1.68%. Unlike intrapopulation *Z*-scores, these CDF percentiles project the model quality on the uniform scale of correctness which makes them comparable not only (1) between the models, but also between (2) different targets and (3) assessment criteria.

Table 1
Cumulative distribution function (CDF) percentiles of GPCR Dock 2010 models in the experimental distribution

	Average CDF		Best CDF	
Protein similarity measure	D3 (%)	CXCR4 (%)	D3 (%)	CXCR4 (%)
Superimposition error	5.24	1.68	8.40	2.40
Virtual C'_β - C'_β contact strength difference	2.06	0.10	3.99	1.20
Ligand heavy atom RMSD	3.65	0.91	17.57	5.02
Ligand-pocket contact strength difference	2.36	0.75	13.46	2.60

Statistics are calculated for the top half of each model set, i.e., models less accurate than average are eliminated

For example, by averaging CDF percentiles over the four comparison criteria in Table 1, we can obtain the CDF score of 3.33% for an average D3 model but only 0.86% for an average CXCR4 model, which is representative of both absolute and relative accuracy of the modeling in the two cases. This result is, of course, expected given the fact that closer homology modeling templates were available in PDB for D3 than for CXCR4 at the time of the assessment. It is quite encouraging, however, that several D3 predictions fell into a significantly populated region of the experimental distribution, with the most accurate D3 model achieving 17.57 and 13.46% CDF values in terms of ligand RMSD and contacts, respectively.

3. Notes

3.1. X-ray Structures as "Golden Standard" in Model Evaluation

Structural variability within sets of protein structures determined for the same parent protein but in different crystal or molecular environments has been acknowledged and quantified in several publications (3, 30, 46). On one hand, such variability may be due to the inherent protein flexibility triggered by a different complex composition or crystallization environment. On the other hand, it may be an artifact of the limited resolution of the structure determination techniques and the inevitable experimental errors. The extent of conformational changes observed between multiple structures of the same protein ranges from minor side-chain rearrangements to large-scale domain and loop movements, and depends on the protein functional class, crystal form and contacts (47), co-crystallized interaction partners (30), and other factors. A large-scale analysis of a redundant set of protein structures was performed in ref. 3 and led the authors to the conclusion about the limited possibility of modeling proteins with multiple conformational states. In this regard, a legitimate question is whether a set of crystallographic coordinates represents an undisputable truth about native, biologically relevant structure of the protein, and whether it is conceptually correct to judge models by the degree of their structural similarity to the X-ray "answer." The question is open-ended, because up to date, X-ray crystallography is the only experimental method capable of elucidating proteins and their interactions at the atomic resolution level. Using crystallographic structures as modeling standards is, therefore, inevitable; however, several measures can be taken to account for arising issues:

- Compare the model to the relevant conformational states and complex compositions.
- Compare the model to the conformational ensemble and not a single structure (choose either the best or the average score).
- Down-weight or eliminate the contribution of flexible or poorly defined regions.
- Report comparison scores in context of their distribution between the multiple structures in the ensemble.

These steps help to translate the knowledge about the "natural" protein variation into an improved comparison measure. For example, in GPCR Dock 2010, all dopamine D3 receptor models were compared to the two noncrystallographic symmetry-related complexes in the reference structure, PDB 3pbl. The CXCR4 models were compared to the ensemble of as many as eight reference complexes. For each combination of criteria, the values were reported in comparison with the most favorable reference in this ensemble. Moreover, the primary focus of the assessment was made on prediction of the ligand binding area and interactions which, in contrast to the intracellular or extracellular loops, are unlikely to be significantly affected by protein flexibility.

3.2. Separating Trivial from Nontrivial: The Naïve Models

In addition to the question of how close a model is to the experimental structure, it is also important to know how far it is from the result of applying a sensible but trivial procedure. The so-called "naïve" models allow evaluation of the contribution of newly developed advanced modeling and refinement procedures in comparison with the most simple and straightforward approaches. In a way, the role of naïve models is similar to the role of placebo in drug clinical evaluation. Quite interestingly, the number of drugs that fail in the clinical trials by the reason of being no more effective than placebos constantly increases (48), leading some to the conclusion that the placebo effect is strengthening. Similarly, the constant method development in protein structure prediction makes the "naïve" models increasingly sophisticated thus shifting the baseline in model evaluation.

The most straightforward way to build a naïve model is threading the target sequence through a homology template without any subsequent optimization, or, in some cases, with fast side-chain optimization aimed at removal of major steric clashes. Even along this simple path, several factors may dramatically affect the quality and the degree of naivety of the models. They include (1) choice of the homologous protein and (2) of the specific structure of that protein to be used as the homology template, as well as (3) choice of the target-template sequence alignment which, with the exception of the extremely high homology cases, usually appears ambiguous. Figure 3d–f presents the scatter plot of such naïve model on the background of the top half of GPCR Dock models. The accuracy range of the naïve models is substantial; in this case, the range is primarily determined only by the choice of the homology template because we used our best knowledge sequence alignment in each case. For homology modeling, we used the six GPCR structures available in PDB prior to the 2010 GPCR Dock assessment: those of bovine rhodopsin in dark (bRho) and light-activated ligand free (opsin) states (49–51), β_1 and β_2 adrenergic receptors (52–54), and adenosine A_{2A} receptor (44). Our naïve models are close to the center of the distribution of the assessment models which may indicate the similarity of the approaches used by the GPCR Dock participants. However, a few models stand out and fall closer to the natural variation zone.

Whenever the modeling process includes not only modeling of the protein structure but also the docking of a protein or a small-molecule ligand, the definition of a naive model becomes even less defined. In rare cases when a homologous complex structure exists, it may be used to build a naïve, non-optimized model of the target complex as long as the target and the template ligands can be unambiguously (structurally) aligned. For protein ligands, the alignment may be based on sequence homology; but small molecules or in some cases short peptides may require finding the maximal common substructure between the target and template ligands, or establishing the correspondence in some other nontrivial way.

As an example, let us consider the challenges of building a "naive" model of the dopamine D3 receptor complex with eticlopride. This molecule belongs to a large class of aminergic antagonists and shares some degree of pharmacophoric similarity with previously crystallized antagonists of β_2 adrenergic receptor, carazolol, and timolol. We performed pharmacophore-based alignment of the three-dimensional eticlopride molecule onto the structures of these two adrenergic antagonists. Because the procedure produced several answers, the top ten chemical alignments were taken for each template, each was combined with the six naïve models generated by sequence threading and locally minimized to eliminate severe side-chain/ligand steric clashes. This produced a population of "naive" D3 complex models presented in Fig. 7b. The accuracy

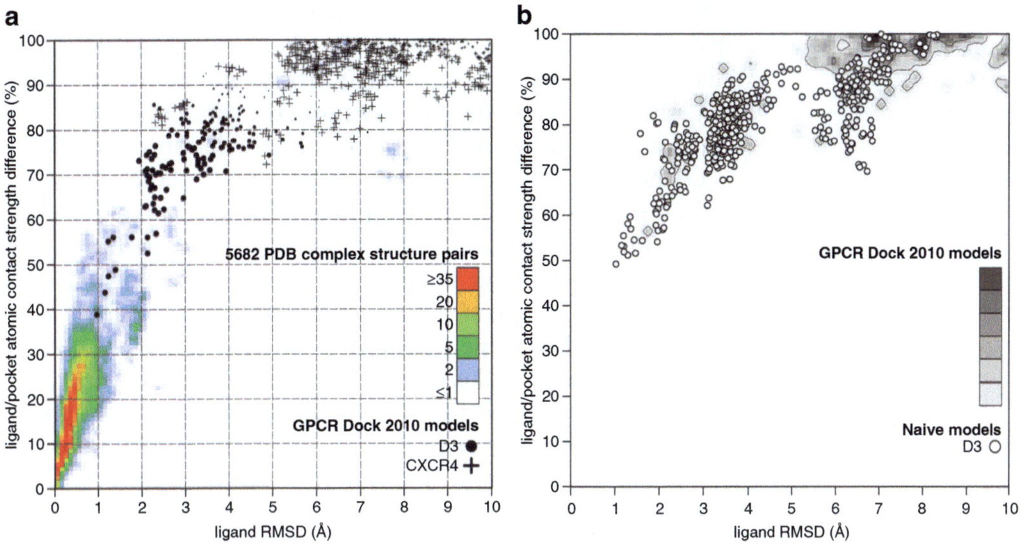

Fig. 7. Distribution of ligand RMSD values and atomic contact strength differences between identical composition complex structures: statistics of a large subset of experimental complex structures pairs in PDB (**a**, heat map), GPCR Dock 2010 models (**a**, filled circle for D3 and plus sign for CXCR4; **b**, heat map), and naïve models of dopamine D3 receptor (**b**, open circle).

range of these models is huge. Some of them approach (though none of them exceeds!) the level of accuracy of the best D3 models in GPCR Dock 2010. Though the step of scoring and selection was not employed in this exercise, it illustrates that (1) the level of model naivety may be highly variable, especially in the case of protein–ligand docking complexes and (2) "naïve" sampling is capable of producing very accurate models.

In summary, the naïve models are useful to separate the actual advances from the trivial sensible approach; however, their definition appears too ambiguous to make them reliable standards of structure comparison and evaluation.

3.3. Evaluation of Model Quality Without Direct Comparison to the Reference Structure

The first question that has to be answered about a model is, in fact, not the degree of its similarity to the reference structure, but its spatial feasibility. This kind of evaluation is widely used to assess local errors in crystallographic coordinates during the refinement process or submissions for a modeling competition. The evaluation may be based on geometrical, stereochemical, or statistical criteria, e.g., WhatCheck (55, 56), PROCHECK (57), or MolProbity (58), while some others, e.g., ICM Protein Health (59), use realistic normalized force field residue energies, where the expected distributions for the energies for each residue are derived from high-quality crystal structures. An alternative approach involves the cumulative residue pseudo-energies or scores calculated as function of local atom, residue, secondary structure, accessibility environment, and trained to predict the deviations from the near native models. Multiple methods (VERIFY3D, PROSA, BALA, ANOLEA, PROVE,

TUNE, REFINER, PROQRES) were integrated into a meta-server called MetaMQAP and trained to predict the residue deviations. While the individual residue predictions may not be accurate, combining different methods, and averaging the residue signal in a five residue window led to impressive quality prediction values (60).

Despite the obvious progress in protein structure prediction methodology and tools, the gain in modeling accuracy, as evaluated by similarity to the experimentally solved answer, has become less prominent in recent years (4). It appears, therefore, that the progress in the protein structure prediction area is reaching a certain plateau and that the question of primary importance at this stage is not how to make models more similar to the experimentally derived structures, but how to make the most use of these models at the given level of prediction accuracy. Because one of major applications of modeling is in rational structure-based drug discovery and optimization, it appears relevant to directly evaluate the drug discovery potential of the models.

In the area of prediction of protein/ligand complex structures, virtual ligand screening (VLS) enrichment by a model represents a clever way of evaluation of the model compliance with the experimental data in the form of small molecule chemical activity against the modeled protein. In this experiment, a large set of chemicals containing known potent binders to the protein of interest (1–10% of the set) and diverse decoys of similar molecular weights and atom counts (90–99% of the set) is docked to the model, and the molecules are ranked by their predicted binding affinity. The model that efficiently and selectively scores the active molecules better than decoys apparently has a good potential for *de novo* drug discovery efforts. Quite interestingly, it appears also that such models often are most accurate in terms of predicted contacts between the ligand and the pocket atoms. For example, in both GPCR Dock 2008 (7) and GPCR Dock 2010 (10) assessments, model selection by VLS enrichment proved to be a successful strategy leading to most accurate predictions.

An important question is how to quantify VLS enrichment by a model. One of the traditional approaches to the problem involves calculation of the area under the so-called receiver operating characteristic curve (ROC curve) which plots the ratio of true positives (TP, y-axis) against the ratio of false positives (FP, x-axis) in the top portion of the hit list ordered by the predicted binding affinity for each value of the affinity cutoff. A variation of the ROC curve is built when the fraction of TP is plotted against the total number of compounds scoring below the given cutoff rather than the FP rate. Both approaches suffer from the inability to distinguish early enrichment from late enrichment, and therefore are often complemented by the specific enrichment factors (EF) at the given FP rate, e.g., EF1 denotes the fraction of correct, active compounds that score better than 1% of the top-scoring decoys.

The normalized square-root area under curve (nsAUC) is the area under the curve that plots the fraction of TP on top of the hit list (*y*-axis) against the *square root* of the total number of compounds scoring below the given cutoff (*x*-axis). Previous studies indicated that this measure is more representative of the true model selectivity than either the regular ROC which understresses the initial compound recognition (Fig. 8) of the log-AUC (61, 62) which overstresses it. With the non-normalized square-root AUC approach, the ideal sAUC (perfect recognition, all actives are ranked better than all inactives) and the random sAUC (actives are retrieved at the same rate as total compounds in the set, no recognition) are given by

$$\mathrm{sAUC}_{\mathrm{ideal}} = \frac{1}{c}\int_0^{\sqrt{c}} x^2\,\mathrm{d}x + (1-\sqrt{c}) = 1 - \frac{2\sqrt{c}}{3}$$

and

$$\mathrm{sAUC}_{\mathrm{rnd}} = \int_0^1 x^2\,\mathrm{d}x = \frac{1}{3},$$

respectively. Here *c* is the fraction of the active compounds in the set. For the purpose of comparing the AUC across different datasets, sAUC is normalized to get:

$$\mathrm{nsAUC} = \frac{\mathrm{sAUC} - \mathrm{sAUC}_{\mathrm{rnd}}}{\mathrm{sAUC}_{\mathrm{ideal}} - \mathrm{sAUC}_{\mathrm{rnd}}} \times 100\%$$

that ranges from 0% (random) to 100% (ideal).

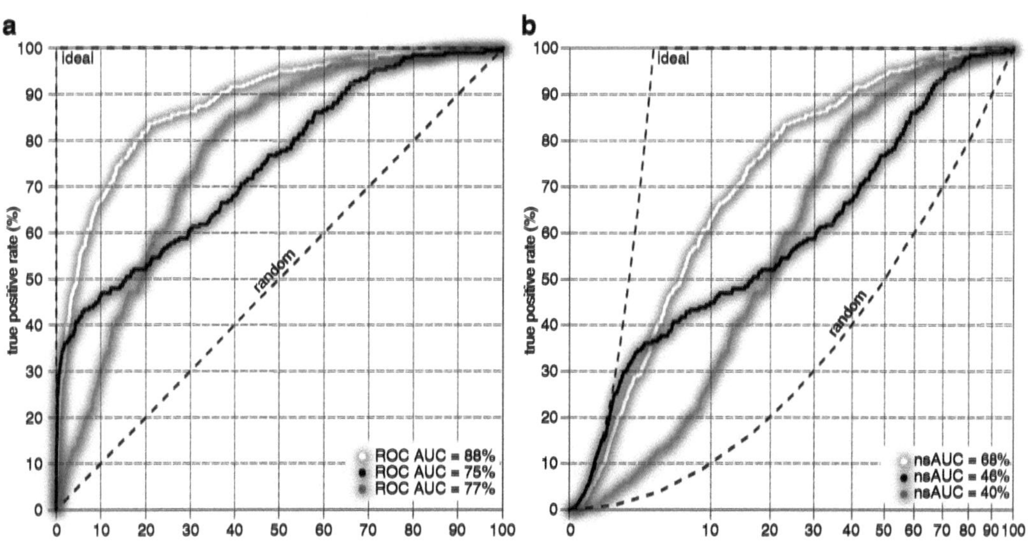

Fig. 8. Unlike the routinely used ROC AUC (**a**), the normalized square-root AUC (**b**) rewards the initial hit recognition in virtual ligand screening. This approach makes the profile in *black* preferable over the one in *gray*.

Finally, the VLS enrichment is not the only possible way to incorporate ligand binding information in the modeling process. Alternative approaches may be based on known active ligand pharmacophores, for example, by the detection of complementarity of such pharmacophores to the model pocket. Though not directly measuring the drug discovery potential of the model, this approach also proved fruitful for increasing the accuracy of the GPCR–ligand complex structure prediction in GPCR Dock 2010 (10).

Acknowledgments

Authors wish to thank the organizers and the participants of the GPCR Dock 2010 assessment for providing the model statistics, Max Totrov and Eugene Raush for implementing some of the core functions in ICM, Manuel Rueda for helpful discussions and Karie Wright for help with manuscript preparation. We would like to acknowledge financial support by NIH, grants # R01 GM071872, U01 GM094612, and U54 GM094618.

References

1. Gabanyi M, Adams P, Arnold K, Bordoli L, Carter L, Flippen-Andersen J, Gifford L, Haas J, Kouranov A, McLaughlin W, *et al.* (2011) *Journal of Structural and Functional Genomics*, 1–10.
2. Rose PW, Beran B, Bi C, Bluhm WF, Dimitropoulos D, Goodsell DS, Prlic A, Quesada M, Quinn GB, Westbrook JD, *et al.* (2011) *Nucleic Acids Research* **39,** D392–D401.
3. Burra PV, Zhang Y, Godzik A, & Stec B (2009) *Proceedings of the National Academy of Sciences* **106,** 10505–10510.
4. Kryshtafovych A, Fidelis K, & Moult J (2009) *Proteins: Structure, Function, and Bioinformatics* **77,** 217–228.
5. Cozzetto D, Kryshtafovych A, Fidelis K, Moult J, Rost B, & Tramontano A (2009) *Proteins: Structure, Function, and Bioinformatics* **77,** 18–28.
6. Wodak SJ (2007) *Proteins: Structure, Function, and Bioinformatics* **69,** 697–698.
7. Michino M, Abola E, participants of GPCR Dock 2008, Brooks CL, Dixon JS, Moult J, & Stevens RC (2009) *Nat Rev Drug Discov* **8,** 455–463.
8. Warren G, Nevins N, & McGaughey G (2011) in *241st ACS National Meeting* (Anaheim, CA).
9. Warren GL, Andrews CW, Capelli A-M, Clarke B, LaLonde J, Lambert MH, Lindvall M, Nevins N, Semus SF, Senger S, *et al.* (2005) *Journal of Medicinal Chemistry* **49,** 5912–5931.
10. Kufareva I, Rueda M, Katritch V, participants of GPCR Dock 2010, Stevens RC, & Abagyan R (2011) *Structure* **19**(8), 1108–1126.
11. Wu B, Chien EYT, Mol CD, Fenalti G, Liu W, Katritch V, Abagyan R, Brooun A, Wells P, Bi FC, *et al.* (2010) *Science* **330,** 1066–1071.
12. Chien EYT, Liu W, Zhao Q, Katritch V, WonHan G, Hanson MA, Shi L, Newman AH, Javitch JA, Cherezov V, *et al.* (2010) *Science* **330,** 1091–1095.
13. Kryshtafovych A, Venclovas, Fidelis K, & Moult J (2005) *Proteins: Structure, Function, and Bioinformatics* **61,** 225–236.
14. Zemla A (2003) *Nucleic Acids Research* **31,** 3370–3374.
15. Shindyalov IN & Bourne PE (1998) *Protein Engineering* **11,** 739–747.
16. Holm L & Sander C (1993) *Journal of Molecular Biology* **233,** 123–138.
17. Kleywegt GJ & Jones AT (1997) in *Methods in Enzymology* (Academic Press), pp. 525–545.
18. Ortiz AR, Strauss CEM, & Olmea O (2002) *Protein Science* **11,** 2606–2621.
19. Levitt M & Gerstein M (1998) *Proceedings of the National Academy of Sciences of the United States of America* **95,** 5913–5920.

20. Shapiro J & Brutlag D (2004) *Nucleic Acids Research* **32,** W536-W541.
21. Szustakowski JD & Weng Z (2000) *Proteins: Structure, Function, and Bioinformatics* **38,** 428–440.
22. Kleywegt GJ (1996) *Acta Crystallogr D Biol Crystallogr* **52,** 842–857.
23. Kawabata T & Nishikawa K (2000) *Proteins* **41,** 108–122.
24. Kawabata T (2003) *Nucleic Acids Res* **31,** 3367–3369.
25. Yang A-S & Honig B (2000) *Journal of Molecular Biology* **301,** 665–678.
26. Lackner P, Koppensteiner WA, Sippl MJ, & Domingues FS (2000) *Protein Engineering* **13,** 745–752.
27. Krissinel E & Henrick K (2004) *Acta Crystallographica Section D* **60,** 2256–2268.
28. Zemla A, Venclovas, Moult J, & Fidelis K (2001) *Proteins* **Suppl 5,** 13–21.
29. Zhang Y & Skolnick J (2004) *Proteins: Structure, Function, and Bioinformatics* **57,** 702–710.
30. Abagyan R & Kufareva I (2009) *Methods Mol Biol* **575,** 249–279.
31. McLachlan AD (1979) *J Mol Biol* **128,** 49–79.
32. Damm KL & Carlson HA (2006) *Biophysical journal* **90,** 4558–4573.
33. Phillips DC (1970) *Biochem Soc Symp* **30,** 11–28.
34. Nishikawa K & Ooi T (1974) *J.Theor.Biol.* **43,** 351–274.
35. Liebman MN (1980) *Biophys. J.* **32,** 213–215.
36. Sippl MJ (1982) *Journal of Molecular Biology* **156,** 359–388.
37. Abagyan RA & Totrov MM (1997) *J Mol Biol* **268,** 678–685.
38. Marsden B & Abagyan R (2004) *Bioinformatics* **20,** 2333–2344.
39. Lensink MF & Wodak SJ (2010) *Proteins: Structure, Function, and Bioinformatics* **78,** 3085–3095.
40. Bottegoni G, Kufareva I, Totrov M, & Abagyan R (2009) *J Med Chem* **52,** 397–406.
41. Totrov M & Abagyan R (2008) *Curr Opin Struct Biol.*
42. Coupez B & Lewis RA (2006) *Curr Med Chem* **13,** 2995–3003.
43. Katritch V, Rueda M, Lam PC-H, Yeager M, & Abagyan R (2010) *Proteins* **78,** 197–211.
44. Jaakola V-P, Griffith MT, Hanson MA, Cherezov V, Chien EYT, Lane JR, Ijzerman AP, & Stevens RC (2008) *Science* **322,** 1211–1217.
45. Rueda M, Katritch V, Raush E, & Abagyan R (2010) *Bioinformatics* **26,** 2784–2785.
46. Stroud RM & Fauman EB (1995) *Protein Science* **4,** 2392–2404.
47. Eyal E, Gerzon S, Potapov V, Edelman M, & Sobolev V (2005) *Journal of Molecular Biology* **351,** 431–442.
48. Golomb BA, Erickson LC, Koperski S, Sack D, Enkin M, & Howick J (2010) *Annals of Internal Medicine* **153,** 532–535.
49. Palczewski K, Kumasaka T, Hori T, Behnke CA, Motoshima H, Fox BA, Trong IL, Teller DC, Okada T, Stenkamp RE, *et al.* (2000) *Science* **289,** 739–745.
50. Scheerer P, Park JH, Hildebrand PW, Kim YJ, Krausz N, Choe H-W, Hofmann KP, & Ernst OP (2008) *Nature* **455,** 497–502.
51. Park JH, Scheerer P, Hofmann KP, Choe H-W, & Ernst OP (2008) *Nature* **454,** 183–187.
52. Warne T, Serrano-Vega MJ, Baker JG, Moukhametzianov R, Edwards PC, Henderson R, Leslie AGW, Tate CG, & Schertler GFX (2008) *Nature* **454,** 486–491.
53. Rosenbaum DM, Cherezov V, Hanson MA, Rasmussen SGF, Thian FS, Kobilka TS, Choi H-J, Yao X-J, Weis WI, Stevens RC, *et al.* (2007) *Science* **318,** 1266–1273.
54. Cherezov V, Rosenbaum DM, Hanson MA, Rasmussen SGF, Thian FS, Kobilka TS, Choi H-J, Kuhn P, Weis WI, Kobilka BK, *et al.* (2007) *Science* **318,** 1258–1265.
55. Hooft RW, Vriend G, Sander C, & Abola EE (1996) *Nature* **381,** 272–272.
56. Vriend G (1990) *J Mol Graph* **8,** 52–56.
57. Laskowski RA, MacArthur MW, Moss DS, & Thornton JM (1993) *Journal of Applied Crystallography* **26,** 283–291.
58. Chen VB, Arendall WB, III, Headd JJ, Keedy DA, Immormino RM, Kapral GJ, Murray LW, Richardson JS, & Richardson DC (2010) *Acta Crystallographica Section D* **66,** 12–21.
59. Maiorov V & Abagyan R (1998) *Fold Des* **3,** 259–269.
60. Pawlowski M, Gajda MJ, Matlak R, & Bujnicki JM (2008) *BMC Bioinformatics* **9,** 403–403.
61. Jain A & Nicholls A (2008) *Journal of Computer-Aided Molecular Design* **22,** 133–139.
62. Clark R & Webster-Clark D (2008) *Journal of Computer-Aided Molecular Design* **22,** 141–146.

Chapter 11

Homology Modeling of Class A G Protein-Coupled Receptors

Stefano Costanzi

Abstract

G protein-coupled receptors (GPCRs) are a large superfamily of membrane bound signaling proteins that hold great pharmaceutical interest. Since experimentally elucidated structures are available only for a very limited number of receptors, homology modeling has become a widespread technique for the construction of GPCR models intended to study the structure–function relationships of the receptors and aid the discovery and development of ligands capable of modulating their activity. Through this chapter, various aspects involved in the constructions of homology models of the serpentine domain of the largest class of GPCRs, known as class A or rhodopsin family, are illustrated. In particular, the chapter provides suggestions, guidelines, and critical thoughts on some of the most crucial aspect of GPCR modeling, including: collection of candidate templates and a structure-based alignment of their sequences; identification and alignment of the transmembrane helices of the query receptor to the corresponding domains of the candidate templates; selection of one or more templates receptor; election of homology or de novo modeling for the construction of specific extracellular and intracellular domains; construction of the 3D models, with special consideration to extracellular regions, disulfide bridges, and interhelical cavity; validation of the models through controlled virtual screening experiments.

Key words: G protein-coupled receptors, Membrane spanning helices, Extracellular loops, Homology modeling, De novo modeling, Multiple sequence alignment, Model validation, Controlled virtual screening

1. Introduction

G protein-coupled receptors (GPCRs), also known as seven transmembrane (7TM) receptors, are proteins expressed on the plasma membrane that mediate the receiving of extracellular stimuli given by a variety of first messengers (1). The latter can be either endogenous molecules secreted by the body, for example neurotransmitters or hormones, or exogenous molecules of external origin, for example odorants. In humans, the superfamily of GPCRs includes over 800 members that, according to the GRAFS classification scheme, can be divided into five main families: the glutamate family

Andrew J.W. Orry and Ruben Abagyan (eds.), *Homology Modeling: Methods and Protocols*, Methods in Molecular Biology, vol. 857, DOI 10.1007/978-1-61779-588-6_11, © Springer Science+Business Media, LLC 2012

(G; also class C or family III), the rhodopsin family (R; also class A or family I), the adhesion family (A; also class B or family 2, together with the secretin family), the frizzled/taste2 family (F), and the Secretin family (S, also class B or family 2, together with the adhesion family) (2). The rhodopsin family, which also comprises numerous odorant receptors, is by far the largest of the five, accounting for about 84% of the entire superfamily (2). Coupling with intracellular proteins, GPCRs transduce extracellular stimuli into biochemical signals that alter the functioning of the cell, with vast physiological and pathophysiological implications (1). Notably, GPCRs signaling can be ad hoc modulated by exogenous molecules that either stimulate the receptors in lieu of their physiological first messengers or block their stimulation. As a result of this opportunity for pharmacological intervention, GPCRs are the target of a large share of the currently marketed drugs (3) and are the object of intense studies aiming at the development of novel therapeutic strategies.

Despite the large size of the superfamily, GPCRs have traditionally been characterized by a paucity of structural information and, for many years, detailed 3D structures were available only for rhodopsin. However, rhodopsin is a peculiar receptor with a very distinctive mechanism of activation: it features a covalently bound ligand, retinal, that triggers the activation of the receptor upon isomerization by the action of light photons—for a synoptic perspective on the role of rhodopsin as a prototypical class A GPCR, *see* Costanzi et al. (4). More recently, breakthroughs in GPCR crystallography led to the solution of the structure of additional receptors, all belonging to class A. Specifically, as shown in Table 1, at the time of this writing the Protein Data Bank (http://www.rcsb.org), enlists structures for: bovine rhodopsin crystallized in the ground state and at early stages of the photoactivation cycle; squid rhodopsin; the unliganded opsin alone and in complex with the C-terminal peptide of the α-subunit of transducin; the β_1 and β_2 adrenergic receptors in complex with a variety of blockers and agonists; the adenosine A_{2A} receptor in complex with a neutral antagonist; the CXCR4 chemokine receptor in complex with a small molecule and a cyclic peptide antagonist; and the dopamine D3 receptor (4–10). Additional structures are very likely to be solved in the near future.

The experimentally elucidated structures confirmed the idea, initially founded on sequence analysis (4), that GPCRs are constituted by a single polypeptide chain that spans the plasma membrane seven times, with seven α-helical structures (numbered from helix 1 to 7) interconnected by three extracellular and three intracellular loops (ELs and ILs, numbered from EL1 to EL3 and from IL1 to IL3), as schematically shown in Fig. 1 (11). The N terminus is in the extracellular milieu. Although usually relatively short, for some receptors—notably those belonging to class B and C and to

Table 1
Crystal structures of GPCRs deposited in the Protein Data Bank (http://www.rcsb.org) at the time of this writing

Receptor	PDB ID
Bovine rhodopsin, ground state	1F88 (40), 1GZM (41), 1HZX (42), 1L9H (43), 1U19 (44), 2I35 (45), 2I36 (45), 2J4Y (46),[a] 3C9L (47),[b] 3C9M (47)[b]
Bovine rhodopsin, early stages of photoactivation	2G87 (48), 2HPY (49), 2I37 (45), 2PED (50)
Squid rhodopsin, ground state	2ZIY (34), 2Z73 (51)
Bovine opsin	3CAP (52), 3DQB (53)[d]
Turkey β_1 adrenergic receptor in complex with antagonists, partial agonists, and full agonists	2VT4 (33),[a,e] 2Y00 (9),[a,f] 2Y01 (9),[a,f] 2Y02 (9),[a,g] 2Y03 (9),[a,g] 2Y04 (9)[a,f]
Human β_2 adrenergic receptor in complex with inverse agonists, antagonists, and agonists	2R4R (54),[h,i,j] 2R4S (54),[h,i,j] 2RH1 (27, 28),[i,k] 3D4S (55),[i,k] 3KJ6 (56),[h,i,j] 3NY8 (57),[i,k] 3NY9 (57),[i,k] 3NYA (57),[e,k] 3P0G (7),[g,k,l] 3PDS (8)[k,m]
Human adenosine A_{2A} receptor in complex with an antagonist	3EML (58)[e,k]
Human CXCR4 chemokine receptor in complex with antagonists	3ODU (6),[e,k] 3OE9 (6),[e,k] 3OE8 (6),[e,k] 3OE6 (6),[e,k] 3OE0 (6)[k,n]
Human dopamine D_3 receptor	3PBL (10)[e,k]

[a] Thermally stable mutant receptor
[b] Alternative model of 1GZM
[c] Alternative model of 2J4Y
[d] In complex with a C-terminal peptide of the α-subunit of transducin
[e] In complex with an antagonist
[f] In complex with a partial agonist
[g] In complex with a full agonist
[h] In complex with a Fab
[i] In complex with an inverse agonist
[j] Ligand not visible
[k] T4-lysozime fusion protein
[l] In complex with a camelid antibody fragment
[m] In complex with an irreversible agonist
[n] In complex with a cyclic peptide antagonist

the glycoprotein hormone subfamily of class A—this region is fused to a large soluble ectodomain responsible for ligand binding. For the protease-activated receptors (PAR), the N terminus plays a very peculiar role: it functions as a tethered ligand that, when unmasked by the action of proteases, activates the receptor. The C terminus, instead, is inside the cytoplasm. Notably, for all the receptors crystallized at the time of this writing, with the exception of the CXCR4 chemokine receptor, the portion of the C-terminal domain immediately following the junction with helix 7 has been shown to adopt

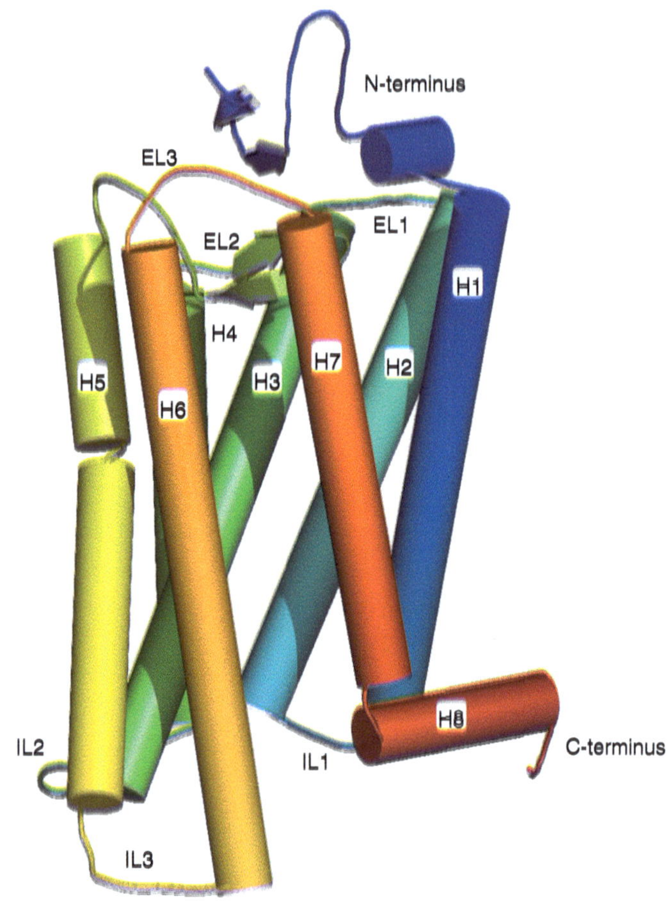

Fig. 1. Schematic representation of the crystal structure of bovine rhodopsin (1GZM), showing the seven transmembrane domain spanning topology characteristic of GPCRs. The structure is rendered with a continuum spectrum of colors going from , at the N terminus, to , at the C terminus.

an α-helical structure parallel to the plane of the membrane, dubbed helix 8. Sequence similarity suggests that many of the receptors belonging to the rhodopsin family may feature this amphipathic helix.

With such a large superfamily of pharmaceutically appealing receptors and so little structural information, homology modeling, initially based exclusively on the structure of rhodopsin, became a widespread technique to get insights into the structure–function relationships of the receptors and facilitate the discovery of chemicals capable of modulating their activity (4, 11, 12). In the most successful examples, the models were generated on the basis of biochemical and medicinal chemistry data, especially for the in silico generation of the complexes between the receptors and the small molecule ligands (13). A particularly powerful approach is the neoceptor/neoligand method developed by Jacobson and coworkers, in which receptor–ligand interactions are probed through mutagenesis experiments coupled to complementary chemical modification of the ligands (14).

In recent times, the above mentioned advancements in GPCR crystallography have significantly changed the landscape of GPCR homology modeling. First of all, multiple template strategies can now be applied to the construction of the models (11, 15, 16)—for a detailed analysis of the impact of the disclosure of new crystal structures to GPCR homology modeling, see Mobarec and coworkers (16). Moreover, comparisons between in silico and experimental models of the same receptor are now possible and can be used not only to evaluate the state of the art but also to develop new and improved modeling strategies. In this context, soon after the β_2 adrenergic receptor became the first GPCR, after rhodopsin, with a crystallographically elucidated structure, I published the first direct evaluation of the accuracy of a GPCR homology model (17). In particular, I compared the crystal structure of the β_2 adrenergic receptor in complex with its inverse agonist carazolol to in silico models of the same receptor–ligand complex constructed through rhodopsin-based homology modeling followed by molecular docking. Notably, not only the structure of the receptor but also the binding mode of the ligand and the receptor–ligand interactions were approximated reasonably well by the models. A wider evaluation of the state of the art was subsequently provided by the first "community-wide assessment of GPCR structure modeling and ligand docking," organized in coordination with the solution of the structure of the adenosine A_{2A} receptor in complex with the neutral antagonist ZM241385 (18). This time, models of the receptor–ligand complex were submitted to the organizers of the assessment by a number of molecular modelers prior to the unveiling of the crystal structure. In line with what I had found for the β_2 adrenergic receptor, this blind test revealed that the seven-helix bundle of the A_{2A} receptor could be built with good accuracy, while the modeling of the interconnecting loops, especially the long ones, was confirmed to be problematic. The docking of the ligand revealed to be a very challenging aspect too, as testified by the wide distribution found for the accuracy of the predictions. However, the top three scoring models (submitted by Costanzi, Abagyan/Katrich and Abagyan/Lam) predicted correctly over 40% of the total number of the receptor–ligand contacts. At the time of this writing, a second community wide assessment is underway (see cmpd.scripps.edu/GPCRDock2010).

This chapter, geared towards researchers already familiar with homology modeling, provides suggestions, guidelines, and critical

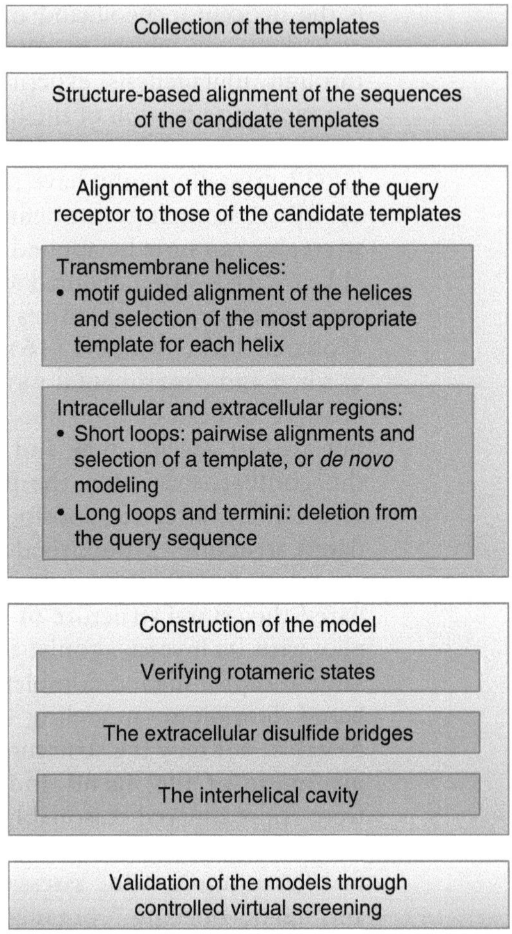

Fig. 2. Schematic overview of the aspects of class A GPCR modeling discussed throughout this chapter.

thoughts on some of the most crucial aspect involved in the constructions of homology models of the serpentine domain of class A receptors (see Fig. 2 for a schematic overview).

2. Materials

The construction and validation of homology models of GPCRs entails performing sequence alignments—including structure-based sequence alignments—generating and refining 3D models, and performing docking-based virtual screening experiments. These operations can be carried out by means a variety of web servers as well as commercial and freely available software. Of note, this chapter is intended for researcher well versed with homology modeling and does not deal with technical aspects relative to the use of specific software packages.

3. Methods

3.1. Collection of the Templates

As mentioned, for a long time rhodopsin has been the only available template for the construction of homology models of class A GPCRs (4). However, this is not the case anymore, as crystal structures for a number additional receptors have been recently solved (4–6).

Files with the coordinates of the crystallized class A GPCRs (see Table 1) can be directly downloaded in PDB format from the Web site of the Protein Data Bank (http://www.rcsb.org). Of note, the availability of additional templates may be verified at any given moment through the "Advanced Search" feature of the Web site, which allows conducting "Sequence Blast" searches based on the amino acid sequence of the query receptor, i.e., the receptor object of the modeling project.

3.2. Structure-Based Alignment of the Sequences of the Templates

Prior to the selection of the most suitable structure—or of multiple structures—to be used as template for the construction of the model of the query receptor, it is convenient to align the amino acid sequences of the candidate templates. Since structures are more conserved than sequences and since, by definition, 3D coordinates are available for all the templates, it is opportune to derive this sequence alignment through a structure-based alignment method. More specifically, it is advisable to derive the multiple sequence alignment only for the seven membrane spanning helices and, when present, for the amphipathic helix 8. In fact, it is, in these domains, that the highest structural conservation is observed in GPCRs, while a much higher variability is observed in the extracellular and the intracellular regions (5).

Before subjecting the PDB files to the structure-based sequence alignment, they should be appropriately edited, as several of their sections need to be expunged (see Notes 1 and 2). In particular, a PDB file often includes multiple receptor molecules contained in the unit cell, each of which with a unique chain name—for example, the β_1 adrenergic receptor structure deposited with the PDB ID of 2VT4 contains four distinct instances of the receptor (chains A, B, C, and D). One of the chains should be selected to serve as a potential template for the construction of the homology model, while the others should be deleted (for a caveat on how to choose the right chain, see Note 3). A PDB file may also contain additional proteins co-crystallized with the receptor—for example, the β_2 adrenergic receptor structure deposited with the PDB ID of 3R4R contains, in addition to the coordinated of the receptor (chain A), those of the light and heavy chains (chains L and H, respectively) of a co-crystallized Fab (fragment antigen binding) that recognizes the IL3 domain of the receptor. All the records pertinent to theses chains should be deleted. For the chain of interest, the ATOM records pertinent to the helical bundle of the receptor are essential for the structure-based sequence alignment and must be preserved (see Note 4). All other records, among which those relative to ligands and cofactors as well as intracellular and extracellular regions are not necessary and may be deleted. Importantly, if the crystal structure has been obtained for a fusion protein of the receptor with the T4-lysozyme, the ATOM records relative to the latter must be deleted too. By way of example, the rhodopsin structure deposited with the PDB ID of 1GZM can be reduced to what represented in Fig. 3.

```
HEADER    SIGNALING PROTEIN                       24-MAY-02   1GZM
ATOM    266  N   PRO A  34      40.293  39.953  21.436  1.00 62.15           N
...
...
...
ATOM    523  NE2 GLN A  64      68.480  13.985  -3.754  1.00 78.20           N
ATOM    578  N   PRO A  71      52.420   8.589 -12.047  1.00 52.70           N
...
...
...
ATOM    813  NE2 HIS A 100      49.042  45.962  10.860  1.00 54.19           N
ATOM    863  N   PRO A 107      39.986  44.959  -0.249  1.00 50.92           N
...
...
...
ATOM   1117  SG  CYS A 140      39.511  -1.820 -11.573  1.00 79.28           S
ATOM   1189  N   GLY A 149      50.821   6.757 -18.797  1.00 91.43           N
...
...
...
ATOM   1366  CG2 VAL A 173      36.152  37.172 -14.261  1.00 41.07           C
ATOM   1578  N   ASN A 200      25.636  36.061  -7.732  1.00 46.81           N
...
...
...
ATOM   1857  OE2 GLU A 232      32.271 -12.814 -10.624  1.00142.33           O
ATOM   1915  N   ALA A 241      50.473 -11.551  -6.685  1.00155.08           N
...
...
...
ATOM   2211  CG2 THR A 277      20.369  26.398   1.054  1.00 38.18           C
ATOM   2264  N   PRO A 285      30.175  35.133  14.236  1.00 53.25           N
...
...
...
ATOM   2574  SG  CYS A 323      73.326  15.731   1.645  1.00 90.64           S
END
```

Fig. 3. Example of a simplified PDB file that can be used to generate a structure-based alignment of the helical bundle of the candidate templates. For each helix, the figure shows only the entries corresponding the first atom of the first residue and the last atom of the last residue, while the entries in between are indicated by *suspension marks*. The simplified PDB file refers to the rhodopsin structure deposited with the PDB ID of 1GZM. The segment from Pro285 to Cys323 refers to both helix 7 and helix 8.

The edited PDB files of the crystallized receptors can then be used to derive a structure-based sequence alignment that, in turn, can serve as a tool for the selection of the template—or of the multiple templates—to be used for the construction of the helical bundle of the query receptor (see Subheading 3.3). Instead, for the selection of the template for the extracellular and intracellular regions, when this is possible, pairwise alignments between each single template and the receptor to be modeled are more appropriate (see Subheading 3.4). As a guide, a structure-based sequence alignment of the seven membrane spanning helices and the amphipathic helix 8 of bovine and squid rhodopsin, the β_1 and β_2 adrenergic receptors, and the adenosine A_{2A} receptors are provided in Fig. 4, together with a 3D view of the resulting structural superimposition.

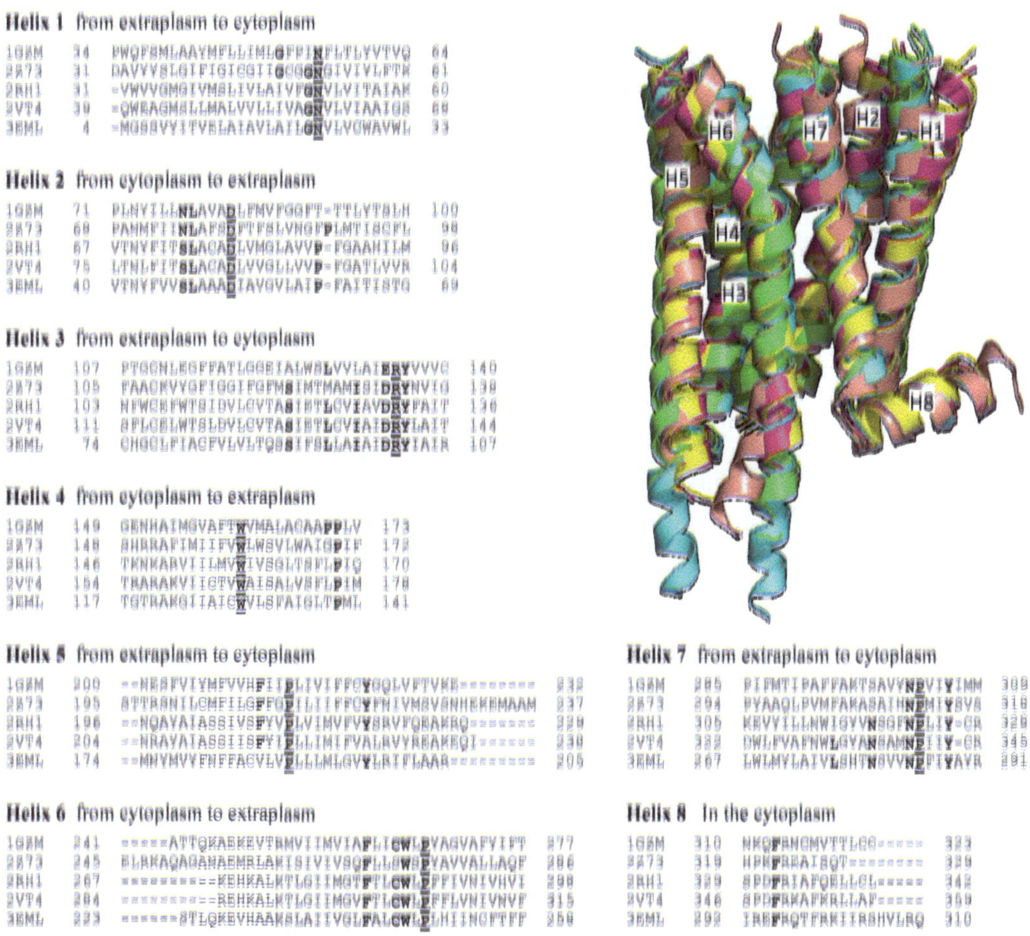

Fig. 4. Structure-based alignment of the sequences of the seven membrane spanning helices and the amphipathic helix 8 of bovine rhodopsin (1GZM), squid rhodopsin (2Z73), human β_2 adrenergic receptor (2RH1), turkey β_1 adrenergic receptor (2VT4), and adenosine A_{2A} receptors (3EML). The most conserved residue of each helix, as defined by Ballesteros and Weinstein (see Note 5), is in **bold** and underlined, while additional significantly conserved residues are in *bold* (see Fig. 5). A 3D structural superimposition is also provided, where bovine and squid rhodopsin are in *green* and *cyan*, the β_1 and β_2 adrenergic receptors in *yellow* and *purple*, and the adenosine A_{2A} receptor in *pink*.

3.3. Alignment of the Query Sequence to the Prealigned Helical Bundle of the Candidate Templates

The alignment of the sequence of the query receptor to the prealigned helical bundle of the candidate templates can be achieved starting with an automatic sequence alignment, performed without allowing the relative alignment of the candidate templates to change. The alignment obtained in this manner, should be subsequently subjected to a careful visual inspection and manual refinement. In particular, the correct identification of the seven membrane spanning helices of the query receptor must be verified on the basis of the presence of specific motifs, also called conservation patterns, that characterize each helix (see Fig. 5) (19). Of particular importance is the identification and the correct alignment of the most conserved residue of each helix (see Fig. 5), defined as residue X.50 according to the GPCR residue indexing system (see Note 5) (20, 21). Of note, these motifs, although frequent, are not present in the membrane spanning helices of all receptors, sometimes making the identification of a certain helix difficult. Once all the helices have been identified, the automatic alignment should be inspected and, if necessary, adjusted to ensure that the motifs of the query are aligned with those of the candidate templates. The presence of gaps in the alignment of the helices should also be avoided (however, see Note 6).

3.3.1. Single Template or Multiple Templates?

Given that the structure of several GPCRs has been solved through X-ray crystallography, GPCR homology models can now be constructed through either a single or a multiple template strategy (16). Single template strategies involve the selection of the crystallized receptor that, overall, seems more likely to be characterized by structural similarity with the query receptor, while multiple template strategies involve the splitting of the query receptor into several domains and the subsequent selection of the most suitable template for each of these domains. In particular, once the sequences of candidate templates and query receptors have been aligned, the selection of the templates can be operated on the basis of sequence similarities, for instance through the calculation of

Helix 1: GX_3**N** or G**N**

Helix 2: N(S,H)LX_3**D**$X_{7,8,9}$P

Helix 3: SX_3LX_2IX_2D(E,H)**R**Y

Helix 4: **W**$X_{8,9}$P

Helix 5: FX_2**P**X7Y

Helix 6: FX_2CW(Y,F)X**P**

Helix 7/Helix 8: LX_3NX_3N(D)**P**X_2Y$X_{5,6}$F

Fig. 5. Motifs relatively common in each of the seven membrane spanning helices and the amphipathic helix 8 of GPCRs. The most conserved residues of each helix, as defined by Ballesteros and Weinstein (see Note 5), are in *bold* and *underlined*; X_n indicates *n* contiguous nonconserved residues; residues in parentheses often replace the preceding residue.

percentages of accepted mutations (PAMs) and/or the presence of specific sequence motifs. Of note is an article published by Worth and coworkers that outlined a detailed integrated workflow for the identification of suitable templates for each of the seven membrane spanning helices and the amphipathic helix 8, based on a thorough structural analysis of the crystallized GPCRs (15). In particular, according to this scheme, the selection criteria should be based not only on sequence similarities but also on the detection of specific features and motifs detected in the sequence of the query receptor, such as the presence of specific glycine and proline residues responsible for helical kinks, or cysteine residues putatively involved in the formation of disulfide bridges (regarding the modeling of helix 7 and helix 8, see Note 7). For advice on how to construct a homology model on the basis of multiple templates, see Note 8.

3.4. The Extracellular and Intracellular Regions: To Align or Not to Align, That is the Question

The extracellular and intracellular domains of class A GPCRs are characterized by very low sequence similarity and great length variability, which make their sequences less straightforward to align than the seven membrane spanning helices. As outlined by the published crystal structures (5, 6), the lack of sequence of similarity detected for these regions is paralleled by a correspondent significant structural diversity, which hampers their modeling by homology. Moreover, further hindering homology modeling, termini and long loops have not been solved for many of the currently crystallized receptors, while in some of the crystal structures IL3 is substituted by a fused T4-lysozyme (5). Thus, not surprisingly, molecular models of class A GPCRs are usually significantly more accurate in the helical bundle than in the extracellular and intracellular regions, if we exclude short interconnecting loops (18). Notably, besides the purely computational methods discussed in this chapter, hybrid experimental and computational approaches have also been proposed, whereby the structures of peptides mimicking the extracellular and intracellular regions of a receptor are determined experimentally, for instance through NMR spectroscopy, and subsequently merged with an in silico generated model of the helical bundle (22). Such hybrid models may offer a very powerful approach to the study of receptors that have not yet been crystallized.

3.4.1. Avoiding the Alignment: De Novo Modeling or Omission of the Loop

A viable solution for the construction of short interconnecting loops can be found in de novo modeling, an approach not based on the use of a template. If this is the chosen route, the corresponding domain can be deleted from the structure of the template. Of note, if cysteine residues are present in the loop of the query receptor, special care deserves the analysis of their possible involvement in the formation of disulfide bridges on the basis of sequence analyses and experimental data (see Subheading 3.5).

In some GPCRs, however, the considerable length of termini and some of the loops—notably IL3—prevents an effective use of de novo modeling for their construction. It is advisable not to model the terminal regions, constructing only the portion of the receptor between the beginning of helix 1 and the end of helix 7 or helix 8, when this thought to be present. Similarly, it is advisable not to model long loops. The omission of a domain from the model can be achieved by deleting the corresponding sequence in the query receptor (for the loops, see Note 9).

3.4.2. Aligning the Loops

Despite the caveats expressed in the previous two subsections, homology modeling can be applied to the construction of interconnecting loops with a length comparable to that of the corresponding regions of the template. In this case, a sequence alignment and the selection of a template are necessary.

Due to the mentioned low sequence similarity and length variability, the alignment of the loops is better performed in a pairwise manner comparing the query receptor to one template at the time, rather than in a multiple sequence alignment context. If a loop has not exactly the same length in the template and the query receptor, a gap will have to be inserted in the sequence of the shorter one. As always in homology modeling, special care needs to be put into the positioning of such gaps, which should be driven not only by the attempt to maximize the similarity score but also by a careful structural analysis of the template. Specifically, it is important to ensure that insertions or deletions are placed in a position compatible with the structure of the template.

If a single template strategy is chosen, it will be sufficient to align the loops of the query receptor to the corresponding loops of the template receptor chosen on the basis of the sequence similarity detected in the helical bundle. Instead, if a multiple strategy template has been chosen, once a loop of the query receptor has been separately aligned with the corresponding loop of each of the candidate templates, the template for the construction of the model can be selected according to sequence similarity or on the basis of the conservation of specific amino acids. Additionally, it is important to carefully analyze the geometric compatibility between the candidate template for the modeling of the loop and the templates chosen for the modeling of the two helices that the loop connects.

3.4.3. Special Considerations Concerning the Second Extracellular Loop

EL2 connects helix 4 and helix 5 and, in the majority of class A GPCRs, is characterized by a highly conserved cysteine residue that connects it to helix 3. Modeling EL2 deserves particular attention since this loop, and in particular the portion downstream of the conserved disulfide bridged cysteine residue, is directly involved in the lining of the interhelical cavity that putatively hosts the orthosteric binding site for all members of class A GPCRs that are activated by small molecules. The crystal structures of class A

GPCRs that have been solved at the time of this writing revealed that EL2 does not feature a common structure shared by all receptors (5, 6, 10) and adopts four different conformations in rhodopsin, β adrenergic, adenosine A_{2A}, dopamine D3, and CXCR4 chemokine receptors. Specifically, in rhodopsin EL2 is characterized by a distinctive β-hairpin conformation that lays over the opening of the interhelical cavity restricting the access of water from the extracellular side, while in the β adrenergic, adenosine A_{2A}, dopamine D3, and CXCR4 chemokine receptors it assumes a significantly more open conformation. These differences are probably attributable to the fact that, while rhodopsin features a covalently bound inverse agonist, 11-*cis*-retinal, that is isomerized in situ to its all-*trans* form by the action of a light photon and consequently triggers the activation of the receptor, the remainder of class A GPCRs are physiologically activated by diffusible agonists (4) (see Note 10).

Despite this common feature that distinguishes receptors for diffusible ligands from rhodopsin, however, a profound structural variability for EL2 has been detected among the various experimentally solved receptors, also due to the different arrays of disulfide bridges detected in their extracellular regions (5). This lack of structural conservation prevents the use of homology modeling for the construction of EL2, unless template and query receptors belong to the same subfamily, and suggests that better results could be achieved through de novo modeling, enforcing the formation of the disulfide bridges that putatively exist in the query receptor (see Subheading 3.5). Accordingly, through a comparison of different rhodopsin-based models of the $β_2$ adrenergic receptor, I have demonstrated that those that featured a de novo-modeled EL2 resulted in lower root mean square deviations in the regions downstream of the disulfide bridge (17). In turn, this yielded the production of significantly more accurate ligand poses as a result of molecular docking (17), as well as better performances when the models were used as platforms for controlled docking-based virtual screening (23).

Alternatively to complete de novo modeling, a short portion around the conserved cysteine residue may be built by homology with one of the templates, while building the remainder of the loop de novo. Notably, I have used this strategy for the construction of C-terminal portion of EL2 in the adenosine A_{2A} receptor model for the above-mentioned "community-wide assessment of GPCR structure modeling and ligand docking"—see supplementary information of ref. 18 for the sequence alignment.

If the models are constructed with the intent of studying the interactions of the receptors with small molecules that bind to their interhelical cavity or conducting docking-based virtual screening experiments targeting said cavity, the segment of EL2 that really matters is the one that is downstream of the above-mentioned conserved disulfide bridge that links the loop to helix 3. The remainder of the loop, if too long to allow robust de novo modeling, may be omitted (see Note 9).

3.5. Construction of the Model

Once a sequence alignment has been obtained and the proper portions of query and/or template sequences have been deleted as outlined in the previous sections, a 3D model of the query receptor can be constructed through homology modeling or a combination of homology and de novo modeling—most modeling packages will directly build de novo those domains of the query receptor that are not aligned with a template.

3.5.1. Verifying Rotameric States

Due to the availability of multiple templates, after the construction of a model, the rotameric state of each residue can be verified and adjusted in light of the whole set of crystallized receptors. Notably, if a residue of the query receptor is not conserved in the template employed to model the domain to which it belongs, nonetheless it may be conserved in one or more of the other crystallized receptors. As the structures of additional GPCRs will be solved, the number of residues of a query receptor that will be conserved in at least one of the templates will increase significantly, with obvious beneficial repercussions on homology modeling (16).

3.5.2. Special Considerations on the Extracellular Disulfide Bridges

As mentioned, the extracellular domains of most class A GPCRs are characterized by the presence cysteine residues involved in the formation of disulfide bridges. Among these, the disulfide bridge that connects EL2 to helix 3 is widely conserved within class A, while additional bridges, when present, are often peculiar to a specific subfamily of receptors, to which they confer a characteristic extracellular architecture functional to ligand binding. As mentioned, it is of utmost importance that the presence of cysteine residues and their putative involvement in the formation of disulfide bridges be identified prior to the construction of the model. In addition to computer-based sequence analyses, the detection and the corroboration of the presence of such bridges can be greatly assisted by biochemical data, either ad hoc generated or retrieved from the literature. For instance, mutagenesis data suggested the presence of a disulfide bridge connecting EL3 to the N terminus of the P2Y receptors (24, 25), while they accurately predicted the presence of a disulfide bridge internal to EL2 of the β_2 adrenergic receptor (26), successively confirmed by the crystal structures (27, 28). Some software for homology modeling allows the enforcement of the formation of disulfide bridges between specified pairs of cysteine residues. This feature is particularly important when the cysteine residues are not conserved in the templates or whenever using de novo loop modeling. However, if this feature is not available within the chosen software, one possible solution is the construction of many alternative loop models and the subsequent selection of those that feature the cysteine pair at a distance compatible with the formation a disulfide bridge, if present. Alternatively, the disulfide bridges can be generated after the construction of the model, for instance through molecular dynamics simulations with a harmonic restraint applied to the distance between the sulfur atoms of the bridged cysteine pairs. After the proper connection of the putative disulfide bridges, a thorough exploration of the conformations accessible to extracellular and intracellular loops, possibly in light of experimental data, is also advisable. Of note, for the extracellular loops, sometimes this operation could be better performed following the docking of a ligand (for instance, see ref. 29).

3.5.3. Special Considerations on the Interhelical Cavity

In general, when the ligand co-crystallized with the template binds also to the query protein, the use of the co-crystallized ligand as environment for the construction of the model significantly helps the modeling of the binding pocket and facilitates the formation of protein–ligand interactions. However, when modeling class A GPCRs, given the wide diversity found within the class and the specificity of each subfamily for a particular set of natural and synthetic ligands, in very rare cases the query receptor will share ligands with any of the available templates. Nonetheless, using the ligand co-crystallized with one of the templates as environment may still be a good practice to grant to the model a binding pocket suitable for molecular docking. Often, in fact, homology modeling procedures tend to occlude internal cavities through subtle backbone movements, especially if the construction of the model involves unrestrained energy minimizations, and through the orientation of the side chains of the residues that line the cavity towards the center of it. However, building the model of a class A GPCR around the ligand co-crystallized with one of the templates can induce artificial rotameric states to some of the residues that line the binding pocket. For example, I have shown that, when building the β_2 adrenergic receptor using rhodopsin as the template and the co-crystallized retinal as the environment (17), Phe290 is prevented from adopting its natural the gauche (+) conformation by the presence of retinal (see Fig. 6). Thus, after the construction of the model a thorough exploration of the rotameric states of the residues that line the binding cavity is needed. This operation can be conveniently performed after the generation of preliminary docking poses of a chosen ligand, possibly guided by experimental constraints, through a variety of differently implemented procedures dubbed "ligand-supported," "ligand-based," or "ligand-steered" or homology modeling (13, 30, 31).

3.6. Validation of the Models Through Virtual Screening Experiments

The ultimate validation of a GPCR homology model can only derive from a direct comparison with its experimentally elucidated structure. However, such a comparison is only possible either when the model of a crystallized receptor is generated so as to probe scope and limitations of the modeling techniques, or, retroactively,

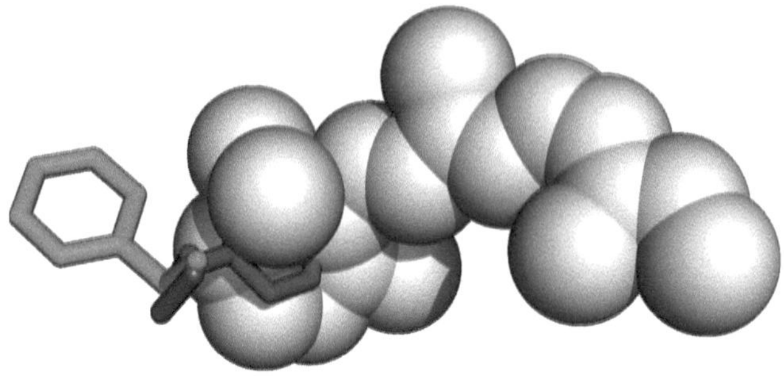

Fig. 6. As indicated by the structural superimposition shown here, Phe290 cannot adopt the right rotameric state in a rhodopsin-based model of the β_2 adrenergic receptor constructed using retinal as the environment: retinal (in *light gray*, from 1GZM) would sterically prevent Phe290 from adopting the gauche(+) conformation revealed by the crystal structure (in *red*, from 2RH1) and would force it in the trans conformation (in *green*, from a rhodopsin-based homology model (17)). Of note, in rhodopsin, the residue corresponding to Phe290 is an alanine, namely Ala269 (in *dark gray*, from 1GZM). The figure appears in color in the online edition.

when the experimental structure of a previously modeled receptor becomes available, possibly many years after the model was generated. In fact, if a computational model of a receptor is generated to shed light into its structure–function relationships and, possibly, to facilitate the discovery of ligands capable of modulating its activity, this very fact implies that experimental structures do not exist for the query receptor. Thus, for all intents and purposes, the only possible way to validate the usefulness of a homology model—if not necessarily its accuracy—is to test the correlation between predictions generated on its basis and experimental results. In particular, if homology models have been built with the purpose of studying receptor–ligand interactions and conducting structure-based drug discovery, the best way to validate their efficacy is to subject them to a series of controlled virtual screening experiments. These are usually performed docking at the receptor a dataset of compounds containing a number of known ligands mixed with a larger number of decoys, i.e., compounds with physicochemical characteristics similar to those of the ligands but presumed to be inactive. Then, the ability of the screening to prioritize ligands over decoys is evaluated by monitoring enrichment factors and/or areas under the receiver operating characteristic (ROC) curve (23, 29, 31, 32). Such controlled experiments constitute very good tools not only for the selection of the initial models but also for the control of the entire optimization process, including the refinement of loops and side chains. Clearly, controlled virtual screening can only be performed if a significant amount of known ligands for the query receptor exists (see Note 11), while can be applied with difficulty to receptors characterized by a marked paucity of known ligands and not applied at all to orphan receptors. Moreover, it is worth keeping in mind that better virtual screening performances do not necessarily parallel higher levels of overall accuracy and may reflect a particularly favorable arrangement, either natural or artificial, of the side chains of the residues that line the binding pocket (16, 17, 29).

4. Notes

1. Text editors can be conveniently used to read and edit PDB files. Alternatively, the files can be directly edited within the specialized modeling package of choice.
2. For a description of the PDB file format, see http://www.pdb.org/docs.html.
3. It is not always safe to blindly opt for the first chain (usually named chain A) and discard the others. The B-factors of the various chains and their completeness are certainly important parameters on which to base the selection. Moreover, to choose the best chain to work with, a careful reading of the main article that describes the crystal is of utmost importance. For example, in the case of the β_1 adrenergic receptor (PDB ID: 2VT4) chain B is to be preferred to chain A, since, as explained by the authors, the latter presents an anomalous 60° kink in helix 1 (33).
4. For a correct interpretation of the secondary structure, some programs require also the portion of the PDB file that defines it (record type: HELIX and SHEET).
5. The GPCR residue identifier system, devised by Ballesteros and Weinstein, is a universal way of numbering GPCR residues on the basis of reference positions that the authors identified for each of the seven membrane spanning helices (20). Specifically, through the analysis of a sequence alignment of Class A receptors, the authors selected a reference position for each of the seven helices, chosen among those featuring one of the most conserved residues in that helix. They then defined a convention by which the identifier X.50—where X is the helix number—is arbitrarily assigned to the reference position, while the remaining residues in the helix are numbered relatively to the reference. Later, van Rhee and Jacobson introduced a modification to the Ballesteros and Weinstein system according to which each residue is indicated with its original sequence number followed by the residue identifier, rather than solely with the residue identifier (21).
6. Although insertion and deletions within the seven helices are not common, structure-based alignments indicate the presence of an insertion in helix 2 of squid rhodopsin (see Fig. 4) (15, 34).

Moreover, the C-terminal region of helix 7, close to the hinge with helix 8, presents a deletion in some receptors, leaving only five rather than six residues between the Tyr and the Phe of the conserved $NPX_2YX_{5,6}F$ motif (see Fig. 4) (35).

7. The presence of either five or six intervening residues between the conserved tyrosine and phenylalanine at the hinge between helix 7 and helix 8 (see Note 6) may guide the selection of the template for this region (15). Importantly, if sequence analysis does not strongly support the presence of an amphipathic helix, the sequence of the query receptor can be truncated at the end of helix 7, leaving the remainder of the receptor unmodeled.
8. While some homology modeling software allows the direct use of multiple templates, others require the use of a single template. A possible workaround to overcome this limitation is the generation of a hybrid template by cutting and pasting the selected portions of the various crystallized receptors into a single PDB file (on the editing of a PDB file, see also Note 1).
9. Some homology modeling software requires that the query be an uninterrupted protein chain. In this case, the loop (or a portion of it) can conveniently be deleted after the construction of the model. If the loop destined to be omitted from the model is particularly long, to avoid the expenditure of excessive computational time in its construction, it may be advisable to delete its central portion from the query sequence, thus constructing only a relatively short loop that will be subsequently removed.
10. As suggested by molecular modeling studies, the egression of the cleaved all-*trans*-retinal consequent the activation of rhodopsin and the following ingression of 11-*cis*-retinal into the unliganded opsin, to reform a functional rhodopsin unit, occur through openings between adjacent membrane spanning helices (36, 37). Instead, the physiological ligands of the β adrenergic receptors, as well as those of all class A GPCRs naturally activated by small molecules, are very likely to enter and exit the receptor through the opening of the interhelical cavity towards the extracellular milieu (38).
11. Known ligands of the query receptor can conveniently be retrieved from the GPCR–ligand database (GLIDA, http://pharminfo.pharm.kyoto-u.ac.jp/services/glida/) (39).

Acknowledgments

This work was supported by the intramural research program of the National Institute of Diabetes and Digestive and Kidney Diseases of the National Institutes of Health.

References

1. Pierce, K., Premont, R., and Lefkowitz, R. (2002) Seven-transmembrane receptors *Nat. Rev. Mol. Cell Biol.* **3,** 639–50.
2. Gloriam, D., Fredriksson, R., and Schiöth, H. (2007) The G protein-coupled receptor subset of the rat genome. *BMC Genomics* **8,** 338.
3. Overington, J. P., Al-Lazikani, B., and Hopkins, A. L. (2006) How many drug targets are there? *Nat. Rev. Drug Discov.* **5,** 993–6.
4. Costanzi, S., Siegel, J., Tikhonova, I., and Jacobson, K. (2009) Rhodopsin and the others: a historical perspective on structural studies of G protein-coupled receptors *Curr. Pharm. Des.* **15,** 3994–4002.
5. Hanson, M. A., and Stevens, R. C. (2009) Discovery of new GPCR biology: one receptor structure at a time *Structure* **17,** 8–14.
6. Wu, B., Chien, E. Y., Mol, C. D., Fenalti, G., Liu, W., Katritch, V., Abagyan, R., Brooun, A., Wells, P., Bi, F. C., Hamel, D. J., Kuhn, P., Handel, T. M., Cherezov, V., and Stevens, R. C. (2010) Structures of the CXCR4 Chemokine GPCR with Small-Molecule and Cyclic Peptide Antagonists *Science.*
7. Rasmussen, S. G., Choi, H. J., Fung, J. J., Pardon, E., Casarosa, P., Chae, P. S., Devree, B. T., Rosenbaum, D. M., Thian, F. S., Kobilka, T. S., Schnapp, A., Konetzki, I., Sunahara, R. K., Gellman, S. H., Pautsch, A., Steyaert, J., Weis, W. I., and Kobilka, B. K. (2011) Structure of a nanobody-stabilized active state of the beta(2) adrenoceptor *Nature* **469,** 175–80.
8. Rosenbaum, D. M., Zhang, C., Lyons, J. A., Holl, R., Aragao, D., Arlow, D. H., Rasmussen, S. G., Choi, H. J., Devree, B. T., Sunahara, R. K., Chae, P. S., Gellman, S. H., Dror, R. O., Shaw, D. E., Weis, W. I., Caffrey, M., Gmeiner, P., and Kobilka, B. K. (2011) Structure and function of an irreversible agonist-beta(2) adrenoceptor complex *Nature* **469,** 236–40.
9. Warne, T., Moukhametzianov, R., Baker, J. G., Nehme, R., Edwards, P. C., Leslie, A. G., Schertler, G. F., and Tate, C. G. (2011) The structural basis for agonist and partial agonist action on a beta(1)-adrenergic receptor *Nature* **469,** 241–4.
10. Chien, E. Y., Liu, W., Zhao, Q., Katritch, V., Han, G. W., Hanson, M. A., Shi, L., Newman, A. H., Javitch, J. A., Cherezov, V., and Stevens, R. C. (2010) Structure of the human dopamine D3 receptor in complex with a D2/D3 selective antagonist *Science* **330,** 1091–5.
11. Costanzi, S. (2010) Modelling G protein-coupled receptors: a concrete possibility *Chimica Oggi-Chemistry Today* **28,** 26–30.
12. Bissantz, C., Bernard, P., Hibert, M., and Rognan, D. (2003) Protein-based virtual screening of chemical databases. II. Are homology models of G-Protein Coupled Receptors suitable targets? *Proteins* **50,** 5–25.
13. Moro, S., Deflorian, F., Bacilieri, M., and Spalluto, G. (2006) Ligand-based homology modeling as attractive tool to inspect GPCR structural plasticity *Curr. Pharm. Des.* **12,** 2175–85.
14. Jacobson, K., Gao, Z., and Liang, B. (2007) Neoceptors: reengineering GPCRs to recognize tailored ligands. *Trends Pharmacol. Sci.* **28,** 111–6.
15. Worth, C., Kleinau, G., and Krause, G. (2009) Comparative sequence and structural analyses of G-protein-coupled receptor crystal structures and implications for molecular models. *PLoS One* **4,** e7011.
16. Mobarec, J., Sanchez, R., and Filizola, M. (2009) Modern Homology Modeling of G-Protein Coupled Receptors: Which Structural Template to Use? *J. Med. Chem.* **52,** 5207–16.
17. Costanzi, S. (2008) On the applicability of GPCR homology models to computer-aided drug discovery: a comparison between in silico and crystal structures of the beta2-adrenergic receptor *J. Med. Chem.* **51,** 2907–14.
18. Michino, M., Abola, E., 2008 Participants, G., Brooks, C. r., Dixon, J., Moult, J., and Stevens, R. (2009) Community-wide assessment of GPCR structure modelling and ligand docking: GPCR Dock 2008 *Nat. Rev. Drug. Discov.* **8,** 455–63.
19. van Rhee, A. M., Fischer, B., van Galen, P. J., and Jacobson, K. A. (1995) Modelling the P2Y purinoceptor using rhodopsin as template *Drug Des. Discov.* **13,** 133–54.
20. Ballesteros, J. A., and Weinstein, H. (1995) Integrated method for the consturction of three dimensional models and computational probing of structure-function relations in G-protein coupled receptors. *Methods Neurosci* **25,** 366–428.
21. van Rhee, A. M., and Jacobson, K. A. (1996) Molecular architecture of G protein-coupled receptors *Drug Develop. Res.* **37,** 1–38.
22. Tikhonova, I., and Costanzi, S. (2009) Unraveling the structure and function of G protein-coupled receptors through NMR spectroscopy. *Curr. Pharm. Des.* **15,** 4003–16.
23. Vilar, S., Ferino, G., Phatak, S. S., Berk, B., Cavasotto, C. N., and Costanzi, S. (2010) Docking-based virtual screening for ligands of G protein-coupled receptors: Not only crystal structures but also in silico models *J. Mol. Graph. Model.,* doi: 10.1016/j.jmgm.2010.11.005.

24. Hoffmann, C., Moro, S., Nicholas, R. A., Harden, T. K., and Jacobson, K. A. (1999) The role of amino acids in extracellular loops of the human P2Y1 receptor in surface expression and activation processes *J. Biol. Chem.* **274,** 14639–47.
25. Costanzi, S., Mamedova, L., Gao, Z., and Jacobson, K. (2004) Architecture of P2Y nucleotide receptors: structural comparison based on sequence analysis, mutagenesis, and homology modeling. *J. Med. Chem.* **47,** 5393–404.
26. Noda, K., Saad, Y., Graham, R. M., and Karnik, S. S. (1994) The high affinity state of the beta 2-adrenergic receptor requires unique interaction between conserved and non-conserved extracellular loop cysteines *J. Biol. Chem.* **269,** 6743–52.
27. Cherezov, V., Rosenbaum, D., Hanson, M., Rasmussen, S., Thian, F., Kobilka, T., Choi, H., Kuhn, P., Weis, W., Kobilka, B., and Stevens, R. (2007) High-resolution crystal structure of an engineered human beta2-adrenergic G protein-coupled receptor *Science* **318,** 1258–65.
28. Rosenbaum, D., Cherezov, V., Hanson, M., Rasmussen, S., Thian, F., Kobilka, T., Choi, H., Yao, X., Weis, W., Stevens, R., and Kobilka, B. (2007) GPCR engineering yields high-resolution structural insights into beta2-adrenergic receptor function *Science* **318,** 1266–73.
29. Katritch, V., Jaakola, V., Lane, J., Lin, J., Ijzerman, A., Yeager, M., Kufareva, I., Stevens, R., and Abagyan, R. (2010) Structure-based discovery of novel chemotypes for adenosine A(2A) receptor antagonists *J. Med. Chem.* **53,** 1799–809.
30. Evers, A., and Klebe, G. (2004) Ligand-supported homology modeling of g-protein-coupled receptor sites: models sufficient for successful virtual screening *Angew. Chem. Int. Ed. Engl.* **43,** 248–51.
31. Cavasotto, C. N., Orry, A. J., Murgolo, N. J., Czarniecki, M. F., Kocsi, S. A., Hawes, B. E., O'Neill, K. A., Hine, H., Burton, M. S., Voigt, J. H., Abagyan, R. A., Bayne, M. L., and Monsma, F. J., Jr. (2008) Discovery of novel chemotypes to a G-protein-coupled receptor through ligand-steered homology modeling and structure-based virtual screening *J. Med. Chem.* **51,** 581–8.
32. Vilar, S., Karpiak, J., and Costanzi, S. (2010) Ligand and structure-based models for the prediction of ligand-receptor affinities and virtual screenings: Development and application to the beta(2)-adrenergic receptor *J. Comput. Chem.* **31,** 707–20.
33. Warne, T., Serrano-Vega, M., Baker, J., Moukhametzianov, R., Edwards, P., Henderson, R., Leslie, A., Tate, C., and Schertler, G. (2008) Structure of a beta1-adrenergic G-protein-coupled receptor. *Nature* **454,** 486–91.
34. Shimamura, T., Hiraki, K., Takahashi, N., Hori, T., Ago, H., Masuda, K., Takio, K., Ishiguro, M., and Miyano, M. (2008) Crystal structure of squid rhodopsin with intracellularly extended cytoplasmic region *J. Biol. Chem.* **283,** 17753–6.
35. Fritze, O., Filipek, S., Kuksa, V., Palczewski, K., Hofmann, K. P., and Ernst, O. P. (2003) Role of the conserved NPxxY(x)5,6F motif in the rhodopsin ground state and during activation *Proc. Natl. Acad. Sci. U. S. A.* **100,** 2290–5.
36. Wang, T., and Duan, Y. (2007) Chromophore channeling in the G-protein coupled receptor rhodopsin *J. Am. Chem. Soc.* **129,** 6970–1.
37. Hildebrand, P. W., Scheerer, P., Park, J. H., Choe, H. W., Piechnick, R., Ernst, O. P., Hofmann, K. P., and Heck, M. (2009) A ligand channel through the G protein coupled receptor opsin *PLoS One* **4,** e4382.
38. Wang, T., and Duan, Y. (2009) Ligand entry and exit pathways in the beta2-adrenergic receptor *J. Mol. Biol.* **392,** 1102–15.
39. Okuno, Y., Tamon, A., Yabuuchi, H., Niijima, S., Minowa, Y., Tonomura, K., Kunimoto, R., and Feng, C. (2008) GLIDA: GPCR--ligand database for chemical genomics drug discovery--database and tools update. *Nucleic Acids Res.* **36,** D907–12.
40. Palczewski, K., Kumasaka, T., Hori, T., Behnke, C. A., Motoshima, H., Fox, B. A., Le Trong, I., Teller, D. C., Okada, T., Stenkamp, R. E., Yamamoto, M., and Miyano, M. (2000) Crystal structure of rhodopsin: A G protein-coupled receptor *Science* **289,** 739–45.
41. Li, J., Edwards, P. C., Burghammer, M., Villa, C., and Schertler, G. F. (2004) Structure of bovine rhodopsin in a trigonal crystal form *J. Mol. Biol.* **343,** 1409–38.
42. Teller, D. C., Okada, T., Behnke, C. A., Palczewski, K., and Stenkamp, R. E. (2001) Advances in determination of a high-resolution three-dimensional structure of rhodopsin, a model of G-protein-coupled receptors (GPCRs) *Biochemistry* **40,** 7761–72.
43. Okada, T., Fujiyoshi, Y., Silow, M., Navarro, J., Landau, E. M., and Shichida, Y. (2002) Functional role of internal water molecules in rhodopsin revealed by X-ray crystallography *Proc. Natl. Acad. Sci. U. S. A.* **99,** 5982–7.
44. Okada, T., Sugihara, M., Bondar, A. N., Elstner, M., Entel, P., and Buss, V. (2004) The retinal conformation and its environment in rhodopsin in light of a new 2.2 A crystal structure *J. Mol. Biol.* **342,** 571–83.

45. Salom, D., Lodowski, D., Stenkamp, R., Le Trong, I., Golczak, M., Jastrzebska, B., Harris, T., Ballesteros, J., and Palczewski, K. (2006) Crystal structure of a photoactivated deprotonated intermediate of rhodopsin. *Proc. Natl. Acad. Sci. U. S. A.* **103,** 16123–8.
46. Standfuss, J., Xie, G., Edwards, P. C., Burghammer, M., Oprian, D. D., and Schertler, G. F. (2007) Crystal structure of a thermally stable rhodopsin mutant *J. Mol. Biol.* **372,** 1179–88.
47. Stenkamp, R. E. (2008) Alternative models for two crystal structures of bovine rhodopsin *Acta Crystallogr. D Biol. Crystallogr.* **D64,** 902–4.
48. Nakamichi, H., and Okada, T. (2006) Crystallographic analysis of primary visual photochemistry *Angew. Chem. Int. Ed. Engl.* **45,** 4270–3.
49. Nakamichi, H., and Okada, T. (2006) Local peptide movement in the photoreaction intermediate of rhodopsin *Proc. Natl. Acad. Sci. U. S. A.* **103,** 12729–34.
50. Nakamichi, H., Buss, V., and Okada, T. (2007) Photoisomerization mechanism of rhodopsin and 9-cis-rhodopsin revealed by x-ray crystallography *Biophys. J.* **92,** L106–8.
51. Murakami, M., and Kouyama, T. (2008) Crystal structure of squid rhodopsin. *Nature* **453,** 363–7.
52. Park, J. H., Scheerer, P., Hofmann, K. P., Choe, H. W., and Ernst, O. P. (2008) Crystal structure of the ligand-free G-protein-coupled receptor opsin *Nature* **454,** 183–7.
53. Scheerer, P., Park, J. H., Hildebrand, P. W., Kim, Y. J., Krauss, N., Choe, H. W., Hofmann, K. P., and Ernst, O. P. (2008) Crystal structure of opsin in its G-protein-interacting conformation *Nature* **455,** 497–502.
54. Rasmussen, S., Choi, H., Rosenbaum, D., Kobilka, T., Thian, F., Edwards, P., Burghammer, M., Ratnala, V., Sanishvili, R., Fischetti, R., Schertler, G., Weis, W., and Kobilka, B. (2007) Crystal structure of the human beta2 adrenergic G-protein-coupled receptor. *Nature* **450,** 383–7.
55. Hanson, M., Cherezov, V., Griffith, M., Roth, C., Jaakola, V., Chien, E., Velasquez, J., Kuhn, P., and Stevens, R. (2008) A specific cholesterol binding site is established by the 2.8 A structure of the human beta2-adrenergic receptor. *Structure* **16,** 897–905.
56. Bokoch, M., Zou, Y., Rasmussen, S., Liu, C., Nygaard, R., Rosenbaum, D., Fung, J., Choi, H., Thian, F., Kobilka, T., Puglisi, J., Weis, W., Pardo, L., Prosser, R., Mueller, L., and Kobilka, B. (2010) Ligand-specific regulation of the extracellular surface of a G-protein-coupled receptor. *Nature* **463,** 108–12.
57. Wacker, D., Fenalti, G., Brown, M. A., Katritch, V., Abagyan, R., Cherezov, V., and Stevens, R. C. (2010) Conserved binding mode of human beta2 adrenergic receptor inverse agonists and antagonist revealed by X-ray crystallography *J. Am. Chem. Soc.* **132,** 11443–5.
58. Jaakola, V., Griffith, M., Hanson, M., Cherezov, V., Chien, E., Lane, J., Ijzerman, A., and Stevens, R. (2008) The 2.6 angstrom crystal structure of a human A2A adenosine receptor bound to an antagonist. *Science* **322,** 1211–7.

Chapter 12

Homology Modeling of Transporter Proteins (Carriers and Ion Channels)

Aina Westrheim Ravna and Ingebrigt Sylte

Abstract

Transporter proteins are divided into channels and carriers and constitute families of membrane proteins of physiological and pharmacological importance. These proteins are targeted by several currently prescribed drugs, and they have a large potential as targets for new drug development. Ion channels and carriers are difficult to express and purify in amounts for X-ray crystallography and nuclear magnetic resonance (NMR) studies, and few carrier and ion channel structures are deposited in the PDB database. The scarcity of atomic resolution 3D structures of carriers and channels is a problem for understanding their molecular mechanisms of action and for designing new compounds with therapeutic potentials. The homology modeling approach is a valuable approach for obtaining structural information about carriers and ion channels when no crystal structure of the protein of interest is available. In this chapter, computational approaches for constructing homology models of carriers and transporters are reviewed.

Key words: Carriers, Ion channels, Drug targets, Homology modeling, Amino acid sequence alignments, Model building and refinements, Model evaluation, ABC transporters, Neurotransmitter transporters

1. Introduction

Membrane proteins are involved in a variety of processes governing cellular functions, and a large partition of presently known drug targets are membrane proteins. Membrane transporter proteins (ion channels and carriers) comprise major functional classes of membrane proteins (1). These proteins are involved in establishing and controlling the voltage gradient across cellular membranes, in transport of nutrients and signal molecules across the cell membrane, and in mediating active excursion of drugs and endotoxins. Their role as major determinants of the pharmacokinetic, safety, and efficacy profiles of drugs has formed the basis for the recommendations of the International Transporter Consortium (2),

Andrew J.W. Orry and Ruben Abagyan (eds.), *Homology Modeling: Methods and Protocols*, Methods in Molecular Biology, vol. 857, DOI 10.1007/978-1-61779-588-6_12, © Springer Science+Business Media, LLC 2012

which elucidates transporter role for drug development, for instance which transporters are clinically important in drug absorption and disposition.

The transporter classification system approved by the transporter nomenclature panel of the International Union of Biochemistry and Molecular Biology (3) states that transporters are either channels or carriers. There are six categories in the transporter classification system: (1) Channels and pores; (2) Electrochemical potential-driven transporters (secondary and tertiary transporters); (3) Primary active transporters; (4) Group translocators; (5) Accessory factors involved in transport; and (6) Incompletely characterized transport proteins. Channels belong to category 1, while categories 2, 3, and 4 are carriers.

Ion channels may be classified by gating, i.e., what opens and closes the channels. The two main types of ion channels are voltage-gated ion channels and ligand-gated ion channels. Ligand-gated ion channels open or close depending on ligand binding and are therefore often classified as receptors, not transporters (4). Voltage-gated ion channels open or close depending on the voltage gradient across the cellular membrane and are involved in nerve impulses. The timescale of channel opening is in milliseconds.

In contrast to channels, carriers feature stereospecific substrate specificities, and their rates of transport are several orders of magnitude lower than those of channels (3). There are carriers for neurotransmitters, amino acids, organic anions, organic cations, vitamins, fatty acids, bicarbonate, peptides, nucleosides, sugars, bile acids, and phosphates.

1.1. Ion Channels and Carriers as Drug Targets

At present, several drugs on the market function by targeting ion channels or carrier proteins. Drugs may exert their effect by binding to carriers and either inhibit transport of the solute or function as a false substrate for the transport process. Examples of drugs that inhibit the transport process, leading to an increase in the concentration of neurotransmitter in the synaptic cleft, are the antidepressants selective serotonin reuptake inhibitors (SSRIs), which inhibit the serotonin transporter (SERT), and cocaine, which inhibit the dopamine transporter (DAT), noradrenaline transporter (NET), and SERT. Other well-known drugs inhibiting transport processes are diuretics like furosemide that inhibit the $Na^+/K^+/Cl^-$ co-transporter; reserpine, ephedrine, and amphetamines that inhibit vesicular monoamine transporters; and omeprazole that inhibits the proton pump (H^+/K^+-ATPase).

Examples of drugs that act as false substrates are chemotherapeutic and antibacterial agents that are transported out of cells by ATP-binding cassette (ABC) transporters including the ABCB1 transporter (P-glycoprotein). P-glycoprotein and other ABC transporters contribute to multidrug resistance by transporting a broad spectrum of structurally distinct drugs out of cells. Around 40% of human tumors develop resistance to chemotherapeutic drugs due to overexpression of ABC transporters (1).

Various clinically important drugs are inhibitors of voltage-gated or ligand-gated ion channels. Examples of drugs acting on ligand-gated ion channels are anxiolytic drugs (benzodiazepines) targeting the γ-aminobutyric acid $(GABA)_A$ receptors, and general anesthetics (e.g., ketamine and phencyclidine) and drugs used in Parkinson's disease (amantadine) and Alzheimer's disease (memantine) targeting ionotropic glutamate receptors. Several local anesthetic drugs (e.g., lidocaine), class 1-antiarrythmics, and antiepileptic drugs target different subtypes of voltage-gated sodium channels. An overview of drugs targeting carriers and ion channels is given by Landry and Gies (1).

1.2. Structural Information

Atomic resolution 3D structures of biologically active molecules provide information about the active site architecture, possible ligand-binding sites, evolutionary relationships between proteins and are also important for the understanding of the molecular mechanisms of protein function. The protein 3D structure may serve as a basis for designing protein engineering experiments exploring structure activity relationships of the protein. When detailed structural data for a target protein is available, computer programs can be used to predict protein–ligand affinities and to screen virtual compound/fragment libraries in the search for hits or leads in drug development. Atomic resolution 3D structures of drug targets also give the possibility of designing new compounds binding to the targets. At present around 65,000 entities of proteins or protein complexes are present in the PDB database (http://www.rcsb.org/pdb/home/home.do). Technical advances in crystallization and structural data collection, notably using synchrotron X-ray beamlines, improvements in membrane protein molecular biology and biochemistry, and the availability of several sequenced genomes, have contributed to progress in the number of transmembrane proteins determined at an atomic level (5–7). However, in spite of recent technical improvements having increased the number of known 3D structures of membrane proteins, including that of carriers and ion channels, only around 700 of the entities in the PDB database are membrane proteins (http://blanco.biomol.uci.edu/Membrane_Proteins_xtal). Of these, only about 260 represent unique membrane protein structures. Membrane proteins are estimated to constitute one-third of all proteins coded for in the human and other genomes, and thus there are estimated to be at least 10,000 membrane proteins encoded in the human genome (8, 9). The huge gap between the total number of membrane proteins and the number with known 3D structure reflects problems with expression in large amounts and in the crystallization of membrane proteins.

The majority of the membrane proteins with known 3D structure are from bacteria, and the lack of atomic resolution 3D structures of human membrane proteins is a problem for new drug discovery. The homology modeling approach is a method that may be used to generate 3D models of human membrane proteins, and thereby contributes with valuable structural information about membrane proteins with unknown 3D structure. The methodology for constructing homology models of carriers and ion channels are reviewed in this chapter.

2. Methods

In the homology modeling approach, a molecular model of a carrier or an ion channel of unknown structure ("Target") may be constructed based on a carrier or an ion channel with known 3D structure ("Template"). The template protein must have a sequence similarity (homology) to the target. Homology between two proteins is determined by sequence similarity, indicating that the two proteins have a common ancestor and similar features such as homologous protein folds.

Three main approaches are used for predicting the structure of proteins. One approach is ab initio (or de novo) methods, which predict the structure of a protein without using structural information from a close homologous protein. The prediction makes use of information from secondary structure prediction and of local sequence and structural relationships to short protein fragments (10). Another approach is threading, which can be used when template structures of distantly homologous proteins exist but are not easily recognized. Each amino acid in the target sequence is "threaded" to a position in the template structure, and thereafter, it is evaluated how well the target sequence fits the template (11). The third approach, homology modeling, is the approach that currently gives the most accurate and reliable structure predictions. The homology modeling approach was originally applied for constructing models of water-soluble proteins. However, the applied methods have been proved to be as applicable to membrane proteins as for water-soluble proteins (12) (see Note 1).

2.1. The Homology Modeling Procedure

The main steps in homology modeling of transporters are (Fig. 1) as follows:

- Find a suitable template
- Target–template alignment
- Model building
- Model validation

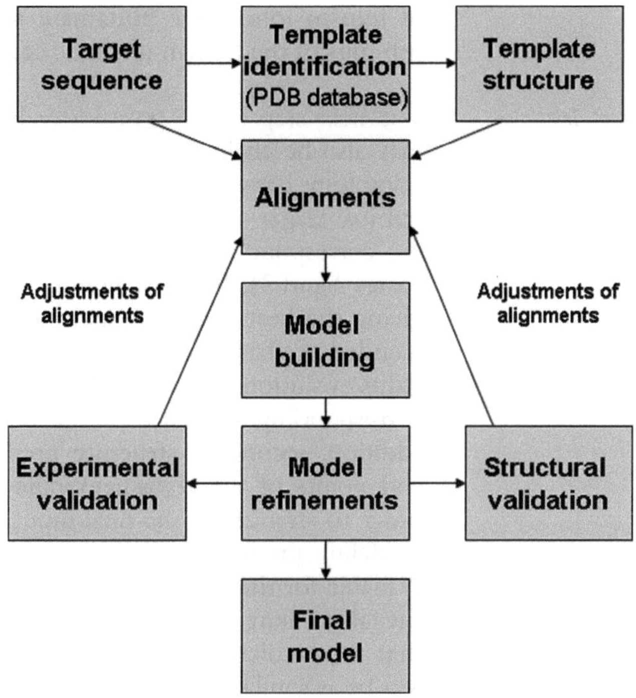

Fig. 1. Flow chart indicating the different steps in a homology modeling procedure of ion channels and carriers.

2.1.1. Template Identification

In order to construct a transporter model based on homology, the transporter structure of interest ("Target") must be matched with experimentally determined structures, the so-called template identification (see Note 2). In general, templates can be obtained by using the target sequence as a query for searching basic local alignment search tool (BLAST). Commonly used methods for template identification represent templates and targets as hidden Markow models (13), or as position-specific substitution profiles such as in PSI-BLAST (14). But since the current knowledge about detailed 3D structures of carriers and ion channels is limited, there may be only one template for your transporter of interest (if any), and consequently, the homology may be very low.

Examples of 3D crystal structures of carriers determined by X-ray crystallography at atomic resolution are the *Mus musculus* ABCB1 (15), the *Staphylococcus aureus* Sav1866 (16), the *Aquifex aeolicus* $\mathrm{LeuT_{Aa}}$ (17), and *Escherichia coli* Lac permease (18). A review concerning the available template structures for carrier modeling is given by Ravna et al. (19). There are also templates present in the PDB database (http://www.rcsb.org/pdb/home/home.do) that can be used to model therapeutically important voltage-gated ion channels (20), and domains of some of the therapeutically important ligand-gated ion channels, like the ligand-binding domain of human ionotropic glutamate receptor 5 (iGluR5) (21) and subunits of the human nicotinic acetylcholine receptor (22).

2.1.2. Target–Template Alignment

The next step in the transporter homology modeling procedure may also be challenging, due to the in many cases relatively low homology between the target transporter and the template. An optimal target–template alignment must be constructed, identifying corresponding positions in the target and the template (see Notes 2 and 3). The best alignment is considered as the alignment giving the best model. A multiple sequence alignment is recommended as a basis for the target–template alignment, since it highlights evolutionary relationships and increases the probability that corresponding sequence positions are correctly aligned (23). In addition, secondary structure predictions that predict start and end points of the transmembrane helices may be important in order to strengthen the final input alignments for the homology modeling procedure. If there are site-directed mutagenesis data available for the target protein, they should also be used to guide the alignment. A correct alignment increases the possibility that the predicted structure of the target, based on the template, will be as similar as possible to an experimental structure of target (see Note 3).

2.1.3. Model Building

In general, transporter model building involves construction of the core areas of the model, based on homology to the template, and construction of loops. The model building procedure may involve three main steps: (1) The core modeling, where transmembrane domains are modeled; (2) loop modeling, where intracellular and extracellular parts of the transporter are constructed de novo; and (3) optimization of side chains (and backbone). One example of core modeling is rigid body superposition (RBS), where the model is constructed from a few core sections defined by the average of Cα atoms in the conserved regions. Examples of homology modeling programs that use RBS are ICM (24) and WHAT IF (25). Other approaches for generating homology models are based on segment matching and modeling by the satisfaction of spatial restraints. The segment matching approach uses the target–template alignment to derive atomic positions which is used to detect matching segments in databases of known structures (26). Modeling by satisfaction of spatial restraints uses a set of restraints derived from the target–template alignment and then generates the model by minimizing the violations of these restrains, as implemented in MODELLER (27).

The lengths of extra- and intracellular loops may differ substantially between the target transporter and the template, introducing uncertainties into the transporter model. In general, existing modeling methods are not reliable for loops longer than 7 residues, and segments of up to 9 residues sometimes have entirely different conformations in different proteins (see Note 4). Consequently, the inclusion of loops in a model may depend on your "aim" with the model. There are several different approaches for loop generation; loop search methods, which can be manual or automatic, combined methods (secondary structure prediction and loop/fragment search), or Monte Carlo/MD methods. In the ICM program (24), loop modeling is part of the homology modeling procedure. Matching loops are searched for from several thousand high-quality pdbs, and the maps around the loops are calculated and scored, selecting the best fitting one.

2.1.4. Model Refinements

After model building, the carrier or ion channel model can be refined using energy minimizations, Monte Carlo simulations, or molecular dynamics calculations. The refinement is often performed as a stepwise process, where the most uncertain parts of the model are refined first. The refinement process depends on the quality of the model generated. If the homology modeling is based upon low homology between template and target, and the quality of the alignment is low, a refinement procedure may not necessarily improve the quality of the model (see Note 5). For molecular dynamics refinements, the transporter model may be embedded in a lipid bilayer to include membrane effects into the calculations.

2.1.5. Model Validation

Since modeling of carriers and ion channels has many elements of uncertainty, model validation is crucial. In the aspect of uncertainty, models should in general be considered as working tools for generating hypotheses and designing further experimental studies related to transporter structure, function, and ligand interactions. Transporter modeling is dependent on an iterating process contributed by experimental studies (e.g., site-directed mutagenesis studies) and molecular modeling, which together may lead toward a better understanding of transporters (Fig. 1). Docking of drug molecules into putative binding sites of carriers or ion channels may identify amino acids that will aid the selection of amino acids for further testing by site-directed mutagenesis studies (see Note 6). If the observations of drug-binding affinities made in the experiments are in accordance with the effects proposed by the modeling study, one may consider the model as partly correct. If not, an adjustment of the model must be performed. Experimental studies based on assumptions made from the models may thus be useful for further model refinements.

In addition to testing the model experimentally, the overall structure of the model should be analyzed for its stereochemical quality. Criteria included may be distribution of backbone ϕ and ψ angels (Ramachandran plots), side-chain packing, secondary structure packing, and side-chain geometry. An example of a structure analysis server is the Structural Analysis and Verification Server (http://nihserver.mbi.ucla.edu/SAVES/), which includes programs such as Procheck (28) and Whatcheck (29). It should be kept in mind that most structure validation programs are developed based on globular, water-soluble protein structures, and that the analysis results may not reflect that transporters have segments traversing the cellular membrane.

Based on model validation the alignments may be adjusted (see Note 3) in order to generate new improved models (Fig. 1). The energetic stability of the model may also be checked by doing molecular dynamics simulations.

2.2. Accuracy and Pitfalls in Homology Modeling of Carriers and Channels

When constructing homology models of carriers and ion channels, there are pitfalls in regard to several of the main steps in the homology modeling procedure. There are few templates available, if any, and the resolution of these templates is generally low. Furthermore, the homology between the target transporter and the template may also be low.

The accuracy of a homology model depends on the functional and sequential similarities between the template protein and the target. These similarities, and available structural information about the protein family of interest, are fundamental for the quality of the generated alignments. For water-soluble proteins, a sequence identity of more than 50% between the template and target are believed to give highly accurate models (about 1 Å Cα root-mean-square deviation from template) (30). Acceptable alignments and thereby also acceptable homology models may be obtained of soluble proteins when the target–template sequence identities are 30% or higher, but the quality sharply decreases when the sequence identity is less than 20% (20).

For water-soluble proteins, an identity between the target protein and the template below 30% may be considered "borderline" of what can be considered as realistic modeling, and structure-based drug design based on low homology models may not be as applicable as for models with identities above 50%. For membrane proteins the overall sequence identity between the target and the template may be quite low, but the structural identity may be high in transmembrane α-helices and active site regions. The overall sequence identity between the G-protein-coupled receptors rhodopsin and β2-adrenergic receptor is less than 20%. However, their X-ray structures indicate that their transmembrane α-helices, which constitute the binding site for endogenous activators and small molecular drugs, are structurally similar. Their X-ray structures show that there are some differences in helical packing, but nevertheless the shape is conserved (31, 32). Thus, in spite of relatively low sequence similarity between template and target, the helical and active site regions of the transporter model may be reliable. Such models provide tools for suggesting candidate residues for mutagenesis experiments, and active sites can be identified when combining molecular modeling and site-directed mutagenesis studies. High-quality models may be used to investigate the molecular interactions between drugs and transporters as an aid in the search to understand the intermolecular forces involved in determining the potency and the specificity of binding compounds (see Note 6). Elucidating structural changes of the drug and the transporter for adopting an energetically favorable complex may indicate how a designed compound will fit into the binding site.

The binding of drugs to carriers is structure- and stereospecific, implying that only drugs with certain chemical groups and spatial orientation has high affinity to a certain transporter. Two homologous carriers may therefore bind different drugs since their amino acid composition in the binding site area may differ from each other, and thus, the differences in pharmacology between template and target may affect the accuracy of the model and thereby the conclusions regarding ligand binding.

The resolution of X-ray crystal structures of transporters is usually low, introducing even more uncertainty to the final model. The amphiphilic nature or membrane proteins cause difficulties in experimental structure determination. The hydrophobic surfaces interact with nonpolar alkyl chains of phospholipids, while the hydrophilic surfaces are exposed to the aqueous medium, and this makes it difficult to obtain stable and homogeneous protein preparations. During crystallization, crystal contacts are formed between hydrophilic and hydrophobic surfaces. Even when crystallization is successful, the protein is no longer in its natural environment and thus the crystallized conformation may not represent a realistic conformation (see Note 2).

2.2.1. Structural Flexibility

Structural flexibility is crucial to take into account when doing homology modeling of transporters. A crystal structure of a carrier is merely a snapshot of a highly flexible protein, and this snapshot may not even be a realistic representation of the transporter in its native form. The majority of the membrane protein structures are determined in a non-membrane environment, and the crystallization is often performed in the presence of detergents or antibodies. Transporters may undergo substantial conformational changes during the transport cycle. Extensive studies of the bacterial carrier Lac Permease (33) have indicated that widespread cooperative conformational changes, including sliding and tilting motions of the TMHs, may occur during substrate transport. X-ray crystal structures of the bacterial ABC transporter lipid flippase, MsbA, trapped in different conformations, have shown that large ranges of motion, changing the accessibility of the transporter from a cytoplasmic (inward) facing to an extracellular (outward)-facing conformation, may be required for substrate transport (34).

When interpreting homology models of transporters and performing docking studies on such models, the structural flexibility of transporters must be considered, as structural changes of both the drug and the drug target for adopting an energetically favorable complex (induced-fit) may be even more important than for drug targets which do not transport their ligands across a translocation pore. Induced-fit and conformational changes due to transport may be an important part of the insight which can help predict how a designed drug will fit into a transporter drug target. As a consequence of structural flexibility, several conformations of the transporter model should be considered in modeling and target-based ligand screening/design approaches (see Note 6).

3. Case Studies

Examples of modeling carrier proteins of pharmacological interests are given below.

3.1. ABC Transporter Modeling

The human ATP-binding cassette (ABC) transporters ABCB1, ABCC4, and ABCC5 belong to the ABC superfamily, a subgroup of primary active transporters that have a common intracellular motif that exhibits ATPase activity (3). The ATPase activity motif cleaves ATP's terminal phosphate to energize the transport of molecules from regions of low concentration to regions of high concentration (3, 35, 36), and the overall topology of ABCB1, ABCC4, and ABCC5 is divided into transmembrane domain 1 (TMD1)—nucleotide-binding domain 1 (NBD1)—TMD2—NBD2.

We have constructed outward-facing molecular models of ABCB1 (37), ABCC4 (38), and ABCC5 (39) based on the *Staphylococcus aureus* ABC transporter Sav1866, which has been crystallized in an outward-facing ATP-bound state (16), and inward-facing models of ABCB1, ABCC4, and ABCC5 (40) based on a wide open inward-facing conformation of *Escherichia coli* MsbA (34). After the models were constructed, we got a unique opportunity to test our methodology when the X-ray crystal structure of the *Mus musculus* ABCB1 in a drug-bound conformation was published (15). The models were also compared with site-directed mutagenesis data on ABCB1 (41–45). Figure 2 shows ABCB1 in three different conformations: In an inward-facing conformation (model) (40), in a drug-bound ABCB1 conformation (X-ray crystal structure) (15), and in an outward-facing conformation (model) (37).

Figure 3 shows that amino acids suggested to participate in ligand recognition from site-directed mutagenesis studies, Ile306 (TMH5) (42, 43, 45), Phe343 (TMH6) (41–43), Phe728 (TMH7) (43), and Val982 (TMH12) (44), form a substrate recognition pocket in the ABCB1 models. The involvement of these amino acid residues is also confirmed by the *Mus musculus* ABCB1

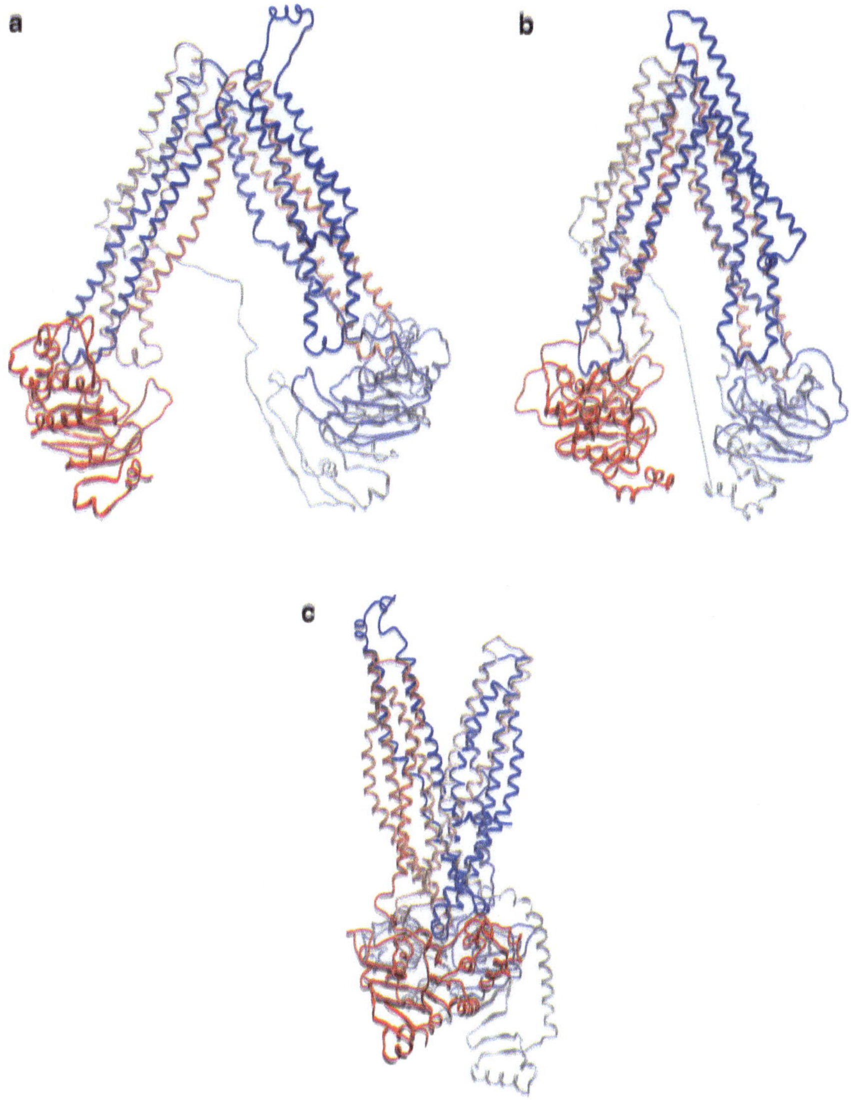

Fig. 2. Backbone Cα-traces of (**a**) inward-facing ABCB1 model (40), (**b**) drug-bound ABCB1 X-ray crystal structure (15), and (**c**) outward-facing ABCB1 model (37), viewed in the membrane plane, cytoplasm downward. Color coding: *blue* via *white* to *red* from N-terminal to C-terminal.

X-ray crystal structure (15) (Fig. 3b). Ile306 (Ile302 in *Mus musculus* ABCB1) points slightly toward the membrane in the X-ray crystal structure, while it points directly toward the translocation pore in the ABCB1 model (Fig. 3a), which may be due to twisting of TMH5 upon changing conformation from a drug recognition conformation to a drug-bound conformation.

ABCB1, ABCC4, and ABCC5 are exporters, pumping substrates out of the cell, and when drugs such as chemotherapeutic agents are expelled from cancer cells as substrates of ABCB1, ABCC4, or ABCC5, the result is multidrug resistance. ABCB1

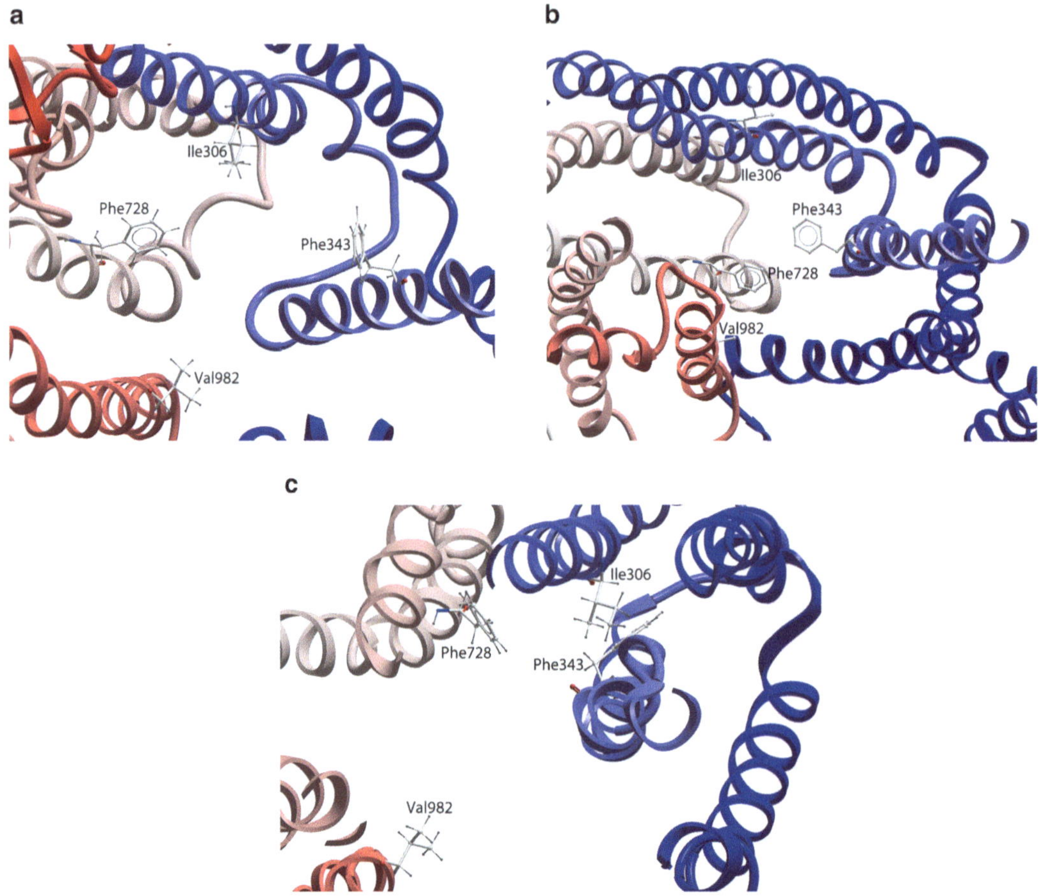

Fig. 3. Drug-binding residues of ABCB1 models and ABCB1 X-ray crystal structure viewed from the intracellular side. Amino acids suggested from site-directed mutagenesis studies to take part in ligand binding are displayed as sticks colored according to atom type (C = *gray*; H = *dark gray*; O = *red*; and N = *blue*); Ile306 (42, 43, 45) (TMH5), Phe343 (41–43) (TMH6), Phe728 (43) (TMH7), and Val982 (44) (TMH12). (**a**) Inward-facing ABCB1 model (40). (**b**) Drug-bound ABCB1 X-ray crystal structure (15). (**c**) Outward-facing ABCB1 model (37). Amino acids in *panel B* are numbered according to human ABCB1. Mus musculus numbering: Ile302, Phe339, Phe724, and Val978. Differences in helix tilting in the panels refer to the different conformations of ABCB1.

transports cationic amphiphilic and lipophilic substrates (46–49), while ABCC4 and ABCC5 transport organic anions (50). The electrostatic potential surface (EPS) of the ABCB1, ABCC4, and ABCC5 models were calculated with the ICM program, and while EPS of the substrate recognition area in the TMDs of ABCB1 was neutral with negative and weakly positive areas, the EPS of the ABCC4 and ABCC5 substrate recognition areas were generally positive (Fig. 4). This serves as an example of how homology modeling of transporters may be used to explain substrate differences between homologous transporters.

The ABCB1, ABCC4, and ABCC5 models are based on low homology templates (21–34%) (37, 38, 40) with low resolution (*Escherichia coli* MsbA (34): 5.30 Å; and *Staphylococcus aureus*

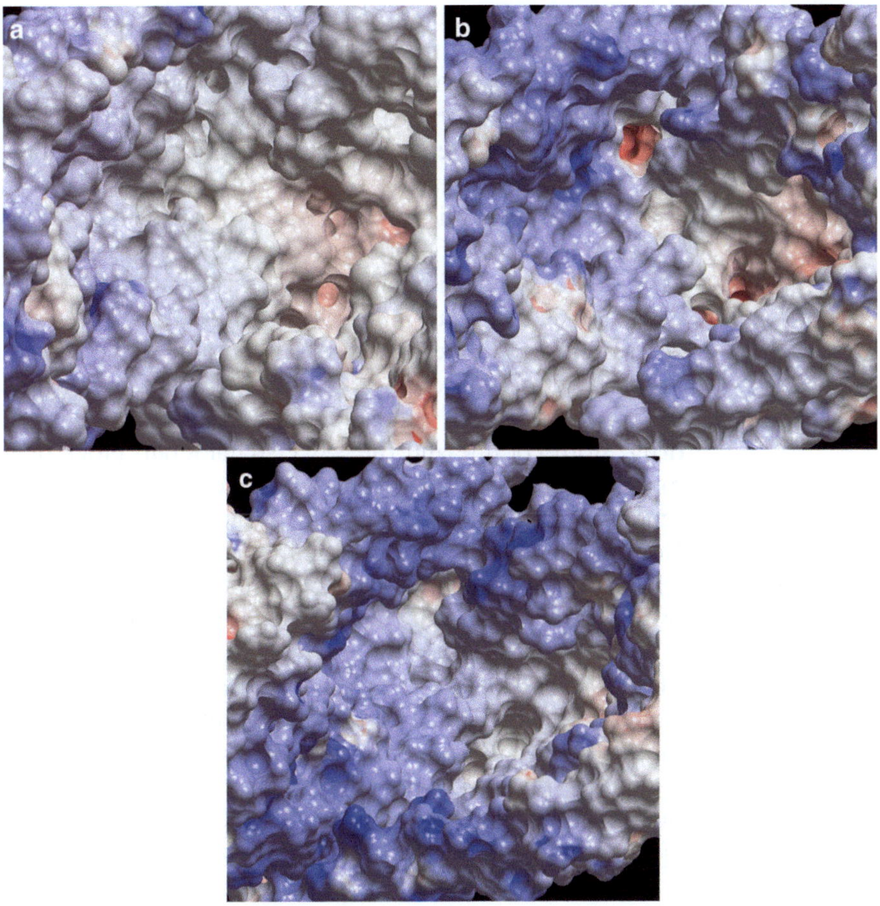

Fig. 4. The water-accessible surfaces of the substrate translocation areas of the ABCB1 model (**a**), the ABCC4 model (**b**), and the ABCC5 model (**c**) viewed from intracellular side color coded according to the electrostatic potentials 1.4 Å outside the surface; negative (−10 kcal/mol), *red* to positive (+10 kcal/mol), *blue*.

ABC transporter Sav1866 (16): 3.00 Å). A 5.3 Å resolution of a template is clearly too low to expect to yield a model of a quality that can be used for, i.e., structure-based drug design. The ABCB1, ABCC4, and ABCC5 models exemplify how structural hypothesis and insights can be obtained even for transporter models which are based on low homology and low resolution templates. These models should be considered as working tools for generating hypotheses and designing further experimental studies related to ABC transporter structure and function, and their limitations due to uncertainties should be kept in mind.

3.2. Neurotransmitter Transporter Modeling

The dopamine transporter (DAT), serotonin transporter (SERT), and noradrenaline transporter (NET) regulate monoamine concentrations at neuronal synapses by carrying monoamines across neuronal membranes into presynaptic nerve cells, using an inwardly directed sodium gradient as an energy source. DAT, SERT, and NET are molecular targets for psychotropic drugs acting in the brain. The dopaminergic system in the brain includes the mesolimbic–mesocortical pathway, which is involved in emotion- and drug-induced reward systems, and the serotonergic and noradrenergic neurons in the brain are associated with mood.

The class of antidepressant drugs termed SSRIs elevates the concentration of serotonin at serotonergic synapses by binding to SERT, and when stimulants such as cocaine bind to DAT, the dopamine concentration is elevated, resulting in a "reward." Interestingly, cocaine and SSRIs have similar molecular mechanisms of action, although SSRIs are therapeutic drugs prescribed for the treatment of depression and cocaine is a highly addictive drug. Both cocaine and the SSRI S-citalopram block neurotransmitter reuptake competitively, but while cocaine is a nonselective reuptake inhibitor, S-citalopram is a selective SERT inhibitor. Cocaine has similar binding affinities for DAT, SERT, and NET, while SSRIs are from 300 to 3,500 times more selective for SERT over NET, and generally have low affinities for DAT (51).

The publication of the *Aquifex aeolicus* LeuTAa crystal structure (17) in 2005 was a major advance in the monoamine transporter modeling field. The sequence identity between LeuTAa and monoamine transporters is relatively low, ~20% (52), for generating models that can be directly used in structure-aided drug design, but still homology models of DAT, NET, and SERT may shed light upon ligand interactions with these transporters.

Homology modeling of DAT, NET, and SERT is an example of how low homology models may be used to aid the selection of amino acids to be mutated in site-directed mutagenesis studies, and also to visualize and interpret results from site-directed mutagenesis data. Such models may also be used for finding binding sites, for instance by using ICMPocketFinder of the ICM program (24), which detects cavities of sufficient size to bind drugs. ICMPocketFinder detected two putative binding sites in our *Aquifex aeolicus* LeuTAa crystal structure (17) (pdbcode 2a65) based DAT, NET, and SERT models (53). The template was in an occluded conformation with leucine bound to its substrate-binding site, and ICMPocketFinder detected the substrate-binding site ("Binding Pocket 1"/"S1") and an additional binding site in the extracellular gateway of the translocation pore of the transporter ("Binding Pocket 2"/"S2") (Fig. 5a). Interestingly, this binding site corresponds to a TCA-binding site reported in two X-ray crystal structures of LeuTAa with TCAs bound in the extracellular-facing cavity (54, 55).

Figure 5b shows cocaine docked into the substrate-binding site of DAT. Cocaine interacts with Asp79, Val152, and Tyr156 in the cocaine–DAT complex. Site-directed mutagenesis data of cocaine binding to DAT also indicate that cocaine interacts with

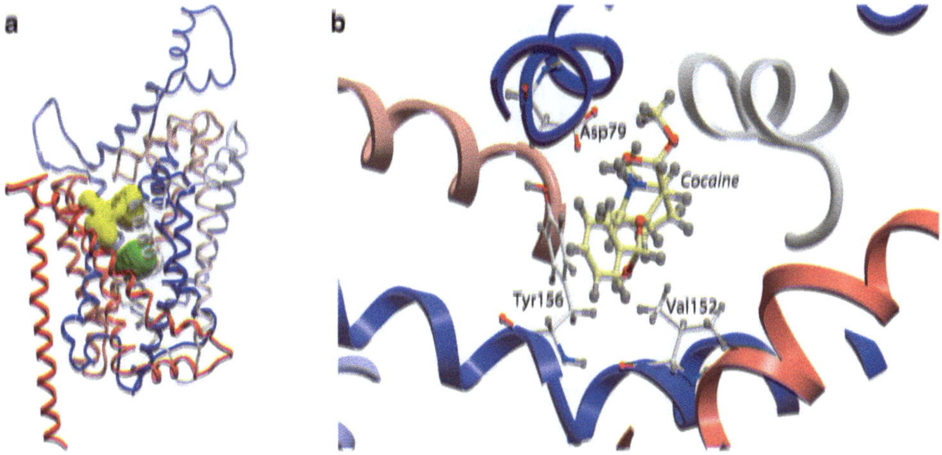

Fig. 5. (**a**) Backbone Cα-traces of DAT model (53) viewed in the membrane plane cytoplasm downward. Binding pocket 1 ("S1") is displayed in *green*, and binding pocket 2 ("S2") is displayed in *yellow*. (**b**) Cocaine docked into the putative substrate-binding area of *DAT* viewed from the extracellular side. Amino acids reported to be part of a cocaine-binding site in site-directed mutagenesis studies: Asp79 (56) (TMH1), Val152 (57) (TMH3), and Tyr156 (58) (TMH3) are displayed as sticks. Color coding as in Figs. 2 and 3.

Asp79 (56) (TMH1), Val152 (57) (TMH3), and Tyr156. Tyr156 corresponds to Tyr176 in SERT, which has been found by site-directed mutagenesis studies to be important for cocaine binding in SERT (58).

4. Notes

1. Please remember that we are dealing with protein models, and the models must be treated as such.
2. The quality of the model depends on the quality of the template and of the template–target amino acid sequence alignments.
3. An incorrect target–template amino acid sequence alignments results in an incorrect model. Manual adjustments of the alignments may therefore be necessary.
4. The lengths and structures of loop segments may differ substantially between the target and the template. It is therefore important to have in mind that loop modeling is uncertain, and overinterpretation of loop structures (if included) must be avoided.
5. Models of transporters and ion channels should be carefully energy refined. Energy refinements using molecular mechanics may result in a more uncorrect model when the structural similarity between the template and target is low.

6. Substrate translocation requires structural flexibility, and the conformation of a transporter model directly obtained by homology modeling may not be correct for substrate and/or inhibitor binding.

5. Summary

In spite of technical improvements in crystallization and structure determination, there is still a huge gap between the number of membrane proteins of known 3D structure and the total number of membrane proteins in the human genome. The homology modeling approach may be used to obtain structural information when detailed experimental structures are lacking (see Note 1). The accuracy of homology-generated models of carriers and ion channels depends mainly on the sequence homology and functional similarities between the template and the target, on the quality of the template–target alignments, and on the resolution of the template (see Notes 2 and 3). Models based on low sequence homology between the template and the target must be regarded as working models for generating new experimental studies, while models based on high homology and functionality between the template and the target may be used for identifying new binders for the target. Carriers must have large conformational flexibility in order to facilitate substrate transport, and inhibitors may bind to different conformations of a carrier (see Note 6). Thus, several conformations of a carrier should be considered in a target-based ligand design approach. The case studies given in this chapter indicate that reliable models of ABC transporters and neurotransmitter transporters may be constructed using presently available structural templates.

Acknowledgments

The molecular modeling group, at the Department of Medical Biology, University of Tromsø, acknowledges the financial support from the Polish-Norwegian Research Fund, the Norwegian Cancer Society, the Research Council of Norway, and the University of Tromsø.

References

1. Landry Y, Gies JP (2008) Drugs and their molecular targets: An updated overview. Fundam Clin Pharmacol 22:1–18
2. Giacomini KM, Huang SM, Tweedie DJ, Benet LZ, Brouwer KL, Chu X, Dahlin A, Evers R, Fischer V, Hillgren KM, Hoffmaster KA, Ishikawa T, Keppler D, Kim RB, Lee CA, Niemi M, Polli JW, Sugiyama Y, Swaan PW, Ware JA, Wright SH, Yee SW, Zamek-Gliszczynski MJ, Zhang L Membrane transporters in drug development. Nat Rev Drug Discov 9:215–236
3. Saier MH, Jr. (2000) A functional-phylogenetic classification system for transmembrane solute transporters. Microbiol Mol Biol Rev 64:354–411
4. Rang HP, Dale MM, Ritter JM, Morre PK (2003) Pharmacology. 5th edn. Churchill Livingstone, ISBN-10 / ASIN: 0443071454
5. Caffrey M (2003) Membrane protein crystallization. J Struct Biol 142:108–132
6. Cherezov V, Clogston J, Papiz MZ, Caffrey M (2006) Room to move: Crystallizing membrane proteins in swollen lipidic mesophases. J Mol Biol 357:1605–1618
7. Cherezov V, Peddi A, Muthusubramaniam L, Zheng YF, Caffrey M (2004) A robotic system for crystallizing membrane and soluble proteins in lipidic mesophases. Acta Crystallogr D Biol Crystallogr 60:1795–1807
8. Frishman D, Mewes HW (1997) Protein structural classes in five complete genomes. Nat Struct Biol 4:626–628
9. Wallin E, von Heijne G (1998) Genome-wide analysis of integral membrane proteins from eubacterial, archaean, and eukaryotic organisms. Protein Sci 7:1029–1038
10. Bradley P, Misura KM, Baker D (2005) Toward high-resolution de novo structure prediction for small proteins. Science 309:1868–1871
11. Casadio R, Fariselli P, Martelli PL, Tasco G (2007) Thinking the impossible: How to solve the protein folding problem with and without homologous structures and more. Methods Mol Biol 350:305–320
12. Forrest LR, Tang CL, Honig B (2006) On the accuracy of homology modeling and sequence alignment methods applied to membrane proteins. Biophys J 91:508–517
13. Eddy SR (1998) Profile hidden markov models. Bioinformatics 14:755–763
14. Altschul SF, Madden TL, Schaffer AA, Zhang J, Zhang Z, Miller W, Lipman DJ (1997) Gapped blast and psi-blast: A new generation of protein database search programs. Nucleic Acids Res 25:3389–3402
15. Aller SG, Yu J, Ward A, Weng Y, Chittaboina S, Zhuo R, Harrell PM, Trinh YT, Zhang Q, Urbatsch IL, Chang G (2009) Structure of p-glycoprotein reveals a molecular basis for poly-specific drug binding. Science 323: 1718–1722
16. Dawson RJ, Locher KP (2006) Structure of a bacterial multidrug abc transporter. Nature
17. Yamashita A, Singh SK, Kawate T, Jin Y, Gouaux E (2005) Crystal structure of a bacterial homologue of na+/cl--dependent neurotransmitter transporters. Nature 437:215–223
18. Abramson J, Smirnova I, Kasho V, Verner G, Kaback HR, Iwata S (2003) Structure and mechanism of the lactose permease of escherichia coli. Science 301:610–615
19. Ravna AW, Sager G, Dahl SG, Sylte I (2009) Membrane transporters: Structure, function and targets for drug design. In: Napier S, Bingham M (eds) Transporters as targets for drugs vol 4. Topics in medicinal chemistry pp 15–51.
20. Tai K, Fowler P, Mokrab Y, Stansfeld P, Sansom MS (2008) Molecular modeling and simulation studies of ion channel structures, dynamics and mechanisms. Methods Cell Biol 90:233–265
21. Frydenvang K, Lash LL, Naur P, Postila PA, Pickering DS, Smith CM, Gajhede M, Sasaki M, Sakai R, Pentikainen OT, Swanson GT, Kastrup JS (2009) Full domain closure of the ligand-binding core of the ionotropic glutamate receptor iglur5 induced by the high affinity agonist dysiherbaine and the functional antagonist 8,9-dideoxyneodysiherbaine. J Biol Chem 284:14219–14229
22. Hibbs RE, Sulzenbacher G, Shi J, Talley TT, Conrod S, Kem WR, Taylor P, Marchot P, Bourne Y (2009) Structural determinants for interaction of partial agonists with acetylcholine binding protein and neuronal alpha7 nicotinic acetylcholine receptor. EMBO J 28: 3040–3051
23. Wieman H, Tondel K, Anderssen E, Drablos F (2004) Homology-based modelling of targets for rational drug design. Mini Rev Med Chem 4:793–804
24. Abagyan R, Totrov M, Kuznetsov DN (1994) Icm - a new method for protein modeling and design. Applications to docking and structure prediction from the distorted native comformation. J Comp Chem 15:488–506

25. Vriend G (1990) What if: A molecular modeling and drug design program. J Mol Graph 8:52–56, 29
26. Levitt M (1992) Accurate modeling of protein conformation by automatic segment matching. J Mol Biol 226:507–533
27. Sali A, Blundell TL (1993) Comparative protein modelling by satisfaction of spatial restraints. J Mol Biol 234:779–815
28. Laskoswki RA, MacArthur MW, Moss DS, Thorton JM (1993) Procheck: A program to check the stereochemical quality of protein structures. J Appl Cryst 26:283–291
29. Hooft RW, Vriend G, Sander C, Abola EE (1996) Errors in protein structures. Nature 381:272
30. Kryshtafovych A, Venclovas C, Fidelis K, Moult J (2005) Progress over the first decade of casp experiments. Proteins 61 Suppl 7:225–236
31. Cherezov V, Rosenbaum DM, Hanson MA, Rasmussen SG, Thian FS, Kobilka TS, Choi HJ, Kuhn P, Weis WI, Kobilka BK, Stevens RC (2007) High-resolution crystal structure of an engineered human beta2-adrenergic g protein-coupled receptor. Science 318:1258–1265
32. Palczewski K, Kumasaka T, Hori T, Behnke CA, Motoshima H, Fox BA, Le Trong I, Teller DC, Okada T, Stenkamp RE, Yamamoto M, Miyano M (2000) Crystal structure of rhodopsin: A g protein-coupled receptor. Science 289:739–745
33. Kaback HR, Wu J (1997) From membrane to molecule to the third amino acid from the left with a membrane transport protein. Q Rev Biophys 30:333–364
34. Ward A, Reyes CL, Yu J, Roth CB, Chang G (2007) Flexibility in the abc transporter msba: Alternating access with a twist. Proc Natl Acad Sci U S A 104:19005–19010
35. Higgins CF, Linton KJ (2001) Structural biology. The xyz of abc transporters. Science 293:1782–1784
36. Oswald C, Holland IB, L. S (2006) The motor domains of abc-transporters - what can structures tell us? Naunyn-Schmiedeberg's Arch Pharmacol 372:385–399
37. Ravna AW, Sylte I, Sager G (2007) Molecular model of the outward facing state of the human p-glycoprotein (abcb1), and comparison to a model of the human mrp5 (abcc5). Theor Biol Med Model 4:33
38. Ravna AW, Sager G (2008) Molecular model of the outward facing state of the human multidrug resistance protein 4 (mrp4/abcc4). Bioorg Med Chem Lett 18:3481–3483
39. Ravna AW, Sylte I, Sager G (2008) A molecular model of a putative substrate releasing conformation of multidrug resistance protein 5 (mrp5). Eur J Med Chem 43:2557–2567
40. Ravna AW, Sylte I, Sager G (2009) Binding site of abc transporter homology models confirmed by abcb1 crystal structure. Theor Biol Med Model 6:20
41. Loo TW, Bartlett MC, Clarke DM (2003) Methanethiosulfonate derivatives of rhodamine and verapamil activate human p-glycoprotein at different sites. J Biol Chem 278: 50136–50141
42. Loo TW, Bartlett MC, Clarke DM (2006) Transmembrane segment 1 of human p-glycoprotein contributes to the drug-binding pocket. Biochem J 396:537–545
43. Loo TW, Bartlett MC, Clarke DM (2006) Transmembrane segment 7 of human p-glycoprotein forms part of the drug-binding pocket. Biochem J
44. Loo TW, Clarke DM (2002) Location of the rhodamine-binding site in the human multidrug resistance p-glycoprotein. J Biol Chem 277:44332–44338
45. Loo TW, Clarke DM (2005) Recent progress in understanding the mechanism of p-glycoprotein-mediated drug efflux. J Membr Biol 206:173–185
46. Muller M, Mayer R, Hero U, Keppler D (1994) Atp-dependent transport of amphiphilic cations across the hepatocyte canalicular membrane mediated by mdr1 p-glycoprotein. FEBS Lett 343:168–172
47. Orlowski S, Garrigos M (1999) Multiple recognition of various amphiphilic molecules by the multidrug resistance p-glycoprotein: Molecular mechanisms and pharmacological consequences coming from functional interactions between various drugs. Anticancer Res 19:3109–3123
48. Smit JW, Duin E, Steen H, Oosting R, Roggeveld J, Meijer DK (1998) Interactions between p-glycoprotein substrates and other cationic drugs at the hepatic excretory level. Br J Pharmacol 123:361–370
49. Wang EJ, Lew K, Casciano CN, Clement RP, Johnson WW (2002) Interaction of common azole antifungals with p glycoprotein. Antimicrob Agents Chemother 46:160–165
50. Borst P, de Wolf C, van de Wetering K (2007) Multidrug resistance-associated proteins 3, 4, and 5. Pflugers Arch 453:661–673
51. Tatsumi M, Groshan K, Blakely RD, Richelson E (1997) Pharmacological profile of antidepressants and related compounds at human monoamine transporters. Eur J Pharmacol 340:249–258
52. Beuming T, Shi L, Javitch JA, Weinstein H (2006) A comprehensive structure-based

alignment of prokaryotic and eukaryotic neurotransmitter/na+ symporters (nss) aids in the use of the leut structure to probe nss structure and function. Mol Pharmacol

53. Ravna AW, Sylte I, Dahl SG (2009) Structure and localisation of drug binding sites on neurotransmitter transporters. J Mol Model
54. Singh SK, Yamashita A, Gouaux E (2007) Antidepressant binding site in a bacterial homologue of neurotransmitter transporters. Nature 448:952–956
55. Zhou Z, Zhen J, Karpowich NK, Goetz RM, Law CJ, Reith ME, Wang DN (2007) Leut-desipramine structure reveals how antidepressants block neurotransmitter reuptake. Science 317:1390–1393
56. Kitayama S, Shimada S, Xu H, Markham L, Donovan DM, Uhl GR (1992) Dopamine transporter site-directed mutations differentially alter substrate transport and cocaine binding. Proc Natl Acad Sci U S A 89:7782–7785
57. Lee SH, Chang MY, Lee KH, Park BS, Lee YS, Chin HR, Lee YS (2000) Importance of valine at position 152 for the substrate transport and 2beta-carbomethoxy-3beta-(4-fluorophenyl) tropane binding of dopamine transporter. Mol Pharmacol 57:883–889
58. Chen JG, Sachpatzidis A, Rudnick G (1997) The third transmembrane domain of the serotonin transporter contains residues associated with substrate and cocaine binding. J Biol Chem 272:28321–28327

Chapter 13

Methods for the Homology Modeling of Antibody Variable Regions

Aroop Sircar

Abstract

Antibodies are one of the critical molecules of our immune system and are unique in their enormous diversity required for recognizing various antigens. Antibodies are protein molecules and their antigen interacting region, the fragment variable (F_V), is typically composed of a light (V_L) and heavy (V_H) chain. In particular, three loops each at the tip of the V_L and the V_H, known as the complementarity determining region (CDR) loops, are responsible for binding to the antigen. While the framework regions of the V_L and V_H are relatively constant across the entire repertoire of antibodies, the conformation of the CDR loops varies extensively to enable the antibody to recognize different antigens. Three-dimensional structures of antibodies illustrating the V_L–V_H relative orientation and the CDR conformations are needed to gain insight into antibody stability, immunogenicity, and antibody–antigen interactions. Computational modeling provides a fast and inexpensive route for generating antibody structural models. This chapter highlights the various features crucial for creating a successful antibody homology model.

Key words: Antibody, Homology, Modeling, RosettaAntibody, PIGS, WAM, Computational, Structure, Prediction, CDR, F_V

1. Introduction

Our immune system comprising billions of different antibodies are equipped to attack any type of antigen that it encounters. On being challenged with an antigen, the immune system selects antibodies against it and subsequently improves the specificity of the selected antibodies by affinity maturation. However, sometimes the response of our immune system is not specific or fast enough to be able to neutralize the antigen. Success of some engineered therapeutic antibodies in curing diseases has demonstrated that we can rationally design antibodies that bind antigens with high specificity and affinity. Three-dimensional structures of antibodies are crucial for

Andrew J.W. Orry and Ruben Abagyan (eds.), *Homology Modeling: Methods and Protocols*, Methods in Molecular Biology, vol. 857, DOI 10.1007/978-1-61779-588-6_13, © Springer Science+Business Media, LLC 2012

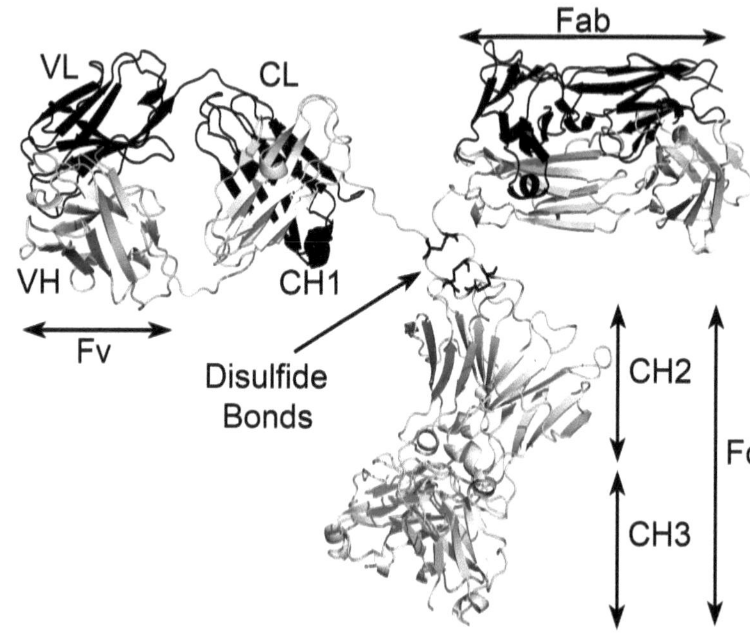

Fig. 1. Cartoon representation of a typical immunoglobulin. (PDB ID: 1IGT) Light (*black*) and heavy (*white*) chains; disulfide bond (*black sticks*).

understanding the precise antibody–antigen interaction, and aid enhancing such interactions. While experimental techniques like X-ray crystallography and nuclear magnetic resonance (NMR) spectroscopy provide accurate and high-resolution three-dimensional structures of proteins such as antibodies, they are laborious, time consuming, and expensive. Computational homology modeling provides a fast alternative method to predict the structure of antibodies, and while computational models are not as accurate as the experimentally determined structures (1) they are still useful in studying protein–protein interactions (2–4).

An understanding of the structural buildup of antibodies is instrumental for successful antibody modeling. Figure 1 shows the usual antibody "Y" shaped molecule comprising four polypeptide chains: two identical light and heavy chains each. The tetramer is made up of a homodimer of light and heavy chain pairs, and the two arms of the "Y" are connected by a disulfide bond between the two heavy chains. Both the heavy and the light chains are comprised of constant and variable domains. The constant domains are the same for all antibodies belonging to the same class, whereas the variable domains differ in different antibodies (but are the same for all antibodies produced by the same B cell). The base of the "Y" responsible for signal transduction is made up of two pairs of heavy chain constant domains (C_H2 and C_H3), and is known as the fragment crystallizable (F_C) region. Each arm of the "Y," referred to as the fragment antibody (F_{ab}), comprises the light chain (variable

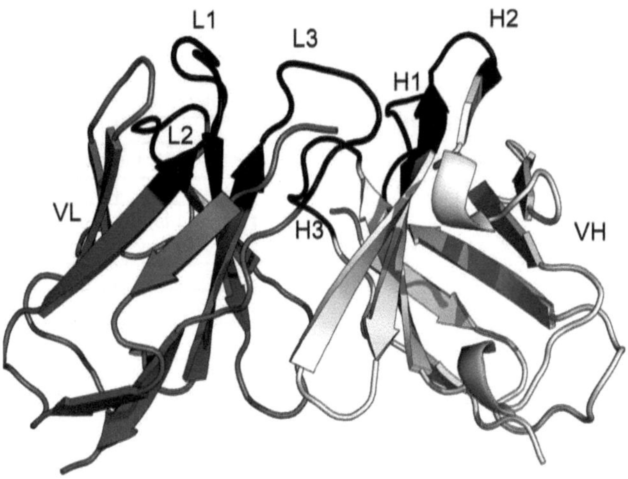

Fig. 2. Cartoon representation of the variable region (F_V) of a typical antibody (PDB ID: 1C08). CDRs (*black*); frameworks of heavy (*white*) and light (*gray*) chains.

(V_L) and constant (C_L) domains) and two domains of the heavy chain (variable (V_H) and constant (C_H1)). The tip of the "Y," i.e., also the tip of the F_{ab}, comprising the variable regions V_L and V_H is referred to as the fragment variable (F_V). F_V interacts with the antigen and is the focus of antibody modeling.

Figure 2 shows that in a typical F_V region the V_L and V_H are oriented to form a conserved β-barrel. Three loops each at the tip of the V_L (L1, L2, L3) and V_H (H1, H2, H3), known as the complementarity determining regions (CDR), exhibit higher sequence diversity among the various antibodies and form the paratope, the actual recognition motif of the antibody. The CDR H3 loop present at the center of the paratope is the most hypervariable loop (both in sequence and length) making it the most difficult to model computationally.

2. Materials and Methods

Figure 3 shows the key components of any antibody modeling algorithm. While the details of each step vary between the different software used, the overall sequence of steps is the same. In particular, the most widely used free antibody modeling protocols will be discussed, viz. RosettaAntibody (1, 5) (http://antibody.graylab.jhu.edu), PIGS (6) (http://arianna.bio.uniroma1.it/pigs/), and WAM (7) (http://antibody.bath.ac.uk/). However, there exist other commercially available antibody modeling software like Accelrys's Discovery Studio and Chemical Computing Group's Molecular Operating Environment (MOE).

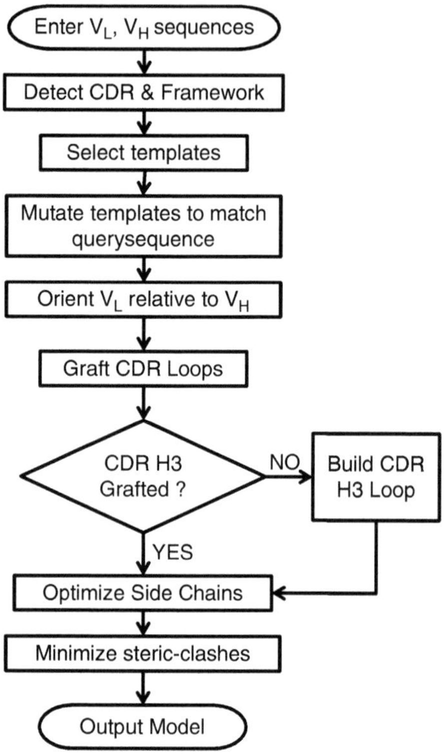

Fig. 3. Flowchart illustrating the key steps of antibody homology modeling.

3. The Input

The V_L and V_H amino acid sequences are required for modeling the F_V region. Most software accept sequences in FASTA format. It has to be ensured that header and linker sequences are removed.

4. Preparing the Input

The first step is to detect the CDR and framework regions in the query sequence. The CDRs are identified by key flanking residues (8) as shown in Table 1. Most software use regular expressions to detect the CDRs.

Once the CDRs have been identified, the sequence has to be numbered using one of the antibody standardized numbering schemes like Kabat (sequence based) (9) or Chothia (structure based) (10). The Abnum (11) antibody numbering server can number sequences by both these conventions. Since we are interested in structural antibody models, we will be using the Chothia numbering system for all subsequent discussions.

Table 1
Key residues for CDR identification

CDR	Residues before	Residues after	Length	Chothia definition
L1	C (starts approximately at residue 24)	W (typically WYQ, WLQ, WFQ, WYL)	10–17	24–34
L2	Generally IY, but also VY, IK, IF (16 residues at the end of L1)	–	7 (mostly)	50–56
L3	C (usually 33 residues at end of L2)	FGXG	7–11	89–97
H1	CXXX (residue 26)	W (mostly WV, but also WI, WA)	10–12	26–32
H2	Typically LEWIG (start always 19 residues at the end of CDRH1)	(KR)(LIVFTA)(TSIA)	9–12	52–56
H3	CXX (typically CAR. Start always 33 residues at end of CDRH2)	WGXG	3–25	95–102

5. CDR Classification

There exist rules (10, 12, 13) that can predict the conformation of the canonical CDRs (L1, L2, L3, H1, H2) based on the respective loop sequence. The loop classes are primarily based on loop length and subclasses are based on key residues at particular sequence positions. The servers WAMPredict (http://antibody.bath.ac.uk/WAMpredict.html) and Canonicals (http://www.bioinf.org.uk/abs/chothia.html) detect and classify CDRs based on the V_L and V_H input sequences. The CDR H3 is a hypervariable loop varying both in amino acid composition and length precludes classification. Still, Shirai et al. have identified sequence-based rules for prediction of *kink* or *extended* conformations of the CDR H3 C-terminal region (14, 15).

6. Template Identification

Once the CDR and framework regions have been identified and properly numbered, structural templates will have to be chosen to assemble the final antibody model. Different antibody modeling software (1, 5–7) have antibody sequence-structure databases, curated from the Protein Data Bank (PDB) (16), from which the template structures are selected. Alternatively, databases can be constructed from available antibody structure databases like SACS (17).

7. Framework Template Selection

The V_L and V_H templates can be selected by one of the following ways:

1. The V_L and V_H sequences are individually scanned against previously created V_L and V_H framework databases respectively for the most sequence homologous match using BLAST (18) (RosettaAntibody and WAM, PIGS *Best H and L chains* option).
2. The combined V_L and V_H sequence is scanned against a previously created database of combined V_L–V_H framework databases using BLAST (18) (PIGS *Same Antibody* option).
3. The V_L and the V_H are individually selected from respective databases based on the maximal match of the canonical classes of the query CDRs and that in the respective template (PIGS *Same Canonical Structures* option).

While WAM and RosettaAntibody web servers do not allow the user to manually select framework templates, PIGS offers a nice interface to manually select desired framework templates. In addition, PIGS also offers users the ability to disallow selected antibody structures from being chosen as framework or CDR templates.

8. CDR Template Selection

The canonical CDR templates are chosen by either of the following two methods:

1. Detecting the canonical class of the query CDR and choosing the representative template from the matching CDR canonical class (PIGS, WAM).
2. Using BLAST (18) to find the most sequence homologous match for the query CDR from a sequence-structure database of the respective CDR (RosettaAntibody). If BLAST does not detect a match, then a template with the same length is chosen from the respective database. However, choosing simply based on length introduces errors and should be avoided as much as possible.

9. Assembling the Templates

Once all the templates for the various segments of the F_V have been selected they are mutated such that the templates now match the residues in the query (input sequence). Finally the mutated templates are assembled to create the complete structural model.

10. β-Barrel Assembly

The relative V_L–V_H orientation results in the formation of a β-barrel, the structure of which clusters very tightly across different antibodies (1). Thus, to position the V_L relative to the V_H or vice versa, one of the following methods is selected:

1. If the V_L and V_H templates are obtained from the same antibody, then the relative V_L–V_H orientation is set as in the template antibody (PIGS *Same Antibody* option).
2. If the V_L and V_H templates are obtained from different antibodies, they can be oriented:
   (a) As in the F_V structure with the highest sequence similarity to the entire query F_V sequence (RosettaAntibody).
   (b) As in the F_V structure from which the V_L template was selected.
   (c) As in the F_V structure from which the V_H template was selected.
   (d) Using certain conserved interfacial residues of known antibody structures (WAM).

If option 2 is selected, the superposition of the V_L and V_H on another template might cause steric clashes. Some software like WAM and PIGS do not attempt to relieve these clashes, but the new antibody modeling protocol RosettaAntibody is the only software that relieves such clashes by optimizing the relative V_L–V_H orientation in a final refinement stage.

11. Grafting the CDRs

The CDRs for which templates have been identified are grafted into the previously assembled V_L and V_H framework. Grafting relies on the fact that while the CDRs themselves have different conformations, the stems flanking the CDRs are part of the conserved immunoglobulin fold. Thus, superimposing the stems flanking the CDR templates on the respective atoms of the stems in the V_L and V_H framework orients the CDRs relative to the framework regions. RosettaAntibody grafts the CDRs by superimposing two C_α atoms on either side of the respective CDR.

While grafting the CDRs captures the structural features of the paratope, sometimes grafting results in intra-loop steric clashes. WAM and PIGS does not attempt to relieve such clashes, but RosettaAntibody optimizes the CDR backbone positions to eliminate such clashes thereby generating more physically realistic models. However, WAM performs steepest descent minimization to smooth the graft location.

12. Building the CDR H3

Predicting the CDR H3 is the most challenging part of generating an antibody homology model. CDR H3s vary in length from 3 to 30 residues and exhibit a huge sequence diversity limiting the possibility of capturing the conformation by mere superposition of an existing template. Additionally, some of the most accurate loop prediction algorithms (19, 20) can model only 13 residue loops and that too is computationally expensive. Finally, modeling CDR H3 in homology models is even harder because of the nonnative environment in which the loop conformation has to be predicted. Given that the CDR H3 is at the center of the paratope and is often the most crucial region for antigen recognition, the usefulness of an F_V homology model depends on the accurate prediction of CDR H3.

While software like PIGS does not even attempt to model the CDR H3 and simply grafts the most sequence homologous CDR H3 loop of the same length, WAM takes an intermediate approach and grafts loops if they are less than 13 residues and builds longer loops using ab inito loop modeling methods. PIGS's simplistic treatment enables it to generate a homology model instantly compared to the few days required by WAM. RosettaAntibody leaves it to the user to make the choice between a fast crude model in which the CDR H3 is grafted from a template or a long protocol that uses loop modeling to generate more accurate models. All CDR H3 loop building-based modeling protocols build multiple models, score each model using a scoring function, and return the model with the best score as the putative predicted structure. RosettaAntibody is the only antibody modeling software that attempts to compensate for the inaccuracies in the scoring function by providing the ten best scoring models (out of 2,000 models) to the user. The usefulness of multiple models has been demonstrated by antibody–antigen docking algorithms like SnugDock (2), which generates more accurate predictions when ten models are used.

13. Side-Chain Optimization

Once the antibody backbone has been generated, the side chains are generated as follows:

1. If residues copied from the template are the same as those in the query sequence, the side-chain orientations of the respective residues can be simply copied. For residues that differ between the template and query sequences, the side-chain orientation can be predicted by screening from standard rotamer libraries (21) (PIGS: Transfer Conserved + SCWRL 3.0 (22) option).

2. Especially if the backbone of the templates has been optimized, it may be necessary to repack the side chains of all residues in the model. For residues that are the same between the template and the query sequence, the side-chain conformation of those residues can be added to the standard rotamer libraries (RosettaAntibody).

14. Using Homology Models

Structural models are useful by themselves as well as in complex with interacting partners. Changes in thermodynamic stability on mutating key residues can be computed by protein stability prediction servers like Eris (23). In conjunction with epitope mapping software like Discotope (24) and Pepitope (25), epitopes on protein or peptide antigen can be identified and subsequently the antibody–antigen complex structure can be predicted using SnugDock (2). The computational pipeline from antibody sequence to increased specificity can be achieved by using computational mutagenesis software like RosettaDesign (26) to increase the binding affinity of the antibody to the antigen.

15. Notes

1. The input sequences should not have any amino acids from the constant (C_H1 or C_L) regions. If the Abnum antibody numbering server (http://www.bioinf.org.uk/abs/abnum/) can successfully renumber the query sequence, then it is a good indicator that the input is valid. If Abnum truncates any upstream or downstream residues, the same should be truncated from the query sequence.
2. The key residues used to identify CDRs are applicable to classical antibodies that have both heavy and light chains. These rules might not hold for heavy chain-only (V_HH) antibodies found in animals like camelids and sharks (27).
3. The canonical CDR classification holds for classical antibodies, but might not be applicable to V_HH antibodies (27). Moreover, as more and more antibodies are being crystallized, it is possible that more conformations are discovered.
4. Unless the query CDR H3 sequence matches exactly with a respective sequence in the database, the CDR H3 has to be modeled using loop modeling to generate physically realistic models. However, for crude models the computational cost can be minimized by either (a) choosing CDR H3 from a database

(PIGS) or (b) for short (<8 residues) CDR H3 loops select from a database, and for longer loops use loop building techniques (WAM).

5. The V_L–V_H relative orientation primarily depends on the subtype of V_L, i.e., V_κ or V_λ. One has to ensure that the F_V selected for deciding the V_L–V_H orientation should have the same V_L subtype (κ or λ) as in the query structure.
6. Personal experience has shown that setting the relative V_L–V_H orientation as in the F_V structure from which the V_H template was selected produces more accurate results than as in the F_V from which the V_L template was selected.
7. Longer CDR H3 loops require larger conformational space to be sampled. Thus, for protocols that use loop modeling for CDR H3 structure prediction, a larger number of models should be built. Ideally, for an n-residue loop, 2^n models should be built.

References

1. Sivasubramanian, A., Sircar, A., Chaudhury, S. and Gray, J.J. (2009) Toward high-resolution homology modeling of antibody Fv regions and application to antibody-antigen docking. *Proteins.* 74(2):497–514.
2. Sircar, A. and Gray, J.J. (2010) SnugDock: paratope structural optimization during antibody-antigen docking compensates for errors in antibody homology models. *PLoS Comput Biol.* 6(1):e1000644.
3. Chaudhury, S., Sircar, A., Sivasubramanian, A., Berrondo, M. and Gray, J.J. (2007) Incorporating biochemical information and backbone flexibility in RosettaDock for CAPRI rounds 6-12. *Proteins.* 69(4):793–800.
4. Sircar, A., Chaudhury, S., Kilambi, K.P., Berrondo, M. and Gray, J.J. (2010) A generalized approach to sampling backbone conformations with RosettaDock for CAPRI rounds 13-19. *Proteins.*
5. Sircar, A., Kim, E.T. and Gray, J.J. (2009) RosettaAntibody: antibody variable region homology modeling server. *Nucleic Acids Res.* 37(Web Server issue):W474–479.
6. Marcatili, P., Rosi, A. and Tramontano, A. (2008) PIGS: automatic prediction of antibody structures. *Bioinformatics.* 24(17):1953-1954.
7. Whitelegg, N.R. and Rees, A.R. (2000) WAM: an improved algorithm for modelling antibodies on the WEB. *Protein Eng.* 13(12):819–824.
8. Martin, A.C.R. 09/11/2010. How to identify the CDRs by looking at a sequence. http://www.bioinf.org.uk/abs/#cdrid. Accessed 09/11/2010.
9. Kabat, E.A., Wu, T.T., Bilofsky, H., Reid-Miller, M. and Perry, H. (1983) Sequence of Proteins of Immunological Interest. *National Institutes of Health, Bethesda*
10. Al-Lazikani, B., Lesk, A.M. and Chothia, C. (1997) Standard conformations for the canonical structures of immunoglobulins. *J Mol Biol.* 273(4):927–948.
11. Abhinandan, K.R. and Martin, A.C. (2008) Analysis and improvements to Kabat and structurally correct numbering of antibody variable domains. *Mol Immunol.* 45(14):3832–3839.
12. Chothia, C. and Lesk, A.M. (1987) Canonical structures for the hypervariable regions of immunoglobulins. *J Mol Biol.* 196(4):901–917.
13. Morea, V., Tramontano, A., Rustici, M., Chothia, C. and Lesk, A.M. (1998) Conformations of the third hypervariable region in the VH domain of immunoglobulins. *J Mol Biol.* 275(2):269–294.
14. Shirai, H., Kidera, A. and Nakamura, H. (1996) Structural classification of CDR-H3 in antibodies. *FEBS Lett.* 399(1-2):1–8.
15. Shirai, H., Kidera, A. and Nakamura, H. (1999) H3-rules: identification of CDR-H3 structures in antibodies. *FEBS Lett.* 455(1-2):188–197.

16. Berman, H.M., Westbrook, J., Feng, Z., Gilliland, G., Bhat, T.N., Weissig, H., Shindyalov, I.N. and Bourne, P.E. (2000) The Protein Data Bank. *Nucleic Acids Res.* 28(1):235–242.
17. Allcorn, L.C. and Martin, A.C. (2002) SACS--self-maintaining database of antibody crystal structure information. *Bioinformatics.* 18(1):175–181.
18. Altschul, S.F., Gish, W., Miller, W., Myers, E.W. and Lipman, D.J. (1990) Basic local alignment search tool. *J Mol Biol.* 215(3):403–410.
19. Zhu, K., Pincus, D.L., Zhao, S. and Friesner, R.A. (2006) Long loop prediction using the protein local optimization program. *Proteins.* 65(2):438–452.
20. Mandell, D.J., Coutsias, E.A. and Kortemme, T. (2009) Sub-angstrom accuracy in protein loop reconstruction by robotics-inspired conformational sampling. *Nat Methods.* 6(8):551-552.
21. Dunbrack, R.L., Jr. and Cohen, F.E. (1997) Bayesian statistical analysis of protein side-chain rotamer preferences. *Protein Sci.* 6(8):1661–1681.
22. Canutescu, A.A., Shelenkov, A.A. and Dunbrack, R.L., Jr. (2003) A graph-theory algorithm for rapid protein side-chain prediction. *Protein Sci.* 12(9):2001–2014.
23. Yin, S., Ding, F. and Dokholyan, N.V. (2007) Eris: an automated estimator of protein stability. *Nat Methods.* 4(6):466–467.
24. Haste Andersen, P., Nielsen, M. and Lund, O. (2006) Prediction of residues in discontinuous B-cell epitopes using protein 3D structures. *Protein Sci.* 15(11):2558–2567.
25. Mayrose, I., Penn, O., Erez, E., Rubinstein, N.D., Shlomi, T., Freund, N.T., Bublil, E.M., Ruppin, E., Sharan, R., Gershoni, J.M., Martz, E. and Pupko, T. (2007) Pepitope: epitope mapping from affinity-selected peptides. *Bioinformatics.* 23(23):3244–3246.
26. Liu, Y. and Kuhlman, B. (2006) RosettaDesign server for protein design. *Nucleic Acids Res.* 34(Web Server issue):W235-238.
27. Sircar, A., Sanni, K.A., Shi, J. and Gray, J.J. Analysis and modeling of the variable region of camelid single-domain antibodies. *J Immunol.* 186(11):6357–6367.

Chapter 14

Investigating Protein Variants Using Structural Calculation Techniques

Jonas Carlsson and Bengt Persson

Abstract

Structure calculation techniques can be very useful to bridge the gap between available sequence information and structural knowledge. In order to understand the molecular mechanisms behind diseases caused by residue exchanges, knowledge about the modified structure is needed. In this chapter, we describe how energy minimizations and molecular dynamics can be useful tools in order to study the structural effects of sequence variation. With these techniques, together with investigation of other properties, it is often possible to obtain a complete picture of the effect and mechanism behind disease-causing mutations. To take this information one step further, we also describe prediction methods that can be used to judge the effects of mutations and how to evaluate these and the interplay between the protein properties.

Key words: Molecular modeling, Energy minimization, Molecular dynamics, Sequence variations, SNP, Disease mechanisms

1. Introduction

1.1. Background

In order to understand the molecular mechanisms of proteins, it is of central importance to have knowledge about the three-dimensional structure. However, the gap between available sequence information and known structures is steadily increasing, since sequencing is currently much more rapid than structure determination. Even though the amount of structural information has increased considerably in recent years, thanks to a number of structural genomics initiatives (1), the avalanche of sequence data from numerous genome-sequencing projects provides an increase that is several magnitudes larger.

The available next-generation sequencing techniques (2) that every year show improved performance at decreasing costs open up for a wide range of biomedically interesting projects. It is now possible at a large scale to investigate multiple human genomes and

Andrew J.W. Orry and Ruben Abagyan (eds.), *Homology Modeling: Methods and Protocols*, Methods in Molecular Biology, vol. 857, DOI 10.1007/978-1-61779-588-6_14, © Springer Science+Business Media, LLC 2012

to study sequence variations for all human protein-coding sequences. One example of such large-scale studies is the 1000 Genome project, aiming at characterization of sequence variation among a thousand individuals from all over the world (3). Similarly, multiple ongoing projects investigating genetic variation are trying to correlate the presence of certain mutations with susceptibility for a particular disease. In order to understand the molecular consequences of residue exchanges, structural knowledge of the modified protein is needed. This can be achieved today using structural calculations techniques. These are steadily improving, thanks to advances in algorithms and improvements of available computational power.

Structural calculations can be used to calculate the optimal structure of the modified protein, using energy minimization techniques, or to characterize the dynamic behavior of the modified protein, using molecular dynamics techniques. The different methods complement each other and provide information on molecular changes that can be useful in understanding the disease-causing mechanisms, which in turn can be used as input for drug development.

In this chapter, we describe how structural calculations can be used in the characterization of protein variants. We also show examples from our work on sets of phenotypically characterized protein variants.

1.2. Strategies

There are several tools available that take a protein sequence and then predict the effect of mutations based on this (cf. Subheading 3.5 below). Some of these also search for known 3D-structures, which will increase the success rate when they are available. However, general predictions are usually of low accuracy and there is often a lack of mechanistic explanation to why a mutation will affect the protein function in a certain way. By doing your own model it is possible to increase the prediction accuracy by integrating knowledge about the protein and also to explain the mechanism. The prediction servers are still useful as a complement.

If the structure of the studied protein is not known, it is possible that a precalculated homology model can be found in a homology model database or can be created from a homologous protein structure. A model based upon a closely homologous structure with high sequence identity yields, in general, better accuracy and thereby better predictions than a model based upon a distantly homologous structure with low sequence identity.

With the help of the protein structure it is now possible to investigate several properties of the protein in addition to those that can be studied based only on the sequence. Using Monte Carlo energy minimization it is possible to calculate stability changes due to residue exchange. Using molecular dynamics simulations, the degree of dynamics of different parts of the protein and how they are affected by mutations can be modeled. The latter is

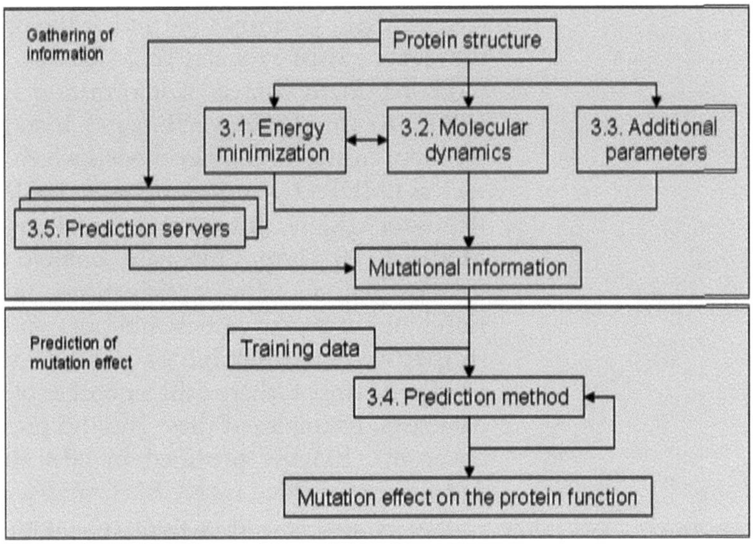

Fig. 1. Flowchart describing the process of investigating protein variants. *Numbers* refer to the relevant sections in the text.

especially useful when the protein is relatively flexible and depends on this flexibility to perform its function. The location of the residue exchange in the structure is also important, i.e., in the core versus on the surface or in the vicinity of active site or binding site. When this structural information is added to conservation and residue exchange analysis, a more complete picture of how the mutations affect the protein can be obtained.

If properties are known, such as activity or clinical severity, for a large number of mutations it is possible to create a prediction model for how hitherto unknown mutations will affect the protein function and structure. Figure 1 describes the general process of investigating protein variants.

2. Materials

2.1. Databases

Information regarding proteins and mutations are stored in numerous biological databases. To be able to obtain knowledge from several sources, there are a number of useful services that provide and connect a large amount of different databases and tools. Two useful Web sites for such services are those of European Bioinformatics Institute (EBI; http://www.ebi.ac.uk), and National Center for Biotechnology Information (NCBI; http://www.ncbi.nlm.nih.gov).

Among databases, the most important in the context of structural calculations are those with sequence information at the DNA level (EMBL, Genbank) (4) and at the protein level (Uniprot with the sections Swiss-Prot and TrEMBL) (5). For protein structures, the most important source of information is the Protein Data Bank (PDB) (6), (http://www.rcsb.org/). If no protein is found here there are precalculated homology models which can be found in databases such as PMDB (7), (http://mi.caspur.it/PMDB/), SWISS-MODEL Repository (8, 9), (http://swissmodel.expasy.org/repository/), and ModBase (10) (http://modbase.compbio.ucsf.edu).

For genome-wide investigations, information is available at Ensembl (http://www.ensembl.org) and NCBI Entrez genome (http://www.ncbi.nlm.nih.gov/sites/entrez?db=genome).

Furthermore, there exist a number of user friendly interfaces to databases. Examples of these are SRS provided by EBI (http://srs.ebi.ac.uk), ExPASy provided by SIB (http://www.expasy.org/), and Entrez provided by NCBI (http://www.ncbi.nlm.nih.gov/).

To be able to add or dock cofactors, substrates, or inhibitors, their structures can be found in molecular databases such as PubChem (http://pubchem.ncbi.nlm.nih.gov/) and ChEBI (http://www.ebi.ac.uk/chebi/). Many of the molecules in these databases have 3D coordinates making it possible to use them without any molecular energy minimization. Useful formats for small molecules are the .mol2 format for single molecules and the .sdf format for multiple molecules.

2.2. Tools

In addition to the databases, there are a number of central tools for analysis of sequence data. For sequence comparisons, there are the FASTA (11), BLAST (12), and PSI-BLAST (12) program suites. These are available as Web servers at EBI (http://www.ebi.ac.uk) and NCBI (http://www.ncbi.nlm.nih.gov).

To create the multiple sequence alignments, MSA, for conservation analysis, a BLAST search against the Uniprot databases Swiss-Prot and TrEMBL is a good start. An MSA can then be created by ClustalW (13) (http://www.ebi.ac.uk/clustalw/), or MUSCLE (14) (http://www.ebi.ac.uk/muscle/), or any other MSA program of choice.

2.3. Software

There exist a large number of programs for energy minimization calculations. One example is ICM from Molsoft LLC, La Jolla, California, USA (15, 16) (http://www.molsoft.com) which is a general purpose molecular modeling program that can perform Monte Carlo-based modeling, docking, and even includes machine learning tools. Other examples of programs that can perform Monte Carlo energy minimizations are Chimera (17) (http://www.cgl.ucsf.edu/chimera/), Boss (Biochemical and Organic Simulation System) (18) from Schrödinger, LLC (http://www.schrodinger.com/) or Cemcomco, LLC (http://www.cemcomco.com/), and MacroModel from Schrödinger LLC, Portland, Oregon, USA (http://www.schrodinger.com/).

The simulation package that we have used for molecular dynamics is GROMACS (19, 20) (http://www.gromacs.org/) as it is fast with linear scaling up to at least 64 cores (21). Other programs that can perform molecular dynamic simulations are AMBER (22) (http://ambermd.org/), CHARMM (23, 24) (http://www.charmm.org/), and MacroModel from Schrödinger LLC, Portland, Oregon, USA (http://www.schrodinger.com/).

3. Methods

3.1. Energy Minimization

Everything in nature strives to reach a position that is as comfortable as possible, i.e., to be in an energy state as low as possible. Proteins are no exceptions to this. This is why the proteins usually fold into a defined structure as this is the lowest energy state given the present environmental factors. Mutations causing amino acid exchanges can negatively influence the structure and even make it unfold partially or completely.

Ideally, one would like to systematically search the complete conformational space to find the global optimal energy. According to Anfinsen's dogma it should theoretically be possible to determine the structure from sequence only (25). However, there are too many possible conformations to be able to test them all. Nevertheless, in reality proteins fold in the order of milliseconds to seconds for small single domain proteins. This paradox is called the Levinthal's paradox (26). Therefore, it is necessary to use heuristic methods that utilize smart strategies to search through the energy landscape. When studying such a small change in a protein as a single residue replacement, Monte Carlo-based energy minimization is a very useful method.

The Monte Carlo energy minimization method is a heuristic technique based on a semi-random walk through the energy landscape. The protein structure is changed locally at a randomly chosen position. While only one residue is in focus in each step, the surrounding residues are included into the minimization. As both the side chains and the backbone surrounding the chosen residue are free to move, a local change can have a propagating effect on the entire protein. If a lower energy conformation is found, the modified structure is kept. Sometimes the structure gets stuck in a local energy minimum, where no locally induced change can improve the energy. To escape these energy traps there is a certain probability that an unfavorable change is kept. The probability decreases exponentially with increasing energy difference. To increase the probability of overcoming local minima the temperature of the system can be raised which induces larger movement (Note 2).

3.1.1. Energy Minimization Applied on Mutations

When an amino acid replacement due to a mutation is first introduced in a protein structural model, there will almost certainly be several clashes between atoms that are too close to each other. This will lead to extreme energies which will tear up the protein structure if not treated carefully. To avoid this, the mutated protein can be minimized using a local to global methodology. First the exchanged residue side chain is positioned optimally, then the side chains of the surrounding residues are energy minimized, followed by allowing local main-chain movements and finally a global Monte Carlo energy minimization.

The suitable number of iterations in the simulations for the global minimization is dependent on the number of degrees of freedom in the protein. As only small changes are introduced in the protein, most of the protein can be approximated as rigid (but still allowed to move in the minimization) and the degrees of freedom will be quite few and therefore also the number of iterations.

As the method is based on random moves, several simulations of the same system are needed to be able to increase the chances of finding the global optimum. The simulations can also be used to evaluate if the simulation was long enough. If several simulation runs obtain similar energies the result should be of higher quality than if they differ to a large extent.

How this can be used to assess the severity of a mutation is described in the Subheading **3.3.4**.

3.1.2. Force Fields

When calculating the total energy of all interactions in a protein some approximations are needed (Note 6). The interactions are divided into different categories called energy terms. The parameters for the energy terms are taken from force fields adapted for biological molecules. The most important of the energy terms are electrostatic interactions, van der Waals forces, hydrogen bonding, and torsion energy. In energy minimization techniques the water molecules are often treated implicitly to speed up calculations, i.e., as an evenly distributed shell around the protein. In molecular dynamics simulations the water molecules must be treated explicitly which is one important reason why this technique often uses more computational time.

The force fields used for proteins are often derived from a combination of experiments and quantum level calculations. The force fields describe both bonded and nonbonded interactions. Besides the general functions that describe the interaction potentials the force fields also provide atom-specific parameters needed to calculate these potentials. Often several different parameters are needed for each element depending on the surrounding atoms, e.g., a carbon in the backbone of a protein or a carbon in a carboxyl group. This makes them approximations of reality and the first level in which errors are introduced.

There are specialized force fields for proteins, like the ECEPP (27) force field used in energy minimization. For molecular dynamics simulations other force fields are used: e.g., the GROMOS force field (28) used in GROMACS (19, 20), the AMBER force field used in AMBER (22), and the widely used CHARMM (23, 24) force fields where CHARMM22 is used for proteins. These latter force fields can also be applied in energy minimizations but are primary for molecular dynamics as they consider all atoms as free variables.

3.2. Molecular Dynamics

As an alternative to energy minimizations, molecular dynamics can be used to investigate protein structures. The drawback with molecular dynamics is that it is very computer intensive in comparison with Monte Carlo energy minimization. The biggest difference of this technique versus energy minimization techniques is that time is introduced, making it possible to study dynamic properties. This can be valuable, since mutations might not only affect the stability of the protein but also the dynamics (Notes 4 and 5).

The dynamics and the conformational space that a protein structure inhabits are found to be more important for the function of the protein than previously anticipated. In fact, recently, an alternative or parallel model to the induced-fit model, where the ligand forces the structure to adapt to a certain conformation, has been proposed (23). Here, instead the conformational space that a folded protein naturally populates is of importance for the binding between ligand and protein or between proteins. A ligand that demands a conformation of the protein that is extremely improbable, for example, caused by a high energy barrier, will not be an effective ligand. Therefore, by doing molecular dynamics simulations the effect of mutations upon the populated conformational space can be studied.

The time in molecular dynamics is not continuous but instead very small time steps are used, usually 1 or 2 fs. The small time step limits the total simulation time to the order of nanoseconds to microseconds, even though using large computer resources, longer simulation times, up to milliseconds can be achieved. Therefore, the conformational shifts must be seen in these rather small timescales for the molecular dynamics simulations to be useful.

3.2.1. Ensembles

Measurement on a real system will result in properties that are an average of all molecules in that system. In a molecular dynamics simulation only one protein molecule is studied. However, for a system in equilibrium, the average of observations over long enough time of a single protein molecule, a statistical ensemble, is equivalent to one observation of a multimolecule system. This means that the properties like temperature and stability can be studied and are in theory equally valid as for measurements in the test tube. In addition to general averaged properties it is also possible to, for example, investigate different states that the protein populates and study the flexibility of different parts of protein structure.

3.2.2. Examples

In our group, we have successfully applied molecular dynamics techniques in two very different projects where the distribution of the conformational space that the protein populates is of importance. The first project is a study of the human amyloid-forming protein islet amyloid polypeptide (IAPP) (29–31), and how mutations in this protein affect the propensity to form amyloids. Here, we observe that the probability at which the protein adapts a beta-sheet conformation is similar to that found in amyloid fibers. These data are then used to predict the amyloid propensity in vitro with high accuracy, showing that the amyloid-forming process in IAPP is dependent on the populated conformational space.

The second project is a study of the antibiotic resistance-associated protein MexR in *Pseudomonas aeruginosa* (32). This protein negatively regulates an efflux pump by binding to DNA. There are several known mutations in this protein that prevent the DNA binding and thereby give rise to antibiotic resistance. Some mutations are directly affecting the DNA binding interface while others have more subtile effects. A few of the mutations not directly affecting the DNA binding probably decreases the stability of the protein while several seem to have no effect or are even stabilizing at the same time as they abolish DNA binding. Here, the data from the molecular dynamics simulations support the fact that these mutations limit or change the populated conformational space so that the probability of the conformations allowing DNA binding is substantially decreased.

3.3. Additional Parameters to Investigate

There are several parameters that can be used in combination to assess the effects a mutation will have on the function of a protein. The most important ones in our opinion are described here.

3.3.1. Evolutionary Conservation

During millions of years, evolution will introduce changes in the genomes that will differentiate proteins in different species from each other. The importance of each individual amino acid residue will affect the probability that a change will be kept in the species. Beneficial mutations will of course have a higher chance of surviving.

Thus, when studying the effect of a mutation the residue conservation is probably the most important aspect. Conservation can be calculated in different ways depending on the goal and available sequences (Note 1). When calculating a multiple sequence alignment, MSA, based on homologous sequences there are a number of issues to take into consideration. If many of the sequences are based on very similar sequences the conservation will be unnaturally biased toward these. In order to avoid this, the sequences could be filtered based on pairwise sequence identity either by hand or by cluster filtering methods such as BLASTCLUST included in the NCBI BLAST package (12) (Note 3). It is also important to remember that even though paralogous proteins are homologous they will have slightly different function and thereby might have different residues at the active sites and binding sites. So in order to capture conserved functional elements it is best to use only orthologs while the structurally important residues can be studied by conservation analysis using a wide range of homologs. The greater the number of unique sequences that the MSA is created from, the better.

There are also different strategies when calculating the conservation score ranging from simply calculating the percentage of the most abundant residue at each position to a conservation score based on a substitution scoring matrix, e.g., PAM (33) or BLOSUM (34). In the latter case, a 20-dimensional centroid vector is calculated for each position based on the average row vector for each residue taken from the substitution score matrix. Then the average distance to the centroid can be calculated as a general measure of the degree of conservation.

To be able to compare the conservation score between proteins, the scores need to be normalized as they are based on different sets and different number of sequences. One way to do this is to adjust the scores based on the relative average conservation.

3.3.2. Surface Accessibility

Amino acid residues that are located on the surface of the protein are in general not as sensitive to changes as those in the core of the protein. There are several reasons for this, i.e., they are less spatially constrained, have a lower number of interaction partners, or are not to the same degree involved in the protein folding process. For a residue to be counted to have access to the surface usually 30% or more of the van der Waals surface must be accessible to the solvent. The accessible surface area is normally calculated by rolling a water molecule sized ball over the surface of the protein.

The accessible surface constitutes a very useful parameter. Mutational sensibility is inversely correlated with surface accessibility in the same manner as it is correlated with evolutionary conservation.

3.3.3. Amino Acid Property

When studying a mutation or investigating potential substitutions the property difference between the native and the new residue is of importance. The simplest measure would be to look up the value in a substitution score matrix. A more accurate score is obtained by taking the conservation profile into account. Here, the same average vector centroid is used as in the evolutionary conservation score for each position. Then the substitution score matrix row vector for the mutation can be used to measure the difference to the centroid. The larger the difference, the higher the probability that the mutation will have a negative effect on the protein functions.

3.3.4. Protein Stability

A mutation that negatively affects the protein function can, for example, do this by directly disturbing the active site or binding sites or altering the stability of the protein. As most proteins are at the very edge of unfolding even small changes in stability can have large effects on the function of the protein. The change in stability upon mutation is therefore a useful indication on the effects of residues not directly involved in the active site or binding sites.

One methodology to calculate the stability is described in the Monte Carlo energy minimization section. There are also servers that make predictions of stability changes upon mutation. One of these is the CUPSAT (Cologne University Protein Stability Analysis Tool) (35) server located at http://cupsat.tu-bs.de/. CUPSAT analyzes the environment around the substitution by calculating several potentials. The change in potentials between native and replaced amino acid residues is used to make a verdict on the change in stability. The most important potentials are the atom potentials precalculated from the PDB structure for atom pairs between 40 different atoms and torsion angle potentials similarly precalculated for the main chain angles of the 20 natural amino acid residues. The resulting energy calculation is used to classify the mutations. Different cutoff values are used depending on secondary structure and surface accessibility.

3.3.5. Proximity to Binding or Active Site

Probably the most obvious parameter is to measure the distance to the active site or a functionally important binding site. If this distance is below a certain threshold, e.g., 5 Å, the mutation will almost certainly negatively affect the function of the protein. There are exceptions when the substituted residue is not critical and the properties of the native and variant residues are very similar but in general this parameter has very high accuracy.

The distance can be calculated by taking the residues that define the active site or binding site and then measure the closest distance to each of the residues defining the site. Residues at the site itself thereby obtain a distance of 0 Å and are therefore always included.

3.3.6. Examples

We have in our group used many of the described prediction parameters to explain the clinical phenotypes of mutations in steroid 21-hydroxylase, CYP21, and then successfully predict the severity of the mutations that were unknown at that time (36). CYP21 has over 60 known mutants found in humans making it a perfect protein to use for mutational investigations. Of these mutations we could explain the clinical severity of all but one mutation.

As no known structure of the protein exists, we first created a homology model based on the closest possible homologue, rabbit cytochrome P450 2C5 with 31% sequence identity. This shows that even when no known structure is found it is often possible to create a homology model that can be used to make more accurate predictions on the effects of mutations than from sequence only.

We have in a similar manner studied p53 (37) to discern severe mutations from non-severe mutations. In p53 there are thousands of known mutations found in human cancer patients with determined properties. By using the activity data as training examples, with 25% activity as a separation between severity classes, and with a total of 12 prediction variables, an automated prediction method was developed.

Different approaches were tested, i.e., PCA, SVM, PLS, and an in-house-developed Monte Carlo-based method. The resulting prediction method manages to predict the 1,148 different residue exchanges with an overall accuracy of 77%. For non-severe mutations, we achieved 74% prediction accuracy and for severe predictions 79% which corresponds to an MCC value of 0.52. Similar MCC values were obtained using SVM and slightly worse with PLS. A subset of cancer mutations found in breast cancer was also evaluated resulting in a prediction accuracy of 88%.

The most important prediction variables in this project were conservation, accessibility, stability calculations, and changes in amino acid property.

3.4. Combining Multiple Parameters

When values for the prediction variables have been collected, how do we determine the effect of the mutation upon the activity? The prediction parameters are not equally informative, some are more important than others. Thus, to be able to determine their mutual importance we need to have training examples, mutations with known effect. These training examples can preferably come from the protein itself or from a protein that is believed to be similar enough.

3.4.1. Principal Component Analysis

Principal component analysis (PCA) (**38**) is a useful mathematical tool that can be used to find patterns in complex data sets with many variables. The input variables, often correlated, are reduced to a few uncorrelated variables, principal components. The first principal component is a vector in the input space where the variability of the data is as large as possible. The second principal component does the same thing for the remaining variability of the data. In this way as much information as possible is captured in very few variables.

As PCA is searching for the highest variability it is important to normalize the input before running the analysis. However, there can still be important variables that are neglected in the first components, because they have low variability in the majority of the data. This can, for example, be the effect of outliers or that the data are nonlinear. The nonlinearity can be corrected by a transformation, for example, by taking the logarithm of the values. It is also important to remember that PCA only finds linear relationships. This can be mitigated somewhat by making combinations of different variables or taking a higher polynomial of one variable and adding these to the input variables.

The advantage of PCA is that it can find patterns in data without any training data. When training data exists, it is often better to use more advanced prediction methods so that this information can be incorporated into the system.

The PCA can be performed using, for example, the free statistics package R (http://www.r-project.org/) or MATLAB from MathWorks.

3.4.2. Support Vector Machines

Support vector machines (SVMs) (39) are the opposite of PCA in the sense that they increase the dimensions of the input space rather than reduce it as in PCA. The method also needs training data to be able to make a classification. By using a kernel function (40) the input space is transformed into a higher dimensional feature space. In this higher dimensional space a linear classification can be found even though the data are not possible to separate linearly in input space. The data are separated by a hyperplane in feature space. However, this hyperplane can be created in an infinite number of ways. This is solved by choosing data points in feature space, support vectors, which maximize the margin between the two groups and place a hyperplane between these support vectors.

The advantages with SVMs are that they can find nonlinear separations between classes using linear separation in feature space, making them fast, besides that they are hard to overtrain and thereby predict well on test data. The disadvantage is that for many of the popular kernels, the importance of the input variables cannot be deduced as the prediction is nonlinear.

The training and prediction of SVMs can be made using, for example, SVM-Light (41), the python library LibSVM (42), and the C++ library Shark (43).

3.4.3. Decision Trees

A decision tree is a rather intuitive way of classifying data where the data are divided into groups, or branches in a tree, at several levels. In every branching a decision is made based on a criterion, most often based on only one variable. A prediction is done when a leaf is reached. The tree can be created automatically or manually, taking advantage of the human experience in the field. Also, the decision tree can be used as a first step where the resulting groups can be further analyzed using different classification techniques.

One way to automatically create a decision tree is to find the variable that best splits the data according to observations (44). The same procedure is then repeated for each of the children of the split until no further improvement can be made or no more splits are possible.

Decision trees capture the fact that the importance of a variable can differ according to the circumstances. In this way a nonlinear classifier is created. The drawback is that the method can be overtrained. This can be avoided to some degree by setting a strict stop criterion for where the decision tree should be pruned.

3.4.4. Random Forest

A random forest (45) is an ensemble of decision trees that bases the classification on the most frequent result from the individual decision trees. All the individual trees are fully extended, i.e., no stop criterion or pruning. One of two differences between the random forest methods lies in how the branching is implemented. The simplest way is to take a random feature at each branch. The second difference lies in what input data are included when building the tree. Either everything is used, or a random subset of the training data is used. The latter seems to yield better accuracy and less generalization errors.

Random forest predictions can be performed using, for example, the open source extension packages to the free statistics package R (http://www.r-project.org/) and MATLAB from MathWorks.

3.4.5. Consensus

When several methods or prediction servers have been applied to the data, it is unnecessary to throw away all but the best method. It might be better to use them all and let the different methods vote in order to form a consensus. If one method is superior, this method's vote can be weighted higher and vice versa for a method that is inferior. In this way several mediocre classifiers can be transformed into a good one, and several good methods into a superior one. This works especially well if the methods work in fundamentally different ways or even better are based on different data.

3.5. Prediction Servers

The different molecular properties described above (energy minimization, molecular dynamics, and other parameters) can together be used to predict the expected properties of a modified protein. There are a number of such tools available today. Several of them also provide user friendly Web sites where the user can enter the sequence to be investigated and as result get a prediction of the expected properties of this modified protein.

There are several prediction servers that perform general predictions. Some of these are SNPs3D (46), SNPs&GO (47), SIFT (48), PANTHER (49, 50), and PolyPhen (51–53). However, when there is in-house knowledge about the protein, a protein-specific prediction can usually outperform the more generalized predictions.

SNPs3D is a resource that can be found at http://www.SNPs3D.org where positive profile scores can be considered as non-severe mutations and negative numbers as severe mutations. In SNPs&GO, found at http://snps-and-go.biocomp.unibo.it/snps-and-go/, mutations are judged as neutral or disease related. SIFT can be found at http://sift.jcvi.org/ where substitutions are annotated as intolerant or tolerant. In PANTHER (http://www.pantherdb.org/tools/csnpScoreForm.jsp) the mutant severity is judged according to the probability of the mutation having functional impact on the protein. The PolyPhen server, located at http://genetics.bwh.harvard.edu/pph/, predicts mutants into three classes: benign, possibly damaging, and probably damaging.

3.6. Evaluation

As many prediction methods exist, it is useful to be able to compare how well they perform. A test is usually performed on data not used in the training procedure. The performance can be evaluated in several ways. The simplest way is to take the method that predicts most data correct. However, when data are not evenly distributed this measure can be misleading. A more objective measure is the MCC value described below. It is also useful to find correlation between variables. Sometimes, the predictions can be improved by removing highly correlated variables as this can decrease overfitting, see cross correlation below.

By taking the best method based on the test data, we have in fact done some training on the test data. Therefore, it is valuable to have a third data set which is never used until at the end, when the performance of the method is evaluated. If enough data exists, this is not a problem, but when data are scarce, the prediction performance can decrease substantially if two different test sets are needed.

3.6.1. Matthews Correlation Coefficient

It can be very useful to get a more objective measure of the prediction quality of a two-state classification than percent correctly predicted, or accuracy. If the two groups of data are unevenly distributed, a prediction that favors the larger group will get good accuracy, but it can still be a bad prediction. Matthews correlation coefficient (MCC) (54) is such an objective measure of prediction quality. The MCC value is calculated as follows:

$$\text{MCC} = \frac{(TP \times TN) - (FP \times FN)}{\sqrt{(TP + FN)(TP + FP)(TN + FP)(TN + FN)}}$$

TP stands for true positive, TN for true negative, FP for false positive, and FN for false negative. A perfect prediction will give the value of 1, a random prediction 0, and a perfect negatively correlated prediction a value of −1.

Very uneven distributions are frequent in bioinformatics, where a common task is to find something specific out of a large sample of data. If we, for example, are looking for genes associated with a disease, we are expecting to find in the order of 10 genes out of 20,000 genes. Even if the FP rate is small, say 1%, and the TP rate is high, say 100%, we would still identify 200 incorrect genes but only 10 correct genes. The MCC value would warn us that this is actually not such a good prediction and give a MCC value of only 0.18.

3.6.2. Cross Correlation

Similarity between parameters can be measured using the Pearson product-moment correlation coefficient (55) described by the following equation

$$r = \frac{\sum\left(x - \bar{x}\right)\left(y - \bar{y}\right)}{\sqrt{\sum\left(x - \bar{x}\right)^2 \sum\left(y - \bar{y}\right)^2}}$$

where x and y are values from the two parameters measured, and $\bar{x}$ and $\bar{y}$ are the mean values for respective parameter. Values of r range from −1 to 1, where 1 means that there is a perfect linear relationship between the two parameters and −1 a perfect negative correlation. Optimal when combining two parameters for prediction are that they have low correlation to each other but high correlation to the prediction variable so that the information they contain is not redundant but instead complement each other. If the real value of the prediction is known, the correlation can be used to see which parameters best describe the effect we are looking for and thereby weigh how much each parameter should contribute to the final prediction.

Limitations with this method are that it does only find linear correlations and that it is sensitive to outliers.

A method that can be used to automatically remove variables, that have low correlation with the predicted variable, is LASSO (56). The method minimizes the sum of square errors using linear regression. In addition, LASSO constrains the sum of the absolute values of the parameter weights. The algorithm starts with zero weights for all variables and increments the weights for the variable with the highest remaining correlation to the predicted variable up until the constraint is met or until all parameters have nonzero weight. This means that, for low constraint values, some parameters will get zero weights. By varying the constraint from zero to the infinity, the best linear regression is found. Unnecessary parameters are as a consequence removed entirely.

4. Notes

1. When building a molecular model the quality of the alignment is very important. If the alignment is not optimal it can lead to problems in the homology modeling, see Subheading 2.1. For example, large gaps might give rise to large loops that will be energetically unfavorable and might cause problems when energy minimizing the structures. If this happens, try to adjust the alignment.

   If there are alternatives in the alignment, multiple models can be created based upon different alignment variants. Subsequently, the energy levels of each model can be compared and the most optimal one chosen. These alternative alignments can either be created manually or even better by using different alignment programs.

2. When building homology models (cf. Subheading 3.1) of multimeric proteins consisting of identical subunits, it is useful to only model one monomer and subsequently copy that in desired numbers and position the copies using a related

multimeric structure as template. This will both speed up the calculations and avoid asymmetry. Subsequently, the interfaces between the subunits needs to be energy minimized to avoid steric clashes.

3. To create an optimal MSA it is crucial to inspect the sequences to include. It is important to use as many sequences as possible but at the same time not to bias the MSA toward a subset of very similar sequences, see Subheadings 2.2 and 3.3.1.
4. Since molecular dynamics simulations are computationally very demanding, it is recommended to perform initial tests on single cases in order to estimate needed computer time and the biologically relevant simulation time. Furthermore, if the simulations can be limited by excluding or fixing irrelevant parts of the protein, much time can be gained, see Subheading 3.2.
5. When doing lengthy simulations of all kind it is a good idea to save checkpoint files from which the simulations can be restarted if anything crashes, for example, by a power outage, see Subheading 3.1. This is standard for most MD simulations but not in all energy minimization programs.
6. Energy values are normally not relevant as absolute values; rather it is the relative differences in values between molecular variants that are reflecting stability changes. Especially at the active site, the energy is of minor importance since this area is optimized for functional properties and not stability. Normally, decreased stability is a sign of a mutation that can affect the protein in a negative fashion. However, for very dynamic proteins, increased stability might also be harmful as this would decrease the protein flexibility and thereby impair the function, see Subheading 3.1.

References

1. Weigelt J. (2010) Structural genomics-impact on biomedicine and drug discovery, *Exp Cell Res* ***316***, 1332–1338.
2. Metzker M L. (2009) Sequencing technologies - the next generation, *Nat Rev Genet* ***11***, 31–46.
3. Durbin R M, Abecasis G R, Altshuler D L, Auton A, Brooks L D, Gibbs R A, Hurles M E, and McVean G A. (2010) A map of human genome variation from population-scale sequencing, *Nature* ***467***, 1061–1073.
4. Benson D A, Karsch-Mizrachi I, Lipman D J, Ostell J, and Wheeler D L. (2005) GenBank, *Nucleic Acids Res* ***33***, D34–38.
5. Boeckmann B, Bairoch A, Apweiler R, Blatter M C, Estreicher A, Gasteiger E, Martin M J, Michoud K, O'Donovan C, Phan I, Pilbout S, and Schneider M. (2003) The SWISS-PROT protein knowledgebase and its supplement TrEMBL in 2003, *Nucleic Acids Res* ***31***, 365–370.
6. Dutta S, Zardecki C, Goodsell D S, and Berman H M. Promoting a structural view of biology for varied audiences: an overview of RCSB PDB resources and experiences, *J Appl Crystallogr* ***43***, 1224–1229.
7. Castrignano T, De Meo P D, Cozzetto D, Talamo I G, and Tramontano A. (2006) The PMDB Protein Model Database, *Nucleic Acids Res* ***34***, D306–309.
8. Arnold K, Bordoli L, Kopp J, and Schwede T. (2006) The SWISS-MODEL workspace: a web-based environment for protein structure homology modelling, *Bioinformatics* ***22***, 195–201.

9. Kiefer F, Arnold K, Kunzli M, Bordoli L, and Schwede T. (2009) The SWISS-MODEL Repository and associated resources, *Nucleic Acids Res 37*, D387–392.
10. Pieper U, Eswar N, Webb B M, Eramian D, Kelly L, Barkan D T, Carter H, Mankoo P, Karchin R, Marti-Renom M A, Davis F P, and Sali A. (2009) MODBASE, a database of annotated comparative protein structure models and associated resources, *Nucleic Acids Res 37*, D347–354.
11. Mackey A J, Haystead T A, and Pearson W R. (2002) Getting more from less: algorithms for rapid protein identification with multiple short peptide sequences, *Mol Cell Proteomics 1*, 139–147.
12. Altschul S F, Madden T L, Schaffer A A, Zhang J, Zhang Z, Miller W, and Lipman D J. (1997) Gapped BLAST and PSI-BLAST: a new generation of protein database search programs, *Nucleic Acids Res 25*, 3389–3402.
13. Larkin M A, Blackshields G, Brown N P, Chenna R, McGettigan P A, McWilliam H, Valentin F, Wallace I M, Wilm A, Lopez R, Thompson J D, Gibson T J, and Higgins D G. (2007) Clustal W and Clustal X version 2.0, *Bioinformatics 23*, 2947–2948.
14. Edgar R C. (2004) MUSCLE: a multiple sequence alignment method with reduced time and space complexity, *BMC Bioinformatics 5*, 113.
15. Abagyan R, and Totrov M. (1994) Biased probability Monte Carlo conformational searches and electrostatic calculations for peptides and proteins, *J Mol Biol 235*, 983–1002.
16. Abagyan R, Totrov M, and Kuznetsov D. (1994) ICM - A new method for protein modeling and design: Applications to docking and structure prediction from the distorted native conformation, *Journal of Computational Chemistry 15*, 488–506.
17. Pettersen E F, Goddard T D, Huang C C, Couch G S, Greenblatt D M, Meng E C, and Ferrin T E. (2004) UCSF Chimera – a visualization system for exploratory research and analysis, *J Comput Chem 25*, 1605–1612.
18. Jorgensen W L, and Tirado-Rives J. (2005) Molecular modeling of organic and biomolecular systems using BOSS and MCPRO, *J Comput Chem 26*, 1689–1700.
19. Lindahl E, Hess B, and van der Spoel D. (2001) GROMACS: A package for molecular simulation and trajectory analysis, *J Mol Mod 7*, 306–317.
20. Van Der Spoel D, Lindahl E, Hess B, Groenhof G, Mark A E, and Berendsen H J. (2005) GROMACS: fast, flexible, and free, *J Comput Chem 26*, 1701–1718.
21. Gruber C C, and Pleiss J. (2011) Systematic benchmarking of large molecular dynamics simulations employing GROMACS on massive multiprocessing facilities, *J Comput Chem 32*, 600–606.
22. Case D A, Cheatham T E, 3rd, Darden T, Gohlke H, Luo R, Merz K M, Jr., Onufriev A, Simmerling C, Wang B, and Woods R J. (2005) The Amber biomolecular simulation programs, *J Comput Chem 26*, 1668–1688.
23. Brooks B R, Bruccoleri R E, Olafson B D, States D J, Swaminathan S, and Karplus M. (1982) CHARMM: A program for macromolecular energy, minimization, and dynamics calculations, *Journal of Computational Chemistry 4*, 187–217.
24. MacKerell A D, J.; Brooks B, Brooks C L, I., Nilsson L, Roux B, Won Y, and Karplus M. (1998) CHARMM: The Energy Function and Its Parameterization with an Overview of the Program., *The Encyclopedia of Computational Chemistry 1*, 271–277.
25. Anfinsen C B, Haber E, Sela M, and White F H. (1961) The kinetics of formation of native ribonuclease during oxidation of the reduced polypeptide chain., *Proc Natl Acad Sci USA 47*, 1309–1314.
26. Levinthal C. (1968) Are there pathways for protein folding?, *Extrait du Journal de Chimie Physique 65*, 44.
27. Momany F, McGuire R, Burgess A, and Scheraga H. (1975) Energy parameters in polypeptides, VII: Geometric parameters, partial atomic charges, nonbonded interactions, hydrogen bond interactions, and intrinsic torsional potentials for the naturally occurring amino acids., *J. Phys. Chem. 79*, 2361–2380.
28. Schuler L D, Daura X, and van Gunsteren W F. (2001) An improved GROMOS96 force field for aliphatic hydrocarbons in the condensed phase., *Journal of Computational Chemistry 11*, 1205–1218.
29. Westermark P. (1972) Quantitative studies on amyloid in the islets of Langerhans, *Ups J Med Sci 77*, 91–94.
30. Kruger D F, Martin C L, and Sadler C E. (2006) New insights into glucose regulation, *Diabetes Educ 32*, 221–228.
31. Paulsson J F, Andersson A, Westermark P, and Westermark G T. (2006) Intracellular amyloid-like deposits contain unprocessed pro-islet amyloid polypeptide (proIAPP) in beta cells of transgenic mice overexpressing the gene for human IAPP and transplanted human islets, *Diabetologia 49*, 1237–1246.
32. Lim D, Poole K, and Strynadka N C. (2002) Crystal structure of the MexR repressor of the mexRAB-oprM multidrug efflux operon of

Pseudomonas aeruginosa, *J Biol Chem 277*, 29253–29259.

33. Dayhoff M O, Schwartz R, and Orcutt B C. (1978) A model of Evolutionary Change in Proteins, *Atlas of protein sequence and structure (volume 5, supplement 3 ed.). Nat. Biomed. Res. Found.*, 345–358.
34. Henikoff S, and Henikoff J G. (1992) Amino Acid Substitution Matrices from Protein Blocks, *PNAS 89*, 10915–10919.
35. Parthiban V, Gromiha M M, and Schomburg D. (2006) CUPSAT: prediction of protein stability upon point mutations, *Nucleic Acids Res 34*, W239–242.
36. Robins T, Carlsson J, Sunnerhagen M, Wedell A, and Persson B. (2006) Molecular model of human CYP21 based on mammalian CYP2C5: structural features correlate with clinical severity of mutations causing congenital adrenal hyperplasia, *Mol Endocrinol 20*, 2946–2964.
37. Carlsson J, Soussi T, and Persson B. (2009) Investigation and prediction of the severity of p53 mutants using parameters from structural calculations, *FEBS J 276*, 4142–4155.
38. Pearson K. (1901) On Lines and Planes of Closest Fit to Systems of Points in Space, *Philosophical Magazine 1901*, 13.
39. Boser B, Guyon I, and Vapnik V. (1992) A training algorithm for optimal margin classifiers., *Fifth Annual Workshop on Computational Learning Theory. ACM Press, Pittsburgh.*
40. Kecman V. (2001) Learning and Soft Computing - Support Vector Machines, Neural Networks, Fuzzy Logic Systems, *The MIT press.*
41. Joachims T. (1999) Making large-Scale SVM Learning Practical. Advances in Kernel Methods - Support Vector Learning, *MIT Press.*
42. Chang C-C, and Lin C-J. (2001) LIBSVM : a library for support vector machines.
43. Igel C, Heidrich-Meisner V, and Glasmachers T. (2008) Shark, *Journal of Machine Learning Research 9*, 993–996.
44. Breiman L, Friedman J, Olshen R, and Stone C. (1984) Classification and Regression Trees, *Wadsworth.*
45. Breiman L. (2001) Random forests, *Random forests 45*, 28–32.
46. Yue P, Melamud E, and Moult J. (2006) SNPs3D: candidate gene and SNP selection for association studies, *BMC Bioinformatics 7*, 166.
47. Calabrese R, Capriotti E, Fariselli P, Martelli P L, and Casadio R. (2009) Functional annotations improve the predictive score of human disease-related mutations in proteins, *Hum Mutat 30*, 1237–1244.
48. Ng P C, and Henikoff S. (2002) Accounting for human polymorphisms predicted to affect protein function, *Genome Res 12*, 436–446.
49. Thomas P D, Campbell M J, Kejariwal A, Mi H, Karlak B, Daverman R, Diemer K, Muruganujan A, and Narechania A. (2003) PANTHER: a library of protein families and subfamilies indexed by function, *Genome Res 13*, 2129–2141.
50. Thomas P D, Kejariwal A, Guo N, Mi H, Campbell M J, Muruganujan A, and Lazareva-Ulitsky B. (2006) Applications for protein sequence-function evolution data: mRNA/protein expression analysis and coding SNP scoring tools, *Nucleic Acids Res 34*, W645–650.
51. Ramensky V, Bork P, and Sunyaev S. (2002) Human non-synonymous SNPs: server and survey, *Nucleic Acids Res 30*, 3894–3900.
52. Sunyaev S, Ramensky V, and Bork P. (2000) Towards a structural basis of human non-synonymous single nucleotide polymorphisms, *Trends Genet 16*, 198–200.
53. Sunyaev S, Ramensky V, Koch I, Lathe W, 3rd, Kondrashov A S, and Bork P. (2001) Prediction of deleterious human alleles, *Hum Mol Genet 10*, 591–597.
54. Matthews B W. (1975) Comparison of the predicted and observed secondary structure of T4 phage lysozyme, *Biochim Biophys Acta 405*, 442–451.
55. Rodgers J L, and Nicewander W A. (1988) Thirteen ways to look at the correlation coefficient, *The American Statistician 42*, 59–66.
56. Tibshirani R. (1996) Regression shrinkage and selection via the lasso, *J. Royal. Statist. Soc B. 58*, 267–288.

Chapter 15

Macromolecular Assembly Structures by Comparative Modeling and Electron Microscopy

Keren Lasker, Javier A. Velázquez-Muriel, Benjamin M. Webb, Zheng Yang, Thomas E. Ferrin, and Andrej Sali

Abstract

Advances in electron microscopy allow for structure determination of large biological machines at increasingly higher resolutions. A key step in this process is fitting component structures into the electron microscopy-derived density map of their assembly. Comparative modeling can contribute by providing atomic models of the components, via fold assignment, sequence–structure alignment, model building, and model assessment. All four stages of comparative modeling can also benefit from consideration of the density map. In this chapter, we describe numerous types of modeling problems restrained by a density map and available protocols for finding solutions. In particular, we provide detailed instructions for density map-guided modeling using the Integrative Modeling Platform (IMP), MODELLER, and UCSF Chimera.

Key words: Macromolecular complexes, Electron microscopy, Fitting, Homology modeling, Comparative modeling, Integrative modeling, Visualization, Chimera, MODELLER, IMP

1. Introduction

Structural description of macromolecular complexes is required for studying their assembly, function, and evolution (1, 2). Although numerous assembly structures have been determined by X-ray crystallography (3) and NMR spectroscopy (4, 5), these techniques are not always applicable. Recent advances established electron microscopy (EM) as a central technique for studying the structures of macromolecular assemblies in different functional states in vitro and in vivo. EM approaches include electron crystallography, single-particle EM, and electron tomography (6–8). EM generally produces a three-dimensional (3D) grid specifying the observed electron density of the system (i.e., the density map). The resolution of this map is typically better than 25 Å and can be as high as approximately 4 Å for highly symmetric structures (9, 10). In most

Andrew J.W. Orry and Ruben Abagyan (eds.), *Homology Modeling: Methods and Protocols*, Methods in Molecular Biology, vol. 857, DOI 10.1007/978-1-61779-588-6_15, © Springer Science+Business Media, LLC 2012

cases, however, the resolution of a density map is insufficient to provide a full atomic description of a protein complex. To this end, computational integration of atomic resolution structures with EM density maps is essential. In particular, the resolution of the density map is often adequate for accurate rigid fitting of atomic structures of the subunits into the density map, resulting in an atomic model of the entire assembly (11–22). Given sufficient resolution, flexible fitting can be used to further refine the model by fitting into the density map while maintaining correct stereochemistry (23–27).

A key requirement for such density-guided structural modeling techniques is the availability of atomic resolution structures of the assembly components. These structures, however, are frequently not available from X-ray crystallography or NMR spectroscopy. Fortunately, it may be possible to construct useful component models by comparative (homology) modeling. Comparative modeling techniques are routinely used to model the structure of a given protein sequence (target) based primarily on its alignment to one or more proteins of known structure (templates) (28–30). The target structure is predicted by identifying one or more related proteins of known structure, aligning the target sequence to the template structures, building a model, and assessing it. Comparative modeling approaches have become frequently applicable in part due to the success of structural genomics initiatives that aim to solve representative structures of most protein families by X-ray crystallography or NMR spectroscopy, such that most of the remaining proteins can be modeled with useful accuracy based on their similarity to a known structure. In fact, at least two orders of magnitude more sequences can be modeled by comparative modeling than have been determined by experiment (31). Therefore, methods for improving fitting into a density map by considering errors in comparative models have been developed (19, 32, 33). Moreover, the availability of a density map opens a possibility of improving the corresponding comparative model, by helping with fold assignment, sequence–structure alignment, model building, and model assessment (14, 20, 22, 34).

In this chapter, we describe various types of density-guided modeling problems and available solutions within the Integrative Modeling Platform (IMP) (35), MODELLER (28), and UCSF Chimera visualization software (36). This description is followed by Subheading 5 that highlights several practical issues in density-guided modeling.

2. Materials

To follow the examples, IMP, MODELLER, Chimera, and a set of input files are required. The IMP software can be downloaded from http://salilab.org/imp, MODELLER from http://salilab.org/

modeller, and Chimera from http://www.cgl.ucsf.edu/chimera. All programs are available in binary format for most common machine types and operating systems. IMP can also be rebuilt from the source code. The example files are found in the *biological_systems/groel* directory in IMP.

3. Methods

Selecting a protocol for density-guided structural modeling depends on the resolution of the density map and the available atomic information. Interpretation of the density map usually begins by identifying the different structural units (e.g., entire protein chains, domains, secondary structure elements, or nucleic acids) by means of segmentation techniques (6, 37). Independently, the availability of atomic structures of the components is determined; when necessary, comparative models are built (29, 38), if a template can be found. Then, an appropriate integrative modeling protocol is selected (Fig. 1).

We describe in detail the modeling of the bacterial molecular chaperone GroEL (39–41). GroEL promotes protein folding in bacterial cells in conjunction with its lid-like co-chaperonin protein complex GroES. GroEL is composed of two heptameric rings of identical 57 kDa subunits stacked back-to-back. The GroEL structure was extensively studied by both X-ray crystallography (42–44) and EM (45–48) across different species, and thus provides a good illustration of approaches that integrate EM data into assembly modeling (49).

The inputs for the GroEL example (Fig. 2) are the sequence of the *E. coli* GroEL chaperone monomeric unit (UniProt id: P0A6F5, file: *data/sequences/groel_ecoli.ali*) and an EM density map of the naked GroEL at 11.5 Å resolution (45) (EMDB id: 1081, file: *data/em_maps/groel-11.5A.mrc*) consisting of 14 subunits. We start by searching for known structures homologous to the GroEL monomeric unit (Subheadings 3.1 and 3.2) and independently segment the density map (Subheading 3.3). We then use the density map to assess the choice of the template(s) (Subheading 3.4). Next, we build a comparative model of the GroEL monomeric unit based on the selected template(s) (Subheadings 3.5 and 3.6) and model the entire GroEL complex by simultaneously fitting 14 rigid copies of the monomer model into the complete density map (Subheading 3.7). Finally, we improve the accuracy of the model by refining it to better fit into its density map (Subheading 3.8).

3.1. Template Identification

Template identification is achieved by scanning the sequence of a monomeric unit of the GroEL against a library of sequences for the known protein structures in the Protein Data Bank (PDB) (http://www.pdb.org, (50)). We use the *profile.build*() command

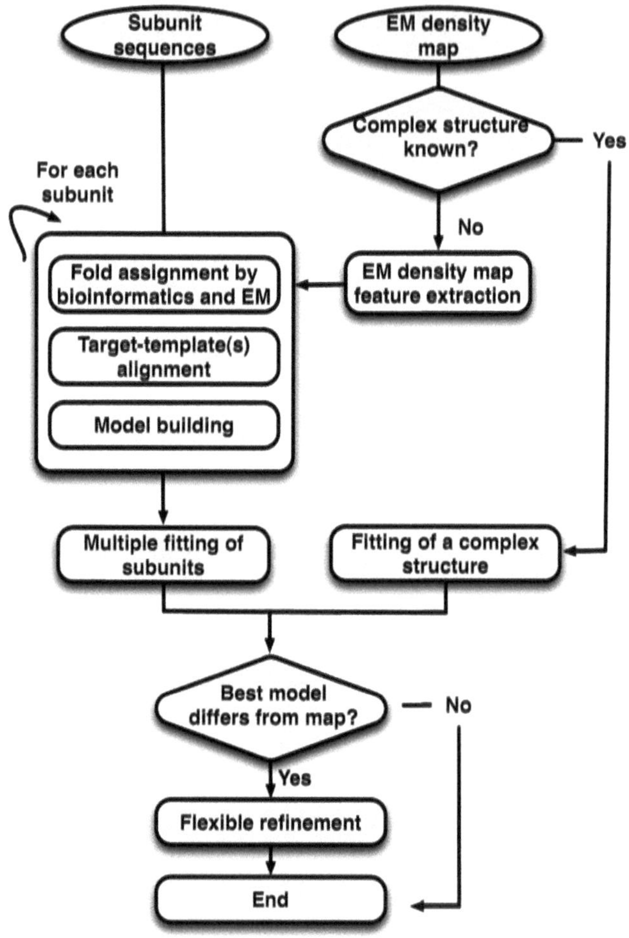

Fig. 1. A flowchart illustrating the steps for modeling a protein complex by comparative modeling and density map fitting.

of MODELLER. The *profile.build*() algorithm uses a local dynamic programming procedure to identify templates with sequences related to the target. In the simplest case, *profile.build*() takes as input the target sequence (file: *data/sequences/groel_ecoli.ali*) and a database of sequences with known structures (file: *data/datasets/pdb_95.pir*), and returns a set of statistically significant alignments (file: *build_profile.prf*). The script and further details can be found in file *scripts/script1_build_profile.py* and Note 1.

3.2. Template(s) Selection by Sequence

Selection of candidate template(s) from known structures found to be homologous to the target is generally a subjective process. Frequently, the selected template(s) share the highest sequence identity to the target. However, additional assessment may be used; in Subheading 3.4, we demonstrate the use of a fit to an EM density map for selecting the most appropriate templates.

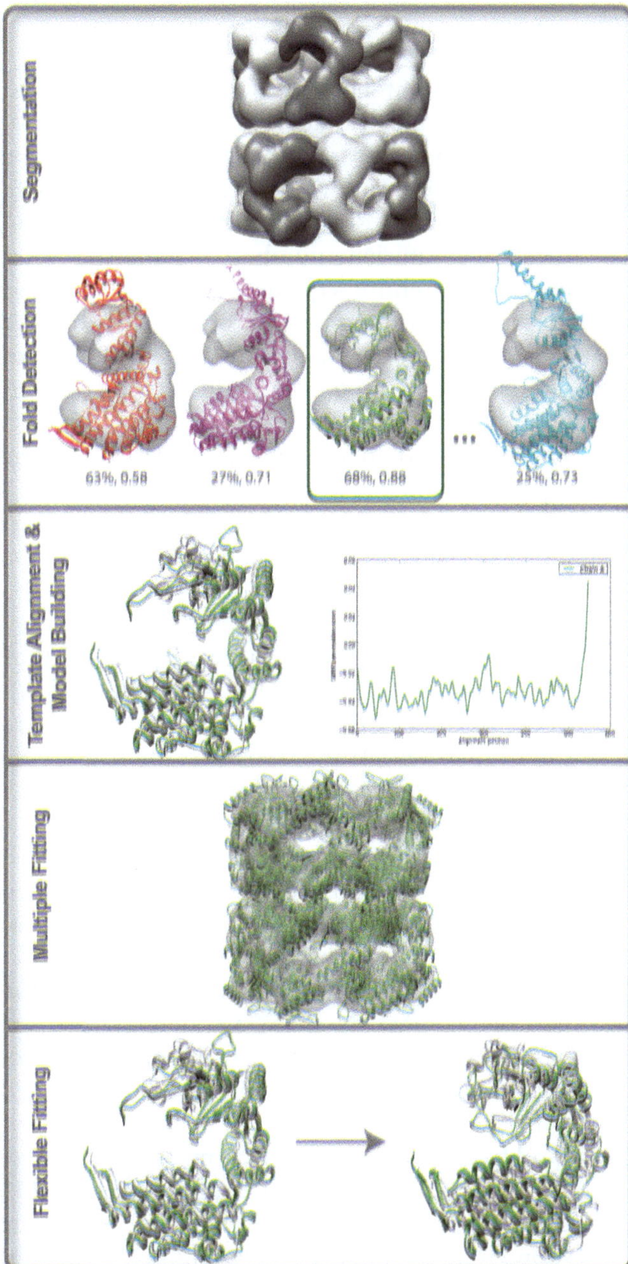

Fig. 2. The steps of EM-guided modeling as applied to the GroEL example. (*Segmentation*) The density map at 11.5 Å resolution is segmented into 14 regions corresponding to the regions occupied by the individual monomers of the assembly. The segments are shown in alternating *shades* of *gray*; (*Fold Detection*) candidate templates are found by scanning the GroEL subunit sequence against the sequences of PDB structures and fitting each of them to the density map. Four of the templates (1we3A, 3kfbA, 1iokA, and 1a6dA), the sequence identity to the target, and the fit into the density map of each of them are shown. The selected template is highlighted in *green* (*Template Alignment and Model Building*); sequence alignment between the target and the selected sequence is generated using a variable gap penalty method. Ten models are constructed and the best model is chosen using the zDOPE, TSVmod, and quality-of-fit scores. A zDOPE profile for the selected model and a superposition of the selected model (*green*) to the reference structure (*gray*) are shown; (*Multiple Fitting*) 14 copies of the target model as simultaneously fitted into the density using the MultiFit method. A model of the complete assembly as generated by MultiFit is shown in *green*; (*Flexible Fitting*) FlexEM is used to refine one of the complex subunits to fit the density map. The starting and refined models (*green*) superposed on the reference structure (*gray*) are shown.

The output file *build_profile.prf* (see Note 2) identifies 13 potential templates, all with high confidence according to their *E*-values, some covering the entire target sequence and others only parts of it. We remove structures matching only a fraction of the target sequence (PDB codes: 1dk7A, 1kidA, 1la1A, and 1srvA), as there is a sufficient number of templates with high confidence covering the entire sequence. To analyze the relationships between the nine remaining structures, we use the *alignment.compare_structures*() command in MODELLER to assess structural and sequence similarity between the structures. This command compares the structures according to the alignment constructed by the *malign3d*() command and produces a clustering tree from the input matrix of pairwise Cα root mean standard deviation (RMSD) distances, helping to visualize differences among the template candidates. The script and further details can be found in file *script2_compare_templates.py* and Notes 2 and 3.

3.3. Density Map Segmentation

Interpretable structural features depend on the resolution of the map and their size. At low resolutions (20–25 Å), the overall shape of the assembly and boundaries of sub-complexes or large proteins can be detected. As the resolution improves, boundaries of smaller proteins or domains can be identified (51–53). At a medium resolution (6–10 Å), secondary structure elements are apparent (37). At a higher resolution, backbone tracing and even side chain conformation may be possible to define (54). Segmentation is, in many cases, performed in a semi-manual manner using visualization tools such as Chimera (21), Amira (http://www.amira.com), Gorgon (http://gorgon.wustl.edu), and Sculptor (http://sculptor.biomachina.org). Notably, a watershed segmentation procedure has been integrated into Chimera (52); secondary structure segmentation and annotation can be performed via the Gorgon visualization software.

Here, we apply a Gaussian mixture model-based segmentation of the density map into 14 regions using the *IMP.multifit.density2anchors* program (55). The resulting segmented regions correspond to the density regions occupied by the subunits. A complete list of commands and further details can be found in file *script3_density_segmentation.py* and Notes 4 and 5.

3.4. Template Selection by Fitting to a Density Map

The density map of the target can aid the process of template selection, by assessing the optimal overlap between a template structure and the density map (14, 19, 20, 34, 56). Such assessment is particularly useful when the templates do not share high sequence similarity with the target or when the conformations of the target and template structures differ (Subheading 3.6). We score the nine remaining candidate templates by fitting each of them into the density map and reporting the EM quality-of-fit score (see Note 6) (25). The score ranges from 0 to 1, with 0 indicating a perfect fit.

Here, the density map is a segmented region corresponding to a monomeric subunit of the GroEL complex density map (file: *groel_subunit_11.mrc*).

Fitting of a component structure into the density map usually optimizes a similarity score between the component and the density map (e.g., the cross-correlation coefficient (CCF)) as a function of the component's translation and rotation relative to the density map (rigid fitting) (49, 57). IMP provides four different methods for performing rigid fitting, based on (1) anchor points matching by geometric hashing (*IMP.multifit.anchor_points_based_rigid_fitting*()) (55), (2) fast Fourier transform (58) (*IMP.multifit.fft_based_rigid_fitting*()), (3) principal component analysis (PCA) (55) (*IMP.multifit.pca_based_rigid_fitting*()), and (4) local Monte Carlo/conjugate gradient search (25) (*IMP.em.local_rigid_fitting*()). Here, we read the profile output into IMP and fit each of the candidate templates into the density map, employing the PCA-based fitting, followed by a local fitting (see Notes 8 and 9). The resulting quality-of-fit scores range from 0.18 to 0.33, indicating that despite the high sequence identity of the target sequence to some of the structures (60% for 1sjpA; 63% for 1we3A), the target structure is in a different conformational state than the templates. Interestingly, some templates with high quality-of-fit scores had lower sequence identity than templates with high sequence identity (e.g., 3kfeA with 27% sequence identity and EM quality-of-fit of 0.3 versus 1we3A with 63% sequence identity and EM quality-of-fit of 0.32), illustrating the potential utility of a density map for improving comparative models. To exemplify advanced flexible fitting techniques, we chose 1iokA as the template. The script and further details can be found in file *scripts/script4_score_templates_by_cc.py*, Notes 6–9, and Figs. 2 and 3.

3.5. Template Alignment

Once template(s) have been selected, the next step of a comparative modeling procedure is aligning the chosen template(s) to the target sequence. Here, sequence–structure alignments are calculated using the *align2d*() command of MODELLER (59). Although *align2d*() relies on a global dynamic programming algorithm (60), it is different from standard sequence–sequence alignment methods because it incorporates structural information from the template when constructing the alignment. This goal is achieved through a variable gap penalty function that tends to place gaps in solvent exposed and curved regions, outside secondary structure segments, and between two positions that are close in space (61). The resulting alignment is written into the file *groel-1iokA.ali* in the PIR format. The script and further details can be found in file *scripts/script5_template_alignment.py*.

In addition, templates and their alignments to the target sequence can be explored using UCSF Chimera (72). Chimera uses BLAST to search the PDB for potential templates, which are

```
import IMP.em
import IMP.atom
import IMP.multifit
IMP.set_log_level(IMP.NONE)
#--- parse the templates file
templates_file="build_profile.prf"
templates=[];seq_ids=[]
for line in open(templates_file):
    if line[0]=="#":
        continue
    s=line.split()
    if int(s[0])==1:
        continue
    if not(int(s[4])-int(s[3])>480):
        print "Not including:",s[0],s[1]
    templates.append([s[1][:4],s[1][-1]])
    seq_ids.append(s[10])
templates_dir="templates/"
#--- load the target density map
dmap=IMP.em.read_map("groel_subunit_8.mrc",IMP.em.MRCReaderWriter())
dmap.get_header_writable().set_resolution(10)
#--- load IMP model
mdl=IMP.Model()
rb_refiner=IMP.core.LeavesRefiner(IMP.atom.Hierarchy.get_traits())
template_fit_sols=[];best_temp=[]
best_fit=1.
dens_threshold=0.02
#--- iterate over the templates and fit each of them
for i,t in enumerate(templates):
    print "fitting template "+t[0]+t[1]
    #load the template
    mh=IMP.atom.read_pdb(templates_dir+t[0]+".pdb",mdl)
    IMP.atom.add_radii(mh)
    #get the right chain
    mh_chain=IMP.atom.get_by_type(mh,IMP.atom.CHAIN_TYPE)[ord(t[1])-ord('A')]
    rb=IMP.atom.setup_as_rigid_body(mh_chain)
    #fit the template to the density map
    sols=IMP.multifit.pca_based_rigid_fitting(rb,rb_refiner,dmap,dens_threshold)
    IMP.core.transform(rb,sols.get_transformation(0))
    #refine the best scoring fit
    mhs=IMP.atom.Hierarchies()
    mhs.append(mh_chain)
    pdb_opt_state=None #IMP.atom.WritePDBOptimizerState(mhs,"refined_temp_%03d.pdb")
    refined_sols = IMP.em.local_rigid_fitting(
        rb,rb_refiner,
        IMP.core.XYZR.get_default_radius_key(),
        IMP.atom.Mass.get_mass_key(),dmap,pdb_opt_state,1,3,1000)
    IMP.core.transform(rb,refined_sols.get_transformation(0))
    IMP.atom.write_pdb(mh_chain,t[0]+t[1]+"_fitted.pdb")
    template_fit_sols.append([
        refined_sols.get_transformation(0)*sols.get_transformation(0),refined_sols.get_score(0)])
    IMP.core.transform(rb,refined_sols.get_transformation(0).get_inverse())
    IMP.core.transform(rb,sols.get_transformation(0).get_inverse())
#--- write the best fitting score for each template
output=open("score_templates_by_cc.log","w")
output.write('{0:<12}{1:<12}{2:<12}{3:<}\n'.format('name','seq id','cc score','transformation'))

for i,t in enumerate(templates):
    rot=template_fit_sols[i][0].get_rotation().get_quaternion()
    v=template_fit_sols[i][0].get_translation()
    pretty_trans='{0:3.6f} {1:3.6f} {2:3.6f} {3:3.6f} {4:3.6f} {5:3.6f} {6:3.6f}\n'.format(
        rot[0],rot[1],rot[2],rot[3],v[0],v[1],v[2])
    output.write('{0:<12}{1:<12}{2:<12}{3:<}\n'.format(
        t[0]+t[1],seq_ids[i],1.-template_fit_sols[i][1],pretty_trans))
output.close()
```

Fig. 3. A Python script used for scoring templates by their quality-of-fit to a segment of a density map.

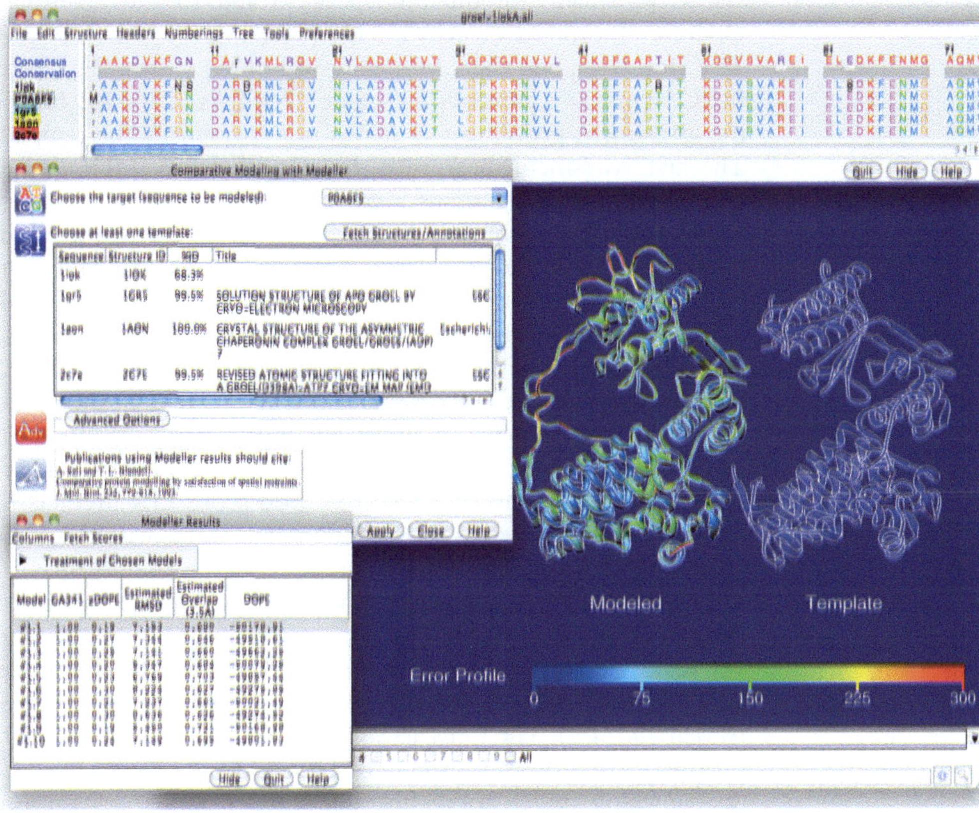

Fig. 4. The Chimera–MODELLER interface. The sequence alignment is displayed in Chimera's Multalign Viewer tool (*top*). In the dialog for running MODELLER (*middle left*), one of the sequences in the alignment is designated as the target (sequence: P0A6F5), and at least one structure (associated with another sequence in the alignment) is designated as the template (structure: 1iok). Structure information is shown to help guide the choice of template. After the run, the resulting models are listed along with various model scores from MODELLER in a table (*bottom left*) and their structures are loaded into Chimera. In this example, the main Chimera window (*right*) shows the template as an outline and one of the model structures as a ribbon colored by error profile.

displayed in the Multalign Viewer tool (Fig. 4, top) (62). The Viewer allows for alignment editing, for example, to remove gaps within an element of regular secondary structure in the template, which frequently contribute to model error. Additional sequences can be added to the alignment, either by typing or extracting from other structures in Chimera.

3.6. Model Building and Assessment

We perform automated comparative model building using the *automodel*() command in MODELLER, generating ten comparative models based on the input target–template alignment (file: *scripts/script6_model_building_and_assessment.py*). Comparison between these ten models reveals structural differences (Cα RMSD between pairs of models range from 4.6 to 8.2 Å, file: *scripts/script7_pairwise_rmsd.py*). To select the most accurate model, we assess the quality of the models according to the normalized Discrete Optimized Protein Energy (zDOPE, see Note 10) (63), TSVMod (64), and the EM quality-of-fit (25) scores. We remove the C terminus region of each model (residues 524–548) prior to the assessment procedure, as it was not covered by the template. The first assessment measure is the zDOPE score (MODELLER command *assess_normalized_dope*()); a value of less than −1 indicates that the distribution of atom pair distances in the model resembles that found in a large sample of known protein structures. The model with the minimum zDOPE score value is model 1 (score of 0.19). However, none of the truncated models got a zDOPE score lower than −0.06, despite the relatively low zDOPE score of the template (−0.6), indicating inaccuracies in the modeling procedure and/or an unusually unfavorable zDOPE score value of the (correct) template structure (see Note 11). The second assessment measure is the TSVMod score that predicts the native overlap (defined as the fraction of Cα atoms within 3.5 Å of the native structure) of a comparative model in the absence of a solved structure using a support vector machine algorithm (64) (http://modbase.compbio.ucsf.edu/evaluation). The predicted Cα RMSD errors are between 5.3 and 8.6 Å for the full models and between 3.4 and 3.9 Å for the truncated models (file: *tsvmod.server.results.txt*). The third assessment measure is the EM quality-of-fit score that measures the fit of a model to the density map. All ten truncated models got comparable scores around 0.2. As according to these criteria all models are of comparable accuracy, we selected model 1 as the starting model for refinement because it scored the best according to zDOPE and EM quality-of-fit scores. A complete list of commands and further details can be found in *scripts/script6_model_building_and_assessment.py*, *scripts/script7_pairwise_rmsd.py*, and Notes 10 and 11.

Alternatively, MODELLER can be called from within Chimera, either as a process run on the user's computer or as a process run remotely via a Web service (72). From the Chimera–MODELLER interface, the user can choose the target sequence, template structure(s), and specify advanced options, e.g., number of output models (Fig. 4, middle left). If the user chooses to run MODELER locally, the MODELLER script file generated by Chimera is accessible and customizable. The MODELLER modeling process runs in the background and can be monitored via Chimera's task manager. Generating 10 comparative models for the GroEL monomer took approximately 20 minutes via the Web service. When the results become available, the models are displayed in Chimera and their scores are shown in a table (Fig. 4, bottom left). The results table lists the GA341 (65), zDOPE, and DOPE scores. Clicking the *Fetch Scores* option triggers a call to TSVMod for accuracy prediction.

3.7. Multiple Fitting into a Density Map

So far we have modeled the structure of the monomeric unit. However, the density map was determined for the entire complex. As a template of the entire complex is not known (for the purpose of this example), we model the whole assembly by fitting 14 copies of the monomeric unit model into the map. We use the symmetric version of the MultiFit program designed to efficiently sample ring complexes. We first split the density into two rings along the *Z*-axis (file: *scripts/script8_split_density.py*). We then run MultiFit separately for each ring (file: *scripts/script9_symmetric_multiple_fitting.py*). The procedure outputs a list of assembly models ranked by their EM quality-of-fit score (files: *multifit.top.output* and *multifit.bottom.output*, see Note 13). The two top-ranking models, one from each ring (files: *model.top.0.pdb* and *model.bottom.0.pdb*), are joined to create a complete model of the assembly with an EM quality-of-fit score of 0.08. A complete list of commands and further details can be found in scripts/script8_split_density.py, scripts/*script9_symmetric_multiple_fitting.py*, and Notes 12 and 13.

Alternatively, MultiFit can be called from within Chimera. From the Chimera–MultiFit interface, the user can choose the monomeric unit model, an EM density map, and specify the map resolution. When MultiFit finishes its calculation in the background, the solutions are displayed and their geometric complementarity scores and EM quality-of-fit scores are shown in a table (72).

3.8. Flexible Fitting into a Density Map

The comparative model generated for the monomeric subunit of GroEL complex is in a different conformational state than the one determined by EM, as indicated by the EM quality-of-fit score (0.2). Conformational differences between a comparative model and its density map can originate from different conditions (e.g., crystallization versus freezing) under which the isolated components and assembly structures were determined, as well as errors in modeling methods (such as mis-assignment of secondary structure elements and their shifts in space caused by target–template misalignment). Flexible fitting can help by refining the conformation of the component, together with its position and orientation. Here, we use the FlexEM method in MODELLER (25) for refining the model to better fit its density. The procedure first adjusts the positions and orientations of its secondary structure segments followed by a full atomic refinement. The increased accuracy of the model is reflected by the EM quality-of-fit score that improved from 0.43 to 0.36. A complete list of commands and further details can be found in file *scripts/script10_flexible_fitting.py* and Notes 14 and 15.

4. Conclusions

EM techniques are becoming increasingly useful for structural characterization of macromolecular assemblies (66). In most cases, however, the resolution of a density map is insufficient to provide a complete atomic description of a protein complex with high confidence. To this end, computational integration of atomic resolution structures with EM density maps is essential. Here, we demonstrate how MODELLER, IMP, and Chimera can be used for modeling structures of such assemblies by a combination of homology modeling, rigid fitting and flexible fitting techniques. These steps are now combined within the Chimera software allowing the user to visualize and control the modeling process (72). We expect such integrative modeling protocols to become increasingly useful and facilitate maximizing the coverage, accuracy, resolution, and efficiency of the structural characterization of macromolecular assemblies.

5. Notes

1. Below we provide a detailed description of *script1_build_profile.py*:
   - *log.verbose*() sets the amount of information that is written out to the log file.
   - *environ*() initializes the "environment" for the current modeling procedure, by creating a new environ object, called env. Almost all MODELLER scripts require this step, as the *environ*() object is needed to build most other objects.
   - *sequence_db*() creates a sequence database object, calling it sdb, which is used to contain large databases of protein sequences.
   - *sdb.read*() reads a text file, containing nonredundant PDB sequences, into the sdb database. The input options to this command specify the name of the database (seq_database_file:'*pdb_95.pir*'), the format (seq_database_format='pir'), whether to read all sequences from the file (chains_list='all'), upper and lower bounds for the lengths of the sequences to be read (minmax_db_seq_len=(30,4000)), and whether to remove nonstandard residues from the sequences (clean_sequences=True).
   - *sdb.write*() writes a binary machine-independent file (seq_database_format='binary') with the specified name (seq_database_file:'pdb_95.bin'), containing all sequences read in the previous step.

- The second call to *sdb.read*() reads the binary format file back in for faster execution.
- *alignment*() creates a new "alignment" object (aln).
- *aln.append*() reads the target sequence groel from the file groel.ali and aln.to_profile() converts it to a profile object (prf). Profiles contain similar information as alignments, but are more compact and better suited for sequence database searching.
- *prf.build*() searches the sequence database (sdb) using the target profile stored in the prf object as the query. Several options, such as the parameters for the alignment algorithm (matrix_offset, rr_file, gap_penalties, etc.), are specified to override the default settings. max_aln_evalue specifies the threshold value to use when reporting statistically significant alignments.
- *prf.write*() writes a new profile containing the target sequence and its homologs into the specified output file (file:*build_profile.prf*).
- The profile is converted back to the standard alignment format and written out using *aln.write*().

2. The results of the *build_profile*() command are stored in the output file *output/build_profile.prf*. The first six lines of this file list the input parameters used to create the alignments between the identified templates and the target sequence. Subsequent lines contain several columns of data, one for each template. For the purposes of this example, the relevant columns are (1) the second column, containing the PDB code of the related template sequences; (2) the tenth column, indicating length of the matched alignment between the GroEL subunit and the template; (3) the 11th column, containing the percentage sequence identity of the alignment; and (4) the twelfth column, containing *E*-values for the statistical significance of the alignments.
3. After a list of all related protein structures and their alignments with the target sequence has been obtained, template structures are usually prioritized depending on the purpose of the comparative model. Template structures may be chosen based purely on the target–template sequence identity or a combination of several other criteria—such as the experimental accuracy of the structures (resolution of X-ray structures, number of restraints per residue for NMR structures), conservation of active site residues, and holo structures that have bound ligands of interest—and fit to other experimental data such as density maps and small angle X-ray scattering curves (67).
4. A segmentation of the EM density map is performed by an adaptation of the Gaussian mixture model (GMM) clustering

technique (55, 68). Geometrically, an assembly of globular proteins can be viewed as a spatial configuration of ellipsoidal components. Each such component can be approximated by a 3D Gaussian, represented by a 3D mean (i.e., its centroid) and a 3D variance (i.e., the square lengths of its principal axes). Thus, a segmentation of an assembly density that corresponds to its molecular configuration can be formulated as finding the most likely linear combination of Gaussian components from which the assembly density was sampled.

5. The script *script3_density_segmentation.py* sets a call to the *IMP.multifit.density2anchors* program; for more options, call the executable directly. *density2anchors* requires specifying of the number of Gaussians (K). It is recommended to set K to the number of proteins (domains) of the assembly, for segmenting a low-resolution (an intermediate resolution) density map, however, different Ks should be tested. To visually inspect segmentation results, add the *seg* option to *density2anchors* run; with this option *density2anchors* writes each segment into a separate MRC file and provides a *load_configurations.cmd* script to load all segments into Chimera.

6. The EM quality-of-fit of a probe (ρ^{P}) to its density (ρ^{EM}) is defined as 1 minus the CCF between them (25). Specifically, CCF is defined as:

$$\mathrm{CCF} = \frac{\sum_{i \in \mathrm{Vox}(p^{\mathrm{P}})} p_i^{\mathrm{EM}} \left(\sum_{j=1}^{N} p_{i,j}^{\mathrm{P}} \right)}{\sqrt{\sum_{i \in \mathrm{Vox}(p^{\mathrm{P}})} \left(p_i^{\mathrm{EM}} \right)^2 \sum_{i \in \mathrm{Vox}(p^{\mathrm{P}})} \left(\sum_{j=1}^{N} p_{i,j}^{\mathrm{P}} \right)^2}},$$ where $Vox(\rho^{\mathrm{P}})$ represents all voxels in the density grid that are within two times the map resolution from any of the atoms of the protein; and where the total density of P at grid point i is $\sum_{j=1}^{N} p_{i,j}^{\mathrm{P}}$. The values of the EM quality-of-fit score range from 0 to 1, where 0 indicates a perfect fit.

7. Below we highlight key commands *in script4_score_templates_by_cc.py* :

   - First few lines parse the *build_profile.prf* file and extract the names of the templates.
   - *IMP.em.read_map*( ) reads the density map. The command gets as input a density map filename and an appropriate reader, which is in this case a *MRCReaderWriter.* IMP supports MRC, Xplor, Spider, and EM formats.
   - The resolution of the density map is not saved in the map and needs to be set using the *set_resolution*( ) command.

- *IMP.Model*() initializes an IMP model which is going to store all templates.
- *IMP.atom.read_pdb*() reads the structure of the template. The function requires a file in PDB format, and a model object that is going to store the molecule. In addition, the function can get as input a *Selector* that specifies which atom types should be read (e.g., *CAlphaPDBSelector* and *NonWaterPDBSelector*).
- *IMP.atom.setup_as_rigid_body*() sets the molecule to a rigid body. The function returns a IMP.core.RigidBody decorator. To learn more on the decorator concept in IMP see http://salilab.org/imp.
- The rigid fitting procedure is performed in two stages. First, coarse fits are explored using the *IMP.multifit.pca_based_rigid_fitting*() command. These fits are then refined by a local Monte Carlo/conjugate gradient (MC/CG) minimization using the *IMP.em.local_rigid_fitting*() command.
- We write the fitted templates using the *IMP.atom.write_pdb*() command. Notice that we used *IMP.core.transform*() to transform the rigid body to its fitted position prior to the writing command.

8. The *IMP.multifit.pca_based_rigid_fitting*() command fits a protein to its density map by aligning their principal components. The principle components of the density are calculated according to all voxels above a density threshold (specified by the user) while the principle components of the density map are calculated according to all atoms. The function returns a list of fits. Each fit is represented by a transformation and a quality-of-fit score.
9. The *IMP.em.local_rigid_fitting*() command locally refines the current fit of a rigid body in a density map by a local MC/CG sampling. At each MC iteration the rigid body is randomly locally transformed followed by a CG minimization. The user can specify the number of MC iterations and the maximum number of CG steps allowed at each iteration.
10. The DOPE score is a pairwise atomic distance statistical potential that assesses atomic distances in a model relative to those observed in many known protein structures. The DOPE potential was derived by comparing the distance statistics from a nonredundant PDB subset of 1,472 high-resolution protein structures with the distance distribution function of the reference state. By default, the DOPE score is not included in the model building routine, and thus can be used as an independent assessment of the accuracy of the output models. In its

normalized version (zDOPE), a score below −1.0 indicates a relatively accurate model, with more than 80% of its Cα atoms within 3.5 Å of their correct positions. However, it might be that the template does not follow a typical shapes found in the PDB, which will result in a high zDOPE for the experimentally determined template. Thus, it is advised to compare the zDOPE profiles of both target and template.

11. The ten models of the GroEL subunit based on 1iok template achieve low zDOPE score (i.e., all models achieved a zDOPE score higher than 0). Visual inspection of the generated models reveals that the C-terminal fragment of the subunit was not covered by the alignment and thus not modeled. After removing this fragment from the models the zDOPE score dropped below 0.

12. MultiFit (55, 69) is a method for modeling the structure of a multi-subunit complex by simultaneously optimizing the fit of the model into its EM density map and the shape complementarity between its interacting subunits (http://www.salilab.org/multifit). It has been shown that the accuracy of both scoring terms is sensitive to errors in comparative modeling (19, 70). Thus, if the target(s) share high sequence identity to their template(s), it is advised to model the assembly based on the template structure(s) and then superpose the target models structure on the corresponding templates. For example, here the accuracy of the subunit homology models were low (as indicated by zDOPE and TSVMod), especially in the loop regions. Thus, we ran MultiFit with the template as input and then replaced the template with the subunit model using a series of transformations commands. A refinement procedure (such as FlexEM) should be next used to fix clashes and improve the fit to the density.

13. Below we highlight key commands *in script9_symmetric_multiple_fitting.py*:
    - *runMSPoints.pl* is a perl script for generating a Connolly surface (71) from the subunit to be fitted.
    - *build_cn_multifit_params.py* is a python script for generating the parameters file to be used by MultiFit. The script initializes the algorithm parameters with its defaults. The user can manually adjust these parameters to allow for an enhanced sampling. Example for one such parameter is the *pca_matching_threshold* parameter. MultiFit filters out ring complexes whose PCA dimensions do not match the ones of the density map. The acceptable difference match size is set by the *pca_matching_threshold* parameter with default value of 3/4 of the EM density map resolution.
    - *symmetric_multifit* is the executable that runs MultiFit given the parameters file. The user can control the number

of output models by the *–n* option. The results are written into a text file consisting, among others, of the following three key fields: (1) The transformation used to build a symmetric complex is written to the *dock rotation* and *dock translation* fields, (2) the transformation used to fit the ring into the density is written to the *fit rotation* and *fit translation* fields, and (3) the cross-correlation score (1 minus the EM quality-of-fit score) is written to the *fitting score* field.

14. A FlexEM refinement procedure is composed of two stages. In the CG stage, the positions and orientations of predefined rigid bodies are resolved via a MC/CG minimization; the rigid bodies usually correspond to secondary structure elements. In the MD stage, positions of all atoms are resolved via a fully atomistic molecular dynamics minimization. A FlexEM tutorial can be found at http://salilab.org/Flex-EM.
15. Below we highlight key commands in *script10_flexible_fitting.py*:
    - Input parameters to be set are as follows: (1) *input_pdb_file*, the name of the comparative model file, already rigidly fitted to the density; (2) *em_map_file*, the name of the density map file; (3) *apix*, the density map voxel size; and (4) *res*, the resolution of the density map.
    - The optimization procedure is controlled by few parameters: (1) *rigid_filename*, the name of the file holding the definition of the rigid bodies (see file *rigid_sses.txt* for the format); (2) *optimization*, the optimization stage to run (CG or MD); (3) *num_of_runs*, the number of models to produce; and (4) *initial_dir*, the initial number for the output directories.
    - This MD optimization stage is controlled by *md_parameters* (i.e., temperatures and number of steps for the simulated annealing algorithm).
    - The *md_return* parameter controls the output model reported as final for each run (*final_mdcg.pdb*). The model can be either the last one sampled (FINAL) or the best scoring one (OPTIMAL).
    - In our example, model #2 got the lowest EM quality-of-fit score.

Acknowledgments

We are grateful to our colleagues Maya Topf, Friedrich Foerster, Jeremy Phillips, and Daniel Russel for their help with EM fitting, MODELLER, and IMP. We also thank Tom Goddard for help with the IMP/Chimera interface. The research of KL was supported by continuous mentorship from Haim J. Wolfson and by the Clore Foundation Ph.D. Scholars program, and carried out her research in partial fulfillment of the requirements for the Ph.D. degree at TAU. This work was also supported by grants from National Institutes of Health [R01 GM54762, U54 GM074945, U54 GM074929, U01 GM61390, P01 GM71790 (AS), P41 RR01081 (TEF)]; the National Science Foundation [0732065 (AS)], and the Sandler Family Supporting Foundation (AS). We are also grateful for computing hardware gifts from Mike Homer, Ron Conway, NetApp, IBM, Hewlett Packard, and Intel.

References

1. Sali A, Glaeser R, Earnest T et al (2003) From words to literature in structural proteomics. Nature 422:216–225
2. Robinson C, Sali A, and Baumeister W (2007) The molecular sociology of the cell. Nature 450:973–982
3. Drenth J (2006) Principles of Protein X-ray Crystallography, 3rd edn. Springer, New York
4. Bonvin AM, Boelens R, and Kaptein R (2005) NMR analysis of protein interactions. Current opinion in chemical biology 9:501–508
5. Neudecker P, Lundstrom P, and Kay LE (2009) Relaxation dispersion NMR spectroscopy as a tool for detailed studies of protein folding. Biophys J 96:2045–2054
6. Frank J (2006) Three-Dimensional Electron Microscopy of Macromolecular Assemblies: Visualization of Biological Molecules in Their Native State 2nd edn. Oxford University Press, New York
7. Stahlberg H, and Walz T (2008) Molecular electron microscopy: state of the art and current challenges. Acs Chemical Biology 3: 268–281
8. Lucic V, Leis A, and Baumeister W (2008) Cryo-electron tomography of cells: connecting structure and function. Histochem Cell Biol 130:185–196
9. Zhang J, Baker ML, Schroder GF et al (2010) Mechanism of folding chamber closure in a group II chaperonin. Nature 463:379–383
10. Chen JZ, Settembre EC, Aoki ST et al (2009) Molecular interactions in rotavirus assembly and uncoating seen by high-resolution cryo-EM. Proc Natl Acad Sci U S A 106: 10644–10648
11. Volkmann N, and Hanein D (1999) Quantitative fitting of atomic models into observed densities derived by electron microscopy. J Struct Biol 125:176–184
12. Roseman AM (2000) Docking structures of domains into maps from cryo-electron microscopy using local correlation. Acta Crystallogr D Biol Crystallogr 56:1332–1340
13. Rossmann MG, Bernal R, and Pletnev SV (2001) Combining electron microscopic with x-ray crystallographic structures. J Struct Biol 136:190–200
14. Jiang W, Baker ML, Ludtke SJ et al (2001) Bridging the information gap: computational tools for intermediate resolution structure interpretation. J Mol Biol 308:1033–1044
15. Chacon P, and Wriggers W (2002) Multi-resolution contour-based fitting of macromolecular structures. J Mol Biol 317:375–384
16. Suhre K, Navaza J, and Sanejouand YH (2006) NORMA: a tool for flexible fitting of high-resolution protein structures into low-resolution electron-microscopy-derived density maps. Acta Crystallogr D Biol Crystallogr 62:1098–1100
17. Birmanns S, and Wriggers W (2007) Multi-resolution anchor-point registration of biomolecular assemblies and their components. J Struct Biol 157:271–280
18. Navaza J, Lepault J, Rey FA et al (2002) On the fitting of model electron densities into EM reconstructions: a reciprocal-space formulation. Acta Crystallogr D Biol Crystallogr 58:1820–1825
19. Topf M, Baker M, John B et al (2005) Structural characterization of components of protein assemblies by comparative modeling and electron cryo-microscopy. J Struct Biol 149:191–203
20. Lasker K, Dror O, Shatsky M et al (2007) EMatch: discovery of high resolution structural homologues of protein domains in intermediate resolution cryo-EM maps. IEEE/ACM Trans Comput Biol Bioinform 4:28–39

21. Goddard TD, Huang CC, and Ferrin TE (2007) Visualizing density maps with UCSF Chimera. J Struct Biol 157:281–287
22. Lindert S, Staritzbichler R, Wotzel N et al (2009) EM-fold: De novo folding of alpha-helical proteins guided by intermediate-resolution electron microscopy density maps. Structure 17:990–1003
23. Hinsen K, Beaumont E, Fournier B et al (2010) From electron microscopy maps to atomic structures using normal mode-based fitting. Methods Mol Biol 654:237–258
24. Orzechowski M, and Tama F (2008) Flexible fitting of high-resolution x-ray structures into cryoelectron microscopy maps using biased molecular dynamics simulations. Biophys J 95:5692–5705
25. Topf M, Lasker K, Webb B et al (2008) Protein structure fitting and refinement guided by cryo-EM density. Structure 16:295–307
26. Trabuco LG, Villa E, Mitra K et al (2008) Flexible fitting of atomic structures into electron microscopy maps using molecular dynamics. Structure 16:673–683
27. Schroder GF, Brunger AT, and Levitt M (2007) Combining efficient conformational sampling with a deformable elastic network model facilitates structure refinement at low resolution. Structure 15:1630–1641
28. Sali A, and Blundell TL (1993) Comparative protein modelling by satisfaction of spatial restraints. J Mol Biol 234:779–815
29. Marti-Renom MA, Stuart AC, Fiser A et al (2000) Comparative protein structure modeling of genes and genomes. Annu Rev Biophys Biomol Struct 29:291–325
30. Ginalski K (2006) Comparative modeling for protein structure prediction. Curr Opin Struct Biol 16:172–177
31. Pieper U, Eswar N, Webb B et al (2009) MODBASE, a database of annotated comparative protein structure models and associated resources. Nucleic Acids Res 37:D347–354
32. Zhu J, Cheng L, Fang Q et al (2010) Building and refining protein models within cryo-electron microscopy density maps based on homology modeling and multiscale structure refinement. J Mol Biol 397:835–851
33. Shacham E, Sheehan B, and Volkmann N (2007) Density-based score for selecting near-native atomic models of unknown structures. J Struct Biol 158:188–195
34. Velazquez-Muriel JA, Sorzano CO, Scheres SH et al (2005) SPI-EM: towards a tool for predicting CATH superfamilies in 3D-EM maps. J Mol Biol 345:759–771
35. Alber F, Dokudovskaya S, Veenhoff L et al (2007) Determining the architectures of macromolecular assemblies. Nature 450:683–694
36. Pettersen EF, Goddard TD, Huang CC et al (2004) UCSF Chimera – a visualization system for exploratory research and analysis. J Comput Chem 25:1605–1612
37. Chiu W, Baker ML, Jiang W et al (2005) Electron cryomicroscopy of biological machines at subnanometer resolution. Structure 13:363–372
38. Baker D, and Sali A (2001) Protein structure prediction and structural genomics. Science 294:93–96
39. Horwich AL, Farr GW, and Fenton WA (2006) GroEL-GroES-mediated protein folding. Chem Rev 106:1917–1930
40. Frydman J (2001) Folding of newly translated proteins in vivo: the role of molecular chaperones. Annu Rev Biochem 70:603–647
41. Sigler PB, Xu Z, Rye HS et al (1998) Structure and function in GroEL-mediated protein folding. Annu Rev Biochem 67:581-608
42. Xu Z, Horwich AL, and Sigler PB (1997) The crystal structure of the asymmetric GroEL-GroES-(ADP)7 chaperonin complex. Nature 388:741–750
43. Braig K, Adams PD, and Brunger AT (1995) Conformational variability in the refined structure of the chaperonin GroEL at 2.8 A resolution. Nat Struct Biol 2:1083–1094
44. Braig K, Otwinowski Z, Hegde R et al (1994) The crystal structure of the bacterial chaperonin GroEL at 2.8 A. Nature 371:578–586
45. Ludtke SJ, Jakana J, Song JL et al (2001) A 11.5 A single particle reconstruction of GroEL using EMAN. J Mol Biol 314:253–262
46. Clare DK, Bakkes PJ, van Heerikhuizen H et al (2009) Chaperonin complex with a newly folded protein encapsulated in the folding chamber. Nature 457:107–110
47. Ludtke SJ, Baker ML, Chen DH et al (2008) De novo backbone trace of GroEL from single particle electron cryomicroscopy. Structure 16:441–448
48. Ranson NA, Farr GW, Roseman AM et al (2001) ATP-bound states of GroEL captured by cryo-electron microscopy. Cell 107:869–879
49. Alber F, Forster F, Korkin D et al (2008) Integrating diverse data for structure determination of macromolecular assemblies. Annu Rev Biochem 77:443–477
50. Berman H, Henrick K, Nakamura H et al (2007) The worldwide Protein Data Bank (wwPDB): ensuring a single, uniform archive of PDB data. Nucleic Acids Res 35:D301–303

51. Baker ML, Ju T, and Chiu W (2007) Identification of secondary structure elements in intermediate-resolution density maps. Structure 15:7–19
52. Pintilie GD, Zhang J, Goddard TD et al (2010) Quantitative analysis of cryo-EM density map segmentation by watershed and scale-space filtering, and fitting of structures by alignment to regions. J Struct Biol 170:427–438
53. Volkmann N (2002) A novel three-dimensional variant of the watershed transform for segmentation of electron density maps. J Struct Biol 138:123–129
54. Baker ML, Baker MR, Hryc CF et al (2010) Analyses of subnanometer resolution cryo-EM density maps. Methods Enzymol 483:1–29
55. Lasker K, Sali A, and Wolfson HJ (2010) Determining macromolecular assembly structures by molecular docking and fitting into an electron density map. Proteins 78:3205–3211
56. Khayat R, Lander GC, and Johnson JE (2010) An automated procedure for detecting protein folds from sub-nanometer resolution electron density. J Struct Biol 170:513–521
57. Wriggers W, and Chacon P (2001) Modeling tricks and fitting techniques for multiresolution structures. Structure 9:779–788
58. Frigo M, and Johnson SG (2005) The Design and Implementation of FFTW3. Proceedings of the IEEE 93:216–231
59. Madhusudhan MS, Webb BM, Marti-Renom MA et al (2009) Alignment of multiple protein structures based on sequence and structure features. Protein Eng Des Sel 22:569–574
60. Needleman SB, and Wunsch CD (1970) A general method applicable to the search for similarities in the amino acid sequence of two proteins. J Mol Biol 48:443–453
61. Madhusudhan MS, Marti-Renom MA, Sanchez R et al (2006) Variable gap penalty for protein sequence-structure alignment. Protein Engineering, Design & Selection 19:129–133
62. Meng EC, Pettersen EF, Couch GS et al (2006) Tools for integrated sequence-structure analysis with UCSF Chimera. BMC Bioinformatics 7:339
63. Shen MY, and Sali A (2006) Statistical potential for assessment and prediction of protein structures. Protein Sci 15:2507–2524
64. Eramian D, Eswar N, Shen M et al (2008) How well can the accuracy of comparative protein structure models be predicted? Protein Sci 17:1881–1893
65. Melo F, Sanchez R, and Sali A (2002) Statistical potentials for fold assessment. Protein Sci 11: 430–448
66. Henrick K, Newman R, Tagari M et al (2003) EMDep: a web-based system for the deposition and validation of high-resolution electron microscopy macromolecular structural information. J Struct Biol 144:228–237
67. Putnam CD, Hammel M, Hura GL et al (2007) X-ray solution scattering (SAXS) combined with crystallography and computation: defining accurate macromolecular structures, conformations and assemblies in solution. Q Rev Biophys 40:191–285
68. Bishop CM (2007) Pattern Recognition and Machine Learning (Information Science and Statistics), 1 edn. Springer, New York
69. Lasker K, Topf M, Sali A et al (2009) Inferential optimization for simultaneous fitting of multiple components into a cryoEM map of their assembly. J Mol Biol 388:180–194
70. Ferrara P, and Jacoby E (2007) Evaluation of the utility of homology models in high throughput docking. J Mol Model 13:897–905
71. Connolly ML (1983) Solvent-accessible surfaces of proteins and nucleic acids. Science 221:709–713
72. Yang Z, Lasker K, Schneidman-Duhovny D, et al (2011) UCSF Chimera, MODELLER, and IMP: an Integrated Modeling System. J Struct Biol. (In press, doi:10.1016/j.jsb.2011.09.006)

Chapter 16

Preparation and Refinement of Model Protein–Ligand Complexes

Andrew J.W. Orry and Ruben Abagyan

Abstract

The formation of ligand–protein complexes are critical for the correct functioning of a cell. The prediction of these interactions is important for our understanding of how the cell works and for the development of new drug molecules. Homology modeling is a method for predicting the structure of a protein based on a crystal structure template. Once a model of the protein is complete, a ligand-docking algorithm predicts the ligand–protein model interaction by searching for the best steric and energetically favorable fit. A refinement of the ligand-binding pocket improves the predicted interactions by considering the flexible nature of the ligand-binding pocket. In this chapter, we describe, from first principles, methods to identify and prepare the ligand-binding pocket in a protein model, to dock the ligand, and refine the resulting complex.

Key words: Homology model, Refinement, Docking, Ligand binding, Drug interaction, Structure-based drug design, Internal coordinate mechanics, Virtual screening, Induced fit, GPCR

1. Introduction

The problem of building models by homology that are accurate enough to predict ligand interactions has long been posed by the modeling community. However, despite definite improvements, the latest homology modeling and docking competitions GPCR Dock 2008 and 2010 (1, 2) clearly demonstrate that the success of the results vary from "almost there" (human dopamine D3 receptor bound to eticlopride) to "nothing even remotely similar" (the CVX15 cyclic peptide with CXCR4 chemokine receptor). The model building process for the best models needed to be enhanced with ligand guidance of some sort. However, standard homology modeling methods do not directly take into account any ligand information during the modeling process (3–6). To incorporate chemical biology data into your protein model, you will

Andrew J.W. Orry and Ruben Abagyan (eds.), *Homology Modeling: Methods and Protocols*, Methods in Molecular Biology, vol. 857, DOI 10.1007/978-1-61779-588-6_16, need to use a ligand docking algorithm to predict how and where small molecules such as drugs, chemical probes, and biological substrates bind to your model (7–11). The main steps include (1) the choice of a crystallographic template or templates, (2) the alignment of the modeled sequence to the X-ray template, and (3) the refinement of the whole model as well as its specific parts needed for small molecule recognition.

The Protein Data Bank (PDB) (12) provides template structures for the construction of the model. Depending on the crystallographic conditions, the template structure can be in a multiplicity of functional states including active or inactive and apo and holo forms. In most modeling cases, the modeled structure will inherit the template's structural state and therefore it is important to select the most suitable template for the ligand interaction problem under-investigation. For example, if the aim were to predict the interaction of a ligand to an orthosteric site then the ideal template would be a structure in which a ligand is already bound to a similar pocket. In many cases, the template structure does not reveal detailed information about the ligand-binding pocket and so published biochemical and structural data can be used to locate the pocket or a prediction algorithm can be used (13, 14). In general, three potentially conflicting considerations should be applied to select the main template as well as the secondary templates. First, is the overall structure resolution and its quality, e.g., a structure with resolution of 1.8 Å is definitely preferable over a 3.5 Å resolution structure where a large fraction of side chains was resolved in density. Second, the main template needs to be the closest to the protein of interest not according to a general sequence identity, but in terms of the model areas and substructures of immediate interest. For example, an open structure of protein A may be more relevant as a model of the open state of protein B, even if a closed form of B′ or even B itself exists. Finally, the ligand binding imposes specific requirements and a bound structure may be preferable than the apo one.

Even the best and the closest crystal structure does not address the fundamental properties of a good ligand binding model such as protonation, tautomerization, and conformational induced fit. It is important to consider the limitations of your model based on the crystallographic data of the template structure (15). Crystallographic data such as B-factors, occupancies, and the crystal packing state of the template structure provide information that may affect the structure of the modeled ligand-binding pocket. Likewise, for the model structure, predictions about the orientation of His, Gln, and Asp residues and the charged states of His, Asp, Glu, Arg, and Lys residues need to be made. Also, a decision to include water or cofactors into the model is important, as this will affect the predicted ligand interactions.

In the simplest scenario, the model has high sequence identity to the template structure and the ligand under investigation has similar chemical properties to the template bound ligand. In this case, no docking algorithm is required; the ligand can be manually placed inside the ligand-binding pocket preserving key ligand–receptor contacts. For example, modeling the ligand–receptor interactions of Type I kinase inhibitors is aided by knowing that the inhibitors form 1–3 hydrogen bonds to the hinge region linking the N- and C-terminal. These interactions mimic those formed by the amino group on the adenine ring of adenosine-5′-triphosphate (ATP).

In most cases, a docking algorithm is required to predict and refine the ligand–receptor interactions. To understand the equilibrium between the solvated ligand and the ligand–receptor complex in silico many different complex energy parameters are considered. For the formation of a ligand–receptor complex, the interactions may include electrostatics, hydrogen bonding, van der Waals interactions, hydrophobic interaction, and the loss of entropy of the ligand upon binding. Most efficient docking algorithms use potential energy 3D grid representations of the receptor in the first implementation. A docking energy function discriminates between many different conformations of the docked ligand in the binding pocket to find the global minimum.

The first published computational ligand docking method used a rigid ligand and a receptor geometric matching approach (16) and Fourier transforms to calculate the degree of molecular surface complementarity between the ligand and receptor (17). More recently, docking methods have evolved to allow the ligand to be treated as flexible and incorporate ways of treating protein flexibility. The search for the global minimum can be undertaken via docking algorithms described in this chapter including the biased probability stochastic search in internal coordinates using collective variables (ICM and BPMC) (18), Monte Carlo (MC) (19–21), molecular dynamics (MD) (22–26), genetic algorithm (GA) (27–29), and fragment based (30, 31).

Once an initial model of the ligand–protein complex has been made and the protonation/tautomerization states established, it may be necessary to further refine the model by predicting possible backbone and side-chain flexibility in the pocket. When a ligand binds there is usually some adaptation of the pocket, an effect known as ligand-induced fit. In recent years, a number of methods have been developed to predict this effect including sampling side-chain rotamers, reducing the penalty for van der Waals clashes, and using the ligand or other modeling tools to generate multiple receptor conformations of the ligand pocket (32–34).

2. Materials

2.1. Computer Specifications

The minimum hardware specifications for most docking and refinement algorithms are in the range of 100–400 MB of disk space and 1 GB of RAM. These specifications are well within those of a reasonably priced modern desktop computer. It is recommended to check the exact specifications, platforms, and graphic cards supported by the vendor before purchasing the software or hardware.

2.2. Available Algorithms

Tables 1–4 describe selected commercially available and open source algorithms for each step of a ligand-docking experiment.

Table 1
Selected algorithms for the prediction of ligand-binding pockets

Software name	Download site	Reference
CastP	http://sts.bioengr.uic.edu/castp/	(101)
ConSurf	http://consurf.tau.ac.il	(102)
FPocket	http://fpocket.sourceforge.net/	(103)
SiteHound	http://scbx.mssm.edu/sitehound/sitehound-web/Input.html	(104)
Q-SiteFinder and PocketFinder	http://www.modelling.leeds.ac.uk/qsitefinder/ http://www.modelling.leeds.ac.uk/pocketfinder/	(105, 106)
ICMPocketFinder	http://www.molsoft.com/icm_pro.html	(43, 44)
Pass	http://www.ccl.net/cca/software/UNIX/pass/overview.shtml	(107)
Surfnet	http://www.biochem.ucl.ac.uk/~roman/surfnet/surfnet.html	(35)

Table 2
Selected chemical databases for retrieving ligands for docking

Database name	Download site	Reference
ChEMBL	https://www.ebi.ac.uk/chembldb/	(108)
DrugBank	http://www.drugbank.ca/	(109–111)
KEGG	http://www.genome.jp/kegg/	(112–114)
MolCart Compound Database	http://www.molsoft.com/molcart-compounds.html	
PubChem	http://pubchem.ncbi.nlm.nih.gov/	(115)
Zinc Database	http://zinc.docking.org/	(116)

Table 3
Selected ligand sketching software which can save molecules in formats suitable for ligand docking (e.g., SDF and Mol format)

Software name	Download site
ChemDoodle	http://www.chemdoodle.com/
ChemDraw	http://www.cambridgesoft.com/software/chemdraw/
ChemWriter	http://chemwriter.com/
ICM-Chemist	http://www.molsoft.com/icm-chemist.html
Marvin	http://www.chemaxon.com/products/marvin/

Table 4
Selected ligand docking methods

Software name	Description	Reference
AutoDock and Vina	AutoDock provides a number of different ligand conformation search options including a genetic algorithm and an MC method and uses a grid-based method for energy evaluation. Vina is a new faster algorithm, which has been shown to be more accurate than AutoDock in predicting ligand-binding pose	(21, 117)
cHits	This method breaks the ligand into rigid fragments and then docks each fragment into the ligand-binding pocket. The fragments are then connected by flexible chains and then scored	(118)
DOCK	The original DOCK method used using rigid body docking and geometric matching algorithms. Spheres are used to describe the ligand- and receptor-binding pocket, the spheres are then matched, positioned, and then scored. Newer versions of DOCK use map representation of the ligand-binding pocket, and can also incorporate representations of receptor flexibility	(16, 31, 61)
FlexX	This algorithm uses an "anchor and grow" method whereby the anchor is docked according to chemical complementarity and then the remainder of the ligand is built up incrementally from other fragments. The flexibility of the ligand is represented by multiple conformations and score based on their interaction with the receptor	(30)
FRED	The FRED algorithm uses a combination of shape complementarity and pharmacophore parameters to search the receptor-binding site. Consensus scoring is then used to rank the ligand-binding poses	(119)
Glide	This algorithm uses a series of filters to search for the best position, orientation and conformation of the ligand. A set of ligand conformations are generated and then clustered and selected conformations are minimized in receptor energy grids. The best energy poses are refined using an MC procedure and scored	(120–122)

(continued)

Table 4 (continued)

Software name	Description	Reference
GOLD	A genetic algorithm is used to represent both rotatable dihedral and ligand–receptor hydrogen bonds. The ligand–receptor hydrogen bonds are optimized and each complex is ranked according to this scoring function	(123, 124)
ICM-Pro	The molecular system is represented using internal coordinates. The receptor can be represented by grids and energy calculations are made in the ECEPP force field. A biased probability Monte Carlo global optimization procedure is used to dock a fully flexible ligand	(18, 62, 73)
Surflex	Surflex searches for morphological similarity between the ligand and receptor using a flexible alignment optimization procedure. The Hammerhead scoring function is used to rank the ligand pose predictions	(125–127)

3. Methods

3.1. Ligand-Binding Pocket Identification

The ligand-binding pocket or active site is straightforward to identify in the protein model if:

- The template upon which you have modeled your structure is in the *holo* form and the ligand is bound in the catalytic site.
- The chemical properties of the ligand you are attempting to bind to the model have characteristics that indicate a particular pocket type is required. For example, if the ligand is a nucleotide such as ATP or nicotinamide adenine dinucleotide (NAD) the template structure and the model should have a characteristic Rossmann fold which will help to pinpoint the binding site.
- The modeled protein of interest has extensive sequence evolutionary information and so the ligand-binding pocket information can be gleaned from studying large sequence family alignments (e.g., kinases, nuclear receptors and Family A G-Protein-Coupled Receptors).

In some cases, the ligand-binding pocket is either unknown or partially known. For example, an allosteric binding pocket is under-investigation or mutational data indicates a particular region of the protein may bind a ligand. In this situation, an algorithm is required to fully identify and define the boundaries of the pocket. Table 1 lists some of the available algorithms for identifying ligand-binding pockets in a protein model.

Methods for identifying pockets can be grouped into two categories: (1) geometric approaches, which analyze the surface of the

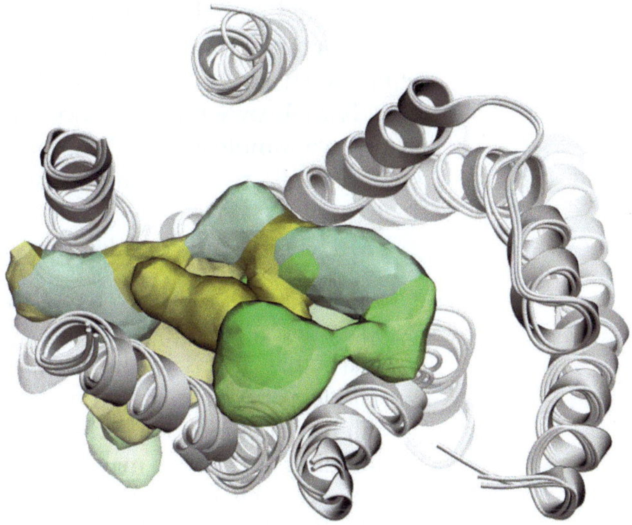

Fig. 1. Predicted ligand binding pockets (displayed as surfaces), generated by icmPocket-Finder (43, 44), for three models of the GPCR Melanin Concentrating Hormone (MCH). The models (displayed in *ribbon* representation) were constructed using ligand-guided modeling and were used for the identification of new MCH inhibitors (98).

protein to find cavities (35–37) and (2) molecular fragment and ligand docking approaches which score the pocket by how well a probe fits into the cavity (38–42). Successful applications of both methods have been reported, but the latter method is computationally expensive, while the geometric approach can sometimes identify pockets that are not drug-like.

The ICM Pocket Finder method in the ICM-Pro software (MolSoft LLC, San Diego) is well validated and straightforward to use (43–45). This method relies solely on the protein structure and can identify cavities and clefts without any prior knowledge of the substrate. The position and size of the ligand-binding pocket are determined based on a transformation of the Lennard-Jones potential, a grid map of a binding potential and construction of equipotential surfaces along the maps. The pockets are displayed graphically as a surface and the dimensions of each pocket are presented in an interactive table and plot (Fig. 1).

The input for pocket identification programs is the model in PDB format and the algorithm will add hydrogen atoms to the structure. Special care should be taken with the software if you are looking for a pocket that is exposed (e.g., protein–protein interaction site) because most of the default parameters are trained to identify buried drug-like pockets (see Note 1).

3.2. Ligand-Binding Pocket Preparation

Before a ligand is docked into a protein model, the inherent inaccuracies or variability associated with the model need to be fully analyzed. This should be addressed at an early stage otherwise the final docked complex will almost certainly be incorrect. The key crystallographic factors which need to be considered about the template structure used to build the model are described below. The ICM-Browser and Browser-Pro software (download here: http://www.molsoft.com/icm_browser.html) provides a useful set of tools to view and analyze the template and model structures.

Model template considerations are:

- The *B-factor*, also referred to as the atomic displacement parameter, will give an indication of the thermal motion of particular atoms in the template structure. Therefore, if the model is based on a region of the template structure which has high B-factors (>50) and this region coincides with the ligand-binding pocket you may want to consider modeling alternative states of this region of the protein (see Note 2). To visualize the B-factors using ICM-Browser:
  - File/Open and choose template PDB file.
  - Select the display tab and display in wire representation.
  - Click and hold the wire representation button and select Color by: B-factor.
- The *occupancy* represents the fraction of atoms that occupy a crystallographic position. So if the electron density of an atom in the template is present the occupancy value will equal to one, but if it is completely absent then the value will be zero. If the occupancy value is zero for side-chain atoms, then the modeling program used to generate the model will build the residues independently of the template and therefore caution should be taken with this region when considering ligand–receptor interactions. To check the occupancy of the template, the electron density file for a PDB structure can be downloaded from the Uppsala Electron Density Server (46) and contoured. The ICM-Browser-Pro software can be used to visualize the electron density map:
  - File/Open and choose template PDB file.
  - File/Load Electron Density and enter the PDB code.
  - Tools/X-Ray/Contour Electron Density.
- The structure of the template ligand-binding pocket might be affected by *crystal-packing interactions* which are only observed due to the crystallization conditions and would not be present in solution. For example, a loop region in a ligand-binding pocket may have a unique conformation only because of its crystal contact neighbors. Therefore, it is important to investigate the template structure to determine where the crystal contacts are located by displaying neighboring molecules in

the template structure. To display the neighboring molecules in ICM-Browser-Pro:

- File/Open and choose template PDB file.
- Tools/X-ray/Crystallographic Neighbors and you can determine whether you want to view the entire molecule or fragments of the neighbors.
- Some template structures, solved at very high resolution, may contain alternative conformations for certain residues. If the residues with alternative conformations are conserved between the template and the model you can make multiple receptor conformations of your model for docking (see Subheading 3.5).

Hydrogen atoms need to be added to the model before a ligand can be docked to the binding pocket, some modeling methods do this automatically, but their placement needs to be checked. The hydrogen positions should ensure that the most favorable hydrogen-bonding networks pattern is achieved. The addH program in the Chimera suite of software (47) is one example of a program that will add and optimize hydrogen atoms. In ICM-Browser, hydrogen atoms can be automatically added to the structure, using an option called Convert PDB which looks at the residue name and adds a full-atom depiction along with full hydrogen optimization.

Once you have built your model the following considerations need to be made:

- The *orientation and protonation states of histidine* residues in your model need to be determined before docking. The histidine residue can be found in two neutral conformations where the positive charge is delocalized between Nδ and Nε at physiological pH or in one charged conformation. A procedure is needed which optimizes the position of the hydrogen to determine the best orientation and protonation state. In the ICM-Browser software, His residues are optimized when converting a PDB file into an ICM object.
  - Right click on the model structure in the ICM workspace.
  - Select convert PDB.
  - Select optimize HisAsnGlnCys.
- The *orientation* at the heavy atom level for Gln and Asn residues in the model need to be determined. There is ambiguity about the positioning of the nitrogen and oxygen atoms in these residues because the electron density for these two atoms looks similar. Maximizing hydrogen bonding and other interactions with neighboring residues in the pocket can achieve the

correct positioning. In ICM-Browser, the Gln and Asn residue are optimized using the same actions as described previously for His residues.

- Assign correct *charges* to Asp, Glu, Lys, and Arg. The basic residues lysine and arginine carry a positive charge at physiological pH and Asp and Glu are negatively charged. There are some situations when these residues may need to be uncharged in the pocket (see Note 3).
- A rule of thumb for docking is that water molecules are removed from the protein and most modeling software do not consider water. In some cases, however, water molecules are modeled into the pocket but this would only be reasonable if the pocket of the model was almost identical to the template structure or the exact location of the water is known and waters were found experimentally to play an important function in ligand binding. The same is generally true for cofactors and metals, which are in the pocket to bind a charged native ligand, so for neutral drugs it would not make sense to model these ions into the pocket.

3.3. Ligand Preparation

There are a number of commercial and academic ligand databases and websites where 2D and 3D sketches of ligands are stored (see Table 2). Alternatively, you can draw the ligand yourself using a molecular editor (see Table 3) or extract the ligand from a PDB file (see Note 4). Many chemical vendors provide their catalog in electronic format on request or you can search their databases online (e.g., ChemDiv's chemical e-Shop http://chemistryondemand.com:8080/eShop/).

Most docking algorithms can read one of the following ligand formats (1) The MOL format (*.mol) developed by MDL (now Symx) (48) is one of the most recognized and used chemical file formats. The main elements of the file is a header containing information about the chemical, and fields for atom, bond connections, and types. A collection of more than one chemical MOL file (separated by $$$$) is called an SDF file, (2) the Mol2 format (*.mol2) developed by Tripos (49) is also a common way to input ligand data into docking algorithms, (3) an easier to read format developed by Daylight is called the Simplified Molecular Input Line Entry Specification (SMILES) (50, 51). The SMILES string is a series of characters representing atoms, bonds, aromaticity, branching, stereochemistry, and isotopes. This is an example of a SMILES string for benzene C1C=CC=CC=1.

Depending on the docking method, the ligand is usually flexible during the docking simulation or conformations of the ligand are generated in the absence of the receptor and then docked into the receptor.

3.4. Docking Method Search Algorithms

Table 4 lists a selection of available docking algorithms. The decision about which docking method to use should be based on published success stories for the protein target receptor family under investigation or by analyzing published performance comparisons (1, 52–56) (see Note 5).

3.4.1. Monte Carlo Docking Methods

A Monte Carlo (MC) docking algorithm docks the ligand by randomly sampling the energy landscape of the ligand-binding pocket (57). Variables in the ligand and/or receptor are randomly changed or the ligand jumps to another region of the pocket. The energy of the system is evaluated and a decision is made whether to accept or reject a conformation based on the energy. If the energy of the new conformation (E_{new}) is lower than the old conformation (E_{old}) then the conformation is accepted if not then the Metropolis criterion is used to determine the outcome of the conformation where k is Boltzman's constant and T is the effective temperature of the simulation.

$$P_{acc} = \exp\left[\frac{-(E_{new} - E_{old})}{kT}\right].$$

The random steps are repeated using adaptive heuristics to determine the termination point. The advantage of MC is that a large rugged energy landscape can be sampled. Monte Carlo-based methods include MCDock (19) and Autodock Vina (21).

3.4.2. Molecular Dynamics Docking Methods

Molecular dynamics (MD) docking simulates the movement of the ligand and/or the receptor atoms as a function of time by integrating Newton's law of motion (58). Each atom within the molecule is considered as a sphere with mass and charge obeying the laws of classical mechanics. The energy of the system is calculated in force fields such as AMBER (25) and CHARMM (26) whereby the acceleration and direction of movement of each atom is determined. A variety of different conformations can be generated by heating and cooling the system over defined periods of time, this allows energy barriers to be overcome by simulating bond stretching and rotation.

The MD approach is very computationally expensive due to the time required to traverse the rugged energy landscape and therefore docking methods that use MD find various ways to overcome this problem. One way to sample the ligand-binding pocket more efficiently using MD is to use a high temperature for translational modes and a lower temperature for the internal degrees of freedom or use hybrid methods that use MD and Brownian dynamics to define a probabilistic distribution of motion to sample the ligand in the pocket (22–24, 59, 60).

3.4.3. Genetic Algorithms

The genetic algorithm (GA) approach to docking takes a set of variables such as rotatable torsion angles of the ligand and then mimics the evolutionary process by placing these into "chromosomes" and evolving them by making "mutations" and "crossovers." The "chromosomes" are then ranked according to a predefined scoring system to determine the most advantageous combination of values and then this spawns a new generation of "fitter" chromosomes which are further ranked and the process is repeated a set number of times. Programs such as GOLD (28), DARWIN (27), and DIVALI (29) use GAs.

3.4.4. Ligand Fragment-Based Methods

Ligand fragment-based docking methods use a piece of the ligand to identify a rigid anchor. This anchor is then docked and then the rest of the ligand is grown from that point. Two of the more popular methods are FlexX (30) and DOCK (16, 31, 61).

- FlexX uses chemical complementarity to dock the anchor fragment and this reduces the number of possible binding orientations of the anchor.
- DOCK uses an algorithm, which identifies the rotatable bonds in a ligand, helping to identify the rigid anchor. The anchor is docked by shape complementarity and then ligand fragments are linked and merged to the anchor. As each fragment is added to the anchor the torsion angles are varied and a collection of best ligand poses are selected.

3.4.5. Internal Coordinate Mechanics and Biased Probability Monte Carlo

Most docking software use standard Cartesian description of the coordinates of each atom (x, y, z). However, you can reduce the number of variables analyzed in the simulation by using internal coordinates (IC), which makes the search for the global energy minimum between the ligand and the receptor more efficient (62). IC takes into account bond lengths, planar angles, and torsion angles and because bond lengths and planar angles are generally rigid under normal conditions, it is only that the torsion angles are variable. The reduction in variables is even greater when you consider that at every branching point in the atom chain there is some sharing of the same torsion angle.

The internal coordinate mechanics (ICM) docking method from MolSoft LLC (San Diego, CA) uses grid potentials to represent the ligand-binding pocket (18, 63). Once the ligand-binding pocket has been identified the grids are setup by using a convenient graphical user interface or via the command line for high throughput docking on a cluster. The docking project is given a name (Docking menu/Set Project) which will label all the files associated with the docking project. The program is then instructed where the ligand-binding pocket is by the selection from ICMPocketFinder or by a ligand bound to the receptor, or defined explicitly by the user (Docking menu/Receptor setup). The program will then ask you to determine the dimensions of the maps (see Note 6) and will

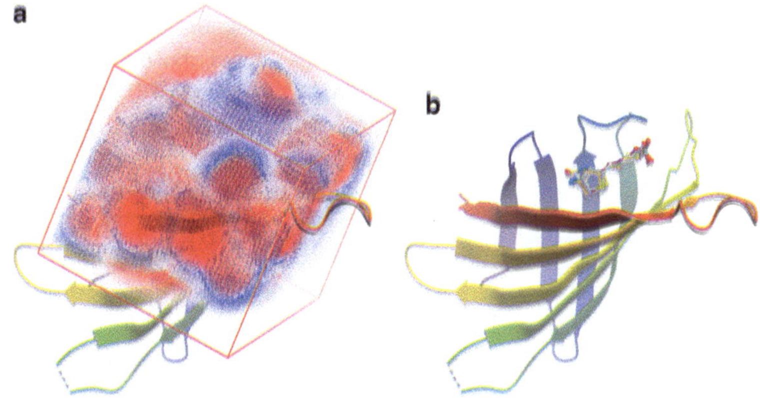

Fig. 2. (**a**) ICM grid potential maps shown as a *box* surrounding the ligand-binding site. Grid maps speed up docking compared to an explicit atom representation of the receptor (displayed in *ribbon* representation). (**b**) During docking, the best energy ligand poses are stored in a stack of conformations. Once docking has completed the stack of ligands ranked by energy or docking score can be displayed in the pocket and the interactions analyzed.

proceed to generate grid maps for the following energy terms (1) hydrogen bond potential energy, (2) van der Waals grid potentials including a smoothed grid potential to allow some flexibility in the receptor, (3) electrostatic potential, and (4) hydrophobic potential (Fig. 2a).

The fully flexible ligand is then docked into the maps using the ICM-biased probability Monte Carlo (BPMC) method (18, 45). The first step in the BPMC global optimization procedure is for the ligand to undergo a random conformation change of free variables according to a defined probability distribution followed by a local gradient energy minimization in torsion angle space. The energy of the complex is then calculated including non-differentiable energy terms such as entropy and solvation and then the conformation is accepted or rejected based on the Metropolis criterion (57). The process is then repeated and terminated using adaptive heuristics based on the ligand size and flexibility.

Once the docking has finished a collection of the most energetically favorable poses of the ligand are collected and can be displayed interactively inside the ligand-binding pocket (Fig. 2b). Further options to incorporate flexibility within the receptor are available (see Subheading 3.5). The ligand–protein model complex can then be saved in PDB format and further analyzed (see Note 7).

3.4.6. Evaluating the Docked Ligand

During the docking procedure, many ligand poses are assessed for their interaction with the receptor. The aim is to discriminate between correct and incorrect ligand poses. Many docked ligand pose predictions can be filtered out because the ligand makes a clash with the receptor. For well-fitting ligands, a scoring function is required to discriminate between a binder and non-binder. The scoring function should give a good approximation of the binding free energy between a ligand and a receptor and is usually a function of different energy terms based on a force-field such as AMBER (25), CHARMM (26), ECEPP (64), and MMFF (65). The scoring function is trained on a large diverse set of ligands and receptors to improve recognition of binders and non-binders. Some docking algorithms use knowledge-based methods such as PMF (66–68) and DrugScore (69–71), while others such as ICM use full atom-based scoring (72, 73).

The ICM scoring function is weighted according to the following parameters (1) internal force-field energy of the ligand, (2) entropy loss of the ligand between bound and unbound states, (3) ligand–receptor hydrogen bond interactions, (4) polar and nonpolar solvation energy differences between bound and unbound states, (5) electrostatic energy, (6) hydrophobic energy, and (7) hydrogen bond donor or acceptor desolvation.

3.5. Ligand-Model Refinement

Once the initial docking is complete, it is necessary to consider refinement of the ligand–protein interactions to ensure an optimal prediction is made. The ligand-model refinement step is required because (1) the protein is flexible and will usually adapt to the ligand upon binding, (2) the side chains of the model surrounding the ligand-binding pocket are likely to be positioned incorrectly, and (3) the ligand-binding pocket may have collapsed partially during modeling (Fig. 3a, b). This section describes methods to overcome these problems and refine the docked complex.

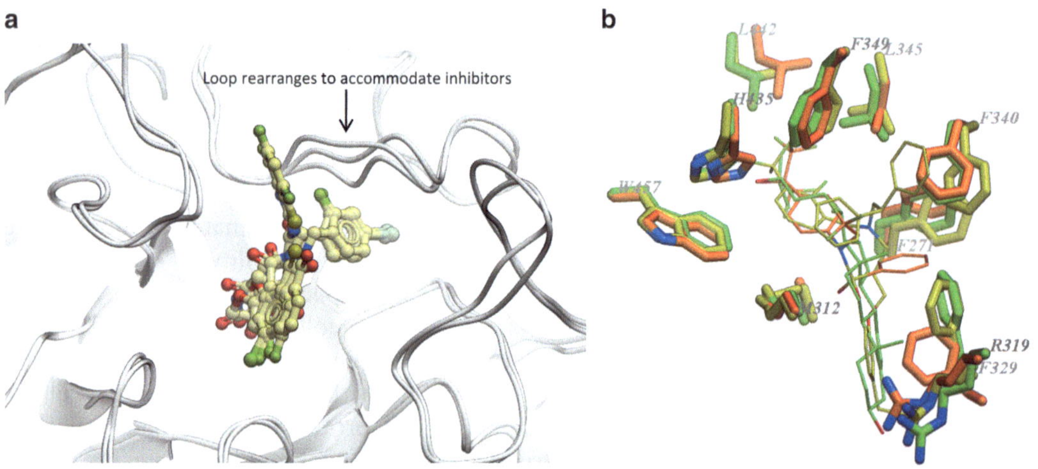

Fig. 3. Examples to demonstrate flexibility in the receptor upon ligand binding: (**a**) Aldose reductase (AR) has a flexible loop in the inhibitor-binding pocket (residues 298–302—*top right hand* corner of image), to show the change in the loop upon inhibitor (*stick* representation) binding two AR X-ray crystal structures (PDB code 1PWM and 1IEI) are superimposed along with a modeled loop (*ribbon* representation). The loop was modeled using ICM (18) and the X-ray and modeled loop conformations can be used in multiple receptor docking. (**b**) The structures of three nuclear receptor (Liver X receptor PDB codes 1PQ6, 1PQC, and 1P8D (99, 100)) are superimposed (*thick sticks*) highlighting the change in side chain positioning when different ligands bind (*thin sticks*). The phenylalanine residues, in particular, provide plasticity to the pocket and highlight the need to consider certain residues as explicit during ligand–receptor refinement. This could be achieved by representing part of the receptor by maps and allowing defined explicit residues to be flexible.

The manner in which a protein receptor adjusts to a ligand, known as "induced fit" is more complicated to model than a simplistic rigid "lock and key" interaction. Modeling induced fit is very computationally expensive and when performed incorrectly or too ambitiously can lead to incorrect ligand–receptor geometries. To refine all possible rotatable torsion angles in the ligand-binding pocket and find a way to identify the lowest energy conformation among many hypothetically generated structures is generally not feasible. Therefore, ways of efficiently sampling different conformations of the receptor that mimic "induced fit" have been developed (34).

To achieve the best refinement you need to thoroughly investigate the ligand-binding pocket to identify regions in your model which may be flexible (e.g., loop regions) and for stabilizing elements such as buried salt bridges and cysteine disulfide bridges and then choose a suitable refinement method (see Note 8).

A method referred to as "soft docking" is one approach, which can account for receptor flexibility upon ligand docking (74–76). This method reduces the penalty for van der Waals interactions between the ligand and receptor and therefore allows the atom radii between the ligand and receptor to overlap slightly. This function can be readily incorporated into docking methods that use grid energy maps for the receptor. The main drawback of this approach is that only minor side-chain rearrangements can be observed.

To refine the receptor side-chain–ligand interactions, the rotameric states of the side chains can be sampled explicitly (77). This approach uses a library of side-chain rotamer conformations and samples the torsion angles of the receptor side chains while predicting the ligand binding energy. In its simplest form, this method can be used to remove any clashes between the ligand and the receptor that you may have in your modeled complex. It can also be a useful approach if you are confident only a small selection of side chains are likely to rearrange upon ligand binding. The method does not take into account any backbone atom rearrangements and is computationally expensive. Most docking algorithms have an option to refine side chains after docking but if the number of degrees of freedom is too high, the approach can lead to incorrectly predicted docking poses.

One method to reduce the number of variables sampled during docking while incorporating flexibility in the receptor is to have a hybrid map/explicit atom grid. Explicit group docking is a recent development in the ICM software that allows selected receptor atoms to be considered explicitly during docking while the rest of the receptor is represented as a grid map. For example, the hydroxyls of Ser, Thr, and Tyr can be allowed to rotate and interact with the ligand during docking.

A computational efficient approach to solving this problem is to use multiple receptor conformations of the receptor. The first step is to generate an ensemble of structures for the ligand-binding pocket. If there are multiple receptor conformations of your template structure available then you can use these structures to build the ensemble by generating multiple models of your protein. If this is not the case then the ensemble can be generated using MC or MD software as described earlier or by using normal modes (NM) (78). NM provides a spring-like representation of the backbone atoms allowing a wide conformational space to be sampled (see Note 9). Alternatively, the ligand is used to mold the binding pocket to generate an ensemble of conformations (see Note 10). The key is to generate a reasonable representative set of structures, which is not too large but focused enough to account for flexibility within the binding pocket as much as possible (79, 80). Many of the leading docking packages listed in Table 4 have been adapted to use multiple receptor conformations, e.g., AutoDock (81), FlexX-Ensemble (82), ICM (78, 83–86), and DOCK (87, 88).

3.6. Benchmarking and Managing Expectations

Several recent modeling and docking competitions established the level of expectations. In 2008, the modeling challenge was to predict the interaction of the antagonist ZM241385 with the A2a human adenosine receptor (1). Only three modeler teams achieved more than 40% of correct ligand–protein interatomic contacts, while subtle rearrangements of the helices is not obvious from the alignment to the b2AR template and were not predicted by any of the groups. The next competition in 2010 had three different GPCR modeling and small molecule docking problems and showed that the best models for the easiest target (human dopamine D3 receptor bound to eticlopride) reached an impressive 58% of correct interatomic contacts (still outside the near-native target of at least 70–80%). The more difficult CXCR4 model based on either b2AR or A2a template with a small molecule antagonist achieved a level of 40% of correct interatomic contacts with over 4 Å RMSD for the best contact model (2).

In a recent separate competition organized by OpenEye, the docking pose prediction accuracy was benchmarked using the modified Astex set of 85 protein–ligand complexes (89). The top score poses were correct (under 2 Å RMSD) in 60 to over 90% of the cases depending on the docking method. The ICM docking method (MolSoft LLC) achieved 78% of the top score poses under 1 Å RMSD and 91% under 2 Å RMSD.

4. Notes

1. Most pocket identification algorithms are trained to find buried "drug-like" pockets. If, however, your pocket of interest is solvent exposed or you are interested in discovering extended regions to the pocket then it is advisable to experiment with different parameters other than the default ones. For example,

methods that use a geometric approach, such as ICM Pocket Finder, the dimensions of the probe used to outline the cavity can be changed.

2. One way to investigate different structural states of your ligand-binding pocket is to search the PDB for similar structures, which may reveal flexible regions (e.g., different loop conformations). The structures can then be used to model different conformations. Alternatively, ab initio methods can be used to predict loop regions but care needs to be taken because the accuracy of loop modeling methods deteriorates with loops longer than 8–13 residues (90, 91).
3. A classic example, where care is needed with setting residue side-chain charge is docking to HIV Protease, which is a dimer with a flexible ligand-binding pocket. One Asp from each chain of the dimer comes together in the active site upon ligand binding. In this case, correct docking can only be achieved if the Asp residues in each chain in the binding pocket are uncharged.
4. Before docking the ligand, check that the ligand has the correct; charges, bond types, bond order, and chirality. The ligand can be corrected using a molecular editor (see Table 3). If the ligand is likely to be covalently bound to the receptor care needs to be taken to choose a docking method that can predict the interaction correctly.
5. One recommended way of testing the ligand docking method is to find a similar ligand–receptor complex to your model in the PDB, then remove the ligand, and redock it. If the docking method is good, the redocked ligand should not have a root mean square deviation (RMSD) of more than 2 Å compared to the crystal structure ligand. If you have more data and sufficient computational facilities you can determine how well each method discriminates between known binders and non-binders. This is undertaken by building a database of chemical decoys (92, 93) and screening the ligands using virtual screening and plotting the score to determine enrichment.
6. Generally, it is fine to use the default map sizes for docking using ICM but if you have an elongated pocket or if you only want to sample a defined region of the pocket you can make the grid size larger or smaller depending on the scenario.
7. LigPlot (94) is a useful program for visualizing the interactions of the ligand with the protein model.
8. The database of molecular motions (95) is a good resource for better understanding the structural flexibility of your protein model.
9. An all heavy atom Elastic Network NM modeling approach was successfully used in the 2008 "blind" G-protein-coupled receptor (GPCR) modeling competition. The method yielded

the best model in terms of ligand–receptor contacts for the Adenosine A2a receptor (1, 86). A useful free resource for generating multiple receptor conformations of a protein using NMs can be found here http://abagyan.ucsd.edu/MRC/.

10. For ligand-guided modeling, a fully flexible seed ligand, which is known to bind, is docked to the protein and the pocket side chain and in some cases backbone atoms are sampled and optimized. This approach generates an ensemble of structures, which can be clustered and filtered down to a few selected conformations. The ability of the model to be able to discriminate binders from non-binders is then tested by screening a database of decoy ligands mixed with known binders (86, 96, 97).

References

1. Michino, M., Abola, E., Brooks, C. L., 3 rd, Dixon, J. S., Moult, J., and Stevens, R. C. (2009) Community-wide assessment of GPCR structure modelling and ligand docking: GPCR Dock 2008, *Nat Rev Drug Discov 8*, 455–463.
2. Kufareva I, Rueda M, Katritch V, Stevens RC, Abagyan R; GPCR Dock 2010 participants. (2011) Status of GPCR modeling and docking as reflected by community-wide GPCR Dock 2010 assessment, *Structure 19*, 1108–1126.
3. Zhang, Y. (2008) Progress and challenges in protein structure prediction, *Curr. Opin. Struct. Biol 18*, 342–348.
4. Martí-Renom, M. A., Stuart, A. C., Fiser, A., Sánchez, R., Melo, F., and Sali, A. (2000) Comparative protein structure modeling of genes and genomes, *Annu Rev Biophys Biomol Struct 29*, 291–325.
5. Moult, J., Fidelis, K., Kryshtafovych, A., Rost, B., and Tramontano, A. (2009) Critical assessment of methods of protein structure prediction - Round VIII, *Proteins 77 Suppl 9*, 1–4.
6. Wallner, B., and Elofsson, A. (2005) All are not equal: a benchmark of different homology modeling programs, *Protein Sci 14*, 1315–1327.
7. Abagyan, R., and Totrov, M. (2001) High-throughput docking for lead generation, *Curr Opin Chem Biol 5*, 375–382.
8. Cavasotto, C. N., and Orry, A. J. W. (2007) Ligand docking and structure-based virtual screening in drug discovery, *Curr Top Med Chem 7*, 1006–1014.
9. Taylor, R. D., Jewsbury, P. J., and Essex, J. W. (2002) A review of protein-small molecule docking methods, *J. Comput. Aided Mol. Des 16*, 151–166.
10. Shoichet, B. K., McGovern, S. L., Wei, B., and Irwin, J. J. (2002) Lead discovery using molecular docking, *Curr Opin Chem Biol 6*, 439–446.
11. Leach, A. R., Shoichet, B. K., and Peishoff, C. E. (2006) Prediction of protein-ligand interactions. Docking and scoring: successes and gaps, *J. Med. Chem 49*, 5851–5855.
12. Berman, H. M., Westbrook, J., Feng, Z., Gilliland, G., Bhat, T. N., Weissig, H., Shindyalov, I. N., and Bourne, P. E. (2000) The Protein Data Bank, *Nucleic Acids Research 28*, 235–242.
13. Leis, S., Schneider, S., and Zacharias, M. (2010) In silico prediction of binding sites on proteins, *Curr. Med. Chem 17*, 1550–1562.
14. Pérot, S., Sperandio, O., Miteva, M. A., Camproux, A.-C., and Villoutreix, B. O. (2010) Druggable pockets and binding site centric chemical space: a paradigm shift in drug discovery, *Drug Discov. Today 15*, 656–667.
15. Davis, A. M., St-Gallay, S. A., and Kleywegt, G. J. (2008) Limitations and lessons in the use of X-ray structural information in drug design, *Drug Discov. Today 13*, 831–841.
16. Kuntz, Blaney, Oatley, Langridge, and Ferrin. (1982) A geometric approach to macromolecule-ligand interactions, *Journal of molecular biology 161*, 269–88.
17. Katchalski-Katzir, E., Shariv, I., Eisenstein, M., Friesem, A. A., Aflalo, C., and Vakser, I. A. (1992) Molecular surface recognition: determination of geometric fit between proteins and their ligands by correlation techniques, *Proc. Natl. Acad. Sci. U.S.A 89*, 2195–2199.
18. Abagyan, R., and Totrov, M. (1994) Biased probability Monte Carlo conformational

searches and electrostatic calculations for peptides and proteins, *J. Mol. Biol 235*, 983–1002.

19. Liu, M., and Wang, S. (1999) MCDOCK: a Monte Carlo simulation approach to the molecular docking problem, *J. Comput. Aided Mol. Des 13*, 435–451.
20. Trosset, J. Y., and Scheraga, H. A. (1998) Reaching the global minimum in docking simulations: a Monte Carlo energy minimization approach using Bezier splines, *Proc. Natl. Acad. Sci. U.S.A 95*, 8011–8015.
21. Trott, O., and Olson, A. J. (2010) AutoDock Vina: Improving the speed and accuracy of docking with a new scoring function, efficient optimization, and multithreading, *Journal of Computational Chemistry 31*, 455–461.
22. Di Nola, A., Roccatano, D., and Berendsen, H. J. (1994) Molecular dynamics simulation of the docking of substrates to proteins, *Proteins 19*, 174–182.
23. Luty, B. A., Wasserman, Z. R., Stouten, P. F. W., Hodge, C. N., Zacharias, M., and McCammon, J. A. (1995) A molecular mechanics/grid method for evaluation of ligand-receptor interactions, *J. Comput. Chem. 16*, 454–464.
24. Kozack, R. E., and Subramaniam, S. (1993) Brownian dynamics simulations of molecular recognition in an antibody-antigen system, *Protein Sci 2*, 915–926.
25. Case, D. A., Cheatham, T. E., 3 rd, Darden, T., Gohlke, H., Luo, R., Merz, K. M., Jr, Onufriev, A., Simmerling, C., Wang, B., and Woods, R. J. (2005) The Amber biomolecular simulation programs, *J Comput Chem 26*, 1668–1688.
26. Brooks, B. R., Brooks, C. L., 3 rd, Mackerell, A. D., Jr, Nilsson, L., Petrella, R. J., Roux, B., Won, Y., Archontis, G., Bartels, C., Boresch, S., Caflisch, A., Caves, L., Cui, Q., Dinner, A. R., Feig, M., Fischer, S., Gao, J., Hodoscek, M., Im, W., Kuczera, K., Lazaridis, T., Ma, J., Ovchinnikov, V., Paci, E., Pastor, R. W., Post, C. B., Pu, J. Z., Schaefer, M., Tidor, B., Venable, R. M., Woodcock, H. L., Wu, X., Yang, W., York, D. M., and Karplus, M. (2009) CHARMM: the biomolecular simulation program, *J Comput Chem 30*, 1545–1614.
27. Taylor, J. S., and Burnett, R. M. (2000) DARWIN: a program for docking flexible molecules, *Proteins 41*, 173–191.
28. Verdonk, M. L., Cole, J. C., Hartshorn, M. J., Murray, C. W., and Taylor, R. D. (2003) Improved protein-ligand docking using GOLD, *Proteins 52*, 609–623.
29. Clark, K. P., and Ajay. (1995) Flexible ligand docking without parameter adjustment across four ligand–receptor complexes, *Journal of Computational Chemistry 16*, 1210–1226.
30. Rarey, M., Kramer, B., Lengauer, T., and Klebe, G. (1996) A fast flexible docking method using an incremental construction algorithm, *J. Mol. Biol 261*, 470–489.
31. Moustakas, D., Lang, P., Pegg, S., Pettersen, E., Kuntz, I., Brooijmans, N., and Rizzo, R. (2006) Development and validation of a modular, extensible docking program: DOCK 5, *Journal of computer-aided molecular design 20*, 601–19.
32. Carlson, H. A. (2002) Protein flexibility and drug design: how to hit a moving target, *Curr Opin Chem Biol 6*, 447–452.
33. Cavasotto, C. N., Orry, A. J. W., and Abagyan, R. A. (2005) The challenge of considering receptor flexibility in ligand docking and virtual screening, *Current Computer-Aided Drug Design 1*, 423–440.
34. Totrov, M., and Abagyan, R. (2008) Flexible ligand docking to multiple receptor conformations: a practical alternative, *Curr. Opin. Struct. Biol 18*, 178–184.
35. Laskowski, R. A. (1995) SURFNET: a program for visualizing molecular surfaces, cavities, and intermolecular interactions, *J Mol Graph 13*, 323–330, 307–308.
36. Levitt, D. G., and Banaszak, L. J. (1992) POCKET: a computer graphics method for identifying and displaying protein cavities and their surrounding amino acids, *J Mol Graph 10*, 229–234.
37. Hendlich, M., Rippmann, F., and Barnickel, G. (1997) LIGSITE: automatic and efficient detection of potential small molecule-binding sites in proteins, *J. Mol. Graph. Model 15*, 359–363, 389.
38. Kortvelyesi, T., Silberstein, M., Dennis, S., and Vajda, S. (2003) Improved mapping of protein binding sites, *J. Comput. Aided Mol. Des 17*, 173–186.
39. Ruppert, J., Welch, W., and Jain, A. N. (1997) Automatic identification and representation of protein binding sites for molecular docking, *Protein Sci 6*, 524–533.
40. Boer, D. R., Kroon, J., Cole, J. C., Smith, B., and Verdonk, M. L. (2001) SuperStar: comparison of CSD and PDB-based interaction fields as a basis for the prediction of protein-ligand interactions, *J. Mol. Biol 312*, 275–287.
41. Verdonk, M. L., Cole, J. C., Watson, P., Gillet, V., and Willett, P. (2001) SuperStar: improved knowledge-based interaction fields

for protein binding sites, *J. Mol. Biol 307*, 841–859.

42. Bliznyuk, A. A., and Gready, J. E. (1998) Identification and energetic ranking of possible docking sites for pterin on dihydrofolate reductase, *J. Comput. Aided Mol. Des 12*, 325–333.
43. An, J., Totrov, M., and Abagyan, R. (2004) Comprehensive identification of "druggable" protein ligand binding sites, *Genome Inform 15*, 31–41.
44. An, J., Totrov, M., and Abagyan, R. (2005) Pocketome via comprehensive identification and classification of ligand binding envelopes, *Molecular & Cellular Proteomics 4*, 752.
45. Orry, A. J. W., Totrov, M., Raush, E., and Abagyan, R. A. (2011) ICM User's Guide, *La Jolla: MolSoft, LLC.*
46. Kleywegt, G. J., Harris, M. R., Zou, J. Y., Taylor, T. C., Wählby, A., and Jones, T. A. (2004) The Uppsala Electron-Density Server, *Acta Crystallogr. D Biol. Crystallogr 60*, 2240–2249.
47. Pettersen, E. F., Goddard, T. D., Huang, C. C., Couch, G. S., Greenblatt, D. M., Meng, E. C., and Ferrin, T. E. (2004) UCSF Chimera--a visualization system for exploratory research and analysis, *J Comput Chem 25*, 1605–1612.
48. Dalby, A., Nourse, J. G., Hounshell, W. D., Gushurst, A. K. I., Grier, D. L., Leland, B. A., and Laufer, J. (1992) Description of several chemical structure file formats used by computer programs developed at Molecular Design Limited, *Journal of Chemical Information and Computer Sciences 32*, 244–255.
49. (2005) Tripos MOL2 format http://tripos.com/data/support/mol2.pdf.
50. Weininger, D. (1988) SMILES, a chemical language and information system. 1. Introduction to methodology and encoding rules, *Journal of Chemical Information and Computer Sciences 28*, 31–36.
51. Weininger, D., Weininger, A., and Weininger, J. L. (1989) SMILES. 2. Algorithm for generation of unique SMILES notation, *Journal of Chemical Information and Computer Sciences 29*, 97–101.
52. Bursulaya, B. D., Totrov, M., Abagyan, R., and Brooks, C. L., 3 rd. (2003) Comparative study of several algorithms for flexible ligand docking, *J. Comput. Aided Mol. Des 17*, 755–763.
53. Chen, H., Lyne, P. D., Giordanetto, F., Lovell, T., and Li, J. (2006) On evaluating molecular-docking methods for pose prediction and enrichment factors, *J Chem Inf Model 46*, 401–415.
54. Cross, J. B., Thompson, D. C., Rai, B. K., Baber, J. C., Fan, K. Y., Hu, Y., and Humblet, C. (2009) Comparison of several molecular docking programs: pose prediction and virtual screening accuracy, *J Chem Inf Model 49*, 1455–1474.
55. Maiorov, V., and Sheridan, R. P. (2005) Enhanced virtual screening by combined use of two docking methods: getting the most on a limited budget, *J Chem Inf Model 45*, 1017–1023.
56. McGaughey, G. B., Sheridan, R. P., Bayly, C. I., Culberson, J. C., Kreatsoulas, C., Lindsley, S., Maiorov, V., Truchon, J.-F., and Cornell, W. D. (2007) Comparison of topological, shape, and docking methods in virtual screening, *J Chem Inf Model 47*, 1504–1519.
57. Metropolis, N., Rosenbluth, A. W., Rosenbluth, M. N., Teller, A. H., and Teller, E. (1953) Equation of State Calculations by Fast Computing Machines, *J. Chem. Phys. 21*, 1087.
58. McCammon, J. A., Gelin, B. R., and Karplus, M. (1977) Dynamics of folded proteins, *Nature 267*, 585–590.
59. Francesca Gerini, M., Roccatano, D., Baciocchi, E., and Di Nola, A. (2003) Molecular dynamics simulations of lignin peroxidase in solution, *Biophys. J 84*, 3883–3893.
60. Mangoni, M., Roccatano, D., and Di Nola, A. (1999) Docking of flexible ligands to flexible receptors in solution by molecular dynamics simulation, *Proteins 35*, 153–162.
61. Ewing, T., Makino, S., Skillman, A., and Kuntz, I. (2001) DOCK 4.0: search strategies for automated molecular docking of flexible molecule databases, *Journal of computer-aided molecular design 15*, 411–28.
62. Abagyan, R., Totrov, M., and Kuznetsov, D. (1994) ICM - a new method for protein modeling and design: applications to docking and structure prediction from the distorted native conformation, *J. Comput. Chem. 15*, 488–506.
63. Totrov, M., and Abagyan, R. (1997) Flexible protein-ligand docking by global energy optimization in internal coordinates, *Proteins Suppl 1*, 215–220.
64. Arnautova, Y. A., Jagielska, A., and Scheraga, H. A. (2006) A new force field (ECEPP-05) for peptides, proteins, and organic molecules, *J Phys Chem B 110*, 5025–5044.
65. Halgren, T. A. (1996) Merck molecular force field. I. Basis, form, scope, parameterization,

and performance of MMFF94, *Journal of Computational Chemistry 17*, 490–519.
66. Muegge, I., and Martin, Y. C. (1999) A general and fast scoring function for protein-ligand interactions: a simplified potential approach, *J. Med. Chem 42*, 791–804.
67. Muegge, I., Martin, Y. C., Hajduk, P. J., and Fesik, S. W. (1999) Evaluation of PMF scoring in docking weak ligands to the FK506 binding protein, *J. Med. Chem 42*, 2498–2503.
68. Ha, S., Andreani, R., Robbins, A., and Muegge, I. (2000) Evaluation of docking/scoring approaches: a comparative study based on MMP3 inhibitors, *J. Comput. Aided Mol. Des 14*, 435–448.
69. Gohlke, H., Hendlich, M., and Klebe, G. (2000) Knowledge-based scoring function to predict protein-ligand interactions, *J. Mol. Biol 295*, 337–356.
70. Sotriffer, C. A., Gohlke, H., and Klebe, G. (2002) Docking into knowledge-based potential fields: a comparative evaluation of DrugScore, *J. Med. Chem 45*, 1967–1970.
71. Velec, H. F. G., Gohlke, H., and Klebe, G. (2005) DrugScore(CSD)-knowledge-based scoring function derived from small molecule crystal data with superior recognition rate of near-native ligand poses and better affinity prediction, *J. Med. Chem 48*, 6296–6303.
72. Schapira, M., Totrov, M., and Abagyan, R. (1999) Prediction of the binding energy for small molecules, peptides and proteins, *J. Mol. Recognit 12*, 177–190.
73. Totrov, M., and Abagyan, R. (1999) Derivation of sensitive discrimination potential for virtual ligand screening, in *Proceedings of the third annual international conference on Computational molecular biology*, pp 312–320. ACM, New York, NY, USA.
74. Gschwend, D. A., Good, A. C., and Kuntz, I. D. (1996) Molecular docking towards drug discovery, *J. Mol. Recognit 9*, 175–186.
75. Jiang, F., and Kim, S. H. (1991) "Soft docking": matching of molecular surface cubes, *J. Mol. Biol 219*, 79–102.
76. Walls, P. H., and Sternberg, M. J. (1992) New algorithm to model protein-protein recognition based on surface complementarity. Applications to antibody-antigen docking, *J. Mol. Biol 228*, 277–297.
77. Leach, A. R. (1994) Ligand docking to proteins with discrete side-chain flexibility, *J. Mol. Biol 235*, 345–356.
78. Rueda, M., Bottegoni, G., and Abagyan, R. (2009) Consistent improvement of cross-docking results using binding site ensembles generated with elastic network normal modes, *J Chem Inf Model 49*, 716–725.
79. Damm, K. L., and Carlson, H. A. (2007) Exploring experimental sources of multiple protein conformations in structure-based drug design, *J. Am. Chem. Soc 129*, 8225–8235.
80. Sperandio, O., Mouawad, L., Pinto, E., Villoutreix, B. O., Perahia, D., and Miteva, M. A. (2010) How to choose relevant multiple receptor conformations for virtual screening: a test case of Cdk2 and normal mode analysis, *Eur. Biophys. J 39*, 1365–1372.
81. Osterberg, F., Morris, G. M., Sanner, M. F., Olson, A. J., and Goodsell, D. S. (2002) Automated docking to multiple target structures: incorporation of protein mobility and structural water heterogeneity in AutoDock, *Proteins 46*, 34–40.
82. Claussen, H., Buning, C., Rarey, M., and Lengauer, T. (2001) FlexE: efficient molecular docking considering protein structure variations, *J. Mol. Biol 308*, 377–395.
83. Schapira, M., Abagyan, R., and Totrov, M. (2003) Nuclear hormone receptor targeted virtual screening, *J. Med. Chem 46*, 3045–3059.
84. Cavasotto, C. N., Kovacs, J. A., and Abagyan, R. A. (2005) Representing receptor flexibility in ligand docking through relevant normal modes, *J. Am. Chem. Soc 127*, 9632–9640.
85. Cavasotto, C. N., and Abagyan, R. A. (2004) Protein flexibility in ligand docking and virtual screening to protein kinases, *J. Mol. Biol 337*, 209–225.
86. Katritch, V., Rueda, M., Lam, P. C.-H., Yeager, M., and Abagyan, R. (2010) GPCR 3D homology models for ligand screening: lessons learned from blind predictions of adenosine A2a receptor complex, *Proteins 78*, 197–211.
87. Ferrari, A. M., Wei, B. Q., Costantino, L., and Shoichet, B. K. (2004) Soft docking and multiple receptor conformations in virtual screening, *J. Med. Chem 47*, 5076–5084.
88. Huang, S.-Y., and Zou, X. (2007) Ensemble docking of multiple protein structures: considering protein structural variations in molecular docking, *Proteins 66*, 399–421.
89. Hartshorn, M. J., Verdonk, M. L., Chessari, G., Brewerton, S. C., Mooij, W. T. M., Mortenson, P. N., and Murray, C. W. (2007) Diverse, High-Quality Test Set for the Validation of Protein - Ligand Docking Performance, *Journal of Medicinal Chemistry 50*, 726–741.

90. Fiser, A., Do, R. K., and Sali, A. (2000) Modeling of loops in protein structures, *Protein Sci 9*, 1753–1773.
91. Soto, C. S., Fasnacht, M., Zhu, J., Forrest, L., and Honig, B. (2008) Loop modeling: Sampling, filtering, and scoring, *Proteins 70*, 834–843.
92. Huang, N., Shoichet, B. K., and Irwin, J. J. (2006) Benchmarking Sets for Molecular Docking, *Journal of Medicinal Chemistry 49*, 6789–6801.
93. Wallach, I., and Lilien, R. (2011) Virtual Decoy Sets for Molecular Docking Benchmarks, *Journal of Chemical Information and Modeling 51*, 196–202.
94. Wallace, A. C., Laskowski, R. A., and Thornton, J. M. (1995) LIGPLOT: a program to generate schematic diagrams of protein-ligand interactions, *Protein Eng 8*, 127–134.
95. Echols, N., Milburn, D., and Gerstein, M. (2003) MolMovDB: analysis and visualization of conformational change and structural flexibility, *Nucleic Acids Res 31*, 478–482.
96. Cavasotto, C. N., Orry, A. J. W., and Abagyan, R. A. (2003) Structure-based identification of binding sites, native ligands and potential inhibitors for G-protein coupled receptors, *Proteins 51*, 423–433.
97. Bisson, W. H., Cheltsov, A. V., Bruey-Sedano, N., Lin, B., Chen, J., Goldberger, N., May, L. T., Christopoulos, A., Dalton, J. T., Sexton, P. M., Zhang, X.-K., and Abagyan, R. (2007) Discovery of antiandrogen activity of nonsteroidal scaffolds of marketed drugs, *Proc. Natl. Acad. Sci. U.S.A 104*, 11927–11932.
98. Cavasotto, C. N., Orry, A. J. W., Murgolo, N. J., Czarniecki, M. F., Kocsi, S. A., Hawes, B. E., O'Neill, K. A., Hine, H., Burton, M. S., Voigt, J. H., Abagyan, R. A., Bayne, M. L., and Monsma, F. J., Jr. (2008) Discovery of novel chemotypes to a G-protein-coupled receptor through ligand-steered homology modeling and structure-based virtual screening, *J. Med. Chem 51*, 581–588.
99. Färnegårdh, M., Bonn, T., Sun, S., Ljunggren, J., Ahola, H., Wilhelmsson, A., Gustafsson, J.-Å., and Carlquist, M. (2003) The Three-dimensional Structure of the Liver X Receptor β Reveals a Flexible Ligand-binding Pocket That Can Accommodate Fundamentally Different Ligands, *Journal of Biological Chemistry 278*, 38821–38828.
100. Williams, S., Bledsoe, R. K., Collins, J. L., Boggs, S., Lambert, M. H., Miller, A. B., Moore, J., McKee, D. D., Moore, L., Nichols, J., Parks, D., Watson, M., Wisely, B., and Willson, T. M. (2003) X-ray crystal structure of the liver X receptor beta ligand binding domain: regulation by a histidine-tryptophan switch, *J. Biol. Chem 278*, 27138–27143.
101. Dundas, J., Ouyang, Z., Tseng, J., Binkowski, A., Turpaz, Y., and Liang, J. (2006) CASTp: computed atlas of surface topography of proteins with structural and topographical mapping of functionally annotated residues, *Nucleic Acids Res 34*, W116-118.
102. Ashkenazy, H., Erez, E., Martz, E., Pupko, T., and Ben-Tal, N. (2010) ConSurf 2010: calculating evolutionary conservation in sequence and structure of proteins and nucleic acids, *Nucleic Acids Res 38*, W529-533.
103. Le Guilloux, V., Schmidtke, P., and Tuffery, P. (2009) Fpocket: an open source platform for ligand pocket detection, *BMC Bioinformatics 10*, 168.
104. Hernandez, M., Ghersi, D., and Sanchez, R. (2009) SITEHOUND-web: a server for ligand binding site identification in protein structures, *Nucleic Acids Res 37*, W413-416.
105. Burgoyne, N. J., and Jackson, R. M. (2006) Predicting protein interaction sites: binding hot-spots in protein-protein and protein-ligand interfaces, *Bioinformatics 22*, 1335–1342.
106. Laurie, A. T. R., and Jackson, R. M. (2005) Q-SiteFinder: an energy-based method for the prediction of protein-ligand binding sites, *Bioinformatics 21*, 1908–1916.
107. Brady, G. P., Jr, and Stouten, P. F. (2000) Fast prediction and visualization of protein binding pockets with PASS, *J. Comput. Aided Mol. Des 14*, 383–401.
108. Overington, J. (2009) ChEMBL. An interview with John Overington, team leader, chemogenomics at the European Bioinformatics Institute Outstation of the European Molecular Biology Laboratory (EMBL-EBI). Interview by Wendy A. Warr, *J. Comput. Aided Mol. Des 23*, 195–198.
109. Knox, C., Law, V., Jewison, T., Liu, P., Ly, S., Frolkis, A., Pon, A., Banco, K., Mak, C., Neveu, V., Djoumbou, Y., Eisner, R., Guo, A. C., and Wishart, D. S. (2011) DrugBank 3.0: a comprehensive resource for "omics" research on drugs, *Nucleic Acids Res 39*, D1035-1041.
110. Wishart, D. S., Knox, C., Guo, A. C., Cheng, D., Shrivastava, S., Tzur, D., Gautam, B., and Hassanali, M. (2008) DrugBank: a knowledgebase for drugs, drug actions and drug targets, *Nucleic Acids Res 36*, D901-906.
111. Wishart, D. S., Knox, C., Guo, A. C., Shrivastava, S., Hassanali, M., Stothard, P., Chang, Z., and Woolsey, J. (2006) DrugBank:

a comprehensive resource for in silico drug discovery and exploration, *Nucleic Acids Res 34*, D668-672.

112. Kanehisa, M., and Goto, S. (2000) KEGG: kyoto encyclopedia of genes and genomes, *Nucleic Acids Res 28*, 27–30.

113. Kanehisa, M., Goto, S., Hattori, M., Aoki-Kinoshita, K. F., Itoh, M., Kawashima, S., Katayama, T., Araki, M., and Hirakawa, M. (2006) From genomics to chemical genomics: new developments in KEGG, *Nucleic Acids Res 34*, D354-357.

114. Kanehisa, M., Goto, S., Furumichi, M., Tanabe, M., and Hirakawa, M. (2010) KEGG for representation and analysis of molecular networks involving diseases and drugs, *Nucleic Acids Res 38*, D355-360.

115. Sayers, E. W., Barrett, T., Benson, D. A., Bolton, E., Bryant, S. H., Canese, K., Chetvernin, V., Church, D. M., DiCuccio, M., Federhen, S., Feolo, M., Fingerman, I. M., Geer, L. Y., Helmberg, W., Kapustin, Y., Landsman, D., Lipman, D. J., Lu, Z., Madden, T. L., Madej, T., Maglott, D. R., Marchler-Bauer, A., Miller, V., Mizrachi, I., Ostell, J., Panchenko, A., Phan, L., Pruitt, K. D., Schuler, G. D., Sequeira, E., Sherry, S. T., Shumway, M., Sirotkin, K., Slotta, D., Souvorov, A., Starchenko, G., Tatusova, T. A., Wagner, L., Wang, Y., Wilbur, W. J., Yaschenko, E., and Ye, J. (2011) Database resources of the National Center for Biotechnology Information, *Nucleic Acids Res 39*, D38-51.

116. Irwin, J. J., and Shoichet, B. K. (2005) ZINC--a free database of commercially available compounds for virtual screening, *J Chem Inf Model 45*, 177–182.

117. Morris, G. M., Goodsell, D. S., Halliday, R. S., Huey, R., Hart, W. E., Belew, R. K., and Olson, A. J. (1998) Automated docking using a Lamarckian genetic algorithm and an empirical binding free energy function, *Journal of Computational Chemistry 19*, 1639–1662.

118. Reid, D., Simon, A., Sadjad, B. S., Johnson, A. P., and Zsoldos, Z. eHiTS: an innovative approach to the docking and scoring function problems., *Current protein peptide science 7*, 421–435.

119. McGann, M. R., Almond, H. R., Nicholls, A., Grant, J. A., and Brown, F. K. (2003) Gaussian docking functions, *Biopolymers 68*, 76–90.

120. Friesner, R. A., Banks, J. L., Murphy, R. B., Halgren, T. A., Klicic, J. J., Mainz, D. T., Repasky, M. P., Knoll, E. H., Shelley, M., Perry, J. K., Shaw, D. E., Francis, P., and Shenkin, P. S. (2004) Glide: A New Approach for Rapid, Accurate Docking and Scoring. 1. Method and Assessment of Docking Accuracy, *Journal of Medicinal Chemistry 47*, 1739–1749.

121. Friesner, R. A., Murphy, R. B., Repasky, M. P., Frye, L. L., Greenwood, J. R., Halgren, T. A., Sanschagrin, P. C., and Mainz, D. T. (2006) Extra Precision Glide: Docking and Scoring Incorporating a Model of Hydrophobic Enclosure for Protein - Ligand Complexes, *Journal of Medicinal Chemistry 49*, 6177–6196.

122. Halgren, T. A., Murphy, R. B., Friesner, R. A., Beard, H. S., Frye, L. L., Pollard, W. T., and Banks, J. L. (2004) Glide: A New Approach for Rapid, Accurate Docking and Scoring. 2. Enrichment Factors in Database Screening, *Journal of Medicinal Chemistry 47*, 1750–1759.

123. Jones, G. (1997) Development and validation of a genetic algorithm for flexible docking, *Journal of Molecular Biology 267*, 727–748.

124. Jones, G., Willett, P., and Glen, R. (1995) Molecular recognition of receptor sites using a genetic algorithm with a description of desolvation, *Journal of Molecular Biology 245*, 43–53.

125. Jain, A. N. (2003) Surflex: fully automatic flexible molecular docking using a molecular similarity-based search engine, *J. Med. Chem 46*, 499–511.

126. Jain, A. N. (2007) Surflex-Dock 2.1: robust performance from ligand energetic modeling, ring flexibility, and knowledge-based search, *J. Comput. Aided Mol. Des 21*, 281–306.

127. Pham, T. A., and Jain, A. N. (2008) Customizing scoring functions for docking, *J. Comput. Aided Mol. Des 22*, 269–286.

Chapter 17

Modeling Peptide–Protein Interactions

Nir London, Barak Raveh, and Ora Schueler-Furman

Abstract

Peptide–protein interactions are prevalent in the living cell and form a key component of the overall protein–protein interaction network. These interactions are drawing increasing interest due to their part in signaling and regulation, and are thus attractive targets for computational structural modeling. Here we report an overview of current techniques for the high resolution modeling of peptide–protein complexes. We dissect this complicated challenge into several smaller subproblems, namely: modeling the receptor protein, predicting the peptide binding site, sampling an initial peptide backbone conformation and the final refinement of the peptide within the receptor binding site. For each of these conceptual stages, we present available tools, approaches, and their reported performance. We summarize with an illustrative example of this process, highlighting the success and current challenges still facing the automated blind modeling of peptide–protein interactions. We believe that the upcoming years will see considerable progress in our ability to create accurate models of peptide–protein interactions, with applications in binding-specificity prediction, rational design of peptide-mediated interactions and the usage of peptides as therapeutic agents.

Key words: Peptide docking, Peptide modeling, Rosetta FlexPepDock, Peptide–protein interactions, Peptide–protein complexes, Peptide binding

1. Introduction

Protein–protein interactions are one of the driving forces of the living cell. A large and important subset of these interactions is mediated by a short, flexible linear peptide that binds to a globular receptor and may form a modular binding motif (1). It has been estimated that between 15 and 40% of all protein–protein interactions are mediated by a short linear peptide (1, 2). Interactions that are mediated by flexible peptides play key roles in major cellular processes, predominantly in signaling and regulatory networks (3), but also in cell localization, protein degradation, and immune response (1, 3). Due to their cardinal role in regulatory interactions, flexible peptides are in many cases implicated in human

Andrew J.W. Orry and Ruben Abagyan (eds.), *Homology Modeling: Methods and Protocols*, Methods in Molecular Biology, vol. 857,
DOI 10.1007/978-1-61779-588-6_17, © Springer Science+Business Media, LLC 2012

disease and cancer (3). Consequently, these peptides provide an attractive starting point as leads for the design of inhibitory peptides and small molecule drugs (4–7).

In vivo, these linear peptides are not necessarily independent molecules, but rather appear within disordered regions at protein termini (8), in-between domains (9), or as flexible loops that bulge out of structured domains and mediate a protein–protein interaction (10). Short peptide molecules may also be created in vivo by proteolytic digestion of precursor molecules (11, 12), or they can be synthesized for in vitro studies or as small drug molecules (13). Flexible peptides, as intrinsically disordered proteins, often lack a distinct fold in their unbound state, and upon encountering their target (the receptor), they go through simultaneous binding and folding (*induced fit* model) (9, 14–16), or go through an equilibrium-shift towards preexisting bound conformations (*conformation sampling* model) (16–18). Their size may vary from short dipeptides that can be likened to small ligand molecules, to flexible peptides dozens of amino acids long, which wrap around the entire perimeter of their receptors (19).

This review aims to summarize the state of the art in modeling the interactions of flexible peptides at high resolution. As this problem involves many degrees of freedom both of the flexible peptide and the receptor, it is conceptually convenient to divide it into several consecutive steps, in line with prevalent approaches for modeling (20) and docking (21) of globular proteins (1) *Model receptor structure*: create an initial model of the receptor (if its structure has not been solved yet); (2) *Predict binding site*: locate potential binding sites on the receptor surface (3) *Build initial model of peptide*: create a set of models of plausible peptide backbone conformations (with or without considering the receptor); (4) *Model and refine peptide–receptor complex structure*: Optimize initial model of the peptide at the receptor binding site (based on steps 1–3) and refine into a high-resolution model. Note that in this last step, the peptide and receptor conformations may change considerably to increase their binding energy. Figure 1 presents an overview of the process, and Table 1 summarizes the different tools available for each step.

The above four steps are not necessarily completely distinct and might rather depend on each other, since the final conformation of the peptide (and sometimes even of the receptor) is stabilized or even induced by the interaction between the two (16). Nonetheless, these rough guidelines make it easier to tackle this complicated problem in a modular fashion. Fortunately, for several well-studied systems (e.g., kinases, MHC proteins, PDZ, SH3, and WW domains), a solved structure of the peptide binding domain in complex with other peptide sequences can be used

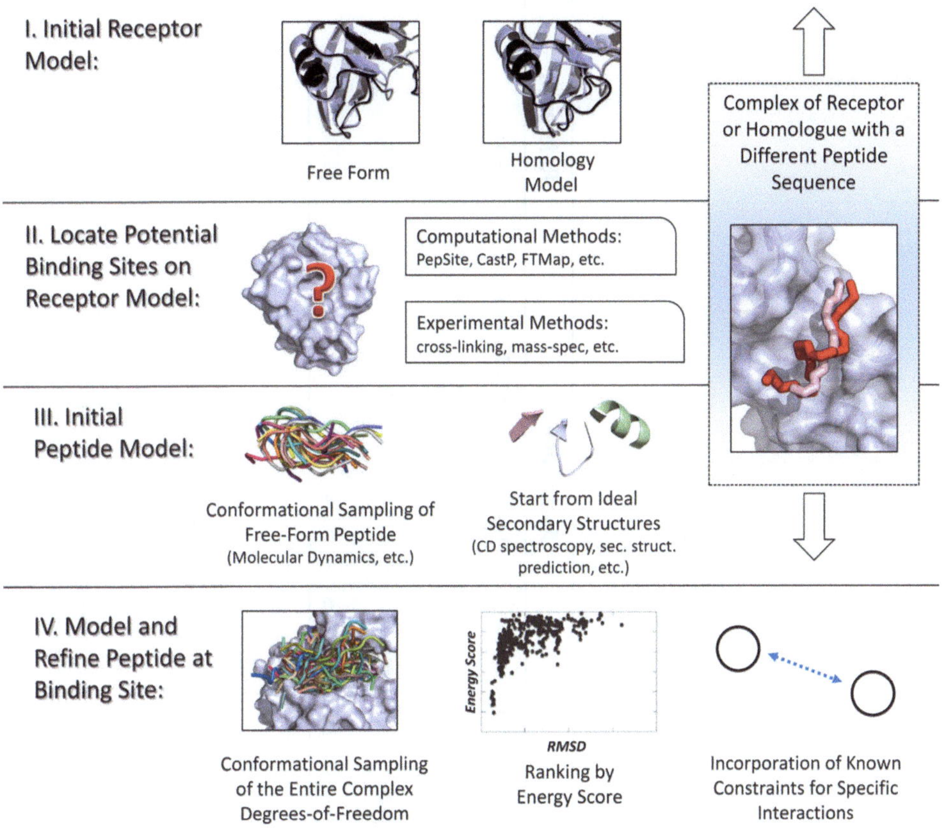

Fig. 1. Modular architecture of modeling peptide–protein interactions. An overview of the four conceptual stages in the high-resolution modeling of peptide–protein interactions.

as a template for subsequent refinement, by simply threading the desired sequence onto the solved peptide backbone. Even in these cases, the last step of refinement is often very important: As in any homology model, the template peptide structure may differ from the target peptide structure to a varying degree, from slight side-chain reorientation (22) to massive backbone rearrangements (23, 24).

Throughout this chapter, we cover the existing approaches for modeling peptide–protein interactions following the steps described above. We include examples of recent applications for the modeling of peptide–protein interactions and discuss some eminent open problems in this field. Finally, we provide the reader with a list of major structural datasets of peptide interactions that have been used to characterize the unique properties of peptide–protein interactions as well as to evaluate existing methods.

Table 1
Summary of methods for modeling peptide–protein interactions

A. Prediction of peptide binding sites

Name	Description	Availability	Reference
PepSite	Peptide binding location predictor; includes partial peptide orientation in the pocket	http://www.russell.embl.de/pepsite/	(42)
FTmap	Solvent mapping of the receptor surface. Correlates well with peptide binding sites	http://ftmap.bu.edu/	(46)
CASTp	Protein surface pocket detector. Peptides tend to bind to the largest pocket	http://sts.bioengr.uic.edu/castp/index.php	(44)
AnchorsMap	Predictor of anchoring residues for peptide or protein binding interfaces	N/A	(47)

B. Peptide backbone conformational sampling approaches

Approach	Description	Reference
Molecular dynamics (MD)	MD has been used to recover the structure of peptides in solution. This works well when the peptide adopts a stable conformation in the absence of the receptor	(53–55)
Monte Carlo (MC)	MC has been used to sample the structure of stable peptides	(56–58)
Fragment-based approaches	Several studies have shown that short peptides have local preferences to adopt a specific conformation based on their sequence. This enables to utilize solved structures of similar sequences in a different context to predict the peptide's conformation	(65, 67)
Extended conformation	When no other data is available, the extended conformation is often a good starting point for the peptide conformation	(24, 27)

C. High-resolution modeling of peptide–protein complexes

Name	Description	Sampling method	Availability	Reference
FlexPepDock	High-resolution refinement of peptide–protein interactions	Monte-Carlo with minimization; implemented in Rosetta	Rosetta 3.2; http://flexpepdock.furmanlab.cs.huji.ac.il/	(27)
DynaDock	High-resolution refinement of peptide–protein interactions	Optimized potential molecular dynamics	Upon request	(28)
AutoDock	Global docking of small molecules and short peptides	Grid based, followed by genetic algorithm-based minimization	http://autodock.scripps.edu/	(75)
MOLS	Global docking and refinement of short peptides	Orthogonal Latin-square sampling	Upon request	(79)

D. Modeling selected systems

System	Constraints	Reference
MHC/peptide	Two peptide anchoring residues bind in specific pockets	(23, 81–86, 100)
PDZ/peptide	The C-terminal residue is anchored at specific location	(24, 88, 89, 102)

Datasets of protein-complex structures

Name	Size	Resolution	Peptide lengths	Availability	Reference
PepX	1,431 (505 unique clusters)	X-ray < 2.5 Å	5–35	http://pepx.switchlab.org	(94)
peptiDB	103 unique clusters	X-ray < 2.0 Å	5–15	London et al. (supplemental information)	(26)
3did	829 (not clustered)	N/A	N/A	http://3did.irbbarcelona.org	(95)

2. Modeling the Receptor Protein

When docking a peptide (or any ligand) to a receptor protein, structures may be available for the receptor protein in its free form (unbound docking), or in complex with other peptide sequences (cross-docking). In more difficult cases, we would have to resort to homology modeling using the methods covered extensively in other chapters of this book or even ab initio modeling. Similar to protein–protein docking and ligand docking, the success of docking to unbound models, cross-docking and homology models, depends on the extent to which the receptor structures changes upon binding, mainly at the binding site (25). In previous work, we have shown that the backbone conformation of the receptor protein does not change substantially (<1 Å backbone root mean square deviation, RMSD) near the binding site, presumably to accommodate the entropic cost incurred by peptides upon binding (26). However, although accurate peptide–protein models were obtained even when starting from unbound backbone models, using methods described below (24, 27, 28), the ranking of the best models was not as good, perhaps due to the susceptibility of full-atom energy scores to small backbone changes that result in local clashes (24, 27).

For specific systems such as MHC receptors and PDZ domains, a rather large set of complex structures is available, and cross-docking, as well as docking of peptides to homology models can result in accurate high-resolution models (see below). In the remainder of this chapter, we assume that a reasonable representation of the receptor protein is available, which might be further optimized in subsequent steps.

We note that the quality of receptor modeling also has implications for structure-based specificity prediction that attempts to define the set of sequences that bind a given receptor. This interesting subject is outside of the scope of this chapter (for examples of such studies, we refer the reader to refs. 29–32, 102, 103).

3. Predicting the Sites for Peptide Binding on the Receptor Surface

As mentioned above, in many (perhaps most) practical cases, the location of the binding site can be inferred from solved structures of similar peptide–receptor complexes, involving the same receptor or its homologues. In other cases, it is at least possible to determine the approximate location of the peptide binding site from cross-linking experiments, mutational analysis, NMR shifts, or any other experimental evidence (33, 34). However, even in those cases in which one has no prior knowledge of the peptide binding site, several approaches have been devised for computational prediction of putative binding sites. Some of these approaches look for those surface regions that may accommodate a specific peptide sequence, while others look for more general, perhaps "promiscuous" regions on protein surfaces. In the latter approach, which follows analogous attempts in the context of globular proteins interactions (e.g., (35–38), reviewed in ref. 39) and small-molecule binding sites (e.g., (40), reviewed in ref. 41), the common characteristics of known peptide binding sites (geometry, amino acids composition, etc.) are used to predict putative binding sites. As a single receptor may include more than one peptide binding site, the correct binding site may be decided upon based on the subsequent steps in which the specific peptide sequence is modeled within the binding site (see illustrative example towards the end of this chapter). In the following, we describe different approaches that may assist in locating peptide binding sites on a given protein structure.

3.1. PepSite (42) (Availability: http://www.russell.embl.de/pepsite/)

Petsalaki et al. (42) have constructed spatial position-specific scoring matrices (PSSMs) to capture the preferred chemical environment for each amino acid in the context of a bound peptide. The 3D matrices were trained based on a database of peptide–protein complex structures (*see* PepX in the datasets section). Given a target protein receptor, these matrices are used to scan the surface of the target protein and score it to find candidate binding sites for each residue of a particular peptide. These predicted binding sites are then combined to suggest the overall binding site, as well as a rough orientation of the binding peptide. This approach might be less accurate for helical peptides, and possibly, also for peptides with sharp turns and coils (see Note 1).

The PepSite method was evaluated on a set of 405 complexes for which an unbound structure of the protein receptor was available, using leave-one-out cross-validation. Conveniently, each prediction is accompanied by a statistical confidence measure in the form of a p value. For instance, predictions with a p value below 0.1 correspond to a true-positive rate (TPR) of about 30% with a false-positive rate (FPR) of only 10%, over the same benchmark set. For very stringent p values below 0.003, the FPR decreases to only 1% with a TPR of about 10%.

PepSite takes into account the specific sequence of the query peptide. This may be of advantage, as protein receptors may contain multiple binding sites (43), but the specific peptide of interest only binds at a certain pocket. On the other hand, this might be too restrictive and miss other sites. Indeed, the reported coverage of this approach is fairly low.

3.2. CASTp (44) (Availability: http://sts.bioengr.uic.edu/castp/index.php)

The original purpose of CASTp is the detection of pockets on protein surfaces, as well as of cavities in the interior of proteins, using an analytical computation that is based on the weighted Delaunay triangulation and the alpha complex for shape measurements (45). The CASTp server provides the user a detailed list of analytic measures, including the area and volume of each pocket or cavity, and further geometric features.

Although CASTp was not developed specifically for detecting peptide binding sites, we have shown that peptides tend to bind at the largest pocket available on the protein surface (26). Over a dataset of 85 peptide–protein complexes (a subset of the peptiDB dataset; *see* Table 1), CASTp detected an average of 15 ± 10 pockets on each protein. We detected two main binding strategies regarding the utilization of pockets (1) *Binding of peptide to a large pocket*: 26% of the peptides in the dataset bind to a very large pocket (pocket accessible surface area (ASA) >100 Å^2; see, for example, Fig. 2). In most of these cases (18/22), this pocket was the largest pocket available on the protein surface. (2) *Binding of specific peptide residue into small hole*: 47% of the peptides in the entire dataset were found to bind to a small pocket instead (pocket area < 100 Å^2); in these cases, one of the peptide's side chains is buried in this pocket in a knob-hole fashion. However, even when the peptide latches onto a small pocket, this is still, in general, the largest pocket available on the protein (29/40 cases). Our analysis further revealed that α-helical peptides tend to bind using the knob-hole strategy, whereas β-strand peptides prefer pockets. Either way, it turns out that finding the largest pockets on a receptor surface can provide useful guidance for peptide binding sites (see Note 2).

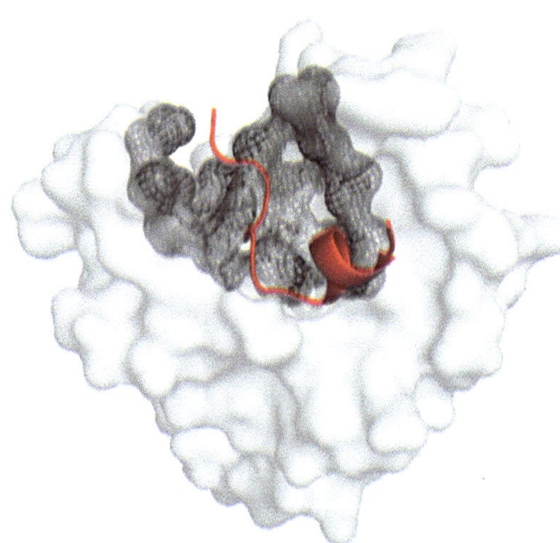

Fig. 2. Peptides tend to bind in large pockets on protein surfaces. An antagonist peptide (in *red cartoon representation*) in complex with the EphB4 receptor (in *white surface representation*; PDB: 2BBA). The largest pocket on the protein surface as detected by CASTp (44) is shown in *dark gray mesh*. Such a pocket can be used to focus the modeling of peptide-protein interactions to the relevant region.

3.3. Small-Molecule Mapping: FTmap (46) (Availability: http://ftmap.bu.edu/) and ANCHORSMAP (47)

The original purpose of FTmap (*Fourier-Transform Maps*) was the mapping of potential solvent binding sites on a protein surface. The server docks small organic molecules on the target protein surface using the Fourier-Transform approach (48), finds favorable binding positions, and clusters the conformations of all predictions. The clusters are then ranked according to their average free energy. Low-energy clusters are grouped into consensus sites, and the largest consensus sites were shown to locate active or ligand binding sites (46). We have recently shown (Raveh et al. (27) and unpublished data) that these clusters can also serve as good predictors of peptide binding sites for peptide anchoring residues. In yet unpublished results, we found that in 82% of the cases, there was at least one molecule cluster that approximately correlated to one of the peptide side chains (at least four atoms were found within 2 Å of the atoms of a single side chain). In 71% of those examples, an even more accurate match was found (at least three atoms were located within 0.7 Å of the atoms of a single side chain).

Another method, which looks for binding sites of peptide anchor residues, is ANCHORSMAP (47), which was shown to locate the peptide anchor binding sites on the PDZ domain and in the protein–peptide complex kinase/PKI, and has recently been applied to characterize the specificity of Thr and Ser kinase binding grooves (104).

We are currently working to combine the different approaches for binding-site prediction (pocket detection, small-molecule mappings, and other features extracted from peptide–protein complexes datasets) to devise an integrated machine learning based classifier that would predict peptide binding sites, in analogy to similar approaches for predicting binding sites for globular proteins and small molecules.

4. Modeling the Initial Backbone Conformation of the Peptide

Most state-of-the-art tools available for modeling and refining the final peptide–receptor complex require an initial conformation of the peptide backbone as part of their input, except for the case of very short peptides made of 2–4 amino acids (49). In the absence of template structures for the target peptide–protein interaction, the initial peptide backbone conformation has to be modeled by other means. We have recently shown that the Rosetta FlexPepDock tool (see below) can model peptide–protein complexes accurately if the initial peptide backbone conformation deviates from the native peptide by at most 50° in terms of φ/ψ torsion angles RMSD (27), meaning that the initial peptide model should at least approximate the correct native secondary structure.

According to an ***induced fit*** model of peptide recognition, a peptide would fold only upon binding to its partner (14) (reviewed in ref. 16). This model suggests that even for building an initial model of the peptide backbone, the effect of the receptor protein on the peptide backbone conformation must be taken into account. In contrast, the *conformational sampling* model rather assumes that the peptide in its free form samples an ensemble of peptide conformations that includes the native, bound peptide conformation. According to this model, the presence of the receptor molecule only shifts the equilibrium further towards the bound form. The conformational sampling model was shown to apply to interactions between intrinsically disordered domains that exist as molten globules in their free state (17, 50) (reviewed in ref. 16). Also, it is known that small peptides that are stabilized by short-range hydrogen bonds, such as β-hairpin peptides (51) and α-helical peptides (52), may adopt a stable secondary structure already in their free form to a varying degree. This suggests that the initial modeling of a set of potential peptide backbone conformations based on sequence preferences alone could well serve as input to consequent peptide refinement within the receptor environment in a subsequent step.

To the best of our knowledge, no generic well-tested tool for conformational sampling of peptide conformations in the context of peptide docking has yet been designed. However, different approaches have been used to address free peptide conformational sampling. Molecular dynamics (MD), for instance, has been used to predict the structure of α-helical and β-hairpin peptides (53, 54) and to study their energy landscape (55). Other sampling methods have also been used for exploring the structures of free peptide molecules. These include Monte-Carlo-based approaches (56–58), which often sample the conformation space more effectively than MD, as well as density-guided importance sampling (59) and simulated annealing-coupled replica exchange molecular dynamics (60).

Sequence-based fragment libraries extracted from PDB structures have been very successful for de novo protein fold prediction (61, 62), loop modeling (63), and other applications (64). Voelz et al. (65) have used replica exchange molecular dynamics (REMD) simulations on 872 different 8-mer, 12-mer, and 16-mer peptide fragments from 13 proteins to examine the extent to which conformations of peptide fragments in water predict native conformations (native contacts) in globular proteins (extending a similar study on a smaller scale by Ho and Dill (66)). Using this scheme, they achieved accuracy of up to 63% in the prediction of native contacts for 8-mers, 71% for 12-mers, and 76% for 16-mers. It seems reasonable that these results would hold also for peptide–protein interaction, as Vanhee et al. (67) recently showed that bound peptides often emulate backbone fragments of monomer proteins. Therefore, already-solved structures can be a good source for estimating the interacting peptide backbone conformation. Preliminary results of an ongoing study in our group show that at least in some specific cases, sequence similarity can be used to detect correct protein segments from structures in the Protein

Data Bank (68), albeit there are many exceptions (see Note 3). Based on these results and on the Rosetta fragment libraries approach (62), we have developed and calibrated ab initio FlexPepDock, an extension of the FlexPepDock refinement protocol described in detail below. FlexPepDock ab initio fully samples the peptide conformations space while docking it to a given site on the protein receptor (105). This protocol has significantly increased the number of peptide-protein interactions that can now be modeled at high accuracy.

Using ideal secondary structure geometry for initial peptide conformation. As the tools used for the final modeling of the peptide–protein complex require only an approximate initial model of the peptide backbone, it might suffice to specify the correct secondary structure composition of the peptide. We have recently shown that for a wide range of peptide–protein interactions, good results can be obtained using the Rosetta FlexPepDock method (27), if we start from an ideally extended initial peptide backbone conformation, even if the native peptide conformation deviates substantially from ideal extended geometry (27). Similar results were shown previously for PDZ domains, which also bind peptides in extended-like conformation (24). It is plausible that if native peptides are, e.g., helical, then an initial conformation with ideal helix geometry would be suitable for the final docking step, although this has not been tested hitherto. We note that the secondary structure propensity of a peptide in its free form can be inferred from experimental methods such as CD spectroscopy (69) or from sequence preferences alone and therefore may provide the necessary information for creating sufficiently good initial peptide models.

Finally, we note that, in some cases, NMR spectroscopy can be used to determine the structure of the *bound* peptide molecule (70, 71), even if for technical reasons the structure of the receptor protein or the relative orientation of the peptide and the receptor cannot be determined (due to, e.g., the size of the receptor).

5. Modeling and Refinement of the Peptide–Protein Complex

Given a known binding site, whether from experimental data or based on prediction, and an estimated conformation for the peptide, be it based on a homologue, predicted as described above, or even a linear representation of the peptide in its binding pocket, we now have reached the last and most critical step of modeling peptide protein interactions: the high-resolution refinement of the peptide within the binding pocket. Again, there is no exact line between "refinement" and "docking" and different tools can reach near-native solutions starting from different representations of the system. This is not a trivial stage, since it has to tackle the sampling of many degrees of freedom. Usually, full flexibility will be given to the peptide backbone and side chains, and some level of flexibility will be sampled for the receptor protein. Moreover, correct selection of the best model is also a hard task, given the large conformational space and rugged energy landscape. In this section, we briefly review methods for the high-resolution modeling of peptide–protein interactions and their performance on various benchmarks.

5.1. Rosetta FlexPepDock (27, 105) (Availability: Rosetta Releases 3.2 and later; Web server at http://flexpepdock.furmanlab.cs.huji.ac.il/(101))

Rosetta FlexPepDock is a high-resolution protocol for refining peptide–protein complexes implemented in the Rosetta modeling suite framework. Given a coarse model of the interaction (either based on homology modeling or generated using the approaches described above), FlexPepDock performs a Monte-Carlo-Minimization-based approach to refine all of the peptide's degrees of freedom (rigid body orientation, backbone and side chain flexibility) as well as the protein receptor side-chain conformations.

FlexPepDock was thoroughly benchmarked against a set of perturbed peptide–protein complexes and an effective range of sampling was defined. For peptides with initial backbone (bb) RMSD of up to 5.5 Å, FlexPepDock is able to create near-native models (peptide bb-RMSD <2 Å) in 91% of the cases for the bound receptor, and rank them as one of the top five models in 78%. In the challenging task of unbound (apo) docking, near-native models were sampled in 85% of the cases and ranked correctly in 59% (for starting structures within 5.5 Å bb-RMSD from the native).

The accuracy of the protocol for high-resolution modeling was tested on consecutive 4-mers, as peptide binding is often mediated by short, highly conserved motifs. Indeed, for starting structures within 3.5 Å bb-RMSD, FlexPepDock managed to sample all-atom sub-angstrom (<1 Å) 4-mers for 82% of the bound cases and 62% of the unbound cases and to rank them among the top five models in 62% and 35% of the cases, respectively.

In cases where no information is available about the conformation of the peptide backbone, docking can be started from an extended conformation of the peptide. In a benchmark in which the peptide was docked starting from an ideal extended backbone conformation (±135° for all φ/ψ angles) based on a single anchor residue, near-native solutions could be sampled in 66% of the 71 non-helical complexes (31% for sub-angstrom models), and ranked among the top five solutions in 49% of the cases (24% for sub-angstrom models).

Recently, FlexPepDock was applied to several "real-world" problems, namely (a) To model the interaction of a bacterial quorum sensing peptide (External Death Factor) with the toxin MazF (72); (b) to model the binding of Dictyostelium myosin II heavy chain kinase A floppy tail at the kinase active site as well as at a putative allosteric site (73); and lastly (c) for the creation of a plausible starting model for a molecular dynamics simulation of a glycogen synthase kinase 3β kinase/substrate peptide interaction (74).

5.2. DynaDock (28) (Availability: Contact Authors)

DynaDock is a three-tiered peptide (small-molecule) docking protocol, which was developed specifically to address the problem of the large number of degrees of freedom that needs to be sampled for peptides. In the first step, broad random sampling of the peptide conformation within the binding pocket is performed to produce 500 starting conformations. In the second step, which is the core of this protocol, an optimized potential molecular dynamics (OPMD) refinement procedure is applied to each of these conformations. This procedure employs a soft-core potential function, which is optimized with respect to the system's energy throughout the simulation and was proven to be superior to standard soft-core potentials. In the last step, a system-specific scoring function is applied to rank the refined models.

DynaDock was benchmarked on a dataset of 15 peptide–protein complexes with peptides that range in length between 2 and 16 amino acids. For starting conformations sampled in the broad sampling stage with >3.5 Å RMSD to the equilibrated native peptide, DynaDock managed to sample a refined structure <2.1 Å for all 15 complexes. For 7/15 complexes, 20–40% of the refined models displayed <2.5 Å RMSD. Similar results were obtained for a set of four unbound peptide docking cases. A scoring function that was reweighted using standard *Z*-score optimization based on this set of 15 complexes was able to rank best a model within 2.1 Å for 11 of these 15 complexes.

5.3. AutoDock (49) (Availability: http://autodock.scripps.edu/) and Other Blind-Docking Methods for Short Peptides

Heteniy et al. showed that AutoDock (49), which was originally developed as a ligand docking tool, is able to "blindly" dock very short peptides (2–4 amino acids) to the bound receptor structure, with high accuracy and with no prior knowledge of the peptide binding site (75). In effect, this approach covers steps (2–4) all at once for very short peptides—from locating the binding site to modeling the peptide backbone within it. Additional studies have used AutoDock to perform docking simulations of even longer peptides, such as a heptapeptide inhibitor binding to the α7-nicotinic receptor (76), a phage-display selected peptide to a ligand-bound antibody (77) and a pentapeptide ligand to the binding site of the MAP kinase ERK2 (78). Another blind-docking approach that was tested on a set of short peptides (3–7 amino acids long) was presented by Prasad and Gautham, using orthogonal Latin-Square sampling (79). However, to date, automated blind-docking of longer peptides remains an open challenge.

5.4. Peptide Modeling Protocols for Specific Systems

While only few approaches for peptide docking have been developed and tested for general, broad applicability (see above), there have been several studies on peptide docking to specific protein receptors, in particular to MHC receptors and to PDZ domains. We describe these methods, in this section, as several of the approaches implemented therein could well be of use and success on a more general scale of peptide docking.

MHC–peptide interactions. A range of structures has been solved for peptide–MHC receptor interactions (80), and consequently, these have served as a test bed for the development and application of different methodologies for peptide docking. This includes biased-probability Monte-Carlo docking (23), peptide backbone library-based predictions coupled with explicit solvent modeling (81), atomistic-level modeling with implicit solvent (82), simulated annealing-driven molecular dynamics (83–85), and docking the peptide's anchor residues into their binding pockets followed by loop closure and peptide backbone refinement (86). Finally, a recent molecular modeling study of MHC-peptide interactions that integrates sampling techniques from protein–protein docking, loop modeling, de novo structure prediction, and protein design, has constructed atomically detailed peptide binding landscapes for a diverse set of MHC proteins which can be used to study the structural details that confer binding specifities of distinct MHC alleles (102).

PDZ domain–peptide interactions. Another biological system that spurred interest from the peptide docking perspective is the binding of peptides to the PDZ domain. Niv et al. devised a protocol for flexible peptide docking based on a simulated annealing molecular dynamics approach (24). The protocol requires one fixed anchoring point to be in the peptide, e.g., the well-conserved position of the Cα atom of the C-terminal residue of the peptide in the PDZ case. The peptide–protein complex conformational space is explored at elevated temperature, followed by cooling and side-chain assignment based on SCWRL 3.0 (87) for each of the hundreds of conformations obtained from the heated trajectory. The resulting models are minimized and scored.

This protocol was benchmarked on a test set of PDZ–peptide complexes. Redocking to native structures (starting from the solved protein–peptide complex) yielded models with RMSD <2 Å for the six tested penta- to octapeptides. When docking to either apo structures of the same protein (unbound docking) or structures of the domain originally solved complexed with another peptide (cross docking) or to homology models of the protein, the best-scoring models displayed RMSD <2.8 Å for all heavy atoms of tetra- to octapeptides in 9 of 12 cases.

Staneva and Wallin (88) developed a procedure that provides limited receptor flexibility using soft constraints while allowing the peptide chain full flexibility. Using an effective all-atom energy function, they perform extensive Monte-Carlo simulations, to achieve full representative conformational ensembles. The procedure was tested on a set of 11 PDZ domain–peptide pairs (bound docking). In 8/11 cases, the minimum-energy conformations displayed all-atom peptide RMSDs <6 Å to the native structure. Similar results were obtained on a test set of nine unbound structures (unbound docking).

Recently Gerek and Ozkan (89) used this system to benchmark another protocol, which is focused on better addressing the backbone flexibility of the receptor. This protocol is based on a dihedral restrained REMD, in which normal modes obtained by an elastic network model (ENM) (90) are incorporated into the molecular dynamics simulations as dihedral restraints to speed up the search. In this way, conformations of the unbound protein receptor are produced along the binding fluctuation mode. Clustering the lowest replica trajectory creates an ensemble of multiple receptor conformations, and peptides are then docked onto these clusters using RosettaLigand (91). The method was tested on a set of PDZ–peptide complexes and indeed was proved to create lower RMSD models, when compared with docking to a fixed backbone unbound receptor (see Note 4).

5.5. Improved Modeling of Peptide–Protein Interactions Using Constraints

Many protein–peptide docking approaches utilize well-characterized structural constraints available for the specific system at hand. Other methods rely on more general constraints. As an example, Liu et al. used an energy scoring function that was explicitly biased towards the native backbone using a coarse-grained Gö potential to dock peptides to their receptors for a dataset of 25 peptide interactions with a large number of rotatable bonds (92). Maurer et al. introduced NMR-derived NOE constraints into an MCM-based docking approach to dock a fibrinogen-like peptide to thrombin (93).

6. Structural Databases of Peptide–Protein Complexes

As mentioned above, only few approaches for peptide–protein docking have been developed, tested, and applied for a large representative range of interactions. Indeed, a crucial step on the path to develop peptide–protein modeling tools was and still is the creation of suitable databases. In addition to their utility for benchmarking purposes, these datasets provide representative templates for homology models, and have enabled large-scale characterization of the features that govern peptide–protein interactions. Below are three collections of peptide–protein complex structures that have emerged recently thanks to the increase in structural information available for these interactions.

6.1. PepX (94) (Availability: http://pepx.switchlab.org)

PepX contains protein–peptide complexes solved by X-ray crystallography with a resolution better than 2.5 Å, with peptides that are between 5 and 35 residues long and that contain natural amino acids only. 1,431 complexes were retained and clustered according to their binding architecture: Any two structures are grouped together if they superpose below 2 Å Cα RMSD for at least 75% of their interface residues. This results in 505 unique protein–peptide interface clusters. It is interesting to note that 64–87% of all clusters are singletons for thresholds of 1–3 Å and 50–95% alignment similarity.

6.2. peptiDB (26) (Availability: London et al. (26) Supplemental Information)

This database was constructed to investigate the binding strategies of peptides to proteins. This is a small, but highly curated database which contains only structures solved by X-ray crystallography with a resolution better than 2.0 Å, without heteroatoms at the interface. Peptide length ranges between 5 and 15 residues, and the structures are clustered at 70% sequence identity for the protein monomer. The resulting dataset contains 103 complexes.

6.3. 3did Peptide-Mediated Interactions (95) (Availability: http://3did.irbbarcelona.org)

The construction of this dataset was based on the idea of detecting structures of interactions involving short linear motifs. Linear motifs are short patterns of around ten residues, which in isolation bind their target proteins with sufficient strength to establish a functional interaction. They are frequently found in disordered or unstructured regions and adopt a well-defined structure only upon binding. The eukaryotic linear motif (ELM) database contains information about many such motifs (96). The PDB was parsed to identify all of the structures of motif binding domains from ELM, followed by the detection of the occurrences of the linear consensus motif within its contacting partners. This was followed by manual visual inspection and at the time of publication 3did contained data on 829 hand-curated peptide-mediated interactions of known 3D structure, from 611 protein pairs, involving 32 globular domains and 51 linear motifs (97) (see Note 5).

7. Towards Automated De Novo Peptide Modeling

After introducing the main challenges and approaches of peptide–protein docking, we conclude our chapter with an illustrative example originally presented by Raveh et al. (27), which exemplifies the different steps described in this chapter, and some of the methods that are available for "real-world" peptide docking. This example highlights the current challenges and limitations in the field of peptide docking.

The HIV-capsid protein interacts in the cell with the human Proline isomerase cyclophilin A (CypA), as part of the virus life cycle. This interaction is mediated by a single peptide (solvent exposed loop) derived from the capsid protein (Sequence: HAGPIA). The structure of the complex between CypA and the peptide was solved (PDB: 1AWR (98)) and is of major interest both as a therapeutic target and for the understanding of HIV. We will try to predict the structure of this complex.

1. As a first step, we delete the peptide partner from the complex and use the FTmap server by Brenke et al. (46), to map potential binding sites for the peptide over the bound receptor surface. The binding position that is ranked second by the

FTMap server roughly correlates with the native position of the central Proline residue (Fig. 3a).

2. We manually pose (using a standard molecular viewer) an extended form of the peptide (±135° for all φ/ψ angles) onto the binding position predicted by FTmap, such that the peptide's central Proline would overlap the predicted fragment location (Fig. 3b).

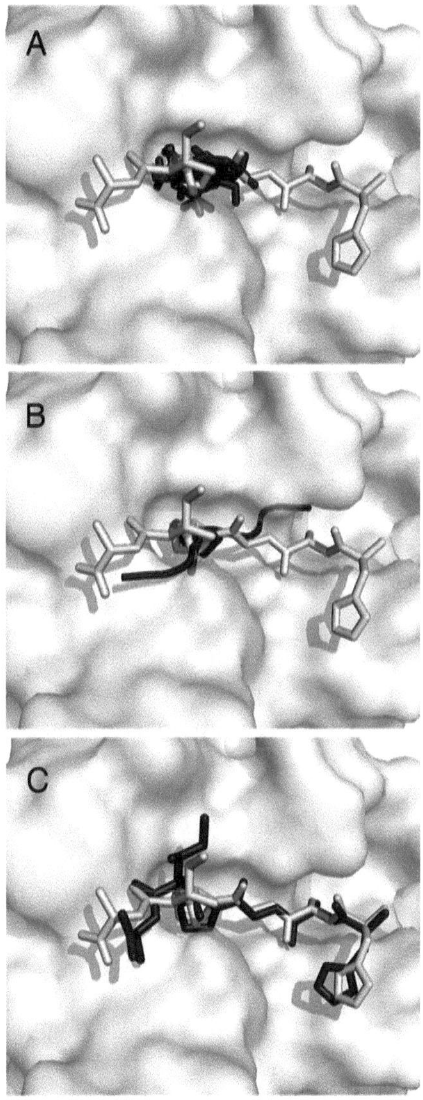

Fig. 3. Peptide-docking example. The CypA protein receptor is depicted in *white surface*. The native bound HIV peptide (HAGPIA) is depicted in *stick representation* (PDB: 1AWR) and was docked using the FlexPepDock protocol as described in Raveh et al. (27). (**a**) The second ranked cluster of FTmap predicts accurately the position of the anchoring Proline residue of the peptide. (**b**) Manual placement of an extended conformation peptide serves as a starting structure for further refinement. (**c**) The final model produced by FlexPepDock is 0.8 Å backbone-RMSD from the native peptide.

3. We use FlexPepDock to refine the complex. In the third ranking solution provided by FlexPepDock, the starting structure was refined from 4.3 Å bb-RMSD to only 0.8 Å bb-RMSD from the native, with sub-angstrom all-atom modeling for most interacting residues (Fig. 3c).

The example describe above, even though successful, highlights several challenges that still need to be addressed before a fully automated general ab initio peptide docking protocol can be used.

First of all, the example was performed using the bound structure of the protein receptor, in real world problems, the bound receptor structure is usually not available but rather an unbound structure or perhaps a structure solved with another partner or an homology model. There are several indications that this is not a major limitation for many peptide–protein interactions. We previously demonstrated that usually the receptor does not undergo major conformational changes upon peptide binding (26) and we showed good performance for our refinement protocol on unbound structures (although not as good as for bound structures) (27). That said, it is clear that receptor flexibility can play a major role in other cases.

The second limitation is the adequate mapping of the *global* peptide binding energy landscape—that is, the correct *ranking* of solutions in different binding sites. For example, our *Rosetta FlexPepDock* protocol was shown to provide accurate ranking of different solutions within the vicinity of the correct binding site. However, in this example, FTMap suggested a set of possible binding sites. When we positioned the initial peptide model in the vicinity of the correct binding site (the binding site ranked second by FTMap), the *Rosetta FlexPepDock* energy function was able to select a sub-angstrom solution as one of its three top-ranking models. However, when starting a similar docking simulation from the FTmap best ranking (but incorrect) binding site, the *Rosetta FlexPepDock* protocol produced models with even better scores than the native—meaning that were we to choose between the two different runs based on the current energy function, we would have selected a false positive in this case. While this is only one specific example, future research should be able to improve the global ranking of solutions within different binding sites. A similar limitation lies in the manual placement of the peptide. This would be an easy step to automate but would have to be coupled with a better scoring scheme as described above. Given an approximate anchor point on the receptor surface there are many different directions a linear peptide can be placed, and therefore the energy landscape of the interaction needs to be accurate enough to select the correct orientation.

Another challenge relates to the sampling in the vast space of both peptide conformations and protein conformations, in fully de novo peptide modeling. The modular approach we outlined in this chapter reduces this bigger problem to a set of smaller subproblems. Vanhee et al. (67) recently made an interesting finding that may be of help in reducing the sampling-space of peptide–protein interactions in the future. In their work, they compared the interfaces of peptide–protein complexes to interactions observed within monomeric proteins and found surprising similarities. Of a dataset of 731 protein–peptide interfaces, over 65% could be reconstructed within 1 Å RMSD using structural fragments of interacting residues within monomeric protein folds. Interestingly, more than 80% of the fragments used for this reconstruction originated from proteins of entirely different structural classification, with an average sequence identity below 15%. This finding suggests that the plethora of available protein structures could be searched to find suitable templates for protein–peptide interactions and, more importantly, that sequence homology is no prerequisite. Indeed, our fragment-based ab initio FlexPepDock protocol has demonstrated that using fragments derived from other, non-related protein structures, near-native models can be created in most of the examined cases (105).

Despite all of these challenges that are still being addressed by various research groups, there are many actual problems that can be tackled already by state-of-the-art approaches. We believe that the upcoming years will see considerable progress in our ability to create accurate models of peptide–protein interactions in an increasingly automated fashion, with applications in binding-specificity prediction and rational design of peptide-mediated interactions, motivated by the pivotal role of peptide interactions in the cellular network of protein–protein interactions and their promise as leads for drug molecules. These are indeed exciting times for the research of peptide–protein interactions.

8. Notes

1. An underlying assumption of PepSite is that flexible peptides bind in roughly extended conformation, which makes it somewhat less suitable for helical peptides, which constitute around 20% of the peptides in peptide–receptor datasets (26) as well as for peptides with sharp coils and turns.
2. We should note that there are many tools available for pocket detection (40), but these have not been evaluated specifically for peptides.
3. In certain cases such as the interaction between protease-cleaved peptides and MHC receptors, the cleaved peptides adapt an extended conformation upon binding to the MHC receptor, regardless of their conformation within their parent proteins, which may vary considerably (99).

4. It should be noted that the native peptide backbones were kept fixed during the simulations—thus avoiding one of the major hurdles of peptide docking.
5. Note that the data in this collection is not clustered and is somewhat redundant.

References

1. Petsalaki, E., and Russell, R. B. (2008) Peptide-mediated interactions in biological systems: new discoveries and applications, *Curr Opin Biotechnol 19*, 344–350.
2. Neduva, V., Linding, R., Su-Angrand, I., Stark, A., de Masi, F., Gibson, T. J., Lewis, J., Serrano, L., and Russell, R. B. (2005) Systematic discovery of new recognition peptides mediating protein interaction networks, *PLoS Biol 3*, e405.
3. Pawson, T., and Nash, P. (2003) Assembly of cell regulatory systems through protein interaction domains, *Science 300*, 445–452.
4. Rubinstein, M., and Niv, M. Y. (2009) Peptidic modulators of protein-protein interactions: progress and challenges in computational design, *Biopolymers 91*, 505–513.
5. Vlieghe, P., Lisowski, V., Martinez, J., and Khrestchatisky, M. (2010) Synthetic therapeutic peptides: science and market, *Drug Discov Today 15*, 40–56.
6. Parthasarathi, L., Casey, F., Stein, A., Aloy, P., and Shields, D. C. (2008) Approved drug mimics of short peptide ligands from protein interaction motifs, *J Chem Inf Model 48*, 1943–1948.
7. London, N., Raveh, B., Movshovitz-Attias, D., and Schueler-Furman, O. (2010) Can Self-Inhibitory Peptides be Derived from the Interfaces of Globular Protein-Protein Interactions?, *Proteins 78*, :3140–3149.
8. Jemth, P., and Gianni, S. (2007) PDZ domains: folding and binding, *Biochemistry 46*, 8701–8708.
9. Vacic, V., Oldfield, C. J., Mohan, A., Radivojac, P., Cortese, M. S., Uversky, V. N., and Dunker, A. K. (2007) Characterization of molecular recognition features, MoRFs, and their binding partners, *J Proteome Res 6*, 2351–2366.
10. Gamble, T. R., Vajdos, F. F., Yoo, S., Worthylake, D. K., Houseweart, M., Sundquist, W. I., and Hill, C. P. (1996) Crystal structure of human cyclophilin A bound to the amino-terminal domain of HIV-1 capsid, *Cell 87*, 1285–1294.
11. Heemels, M. T., and Ploegh, H. (1995) Generation, translocation, and presentation of MHC class I-restricted peptides, *Annu Rev Biochem 64*, 463–491.
12. Zhou, A., Webb, G., Zhu, X., and Steiner, D. F. (1999) Proteolytic processing in the secretory pathway, *J Biol Chem 274*, 20745–20748.
13. Schweizer, A., Briand, C., and Grutter, M. G. (2003) Crystal structure of caspase-2, apical initiator of the intrinsic apoptotic pathway, *J Biol Chem 278*, 42441–42447.
14. Sugase, K., Dyson, H. J., and Wright, P. E. (2007) Mechanism of coupled folding and binding of an intrinsically disordered protein, *Nature 447*, 1021–1025.
15. Fuxreiter, M., Tompa, P., and Simon, I. (2007) Local structural disorder imparts plasticity on linear motifs, *Bioinformatics 23*, 950–956.
16. Wright, P. E., and Dyson, H. J. (2009) Linking folding and binding, *Curr Opin Struct Biol 19*, 31–38.
17. Kjaergaard, M., Teilum, K., and Poulsen, F. M. (2010) Conformational selection in the molten globule state of the nuclear coactivator binding domain of CBP, *Proc Natl Acad Sci U S A 107*, 12535–12540.
18. Rosal, R., Pincus, M. R., Brandt-Rauf, P. W., Fine, R. L., Michl, J., and Wang, H. (2004) NMR solution structure of a peptide from the mdm-2 binding domain of the p53 protein that is selectively cytotoxic to cancer cells, *Biochemistry 43*, 1854–1861.
19. Wu, G., Chen, Y. G., Ozdamar, B., Gyuricza, C. A., Chong, P. A., Wrana, J. L., Massague, J., and Shi, Y. (2000) Structural basis of Smad2 recognition by the Smad anchor for receptor activation, *Science 287*, 92–97.
20. Zhang, Y. (2009) Protein structure prediction: when is it useful?, *Curr Opin Struct Biol 19*, 145–155.
21. Vajda, S., and Kozakov, D. (2009) Convergence and combination of methods in protein-protein docking, *Curr Opin Struct Biol 19*, 164–170.
22. Lane, K. T., and Beese, L. S. (2006) Thematic review series: lipid posttranslational modifications. Structural biology of protein farnesyl-

transferase and geranylgeranyltransferase type I, *J Lipid Res 47*, 681–699.

23. Bordner, A. J., and Abagyan, R. (2006) Ab initio prediction of peptide-MHC binding geometry for diverse class I MHC allotypes, *Proteins 63*, 512–526.
24. Niv, M. Y., and Weinstein, H. (2005) A flexible docking procedure for the exploration of peptide binding selectivity to known structures and homology models of PDZ domains, *J Am Chem Soc 127*, 14072–14079.
25. Hwang, H., Pierce, B., Mintseris, J., Janin, J., and Weng, Z. (2008) Protein-protein docking benchmark version 3.0, *Proteins 73*, 705–709.
26. London, N., Movshovitz-Attias, D., and Schueler-Furman, O. (2010) The structural basis of peptide-protein binding strategies, *Structure 18*, 188–199.
27. Raveh, B., London, N., and Schueler-Furman, O. (2010) Sub-angstrom modeling of complexes between flexible peptides and globular proteins, *Proteins 78*, 2029–2040.
28. Antes, I. (2010) DynaDock: A new molecular dynamics-based algorithm for protein-peptide docking including receptor flexibility, *Proteins 78*, 1084–1104.
29. Smith, C. A., and Kortemme, T. (2010) Structure-Based Prediction of the Peptide Sequence Space Recognized by Natural and Synthetic PDZ Domains, *J Mol Biol 402*, 460–474.
30. Kaufmann, K., Shen, N., Mizoue, L., and Meiler, J. (2010) A physical model for PDZ-domain/peptide interactions, *J Mol Model 17*, 315–324.
31. Chaudhury, S., and Gray, J. J. (2009) Identification of structural mechanisms of HIV-1 protease specificity using computational peptide docking: implications for drug resistance, *Structure 17*, 1636–1648.
32. King, C. A., and Bradley, P. Structure-based prediction of protein-peptide specificity in Rosetta, *Proteins 78*, 3437–3449.
33. Morrison, K. L., and Weiss, G. A. (2001) Combinatorial alanine-scanning, *Curr Opin Chem Biol 5*, 302–307.
34. Mandell, J. G., Falick, A. M., and Komives, E. A. (1998) Identification of protein-protein interfaces by decreased amide proton solvent accessibility, *Proc Natl Acad Sci U S A 95*, 14705–14710.
35. Bradford, J. R., and Westhead, D. R. (2005) Improved prediction of protein-protein binding sites using a support vector machines approach, *Bioinformatics 21*, 1487–1494.
36. Neuvirth, H., Raz, R., and Schreiber, G. (2004) ProMate: a structure based prediction program to identify the location of protein-protein binding sites, *J Mol Biol 338*, 181–199.
37. Qin, S., and Zhou, H. X. (2007) meta-PPISP: a meta web server for protein-protein interaction site prediction, *Bioinformatics 23*, 3386–3387.
38. de Vries, S. J., van Dijk, A. D., and Bonvin, A. M. (2006) WHISCY: what information does surface conservation yield? Application to data-driven docking, *Proteins 63*, 479–489.
39. Zhou, H. X., and Qin, S. (2007) Interaction-site prediction for protein complexes: a critical assessment, *Bioinformatics 23*, 2203–2209.
40. Capra, J. A., Laskowski, R. A., Thornton, J. M., Singh, M., and Funkhouser, T. A. (2009) Predicting protein ligand binding sites by combining evolutionary sequence conservation and 3D structure, *PLoS Comput Biol 5*, e1000585.
41. Laurie, A. T., and Jackson, R. M. (2006) Methods for the prediction of protein-ligand binding sites for structure-based drug design and virtual ligand screening, *Curr Protein Pept Sci 7*, 395–406.
42. Petsalaki, E., Stark, A., Garcia-Urdiales, E., and Russell, R. B. (2009) Accurate prediction of peptide binding sites on protein surfaces, *PLoS Comput Biol 5*, e1000335.
43. Liu, X., and Marmorstein, R. (2007) Structure of the retinoblastoma protein bound to adenovirus E1A reveals the molecular basis for viral oncoprotein inactivation of a tumor suppressor, *Genes Dev 21*, 2711–2716.
44. Dundas, J., Ouyang, Z., Tseng, J., Binkowski, A., Turpaz, Y., and Liang, J. (2006) CASTp: computed atlas of surface topography of proteins with structural and topographical mapping of functionally annotated residues, *Nucleic Acids Res 34*, W116-118.
45. Binkowski, T. A., Naghibzadeh, S., and Liang, J. (2003) CASTp: Computed Atlas of Surface Topography of proteins, *Nucleic Acids Res 31*, 3352–3355.
46. Brenke, R., Kozakov, D., Chuang, G. Y., Beglov, D., Hall, D., Landon, M. R., Mattos, C., and Vajda, S. (2009) Fragment-based identification of druggable 'hot spots' of proteins using Fourier domain correlation techniques, *Bioinformatics 25*, 621–627.
47. Ben-Shimon, A., and Eisenstein, M. (2010) Computational mapping of anchoring spots on protein surfaces, *J Mol Biol 402*, 259–277.
48. Katchalski-Katzir, E., Shariv, I., Eisenstein, M., Friesem, A. A., Aflalo, C., and Vakser, I. A. (1992) Molecular surface recognition: determination of geometric fit between pro-

teins and their ligands by correlation techniques, *Proc Natl Acad Sci U S A 89*, 2195–2199.
49. Goodsell, D. S., Morris, G. M., and Olson, A. J. (1996) Automated docking of flexible ligands: applications of AutoDock, *J Mol Recognit 9*, 1–5.
50. Song, J., Guo, L. W., Muradov, H., Artemyev, N. O., Ruoho, A. E., and Markley, J. L. (2008) Intrinsically disordered gamma-subunit of cGMP phosphodiesterase encodes functionally relevant transient secondary and tertiary structure, *Proc Natl Acad Sci U S A 105*, 1505–1510.
51. Blandl, T., Cochran, A. G., and Skelton, N. J. (2003) Turn stability in beta-hairpin peptides: Investigation of peptides containing 3:5 type I G1 bulge turns, *Protein Sci 12*, 237–247.
52. Andrews, M. J. I., and Tabor, A. B. (1999) Forming stable helical peptides using natural and artificial amino acids, *Tetrahedron 55*, 11711–11743.
53. Schaefer, M., Bartels, C., and Karplus, M. (1998) Solution conformations and thermodynamics of structured peptides: molecular dynamics simulation with an implicit solvation model, *J Mol Biol 284*, 835–848.
54. Fuchs, P. F., Bonvin, A. M., Bochicchio, B., Pepe, A., Alix, A. J., and Tamburro, A. M. (2006) Kinetics and thermodynamics of type VIII beta-turn formation: a CD, NMR, and microsecond explicit molecular dynamics study of the GDNP tetrapeptide, *Biophys J 90*, 2745–2759.
55. Higo, J., Ito, N., Kuroda, M., Ono, S., Nakajima, N., and Nakamura, H. (2001) Energy landscape of a peptide consisting of alpha-helix, 3(10)-helix, beta-turn, beta-hairpin, and other disordered conformations, *Protein Sci 10*, 1160–1171.
56. Kidera, A. (1995) Enhanced conformational sampling in Monte Carlo simulations of proteins: application to a constrained peptide, *Proc Natl Acad Sci U S A 92*, 9886–9889.
57. Abagyan, R., and Totrov, M. (1994) Biased probability Monte Carlo conformational searches and electrostatic calculations for peptides and proteins, *J Mol Biol 235*, 983–1002.
58. Ulmschneider, J. P., and Jorgensen, W. L. (2004) Polypeptide folding using Monte Carlo sampling, concerted rotation, and continuum solvation, *J Am Chem Soc 126*, 1849–1857.
59. Thomas, G. L., Sessions, R. B., and Parker, M. J. (2005) Density guided importance sampling: application to a reduced model of protein folding, *Bioinformatics 21*, 2839–2843.
60. Kannan, S., and Zacharias, M. (2009) Simulated annealing coupled replica exchange molecular dynamics--an efficient conformational sampling method, *J Struct Biol 166*, 288–294.
61. Camproux, A. C., Gautier, R., and Tuffery, P. (2004) A hidden markov model derived structural alphabet for proteins, *J Mol Biol 339*, 591–605.
62. Simons, K. T., Bonneau, R., Ruczinski, I., and Baker, D. (1999) Ab initio protein structure prediction of CASP III targets using ROSETTA, *Proteins Suppl 3*, 171–176.
63. Wang, C., Bradley, P., and Baker, D. (2007) Protein-protein docking with backbone flexibility, *J Mol Biol 373*, 503–519.
64. Budowski-Tal, I., Nov, Y., and Kolodny, R. (2010) FragBag, an accurate representation of protein structure, retrieves structural neighbors from the entire PDB quickly and accurately, *Proc Natl Acad Sci U S A 107*, 3481–3486.
65. Voelz, V. A., Shell, M. S., and Dill, K. A. (2009) Predicting peptide structures in native proteins from physical simulations of fragments, *PLoS Comput Biol 5*, e1000281.
66. Ho, B. K., and Dill, K. A. (2006) Folding very short peptides using molecular dynamics, *PLoS Comput Biol 2*, e27.
67. Vanhee, P., Stricher, F., Baeten, L., Verschueren, E., Lenaerts, T., Serrano, L., Rousseau, F., and Schymkowitz, J. (2009) Protein-peptide interactions adopt the same structural motifs as monomeric protein folds, *Structure 17*, 1128–1136.
68. Berman, H. M., Westbrook, J., Feng, Z., Gilliland, G., Bhat, T. N., Weissig, H., Shindyalov, I. N., and Bourne, P. E. (2000) The Protein Data Bank, *Nucleic Acids Res 28*, 235–242.
69. Greenfield, N., and Fasman, G. D. (1969) Computed circular dichroism spectra for the evaluation of protein conformation, *Biochemistry 8*, 4108–4116.
70. Hayouka, Z., Levin, A., Maes, M., Hadas, E., Shalev, D. E., Volsky, D. J., Loyter, A., and Friedler, A. (2010) Mechanism of action of the HIV-1 integrase inhibitory peptide LEDGF 361–370, *Biochem Biophys Res Commun 394*, 260–265.
71. Moller, H., Serttas, N., Paulsen, H., Burchell, J. M., and Taylor-Papadimitriou, J. (2002) NMR-based determination of the binding epitope and conformational analysis of MUC-1 glycopeptides and peptides bound to the breast cancer-selective monoclonal antibody SM3, *Eur J Biochem 269*, 1444–1455.

72. Belitsky M, A. H., Yelin I, London N, Shperber M, Schueler-Furman , and O, E.-K. H. (2011) The Escherichia coli Extracellular Death Factor EDF induces the endoribonucleolytic activities of the toxins MazF and ChpBK, *Molecular Cell 41*, 625–635.
73. Buch, I., Fishelovitch, D., London, N., Raveh, B., Wolfson, H. J., and Nussinov, R. Allosteric regulation of glycogen synthase kinase 3beta: a theoretical study, *Biochemistry 49*, 10890–10901.
74. Crawley, S. W., Samimi Gharaei, M., Ye, Q., Yang, Y., Raveh, B., London, N., Schueler-Furman, O., Jia, Z., and Cote, G. P. Autophosphorylation activates Dictyostelium myosin II heavy chain kinase A by providing a ligand for an allosteric binding site in the {alpha}-kinase domain, *J Biol Chem 286*, 2607–2616.
75. Hetenyi, C., and van der Spoel, D. (2002) Efficient docking of peptides to proteins without prior knowledge of the binding site, *Protein Sci 11*, 1729–1737.
76. Espinoza-Fonseca, L. M., and Trujillo-Ferrara, J. G. (2006) Fully flexible docking models of the complex between alpha7 nicotinic receptor and a potent heptapeptide inhibitor of the beta-amyloid peptide binding, *Bioorg Med Chem Lett 16*, 3519–3523.
77. Tanaka, F., Hu, Y., Sutton, J., Asawapornmongkol, L., Fuller, R., Olson, A. J., Barbas, C. F., 3rd, and Lerner, R. A. (2008) Selection of phage-displayed peptides that bind to a particular ligand-bound antibody, *Bioorg Med Chem 16*, 5926–5931.
78. Sheridan, D. L., Kong, Y., Parker, S. A., Dalby, K. N., and Turk, B. E. (2008) Substrate discrimination among mitogen-activated protein kinases through distinct docking sequence motifs, *J Biol Chem 283*, 19511–19520.
79. Arun Prasad, P., and Gautham, N. (2008) A new peptide docking strategy using a mean field technique with mutually orthogonal Latin square sampling, *J Comput Aided Mol Des 22*, 815–829.
80. Yaneva, R., Schneeweiss, C., Zacharias, M., and Springer, S. (2010) Peptide binding to MHC class I and II proteins: new avenues from new methods, *Mol Immunol 47*, 649–657.
81. Bui, H. H., Schiewe, A. J., von Grafenstein, H., and Haworth, I. S. (2006) Structural prediction of peptides binding to MHC class I molecules, *Proteins 63*, 43–52.
82. Schafroth, H. D., and Floudas, C. A. (2004) Predicting peptide binding to MHC pockets via molecular modeling, implicit solvation, and global optimization, *Proteins 54*, 534–556.
83. Fagerberg, T., Cerottini, J. C., and Michielin, O. (2006) Structural prediction of peptides bound to MHC class I, *J Mol Biol 356*, 521–546.
84. Davies, M. N., Sansom, C. E., Beazley, C., and Moss, D. S. (2003) A novel predictive technique for the MHC class II peptide-binding interaction, *Mol Med 9*, 220–225.
85. Antes, I., Siu, S. W., and Lengauer, T. (2006) DynaPred: a structure and sequence based method for the prediction of MHC class I binding peptide sequences and conformations, *Bioinformatics 22*, e16-24.
86. Tong, J. C., Tan, T. W., and Ranganathan, S. (2004) Modeling the structure of bound peptide ligands to major histocompatibility complex, *Protein Sci 13*, 2523–2532.
87. Xie, W., and Sahinidis, N. V. (2006) Residue-rotamer-reduction algorithm for the protein side-chain conformation problem, *Bioinformatics 22*, 188–194.
88. Staneva, I., and Wallin, S. (2009) All-atom Monte Carlo approach to protein-peptide binding, *J Mol Biol 393*, 1118–1128.
89. Gerek, Z. N., and Ozkan, S. B. (2010) A flexible docking scheme to explore the binding selectivity of PDZ domains, *Protein Sci 19*, 914–928.
90. Bahar, I., and Rader, A. J. (2005) Coarse-grained normal mode analysis in structural biology, *Curr Opin Struct Biol 15*, 586–592.
91. Meiler, J., and Baker, D. (2006) ROSETTALIGAND: protein-small molecule docking with full side-chain flexibility, *Proteins 65*, 538–548.
92. Liu, Z., Dominy, B. N., and Shakhnovich, E. I. (2004) Structural mining: self-consistent design on flexible protein-peptide docking and transferable binding affinity potential, *J Am Chem Soc 126*, 8515–8528.
93. Maurer, M. C., Trosset, J. Y., Lester, C. C., DiBella, E. E., and Scheraga, H. A. (1999) New general approach for determining the solution structure of a ligand bound weakly to a receptor: structure of a fibrinogen Aalpha-like peptide bound to thrombin (S195A) obtained using NOE distance constraints and an ECEPP/3 flexible docking program, *Proteins 34*, 29–48.
94. Vanhee, P., Reumers, J., Stricher, F., Baeten, L., Serrano, L., Schymkowitz, J., and Rousseau, F. (2010) PepX: a structural database of non-redundant protein-peptide complexes, *Nucleic Acids Res 38*, D545-551.
95. Stein, A., Panjkovich, A., and Aloy, P. (2009) 3did Update: domain-domain and peptide-mediated interactions of known 3D structure, *Nucleic Acids Res 37*, D300-304.

96. Puntervoll, P., Linding, R., Gemund, C., Chabanis-Davidson, S., Mattingsdal, M., Cameron, S., Martin, D. M., Ausiello, G., Brannetti, B., Costantini, A., Ferre, F., Maselli, V., Via, A., Cesareni, G., Diella, F., Superti-Furga, G., Wyrwicz, L., Ramu, C., McGuigan, C., Gudavalli, R., Letunic, I., Bork, P., Rychlewski, L., Kuster, B., Helmer-Citterich, M., Hunter, W. N., Aasland, R., and Gibson, T. J. (2003) ELM server: A new resource for investigating short functional sites in modular eukaryotic proteins, *Nucleic Acids Res 31*, 3625–3630.
97. Stein, A., and Aloy, P. (2008) Contextual specificity in peptide-mediated protein interactions, *PLoS One 3*, e2524.
98. Vajdos, F. F., Yoo, S., Houseweart, M., Sundquist, W. I., and Hill, C. P. (1997) Crystal structure of cyclophilin A complexed with a binding site peptide from the HIV-1 capsid protein, *Protein Sci 6*, 2297–2307.
99. Schueler-Furman, O., Altuvia, Y., and Margalit, H. (2001) Examination of possible structural constraints of MHC-binding peptides by assessment of their native structure within their source proteins, *Proteins 45*, 47–54.
100. Sezerman, U., Vajda, S., Cornette, J., and DeLisi, C. (1993) Toward computational determination of peptide-receptor structure, *Protein Sci 2*, 1827–1843.
101. London, N., Raveh, B., Cohen, E., Fathi, G., & Schueler-Furman, O. (2011) Rosetta FlexPepDock web server-high resolution modeling of peptide-protein interactions. Nucleic Acids Res 39, W249–53. doi:10.1093/nar/gkr431.
102. Yanover, C., & Bradley, P. (2011). Large-scale characterization of peptide-MHC binding landscapes with structural simulations. *Proc Natl Acad Sci USA 108*, 6981–6986. doi:10.1073/pnas.1018165108.
103. London, N., Lamphear, C. L., Hougland, J. L., Fierke, C. A., & Schueler-Furman, O. (2011). Identification of a novel class of farnesylation targets by structure-based modeling of binding specificity, *PLoS Comput Biol 7*, e1002170.
104. Ben-Shimon, A., and Niv, M. Y. (2011). Deciphering the arginine-binding preferences at the substrate-binding groove of ser/thr kinases by computational surface mapping, *PLoS Comput Biol 7*, e1002288. doi:10.1371/journal.pcbi.1002288.
105. Raveh, B., London, N., Zimmerman, L., & Schueler-Furman, O. (2011). Rosetta FlexPepDockab-initio: Simultaneous folding, docking and refinement of peptides onto their receptors. *PLoS One 6*, e18934.

Chapter 18

Comparison of Common Homology Modeling Algorithms: Application of User-Defined Alignments

Michael A. Dolan, James W. Noah, and Darrell Hurt

Abstract

The number of known three-dimensional protein sequences is orders of magnitude higher than the number of known protein structures. This is a result of an increase in large-scale genomic sequencing projects, the inability of proteins to crystallize or crystals to diffract well, or a simple lack of resources. An alternative is to use one of a variety of available homology modeling programs to produce a computational model of a protein. Protein models are produced using information from known protein structures found to be similar. Here, we compare the ability of a number of popular homology modeling programs to produce quality models from user-defined target–template sequence alignments over a range of circumstances including low sequence identity, variable sequence length, and when interfaced with a protein or small molecule. Programs evaluated include Prime, SWISS-MODEL, MOE, MODELLER, ROSETTA, Composer, ORCHESTRAR, and I-TASSER. Proteins to be modeled were chosen to test a range of sequence identities, sequence lengths, and protein motifs and all are of scientific importance. These include HIV-1 protease, kinases, dihydrofolate reductase, a viral capsid protein, and factor Xa among others. For the most part, the programs produce results that are similar. For example, all programs are able to produce reasonable models when sequence identities are >30% and all programs have difficulties producing complete models when sequence identities are lower. However, certain programs fare slightly better than others in certain situations and we attempt to provide insight on this topic.

Key words: Homology modeling, Comparative modeling, Sequence alignments, Protein modeling software, Loop modeling

1. Introduction

Obtaining the three-dimensional structure of a protein often proves to be challenging, employing techniques such as X-ray crystallography and NMR, sometimes taking years to yield results. Frequently, the structure of a protein cannot be determined by X-ray crystallography because it cannot be crystallized or if coaxed into crystallizing, will not diffract well. Similarly, a protein may be

Andrew J.W. Orry and Ruben Abagyan (eds.), *Homology Modeling: Methods and Protocols*, Methods in Molecular Biology, vol. 857, DOI 10.1007/978-1-61779-588-6_18, © Springer Science+Business Media, LLC 2012

unsuitable for NMR experiments due to relatively large size or because of aggregation. One example is that of the membrane-bound G-protein-coupled receptor (GPCR) family of proteins where crystal structures traditionally have been difficult to obtain (1, 2), although recent efforts resulting in determination of the human β2-adrenergic GPCR structure should be noted (3–5). Experimental difficulties coupled with the availability of approximately five million protein sequences (6) and limited amount of resources to experimentally derive three-dimensional structures make an alternative method of structure determination desirable.

Creating a three-dimensional protein model based on information from similar or "homologous" proteins whose structures are known is a faster way of gaining structural insight compared to experimental methods and is often the only way to obtain a three-dimensional view of a protein. The classic paradigm of constructing a homology model is to first find proteins that are homologous to a query or "target" sequence and align them according to common sequence and structural features. The next step is to construct a backbone model consisting of regions that are structurally conserved across the homologs followed by building regions that vary structurally, often comprising loops, insertions, or deletions ("gaps") relative to homologous regions. The final step is to add side chains to the backbone followed by a minimization or molecular dynamics protocol to lower the overall energy of the structure by correcting any bad geometries or steric problems.

Over several decades, a number of homology modeling packages have been developed that rely on knowledge-based methods, *ab initio* methods or a combination of the two to produce a protein model. Knowledge-based programs such as SWISS-MODEL (7), PROFIT (8), ICM (9), and ROSETTA (10) use information from known structures, often represented as a library of fragments to construct a three-dimensional model from a target sequence. Homology modeling programs such as MODELLER (11) use *ab initio* methods producing solutions that satisfy a set of spatial rules derived from probability density functions and statistical analysis of a protein structure as a whole. ORCHESTRAR (12–16), Composer (17, 18), GENEMINE/LOOK (19), MOE (20), and Prime (21) use a combination of ab initio and knowledge-based approaches.

A difficult question then arises: how does one evaluate the quality of a model? One obtains a different answer depending on the nature of the question asked and the method used for evaluation. For example, if the overall fold of a large protein (~500 residues) is compared by measuring the root-mean-square deviation (RMSD) between the backbone atoms of the model to a solved structure, the resulting value may not be as good as if one compared individual domains in the same way, due to differences in overall domain orientations between the model and the solved structure. In this case, one would better understand model quality by comparing the individual model domains to the solved structure domains, and looking at domain orientation separately. The message to the reader is to take comparative results with a grain of salt: look very closely at the methods used to make comparisons and what was compared, whether it is part of or the entire model, which atoms were used in the comparison, what stage of the modeling process is being compared, and the quality of the template to which the model is being compared.

A wide variety of protein homology modeling algorithms have participated over the years in the Critical Assessment for Structure Prediction (CASP) (22) where researchers are given a set of sequences that have known, but yet to be released three-dimensional structures. Three-dimensional solutions are submitted, evaluated and compared to the known protein structures, once the contest ends. Like CASP, this study compares the capability of popular homology modeling packages to produce models of proteins whose three-dimensional structures are known with an exception being that each program is provided identical, specific, user-defined alignments as input. Unlike CASP, it attempts to produce models that use only the default settings of the programs and does not include any additional energy refinement procedure at the end of the modeling process. An attempt is made therefore, to assess only the structure building capabilities of each program. Importantly, modeling using multiple homologs was not examined in this study as not all programs evaluated are able to use information from multiple templates across all parts of a model. In order to include a wider variety of programs, we opted to produce homology models based only on a single template. Of note, although other comparisons have been performed (23–25), this is the first study to evaluate ORCHESTRAR, a more recently developed homology modeling package, when compared to a number of different programs. Finally, we make little attempt to gauge the user-friendliness of the software as this can be subjective between researchers, but instead refer the reader to usability information found in other studies (23–25).

2. Materials

2.1. Sequence Selection

A total of 18 protein sequences were chosen that provided a range of sequence lengths and sequence identities as well as a wide variety of protein folds. Sequences range from 46 to 504 residues and have identities to templates of between 17 and 94%. A number of pharmaceutically relevant proteins were examined including several kinases, dihydrofolate reductase (DHFR), HIV-1 protease, and factor Xa, among others. Protein models are often produced with the intent of using the model for peptide or ligand-binding studies or for examining protein–protein interactions. Therefore, we examined in detail those models produced from homologs containing a protein–protein interface, peptide, or small molecule-binding site and determined how well each program reproduced these regions. Specifically, we examined backbone atom and all-atom positions within 5 Å of these regions.

2.2. Software

Default settings were used for all software except for those that modeled termini and those that allowed additional minimization of the final model with the exception of SWISS-MODEL where it is not possible to produce models without modeling the termini or minimizing the final structure. For all other programs, an all-atom minimization is not performed, but each program has internal optimization strategies for modeling including those that add and optimize side-chain positions.

1. *ORCHESTRAR*
   ORCHESTRAR (distributed by Tripos) is comprised of a group of algorithms including programs to structurally align homologs (Baton) (15, 16), generate conserved region models (CHORAL) (12), find structurally variable regions or "loops" using knowledge-based and *ab initio* methods (PETRA and FREAD) (14), and add side chains (ANDANTE) (13).
2. *Prime*
   Prime (developed and distributed by Schrödinger, LLC) constructs a model using aligned atom positions of homologs. Default settings use the OPLS force field (26, 27) and a surface-generalized Born solvent model (28). Prime constructs model regions not derived from the templates by an *ab initio* method (29) while side-chain conformations are taken from a rotamer library. In this study, we used default settings with the exception of building terminal tails beyond secondary structure elements and minimizing residues.
3. *MOE*
   MOE-Homology (developed by Chemical Computing Group, Inc.) combines the methods of *segment-matching* procedure (19) and the approach to the modeling of insertion/deletion regions (30). MOE-Homology creates ten models by default using a knowledge-based loop searching method and side-chain rotamer selection method after which an average model is created and then submitted to a user-controlled energy minimization. In our study, the "Best Intermediate" model was chosen using the default settings with the exception of a minimization.
4. *SWISS-MODEL*
   Differing from the other modeling methods in the study, SWISS-MODEL (7) is a fully automated comparative protein modeling server (http://swissmodel.expasy.org/). The Alignment

Mode was used which takes an aligned query–template sequence as input and uses the knowledge-based ProModelI I (31) program to produce a model. SWISS-MODEL attempts to produce a complete, minimized model using the Gromos96 force field (32).

5. *Composer*
The Composer program (17, 18) was integrated into SYBYL (distributed by Tripos) prior to version 8.0. The alignment portion of the program was bypassed to preserve the alignment of the input. In default mode, Composer uses structural alignment information from multiple templates to first define structurally conserved regions (SCRs) across all homologs which it then uses to construct a partial model. Any remaining gaps or structurally variable regions (SVRs) between SCRs are modeled using a loop modeling algorithm. When only a single template is used for model construction as in this study, Composer defines an SCR as those regions where no gaps occur between the alignment of the target and template sequences.

6. *MODELLER*
MODELLER uses the "automodel" class to construct a three-dimensional model of the target protein. Model building is implemented by satisfaction of spatial constraints (11). Target/templates were submitted to the program and five models were generated and evaluated. Top models were chosen based on discrete optimized protein energy (DOPE) score (33, 34).

7. *Rosetta*
Homology models were constructed using Rosetta version 3.1 which leverages the loop modeling algorithm within the Rosetta software suite. For each target, 10K models (referred to as "decoys") were generated using the Biowulf Linux cluster (National Institutes of Health, Bethesda, MD; http://biowulf.nih.gov). The top 1,000 decoys in terms of lowest energy were clustered using an RMSD of 5 Å between decoys. The energies of representative decoys from each cluster were obtained and the representative decoy having the lowest overall energy was taken as the "correct" solution.

8. *I-TASSER*
Sequence alignments were submitted to the I-TASSER server (35) after selecting the option "Specify template with alignment." This option allows one to specify both the template structure and the target–template sequence alignment. This differs from the default mode where one submits the target sequence only and allows the program to provide templates and sequence alignments.

3. Methods

3.1. Sequence Selection

Target sequences were chosen (a) based on availability of their 3D coordinates having a resolution of <3 Å, (b) based on general interest to the scientific community, (c) to provide a wide a range of sequence lengths, (d) to cover a range of morphologies, and (e) to provide a wide range of target–template sequence identities, in an effort to test a wide variety of input. N- or C-terminal tags were not included in modeling. Sequences were obtained in FASTA format from the Protein Data Bank (36). Studies using Prime, ORCHESTRAR, Composer, and Rosetta were performed using the Red Hat Enterprise Linux 5.3 operating system. All other software used Windows XP or was run through an associated Web server.

3.2. Sequence Alignment and Template Selection

For each target sequence in the study, a PSI-BLAST (37) search was run to produce an initial sequence alignment which served as input for the sequence–structure homology recognition algorithm FUGUE (38), which identified structural homolog families within the HOMSTRAD database (release date 08/12/2006) (39, 40). No two structures in HOMSTRAD have greater than 90% identity. From each FUGUE search, the top HOMSTRAD multimember family with the rank of CERTAIN (Z score > 6.0) was chosen and from this family, the top homolog based on sequence identity to the target was chosen for modeling. FUGUE was used to realign the target and homolog sequence. This sequence alignment was used as input into all programs, thereby providing a common starting point for subsequent modeling. A list of the homolog families from which a single template was chosen along with the name of the single template and the percent sequence identity to the target is listed (Table 1). Target sequence lengths range from 46 residues for crambin to 504 residues for the protoporphyrinogen IX oxidase. Template/target sequence identities ranged from 17.2 to 96.8% after realigning using FUGUE.

3.3. Evaluation of All-Atom Homology Models

Homology models were evaluated using the Align Structures by Homology tool in the SYBYL7.3 Biopolymer module (Tripos). This tool first aligns a homology model to the known structure derived from X-ray crystallography or NMR by performing a least squares fit between the backbone or all atoms of the homology model followed by calculating the root-mean-square deviation (RMSD) between the model and known structure. RMSD is the square root of the mean of the square of the distances between matched atoms. In other words, an RMSD calculation sums the Cartesian distances between each atom in the model and the corresponding atom in the known structure for a group of atoms. The end result is an aggregation of these distances into a single value

Table 1
Top scoring homologs and associated HOMSTRAD family for each target sequence

Target PDB ID (chain)	Number of residues in target	HOMSTRAD family (*Z*score)	Template PDB ID (chain)	% Seq identity of homolog to target[a]
3CLA	213	cat3 (35.08)	1E2O	17.2
1SEZ(A)	504	Amino_oxidase (29.05)	1H83(A)	18.2
1S9J	335	kinase (28.83)	1BLX(A)	29.6
4DFR	159	dhfr (38.69)	1DHF(A)	30.4
1FDR(C)	245	reductases (25.43)	1A8P	32.6
1CBN	46	thionin (14.55)	1BHP	35.6
3EST	240	sermam (39.76)	1A0L(A)	41.1
1P38	360	kinase (45.34)	1JNK	49.7
2BPY(A)	99	rvp (18.64)	1YTI(A)	50.5
1AAP(A)	58	kunitz (12.73)	1SHP	50.9
1BET	107	ngf (19.52)	1BND(B)	60.4
1HCS (H)	107	sh2 (23.42)	1AOU(F)	65.7
1AYM(A)	285	rhv (37.68)	1R1A	71.4
2BOK(A)	241	sermam (37.67)	1KIG(H)	81.7
1VLC	354	icd (62.11)	1CNZ(A)	87.3
2CTC	307	cpa (57.16)	1PCA	87.3
1PPB(H)	259	sermam (43.56)	1BBR(H)	87.3
1APM	350	kinase (40.20)	1CDK(A)	96.8

[a] Sequence identity to target calculated after sequence realignment using FUGUE

used as a measure of modeling precision. A number of programs offer RMSD calculations including VMD, PyMOL, and Chimera. In addition, all models where examined for the presence of incorrect geometries such as d-amino acids using the ProTable module in SYBYL.

4. Notes

4.1. Model Evaluation

The RMSDs between the backbone atoms of models and known structures are shown, as well as the RMSDs between all atoms (Table 2). Models having the lowest backbone atom RMSD to the

Table 2
Comparison of backbone atoms and all-atoms between models and known structures.

PDB (chain)	% ID	RMSD of backbone atoms between model and known structure (Å)								RMSD of all atoms between model and known structure (Å)								% residues modeled							
		O	P	M	C	S	R	I	MD	O	P	M	C	S	R	I	MD	O	P	M	C	S	R	I	MD
3CLA	17.2	15.65	17.4	15.71	14.7	16.50	16.81	13.43	14.44	16.14	17.8	16.2	15.2	17.02	17.26	13.90	15.01	63.9	100.0	100.0	93.0	100.0	80.1	100.0	100.0
1SEZ	18.2	12.43	20.58	12.93	12.20	---[a]	12.48	10.14	11.97	12.72	21.18	13.21	12.52	---[a]	12.76	10.47	12.30	86.1	90.1	97.4	97.4	---[a]	93.9	100.0	100.0
1S9J	29.6	7.10	8.27	8.35	7.85	8.73	6.56	6.98	8.86	7.72	8.91	8.81	8.34	9.23	7.16	7.51	9.21	88.4	89.9	92.5	92.5	92.5	86.2	100.0	100.0
4DFR	30.4	2.82	2.99	2.90	3.05	2.72	2.59	2.60	2.68	3.64	3.83	3.86	3.82	3.68	3.28	3.36	3.54	92.6	98.7	99.4	99.4	99.4	99.4	100.0	100.0
1FDR (C)	32.6	1.75	2.63	2.15	2.27	2.21	2.07	2.01	1.99	2.41	3.65	3.13	3.22	3.20	3.00	2.97	3.00	78.8	98.0	99.6	99.6	99.6	99.6	100.0	100.0
1CBN	35.6	0.83	1.36	0.94	0.92	0.94	0.62	0.78	0.88	1.45	1.89	1.54	1.60	1.55	1.28	1.19	1.40	97.8	80.4	100.0	100.0	100.0	100.0	100.0	100.0
3EST	41.1	2.49	2.28	2.31	2.67	2.19	2.71	1.34	2.45	3.21	3.14	3.17	3.43	3.05	3.41	2.07	3.28	98.8	94.6	100.0	94.2	100.0	100.0	100.0	100.0
1P38	49.7	3.49	3.44	3.52	6.78	3.57	6.33	4.50	3.84	4.12	3.99	4.13	7.25	4.16	6.71	4.94	4.33	94.4	93.9	95.3	92.5	95.3	84.7	100.0	100.0
2BPY (A)	50.5	1.05	1.09	1.05	1.09	1.06	1.05	1.49	1.10	1.89	2.10	1.93	2.13	1.94	1.96	2.19	2.07	100.0	83.8	100.0	55.6	100.0	100.0	100.0	100.0
1AAP (A)	50.9	1.24	1.23	1.25	1.22	1.23	1.25	1.05	1.24	2.05	2.26	2.30	2.15	2.31	2.04	2.39	2.22	93.1	93.1	94.8	91.4	94.8	94.7	100.0	100.0
1BET	60.4	1.46	1.05	1.11	1.13	1.39	1.19	1.16	1.24	2.50	1.81	1.97	2.01	2.24	2.05	2.06	1.96	97.2	91.6	99.1	95.3	99.1	99.1	100.0	100.0
1HCS (B)	65.7	2.60	2.36	3.07	2.38	3.07	3.06	1.63	3.17	3.30	3.08	3.60	3.05	3.54	3.90	2.87	3.73	95.3	100.0	98.1	85.0	98.1	98.1	100.0	100.0

1AYM (A)	71.4	1.57	0.85	1.36	2.63	1.34	2.33	0.84	5.06	2.28	1.37	2.00	3.11	1.95	2.84	1.80	5.19	97.5	86.0	98.6	85.6	98.6	98.6	100.0	100.0
2BOK (A)	81.7	0.79	0.73	0.79	2.07	0.77	0.79	0.78	0.76	1.65	1.62	1.65	2.84	1.67	1.60	1.80	1.57	99.6	90.0	99.6	90.0	100.0	100.0	100.0	100.0
1VLC	87.3	2.16	2.38	2.36	2.97	2.23	2.12	2.09	2.33	2.52	2.73	2.87	3.37	2.64	2.30	2.61	2.78	99.4	99.4	100.0	95.8	100.0	99.4	100.0	100.0
2CTC	87.3	0.38	0.38	0.38	0.38	0.38	0.38	0.53	0.40	0.96	0.88	0.95	0.94	0.93	0.86	1.44	0.95	99.7	94.8	100	95.1	100.0	100.0	84.4	100.0
1PPB (H)	87.3	1.47	0.43	1.03	0.42	1.82	1.03	1.82	2.16	1.88	1.03	1.56	0.90	2.16	1.68	2.68	2.49	99.6	45.2	57.9	46.7	100.0	100.0	100.0	100.0
1APM	96.8	0.40	0.40	0.41	0.47	0.41	0.41	0.61	0.43	0.42	0.85	0.86	0.94	0.85	0.88	1.45	0.95	96.9	98.3	98.0	97.1	98.0	98.0	100.0	100.0
Total		9	8	6	5	7	8	11	6																

Models were compared to known structures by first aligning structures using backbone atoms (or all atoms) followed by RMSD determination. ***Filled boxes*** indicate models with the lowest RMSD value or within 10% of the lowest RMSD value. The ability to model termini was not selected for these programs except in the case of SWISS-MODEL. O=ORCHESTRAR, P=Prime, M=MOE, C=Composer, S=SWISS-MODEL, R=Rosetta, MD=MODELLER, and I=I-Tasser.

[a] SWISS-MODEL did not produce a model for protoporphyrinogen IX oxidase (1SEZ).

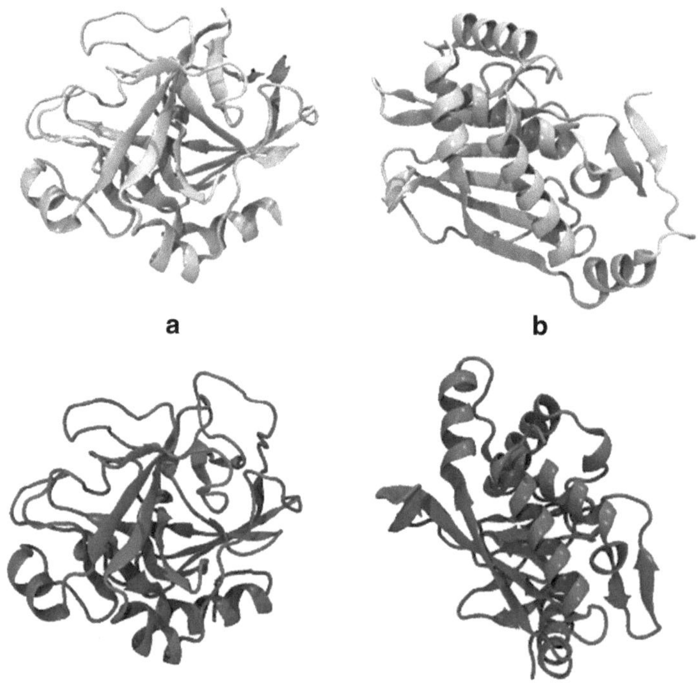

Fig. 1. Comparison of an acceptable homology model to one that was poorly modeled. (**a**) The crystal structure of prothrombinase (PDB ID 2BOK) is shown (*top panel*) along with a homology model (*bottom panel*). The RMSD between backbone atoms is 0.78 Å. (**b**) The crystal structure of type III chloramphenicol acetyltransferase (PDB ID 3CLA) shown (*top panel*) with a poorly modeled structure (*bottom panel*). The RMSD between backbone atoms is 15.7 Å.

known structure are indicated as well as those models within 10% of the lowest RMSD value. Lower RMSD values indicate better modeling precision. RMSD values of <3 Å are generally considered to be good models, whereas models with RMSD values >7 or 8 Å are considered to be poorer models. An example of a good and a poor model is shown in Fig. 1. Overall all programs performed similarly, building good quality homology models with higher sequence identity, and constructing progressively poorer models with lower sequence identity. When examining backbone RMSD data only, I-TASSER performed best overall generating 11 models within 10% of the lowest RMSD, followed by ORCHESTAR with 9, and Rosetta and Prime with 8 each.

4.2. Low Target–Template Sequence Identity

Models of targets having relatively low sequence identity to a template (<25%) are notoriously difficult to obtain. Two targets in this low sequence identity "twilight zone" were modeled and evaluated. The first is type III chloramphenicol acetyltransferase (PDB ID 3CLA) using the catalytic domain from dihydrolipoamide succinyltransferase (PDB ID 1E2O) as a template having sequence identity of 17.2%. The second is protoporphyrinogen IX oxidase (PDB ID 1SEZ) using polyamine oxidase as a template (PDB ID 1H83) with sequence identity of 18.2%. For the first, all programs produced models that were poor, with backbone atom RMSD values between 14 and 18 Å. For the second, all programs produced models with the exception of SWISS-MODEL. The inability of SWISS-MODEL to produce a model for protoporphyrinogen IX oxidase (1SEZ) may be due to the length of the sequence (504 residues) which is the longest in this study, but is most likely due the low sequence identity between the target and template. Models had backbone atom RMSD values of ~12 Å with the exception of PRIME having a backbone atom RMSD value of ~20 Å. Not surprisingly, no program evaluated was able to build a satisfactory model with these targets and templates, but I-TASSER was the only program to produce models for both low sequence identity targets that had backbone RMSDs within 10% of the actual structure. It has been shown in another study that Prime and Profit are able to produce quality models at lower sequence identities (23). Also, ORCHESTRAR makes use of FUGUE which has the ability to find and align to more distant homologs (38). What does one do if no homology modeling program is able to construct a model due to low overall sequence identity? In these cases, it may be worthwhile to perform fold recognition, replica exchange molecular dynamics (REMD) or *in silico* protein folding, such as with the Rosetta program, in an effort to obtain secondary and tertiary structure clues.

4.3. Sequence Size

Six targets were chosen for this study based on their relatively long sequence lengths which range from 307 to 504 residues (Table 1). The longest (protoporphyrinogen IX oxidase, PDB ID 1SEZ) was poorly modeled by all programs most likely due to its relatively low target–template sequence identity (<18.2%) and not to its length (Table 2). This was also the case for human mitogen-activated protein kinase kinase 1, MEK1 (PDB ID 1S9J). Of the remainder, all programs produced comparable, high-quality models of those sequences with the highest target–template sequence identity (PDB IDs 1VLC, 2CTC, and 1APM) with the exception of the MAP kinase P38 (PDB ID 1P38) having sequence identity of 50% and a sequence length of 360 residues. Composer and Rosetta had difficulty modeling this protein while the other programs had a lower backbone RMSD of ~3.5 Å. These results overall suggest that long sequence length is much less of a factor than that of sequence identity. Three targets had sequence lengths of <100 residues ranging from 46 to 99 residues with good target–template sequence identity (range 35.6–50.9%), and all programs produced high quality models.

4.4. Protein–Protein Interfaces

Two sequences were chosen in part because their structures interface with another protein. The first is the factor Xa catalytic domain which is bound to an EGF2-like domain (Stuart–Prower factor, PDB ID 2BOK) for which all programs produced high quality models. Not surprisingly, all programs modeled residues within 5 Å of the interface with high accuracy, having backbone and all-atom RMSD between models and known structures of ~0.5 and ~1.1 Å, respectively (Table 3). The second is the large subunit of human α-thrombin with the small subunit of α-thrombin (PDB ID 1PPB). Similarly, all programs were able to model residue backbone atoms within 5 Å of the protein–protein interface with high accuracy (~0.6 Å RMSD) as well as sidechains (all-atom RMSD range 1.1–2.0 Å).

4.5. Small Molecule and Peptide-Binding Sites

When examining the residues of models located within 5 Å of a known protein interface or a bound small molecule or peptide, Prime produced more models within 10% of the lowest backbone atom RMSD with 7, followed by Composer and SWISS-MODEL with 6, and Rosetta and ORCHESTRAR producing 5 each. In some cases such as with models of dihydrofolate reductase (PDB ID 4DFR), large deviations occurred between programs when comparing backbone atoms and all atoms within 5 Å of methotrexate. This may be a reflection of the differences of side chain and loop modeling algorithms as many ligands bind at protein loops.

4.6. Caveats

A fair amount of data is presented in this study, but it should be made clear that in order to better understand how homology programs handle unconventional modeling situations such as sequences with low identity, one needs to include more examples. For instance, perhaps one or more programs are better at modeling kinases having low sequence identity (see 1P38, Table 2), but another is better at modeling certain viral proteins (see 1AYM, Table 2). Also, it is important to mention that model evaluation as we have done it (comparing RMSDs between atom sets) cannot be presented without revealing the number of atoms that are being compared. For example, one may see that a program produces a relatively low RMSD, but has modeled only part of the structure. A more detailed study might compare different modeled regions between programs to better gauge performance. Also, differences in the modeling of structurally variable termini (SVT) were determined to be substantial across programs evaluated in this study and therefore, the modeling of variable termini was not purposefully conducted except with the Web server modeling programs whereby explicitly excluding certain regions was not possible. Including termini modeling in this study would, therefore, eclipse how well certain programs constructed the nonterminal portions of models. Instead, the authors propose that a future investigation be conducted to evaluate and rank the termini modeling algorithms of each of these programs. Finally, it should be mentioned that an all-atom minimization followed by a simulated annealing procedure

Table 3
Comparison of residues within 5Å of a ligand binding site or protein-protein interface between models and known structures.

PDB ID (chain)	Ligand or protein	Backbone RMSD (Å)								*All atom* RMSD (Å)							
		O	P	M	C	S	R	I	MD	O	P	M	C	S	R	I	MD
2BOK(A)	heterocyclic ligand	2.99	0.45	1.45	1.45	1.06	0.57	0.48	0.57	2.49	0.68	1.28	1.28	0.97	1.70	1.93	1.44
2BOK(A)	EGF-like domain	0.58	0.54	0.58	0.54	0.54	0.59	0.56	0.51	0.97	1.13	1.31	1.19	1.09	1.00	1.45	1.06
4DFR	methotrexate	2.87	1.48	0.63	0.46	0.59	0.38	0.87	0.39	3.56	2.02	1.35	1.13	1.24	0.80	2.07	0.78
2BPY(A)	heterocyclic ligand	0.58	0.58	0.61	0.58	0.62	0.55	0.76	0.66	1.07	1.11	1.18	1.15	1.19	2.00	1.73	1.81
1PPB(H)	small subunit	0.38	0.36	0.28	0.31	0.35	0.33	0.41	0.44	0.71	0.55	1.07	0.45	0.58	0.52	0.62	0.78
1PPB(H)	chloromethylke-tone peptide	2.14	2.06	3.73	3.73	2.07	2.12	2.56	2.44	2.96	2.84	4.05	4.05	2.73	2.95	3.12	2.76
1HCS(B)	hexapeptide	1.48	2.61	2.66	3.01	1.48	3.39	1.11	1.71	1.89	2.79	2.81	3.01	1.91	4.73	2.05	2.81
1AYM(A)	lauric acid	1.82	0.53	0.53	0.53	0.54	0.54	0.57	0.53	1.71	0.92	1.00	1.00	0.93	1.95	1.98	0.98
1APM	peptide inhibitor	2.27	0.39	0.39	0.39	0.39	0.28	0.77	0.42	2.19	0.65	0.65	0.83	0.65	0.57	1.76	0.85
1FDR©	FAD	0.96	0.97	1.24	1.21	1.17	1.27	1.16	1.22	1.59	1.44	2.18	1.92	2.29	2.71	2.63	1.73
2CTC	Zn + L-phenyl lactate	0.22	0.21	0.22	0.22	0.23	0.22	0.97	0.22	0.52	0.47	0.55	0.54	0.55	0.48	1.91	0.64
Total		5	7	4	6	6	5	3	3								

Filled boxes indicate with the lowest RMSD value or within 10% of the lowest RMSD value.

be conducted following the construction of a homology model in an effort to move the model to a lower energy and assumedly more "correct" structure. Such a protocol would have the effect of optimizing side-chain geometries, although most of the programs studied here contain an algorithm that adds and optimizes side-chain geometries during model construction. Knowing this, we have confidence in the all-atom RMSD values obtained (Table 2).

4.7. Summary

At the very least, this study reinforces the idea that all homology programs will produce similar results under most circumstances, using similar settings. If this is the case, then one should find a low cost and user-friendly program for producing homology models. Although usability is often subjective, we find the I-TASSER server to be the best choice overall. Other programs such as Rosetta produce good results, but command line usage can be daunting. Also, with the number of free programs available such as I-TASSER and SWISS-MODEL, one may find it difficult to rationalize the high cost of some proprietary software.

It also highlights the importance of additional measures that must be taken either within a homology modeling program or post-model construction in order to obtain a more accurate model, such as minimizing energy or performing a molecular dynamics simulation to overcome any kinetic barriers leading to a lower energy and assumedly more accurate structure. Construction of a model using homology should be seen as only an initial step in understanding structure and function. This is especially true for lower target–template sequence identities and for models that incorporate a small molecule or protein interface that differs from the template on which it is modeled. Several programs incorporate minimization, molecular dynamics, or "induced-fit" docking methods such as Prime with Glide (41) that effectively increase the accuracy of modeling residues around incorporated ligands during model construction.

Acknowledgments

The authors would like to thank Dr. Judith Hobrath for her technical assistance.

References

1. Evers A and Klebe G (2004) Successful virtual screening for a submicromolar antagonist of the neurokinin-1 receptor base on a ligand-supported homology model. J Med Chem 47:5381–5392
2. Evers A and Klabunde T (2005) Structure-based drug discovery using GPCR homology modeling: Successful virtual screening for antagonists of the alpha1A androgenic receptor. J Med Chem 48:1088–1097

3. Rasmussen SG, Choi HJ, Rosenbaum DM, Kobilka TS, Thian FS, Edwards PC, Burghammer M, Ratnala VR, Sanishvili R, Fischetti RF, Schertler GF, Weis WI, and Kobilka BK (2007) Crystal structure of the human β2-adrenergic G-protein-coupled receptor. Nature 450:383–7
4. Cherezov V, Rosenbaum DM, Hanson MA, Rasmussen SG, Thian FS, Kobilka TS, Choi HJ, Kuhn P, Weis WI, Kobilka BK, and Stevens RC (2007) High-resolution crystal structure of an engineered human β2-adrenergic G protein-coupled receptor. Science 318:1258–65
5. Rosenbaum DM, Cherezov V, Hanson MA, Rasmussen SG, Thian FS, Kobilka TS, Choi HJ, Yao XJ, Weis WI, Stevens RC and Kobilka BK (2007) GPCR engineering yields high-resolution structural insights into β2-adrenergic receptor function. Science 318 (5854):1266–73
6. Wu CH, Apweiler R, Bairoch A, Natale DA et al (2006) The Universal Protein Resource (UniProt): An expanding universe of protein information. Nucl Acids Res 34:Database issue D187-D191
7. Schwede T, Kopp J, Guex N, and Peitsch MC (2003) SWISS-MODEL: An automated protein homology-modeling server. Nucl Acids Res 31:3381–3385
8. Sippl MJ and Weitckus S (1992) Detection of native-like models for amino acid sequences of unknown three-dimensional structure in a database of known protein conformations. Proteins 13:258–271
9. Abagyan RA, Totrov MM, and Kuznetsov DA (1994) ICM: a new method for protein modeling and design: applications to docking and structure prediction from the distorted native conformation. J Comp Chem 15:488–506
10. Misura KM, Chivian D, Rohl CA, Kim DE, Baker D (2006) Physically realistic homology models built with ROSETTA can be more accurate than their templates. PNAS 103(14):5361–6
11. Sali A and Blundell TL (1993) Comparative protein modelling by satisfaction of spatial restraints. J Mol Biol 234:779–815
12. Montalvao RW, Smith RE, Lovell SC and Blundell TL (2005) CHORAL: A differential geometry approach to the prediction of the cores of protein structures. Bioinformatics 21:3719–3725
13. Smith RE, Lovell SC, Burke DF, Montalvao RW and Blundell TL (2007) Andante: reducing side-chain rotamer search space during comparative modeling using environment-specific substitution probabilities. Bioinformatics 23:1099–105
14. Deane CM and Blundell TL (2001) CODA: A combined algorithm for predicting the structurally variable regions of protein models. Protein Sci 10:599–612
15. Sali A and Blundell TL (1990) Definition of general topological equivalence in protein structures. A procedure involving comparison of properties and relationships through simulated annealing and dynamic programming. J Mol Biol 212:403–28
16. Zhu ZY, Sali A and Blundell TL (1992) A variable gap penalty function and feature weights for protein 3-D structure comparisons. Protein Eng 5:43–51
17. Sutcliffe MJ, Haneef I, Carney D, Blundell TL (1987a) Knowledge-based modeling of homologous proteins, Part 1: Three-dimensional frameworks derived from the simultaneous superposition of multiple structures. Protein Eng 1:377–384
18. Sutcliffe MJ, Hayes FR, Blundell TL (1987b) Knowledge-based modeling of homologous proteins, Part 2: Rules for the conformations of substituted sidechains. Protein Eng. 1:385
19. Levitt M (1992) Accurate modeling of protein conformation by automatic segment matching. J Mol Biol 226:507–533
20. MOE. Chemical Computing Group, Montreal, Quebec, Canada.
21. Prime. Schrödinger, LLC, Portland, OR
22. Tramontano A, Cozzetto D, Giorgetti A, Raimondo D (2007) The assessment of methods for protein structure prediction. Methods Mol Biol 413:43–58
23. Nayeem A, Sitkoff D, Krystek S (2006) A comparative study of available software for high-accuracy homology modeling: from sequence alignments to structural models. Protein Sci 15:808–24
24. Wallner B, Elofsson A (2005) All are not equal: A benchmark of different homology modeling programs. Protein Sci 14:1315–1327
25. Dolan MA, Keil M, Baker DS (2008) Comparison of Composer and ORCHESTRAR. Proteins 72:1243–58
26. Jorgensen WL, Maxwell DS and Tirado-Rives J (1996) Development and testing of the OPLS all-atom force field on conformational energetics and properties of organic liquids. J Am Chem Soc 118:11225–11236
27. Kaminski GA, Friesner RA, Tirado-Rives J and Jorgensen WL (2001) Evaluation and reparametrization of the OPLS-AA force field for proteins via comparison with accurate quantum chemical calculations on peptides. J Phys Chem B 105:6474–6487

28. Gallicchio E, Zhang LY and Levy RM (2002) The SGB/NP hydration free energy model based on the surface generalized born solvent reaction field and novel nonpolar hydration free energy estimators. J Comp Chem 23:517–529
29. Jacobson MP, Pincus DL, Rapp CS, Day TJF, Honig B, Shaw DE, Friesner RA (2004) A hierarchical approach to all-atom protein loop prediction Proteins 55:351–367
30. Fechteler T, Dengler U, and Schomburg D (1995) Prediction of protein three-dimensional structures in insertion and deletion regions: A procedure for searching data bases of representative protein fragments using geometric scoring criteria. J Mol Biol 253:114–131
31. Peitsch MC (1996) ProMod and Swiss-Model: Internet-based tools for automated comparative protein modeling. Biochem Soc Trans 24(1):274–279
32. Van Gunsteren WF, Billeter SR, Eising AA, Hünenberger PH, Krüger P, Mark AE, Scott WRP, and Tironi IG (1996) Biomolecular Simulation: The GROMOS96 Manual and User Guide, pp. 1–1042. Vdf Hochschulverlag AG an der ETH Zürich, Zürich, Switzerland
33. Shen M-y, Sali A (2006) Statistical potential for assessment and prediction of protein structures. Protein Science 15:2507–2524
34. Eramian D, Shen M-y, Devos D, Melo F, Sali A and Marti-Renom MA (2006) A composite score for predicting errors in protein structure models. Protein Science 15:1653–1666
35. Roy A, Kucukural A, Zhang Y (2010) I-TASSER: a unified platform for automated protein structure and function prediction. Nature Protocols 5:725–738
36. Berman HM, Westbrook J, Feng Z, Gilliland G, Bhat TN, Weissig H, Shindyalov IN, and Bourne PE (2000) The Protein Data Bank. Nucl Acids Res 28:235–242
37. Altschul SF, Madden TL, Schäffer AA, Zhang J, Zhang Z, Miller W and Lipman DJ (1997) Gapped BLAST and PSI-BLAST: a new generation of protein database search programs. Nucl Acids Res 25:3389–3402
38. Shi J, Blundell TL, and Mizuguchi K (2001) FUGUE: Sequence-structure homology recognition using environment-specific substitution tables and structure-dependent gap penalties. J Mol Biol 310:243–257
39. de Bakker PIW, Bateman A, Burke DF, Miguel RN, Mizuguchi K, Shi J, Shirai H, and Blundell TL (2001) HOMSTRAD: Adding sequence information to structure-based alignments of homologous protein families. Bioinformatics 17:748–749
40. Mizuguchi K, Deane C, Blundell T, and Overington J (1998) HOMSTRAD: A database of protein structure alignments for homologous families. Protein Sci 7:2469–2471
41. Sherman W, Day T, Jacobson MP, Friesner RA, Farid R (2006) Novel procedure for modeling ligand/receptor induced fit effects. J Med Chem 49:534–553

INDEX

Andrew J.W. Orry and Ruben Abagyan (eds.), *Homology Modeling: Methods and Protocols*, Methods in Molecular Biology, vol. 857, DOI 10.1007/978-1-61779-588-6, © Springer Science+Business Media, LLC 2012

L

M

N

O

P

Q

R

S

MIX
Papier aus verantwortungsvollen Quellen
Paper from responsible sources
FSC® C105338

If you have any concerns about our products,
you can contact us on
ProductSafety@springernature.com

In case Publisher is established outside the EU,
the EU authorized representative is:
Springer Nature Customer Service Center GmbH
Europaplatz 3, 69115 Heidelberg, Germany

Printed by Libri Plureos GmbH
in Hamburg, Germany